T0325721

COASTAL HYDROGEOLOGY

Water resources are of enormous societal and ecological importance. In coastal areas, they are under ever greater pressure due to population growth, more affluent lifestyles, food production and the growing tourism industry. Changes to the coastal landscape, through urbanisation and land reclamation, and by natural processes such as climate change and sea level rise, modify the interaction between seawater and groundwater and put water resources at risk. This comprehensive volume covers both theory and practice of coastal hydrogeology. It discusses hydrochemistry; submarine groundwater discharge; groundwater management; palaeo-hydrology; land reclamation; climate change and sea level rise; and mathematical models of variable-density flow. With its up-to-date coverage and numerous case studies that illustrate practical implications, it is perfect for students, practitioners, managers and researchers who wish to develop an in-depth understanding of topics relevant to sustainably managing coastal groundwater resources.

JIMMY JIAO is a professor of hydrogeology in the Department of Earth Sciences at the University of Hong Kong and has more than 20 years of teaching and research experience in various topics in coastal hydrogeology. He was the 2011 recipient of the John Hem Excellence in Science and Engineering Award from the National Groundwater Association. Dr Jiao is a Fellow of the Geological Society of London, the Geological Society of America and the American Society of Civil Engineers. He has provided consultancy services to industry and government organisations regarding various coastal groundwater issues. Dr Jiao was an Associate Editor (2004–2008) and Editor (2008–2015) of the *Hydrogeology Journal* and Associate Editor (2002–2008) for *Groundwater*.

VINCENT POST has nearly 20 years of professional experience in coastal hydrogeology. He completed his PhD on groundwater salinisation in the Netherlands and has worked on projects in coastal zones around the world, including Portugal, Kiribati and Australia. His expertise spans a broad range of topics, which include numerical modelling, hydrochemistry, measurement techniques and freshwater resource management. Dr Post has published extensively on coastal groundwater flow and chemical processes. He was an Editor of the *Hydrogeology Journal* (from 2010 to 2015) and is an Associate Editor for the *Journal of Hydrology*. He is also actively involved in the organisation of the Salt Water Intrusion Meeting series that saw its fiftieth anniversary in 2018.

"This book presents a systematic and comprehensive approach to understanding coastal groundwater and gives readers a balanced examination of almost all important facets of coastal hydrogeology, ranging from palaeo groundwater issues in the recent geological past to future behaviour of coastal aquifers in response to climate change and sea level rise. Numerous international case studies from coastal aquifers cultivate reader understanding of the occurrence, movement and hydrochemistry of coastal groundwater in a variety of geologic settings. I am delighted to have this excellent book on my shelf and believe it will become a classic. I have no hesitation in recommending it as a "must use" book to all those who work in any aspect of groundwater in the coastal environment."

John Cherry, University of Guelph

"The reader of Jiao and Post's book will find excellent chapters from the principles of hydrology and hydrodynamics to highly practical issues of coastal hydrogeology, complementary to the genetic knowledge of aquifers. The book provides the knowledge and background necessary for all researchers, engineers and practitioners dealing with coastal issues, including groundwater resources, land use, environmental values and ecological services."

Emilio Custodio, Polytechnic University of Catalonia, Barcelona

"At last – a readable and comprehensive compilation of the principles, analytical solutions and literature relevant to coastal groundwater. Jiao and Post have done an incredible job of summarising and discussing the diverse and voluminous literature on coastal hydrogeology in a highly readable book. Basic principles such as equivalent freshwater head and Darcy's law in variable density flow are clearly explained and illustrated. Topics covered range from salt water intrusion, submarine groundwater discharge, tidal dynamics and geochemistry to land reclamation, sea level change and relict (palaeo) salt water in coastal aquifers, with interesting historical insights included throughout. A final chapter pulls together many of the concepts in a discussion of coastal aquifer management and seawater intrusion control. This book is essential for anyone interested in groundwater in coastal areas."

Mary P. Anderson, University of Wisconsin-Madison

"Coastal groundwater presents unique challenges to hydrogeologists, both in its vulnerable position squeezed between land and sea and in the physical and chemical complexities created by variations in density, sources of contamination and characteristic coastal forcings. As pressures on coastal water resources rise with increasing demand, changes in climate and sea-level rise, careful management of coastal groundwater is a greater priority than ever. This book is a timely contribution that provides both a broad overview and a thorough dive into fundamental and emerging topics in the field. With a mix of theory and application, this will be a valuable resource for researchers, practitioners and managers looking to address the coastal water challenges of the coming decades."

Holly Michael, University of Delaware

COASTAL HYDROGEOLOGY

JIMMY JIAO
Hong Kong

VINCENT POST
Amsterdam

CAMBRIDGE
UNIVERSITY PRESS

University Printing House, Cambridge CB2 8BS, United Kingdom

One Liberty Plaza, 20th Floor, New York, NY 10006, USA

477 Williamstown Road, Port Melbourne, VIC 3207, Australia

314-321, 3rd Floor, Plot 3, Splendor Forum, Jasola District Centre, New Delhi - 110025, India

79 Anson Road, #06-04/06, Singapore 079906

Cambridge University Press is part of the University of Cambridge.

It furthers the University's mission by disseminating knowledge in the pursuit of education, learning and research at the highest international levels of excellence.

www.cambridge.org
Information on this title: www.cambridge.org/9781107030596
DOI: 10.1017/9781139344142

First published 2019

A catalogue record for this publication is available from the British Library

Library of Congress Cataloging in Publication data
Names: Jiao, Jimmy, author. | Post, Vincent, author.
Title: Coastal hydrogeology / Jimmy Jiao, Vincent Post.
Description: New York, NY: Cambridge University Press, 2019.
Identifiers: LCCN 2018050819 | ISBN 9781107030596
Subjects: LCSH: Hydrogeology. | Coasts.
Classification: LCC GB1005.J536 2019 | DDC 551.49–dc23
LC record available at https://lccn.loc.gov/2018050819

ISBN 978-1-107-03059-6 Hardback

The highest goodness is like water. Water benefits all things and does not compete. It stays in the lowly places which others despise. Therefore it is near The Eternal.

Laozi (1368–1644)

It will not be a case of a check returned with 'No Funds' written across it, but a case of drawing undrinkable water from our faucets.

It behooves us, then, to keep our expenditure within our income, to draw no more water from the artesian system than nature puts into it.

Palmer (1927)

Contents

Foreword

Most of the global population now lives in coastal areas that include megacities such as Los Angeles, New York, Tokyo, Jakarta, Shenzhen, Hong Kong, Shanghai, Singapore and many more. Important groundwater-related problems occur nearly everywhere around the globe, and coastal areas are no exception. The increasing concentration of human settlements in coastal regions, associated with expanding agricultural, industrial and urbanisation activities, places great stress on the water resources, resulting in seawater intrusion and related degradation of water quality. Human activities along coasts, such as land reclamation, and natural factors, such as sea level rise driven by climate change, also modify the natural coastal groundwater flow system and the interaction between seawater and groundwater. The pressure on coastal groundwater can only become more severe.

As a result of investigations of theoretical and practical problems related to coastal groundwater, much new knowledge has been gathered, and advanced technologies have been developed for coastal groundwater investigations. Coastal hydrogeology is an emerging science encompassing the theory and practice of groundwater in the context of issues such as variable-density flow, tidal fluctuations, mixing between freshwater and seawater, submarine groundwater discharge, land reclamation and climate change.

This book *Coastal Hydrogeology* is needed as an assemblage of these advancements for informing local communities, the technical professions and the water supply industry, as well as government regulators and policy makers. The book is a timely response to the academic developments and is much needed for students, engineers, scientists, environmentalists and coastal managers.

Coastal hydrogeology has not been extensively covered in previous textbooks or monographies. Commonly, coverage has been focused mostly on seawater intrusion. This first book focused entirely on coastal hydrogeology is a welcome contribution to groundwater science. It has been prepared by two hydrogeologists who have been working for more than 20 years exclusively and extensively on coastal groundwater environments in the Asia-Pacific and Europe. Both authors have served as editors of the *Hydrogeology Journal* and so are well aware of hydrogeological problems in coastal areas around the world. Many groundwater problems and issues that occur inland also exist in coastal areas, where they take on different shades of complexity commensurate with the highly dynamic physical and chemical interactions between seawater and groundwater and the vulnerability of coastal areas to sea level rise driven by global warming.

Most of the chapters in this book are based primarily on the authors' own research, but they show awareness of what is important in the literature. Between the two, the authors read literature in five languages. This book covers new topics that have emerged over the past two decades but were omitted from or covered only lightly in other books. Chapters contributing most to the uniqueness of this book include Chapter 4, 'Groundwater Tidal Dynamics'; Chapter 7, 'Submarine Groundwater Discharge'; Chapter 8, 'Coastal Palaeo-Hydrogeology'; Chapter 9, 'Impact of Land Reclamation on Coastal Groundwater Systems'; Chapter 10, 'Sea Level Change and Coastal Aquifers'; and Chapter 11, 'Tide-Induced Airflow in Unsaturated Zones'.

This book also covers classic topics related to coastal aquifers, such as seawater intrusion, hydrogeochemistry and aquifer management with most recent research findings and insights from the authors and other researchers. Both authors have an interest in the history of hydrogeology, and so a number of chapters have notes to position scientific advances within the historical context, which makes for even more interesting reading.

As the expanding field of coastal hydrogeology attracts more students, researchers, coastal engineers and water resources administrators, all will benefit from this book. With the subject of coastal hydrogeology becoming more relevant, interactions between coastal hydrologists and other professionals, such as marine scientists, environmental engineers and water managers, will expand. There will be more demand for training in coastal hydrogeology by consulting organisations, state and federal regulatory agencies, and industrial firms. I believe that teachers, students, researchers and practitioners concerned with coastal water issues will find this book to be informative, instructive, useful and timely.

John Cherry
Director, University Consortium
Adjunct Professor, University of Guelph
Distinguished Professor Emeritus, University of Waterloo
13 November 2018

Preface

We live in a time that is unprecedented in our planet's 4.5 billion year history as humans are transforming the natural environment everywhere on Earth. While the number of people continues to grow, so does their impact on the landscape as well as the rate at which changes are occurring. Among the most affected areas are coastal zones, not only because these are among the most densely populated parts of the world but also because they are among the first to bear the brunt of climate change and sea level rise.

This book brings together the available science about the subsurface part of the hydrological cycle in coastal zones. Coastal aquifers have unique issues compared to other aquifers. A particular problem is that of water supply, as subsurface water resources near the sea are more susceptible to salinisation than those in inland regions. Water levels are influenced by tides, which, as this book will show, can have important engineering implications. Moreover, the simultaneous occurrence of fresh- and saltwater in coastal aquifers leads to complex flow dynamics, as well as water quality changes. Only with comprehensive, science-based understanding of these processes can water and natural resources in coastal areas be managed sustainably.

The book's title, *Coastal Hydrogeology*, is also the name of the scientific specialisation that deals with aquifer systems that are under the influence of the sea, and where variable-density flow and tidal effects influence groundwater flow patterns. This specialist area of hydrogeology became established at the start of the twentieth century and has come of age since. It is a dynamic research field, as new discoveries are being made that shape our understanding of the functioning of coastal groundwater systems. A prominent example is the topic of subsurface discharge pathways to the oceans, which has seen a leap in research activity over the past two decades since it became apparent that large quantities of land-derived nutrients and other chemical substances can be delivered to the marine environment via groundwater. This growing interest is part of a broader appreciation of the importance of the connection between onshore and offshore aquifer systems more generally, which has remained understudied to date due to a scarcity of observational data.

Societal change also provides an impetus for coastal hydrogeology research. As coastal cities are expanding, more and more land is being reclaimed, giving rise to a suite of specific groundwater problems. Also, a growing proportion of the rising water demand is nowadays being met by desalination, often using groundwater as the source. The economic value of brackish and saline groundwater, previously considered uninteresting as a resource, is thus increasing. These developments require new and better knowledge of the consequences for coastal groundwater resources and the corollary effects of human activities on ecosystems.

One reason for writing this book was to provide the state of the art of the theory and practice in coastal hydrogeology. Another motivation for writing the book was to cover all the important aspects of coastal hydrogeology. The book's focus is therefore not restricted to seawater intrusion research; it also aims to complement other works by including subjects like land reclamation effects, tidal airflow, palaeo-hydrology and offshore groundwater reserves. No book to date has covered these topics extensively within the context of coastal groundwater research, yet they are important, as they have direct relevance to coastal zone management.

This book is aimed at students, academics, engineers and managers who wish to develop an in-depth understanding of various topics relevant to sustainably managing coastal groundwater resources. It discusses a wide variety of topics, including mathematical models, hydrochemistry, submarine groundwater discharge, coastal aquifer management, palaeo-hydrology, land reclamation and of course climate change and sea level rise. It is assumed that the reader has a basic knowledge of groundwater hydrology and is familiar with the elementary principles of geology, mathematics, physics and chemistry. Its intended use for students is therefore mainly at the graduate level. The book is also a resource for researchers and engineers working on groundwater-related issues in coastal areas. It serves as an access point to the scientific literature, as it provides ample references to other works to help the reader find additional information.

The research in the field of coastal hydrogeology is evolving quickly, and a problem we faced as authors was the vast number of journal articles and other scientific publications that exist. Nowadays, there are so many that it has become impossible to read all of them. Although this necessitated making a selection and, inevitably, leaving out even some good papers, we hope we have done justice to all those who have made important contributions to the science and practice of coastal hydrogeology. We are sorry if you miss your work in this book.

Numerous people have provided valuable contributions to this book. We would first of all like to thank our international colleagues who have reviewed one or more chapters of the book: Maike Gröschke, Yoseph Yechieli, Jacobus Groen, Romain Chesnaux, Maria Pool, Georg Houben, Gu Oude Essink, Willard Moore, Hailong Li and Xingxing Kuang. Their reviews have led to substantial improvements. We are very grateful to Xin Luo and Yi Liu, who drew most of the figures, which has considerably increased the clarity of the scientific message. We would like to acknowledge Jaouher Kerrou, Philippe Renard, Holly Michael, Xuan Yu, Pieter Stuyfzand, Ya Wang, Haipeng Guo, Elad Levanon and Antonio Bosch, who provided data or helped us reproduce figures from their work, as well as Atsushi Kawachi for locating the hard-to-find papers by Nomitsu et al. (1927) and Toyohara (1935). We are grateful to Zoë Pruce and Matt Lloyd at Cambridge University Press for their support during this long project.

It seems that no scientific textbook can be written without the authors sacrificing precious time together with family and friends. We are forever indebted to Tong Chen, Bilin Jiao, Bikun Jiao and Francis Boogaerdt for their understanding, support and, above all, patience.

Figure Credits

The following institutions and publishers are gratefully acknowledged for their kind permission to use figures based on illustrations in journals, books and other publications for which they hold copyright. We have cited the original sources in our figure captions. We have made every effort to obtain permissions to make use of copyrighted materials and apologise for any errors or omissions. The publishers welcome errors and omissions being brought to their attention.

Institutions and publishers	Figure number(s)
American Chemical Society	8.16
American Society of Limnology and Oceanography, Inc.	7.10
British Geological Survey	5.4
C.X. Chen, M. Lin and J.M. Cheng	4.3
D. Williams	5.16
Elsevier	
Advances in Water Resources	6.14, 7.4
Applied Geochemistry	5.18, 8.7, 8.14
Chemical Geology	8.15
Desalination	12.19
Earth-Science Reviews	7.9
Geochimica et Cosmochimica Acta	5.27, 5.31b
Journal of Applied Geophysics	12.4
Journal of Hydrology	3.8, 4.8, 5.13, 5.21, 5.22, 5.28, 6.4, 6.9, 6.10, 6.19, 7.5, 8.17, 9.3, 12.2, 12.9, 12.13
Marine Chemistry	7.14
Science of The Total Environment	5.30, 7.11
F.L. Li	12.12
G.H.P. Oude Essink	3.12, 10.4, 10.9
Geological Society of Hong Kong	9.4, 9.5, 9.6, 9.14, 9.15, 9.16
Geological Society of London	8.8, 12.5
International Association of Hydrological Sciences (IAHS)	2.6, 6.11, 6.16
International Atomic Energy Agency	6.3
IWA Publishing	12.11
J. Groen	8.23, 8.24, 8.25

(cont.)

Institutions and publishers	Figure number(s)
John Wiley & Sons, Inc.	
Applied Contaminant Transport Modeling, 2nd Edition	2.8
Geophysical Research Letters	11.6, 11.8, 11.9
Ground Water (National Ground Water Association)	5.17, 5.31a, 9.8, 9.9
Ground Water Microbiology and Geochemistry, 2nd Edition	5.5, 5.26
Groundwater Hydrology. 2nd ed.	1.1, 12.10
Hydrological Processes	5.32, 6.13, 7.1
Quarterly Journal of Royal Meteorological Society	10.6
Regional Ground Water Quality	5.8a, 5.20
Van Nostrand Reinhold	5.8b, 5.20
Water Resources Research (American Geophysical Union)	1.2, 3.14, 3.15, 3.16, 3.17, 4.5, 4.7, 5.36, 6.15, 9.7, 9.10, 9.13, 11.1, 11.2, 11.3, 11.4, 11.7
P.J. Stuyfzand	5.14, 12.14
Royal Society of South Australia	5.33
Soil Science Society of America	5.24
Springer Nature	
Groundwater in the Coastal Zones of Asia-Pacific	7.2
Environmental Geology/Environmental Earth Sciences	3.6, 5.19, 6.18
Hydrogeology Journal	3.2b, 6.5, 6.8, 8.5, 8.6, 8.9, 8.27
Irrigation and Drainage Systems	12.6
Nature	8.21
The Llobregat	8.10, 8.11, 8.12
Taylor & Francis Group	
Geodinamica Acta	8.28
A.A. Balkema Publishers	5.11
LLC Books	5.3, 5.7
United States Geological Survey	1.3, 1.4, 5.15, 6.2, 6.12, 8.22, 10.5, 12.16

1
Introduction to Coastal Groundwater Systems

1.1 Coastal Zones

Coastal zones are the areas between land and sea that are influenced by marine as well as terrestrial processes (Crossland et al., 2005). The coastal zone is a fuzzy concept for which various definitions have been proposed (Custodio and Bruggeman, 1987; Crossland et al., 2005). For example, it has been defined as the land within 100 km of the shoreline (Small and Nicholls, 2003; SEDAC, 2007; Lange et al., 2010) or the area between 200 m elevation on land and 200 m water depth offshore (Crossland et al., 2005). Even more specific is the definition of the low-elevation coastal zone (LECZ), which is understood to be the contiguous and hydrologically connected zone of land along the coast and below 10 m of elevation (McGranahan et al., 2007). This zone is of particular importance because it is susceptible to flooding by storm surges and tsunamis as well as by sea level rise.

Estimates about how many people live in coastal zones vary depending on the adopted definition and the global population data sets used. Globally, the LECZ covers 2.599×10^6 km^2, which equates to only 2.3% of the total land surface area of all coastal countries. Yet, in the year 2000, 10.9% (625 million people) of their population lived there. China, India, Bangladesh, Indonesia and Vietnam together accounted for more than half of the global LECZ population. Moreover, with 241 people km^{-2}, the population density of the LECZ was about five times the global average (47 people km^{-2}), and the population growth rates in coastal zones are significantly higher than they are in non-coastal zones (Neumann et al., 2015). Various other statistics also show that humans gravitate towards the coast. For example, based on 2003 data, Martínez et al. (2007) found that 41% of the world population, approximately 2.4 billion people, lived within 100 km from the coastline. Within this zone, the population density is highest below an elevation of 20 m (Small and Nicholls, 2003). Their attractiveness stems from the fact that coastal zones offer natural resources, notably fisheries; provide embarkation points for trade and possess endless possibilities for recreation and tourism (Merkens et al., 2016).

The geological, geomorphological and climatological conditions vary tremendously along the world's coastlines. Different combinations of these conditions make up for a wide variety of coastal zone types. Shorelines can consist of solid rocks or loose sediments, and their topography may range from low-lying flat areas to high, steep cliffs. In dry areas, deserts line the coast, but where rain is abundant or rivers bring enough water from the hinterland, lagoons and wetlands form, especially in deltas and low-lying coastal zones. Large areas may be under tidal influence, and in estuaries, seawater can penetrate tens of kilometres upstream, forming a so-called salty tide in the river channel. The shoreline may be dominated by beaches, dunes, mangroves, salt marshes, hills or cliffs.

Because they are at the interface between continental freshwater and marine saline ecosystems, coastal zones are characterised by a high biological diversity. Natural coastlines are increasingly influenced by human activities. The current rapid population growth, land use change and urbanisation are creating severe environmental problems, such as ecosystem and biodiversity loss, land degradation and pollution (Martínez et al., 2007; Shi and Jiao, 2014). It is expected that the pressures on coastal aquifer systems will become more severe as the global population continues to grow (Neumann et al., 2015).

The growth of the population in coastal areas is a main driver for a rise in water demand. Not only is the number of people rising but higher living standards also bring with them a higher consumption per person. An even stronger driver is increased food production, as irrigated agriculture accounts for most of the global freshwater withdrawal (Poore and Nemecek, 2018). A high localised water demand occurs in urban centres or where there are major industrial centres or tourist destinations.

Freshwater is a limited and threatened resource everywhere in the world, but this is particularly true for coastal areas, because their vicinity to seawater entails a high risk of salinisation of available water resources. Groundwater abstraction to meet the rising water demand often exceeds the regenerative capacity of coastal aquifers, and this causes seawater intrusion. At the same time, human activities are causing groundwater quality deterioration. Sources of anthropogenic contaminants include agriculture, landfills, wastewater and chemical spills. The combined threat of contamination from the land surface and by seawater intrusion that leads to a shrinkage of usable fresh groundwater reserves has been labelled the *coastal groundwater squeeze* (Michael et al., 2017).

From a hydrogeological perspective, coastal zones are usually understood to be the regions where fresh groundwater of meteoric origin and saline waters of marine origin meet and interact. The extent of a coastal zone varies according to the type of problem being addressed and the study objectives, so it depends on the relevant groundwater processes and is not defined on the basis of a general distance, elevation or boundary. If the tidal effect is the topic of investigation, then the study area comprises the onshore area where the tidal effect is significant, which can be a few hundred metres to a few kilometres away from the coast (Merritt, 2004). To understand the regional flow in a complicated coastal multilayered aquifer–aquitard system, however, usually a comprehensive groundwater regime, which stretches from the recharge to the discharge areas, has to be included (Figure 1.1). In some cases, the discharge area may be offshore and could even extend all the way to the edge of the continental shelf (e.g. Wilson, 2005).

Coastal zones evolve by the interplay of geological processes such as tectonic movement and sea level change, as well as erosion and sedimentation. In more recent times, humans have become a dominant geological agent by executing large-scale land reclamations,

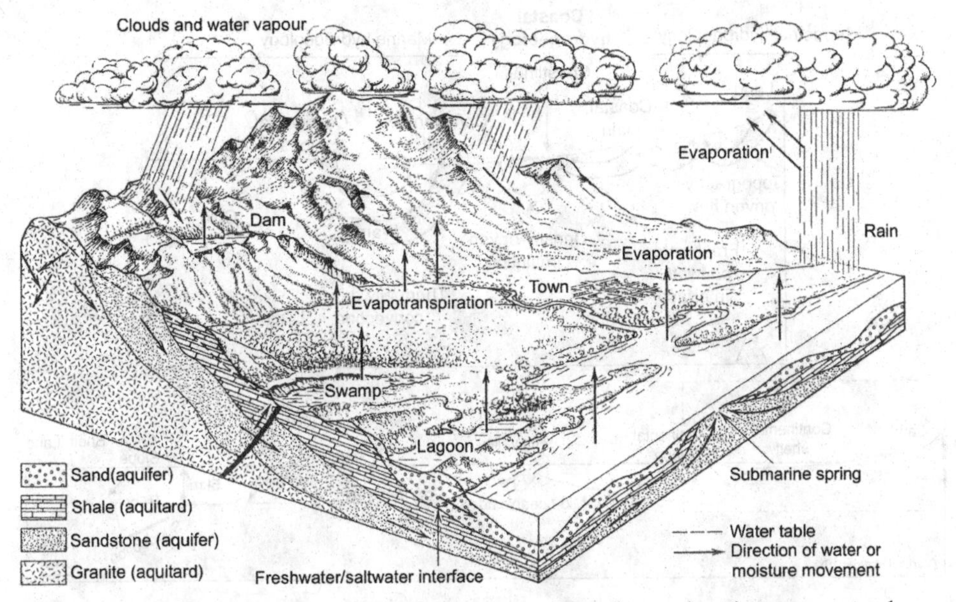

Figure 1.1 The hydrological cycle in a coastal zone. In a natural coastal environment, groundwater flows into the sea by diffuse discharge or concentrated springs, located either along the coastline or in the seabed. The focal points for discharge are outcrops of a confined aquifer or fracture or fault zones. Modified from Todd (1980).

dredging, damming rivers and building coastal defence structures. These processes change the coastal landscape directly or indirectly by impacting the natural erosion and deposition processes. The rate of change by anthropogenic interventions is typically much faster than it is for geological processes, although natural catastrophic events like earthquakes or flood waves can have almost instantaneous impacts.

1.2 Coastal Hydrogeology

Hydrogeology is a branch of geology that deals with the occurrence, movement and chemical state of groundwater, as well as the chemical and physical reactions between water and rock. The study of groundwater and fluid-rock interaction below the seafloor is called *marine hydrogeology* (Fisher, 2005). Traditional hydrogeology excludes the ocean realm and might therefore be equivalently called *terrestrial hydrogeology*. Because marine hydrogeology is not concerned with freshwater resources for human use, its scope is less applied than terrestrial hydrogeology. Instead, it focusses more on topics such as global geochemical cycles, the role of groundwater in ocean floor geochemical processes and the importance of fluid flow for the energy budget of the oceanic crust (Fisher, 2005; Judd and Hovland, 2007; Becker and Fisher, 2008; Kummer and Spinelli, 2009).

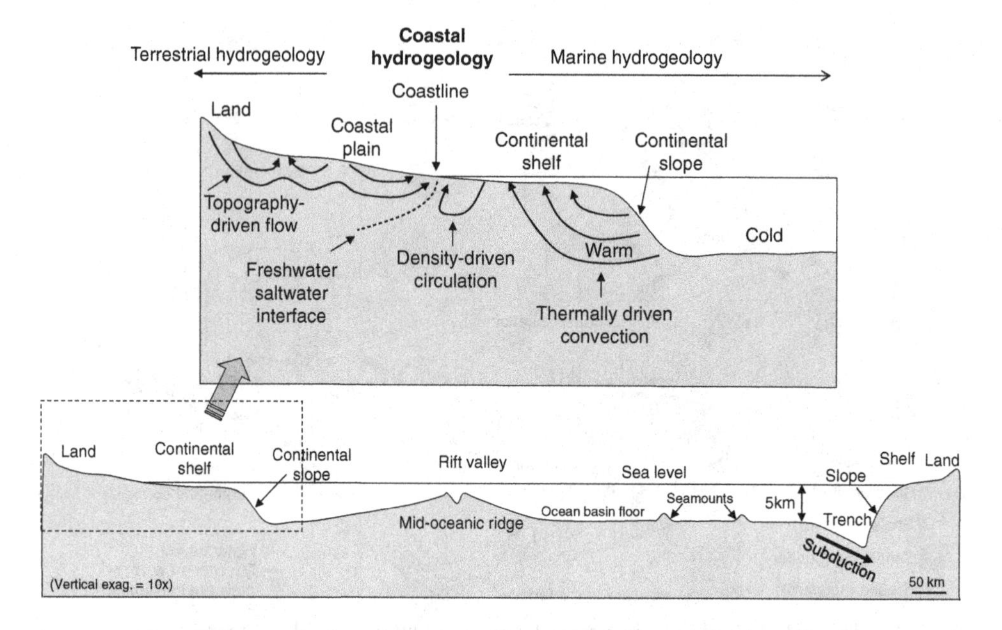

Figure 1.2 Sketch of an ocean basin with inset showing the geographic areas covered by terrestrial hydrogeology, marine hydrogeology and coastal hydrogeology. Inset modified from Wilson (2005).

The physiography of coastal areas was important in determining early human migration patterns (Armitage et al., 2011), and numerous archaeological excavation records testify to human occupation dating back thousands of years (Gaur and Vora, 1999; Zong et al., 2009). These human activities have led to early groundwater exploitation in coastal aquifers. At an early Neolithic Hemudu culture site near the mouth of the Yangtze River in China, a 5600-year-old wooden well was unearthed (Jiao, 2007). This pre-historic well suggests that the people who lived there knew that groundwater had a better and more stable quality than the river water influenced by the tides. Coastal hydrogeology did not become established as a scientific specialisation, though, until the first decade of the twentieth century (Houben and Post, 2017). Nowadays coastal hydrogeology forms one of the most important frontier topic areas in groundwater science, one reason for this being that such a large portion of the world's population lives in coastal areas, giving rise to excessive pressure on water resources, particularly on groundwater. Further human interventions along the coast, such as land reclamation and deep foundations of buildings, as well as natural factors, such as long-term sea level changes, also modify the natural coastal groundwater flow system and the interaction between seawater and groundwater. The resulting groundwater-related resource, environmental and engineering challenges call for enhanced understanding of coastal groundwater systems, as well as solutions to the emerging problems (Michael et al., 2017; Post et al., 2018c). This offers unprecedented novel research opportunities.

There are no clear-cut boundaries between terrestrial, coastal and marine hydrogeology. Geographically, the scope of coastal hydrogeology may stretch offshore to the continental shelf, where the coastal aquifers discharge, or landward to mountainous areas, which provide recharge to the coastal aquifers (Figure 1.2). Coastal hydrogeology comprises a wide variety of topics, including sea level–induced groundwater level fluctuation, submarine groundwater discharge, seawater intrusion, groundwater-dependent coastal ecology, impact of anthropogenic activities on coastal groundwater regimes, palaeo-hydrogeology and groundwater management.

1.3 Coastal Groundwater Systems

Coastal aquifers have some unique characteristics compared to other aquifers. These include variable-density flow with freshwater and seawater as two end-members, groundwater level fluctuations in saturated zones and airflow in unsaturated zones induced by tidal fluctuations, groundwater chemical zoning due to seawater and fresh groundwater mixing and associated chemical processes, submarine groundwater discharge as an important flux and chemical exchange between the land-sea interface and transient but slow changes of the system in response to natural factors like sea level change and anthropogenic factors such as large-scale land reclamation along the coast.

Because of the great diversity of coastal landscapes and geology, as well as the anthropogenic influences impacting them, there is no such thing as a typical coastal groundwater system. Nonetheless, there are certain communalities. The first is that, somewhere in the subsurface, terrestrial groundwater meets intruded seawater. The distinct salinity difference between the two water types, albeit that the details vary per region (Section 5.2), is a feature of all coastal groundwater systems. The second is that, because of its higher salinity, the seawater has a higher density than freshwater; generally, it is greater by approximately 2.5%. Albeit modest, this difference has extremely important consequences for the groundwater flow processes. It gives rise to the typical wedge shape of a body of intruded seawater and makes it so that under equilibrium conditions, the seawater protrudes inland from the coastline (Chapter 6).

The quintessential depiction of a seawater body in a coastal aquifer is displayed in Figure 1.3. Because of its higher density, the seawater extends inland from the coastline and is wedged between the bottom of the aquifer and the overlying freshwater. The ratio of the densities of freshwater and saltwater is one of the primary controls on the position of the boundary between the two water bodies. A dynamic equilibrium exists when the position of the saltwater wedge does not change. The word *dynamic* is meant to indicate that the groundwater is not stagnant: the freshwater flows towards the sea, and as is explained in Chapter 6, a (slow) circulatory flow exists within the saltwater wedge (Figure 1.3).

Figure 1.3 also makes clear that the separation between fresh and saline groundwater is never completely sharp; instead, the two are separated by a zone with gradually varying

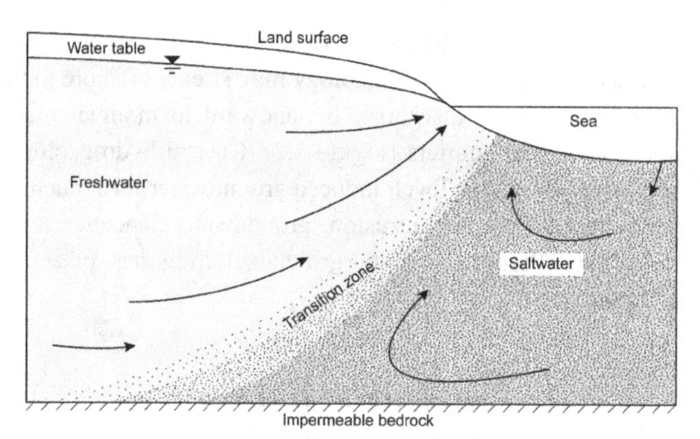

Figure 1.3 Idealised depiction of a wedge of seawater intruded into a coastal unconfined aquifer. The saltwater is separated from the fresh groundwater by a transition zone in which the salinities gradually change from freshwater to seawater values. From Cooper et al. (1964).

intermediate salinities, which is called the *transition zone*, *mixing zone* or *zone of diffusion*. Transition zones can be only a few decimetres wide, but some can reach up to tens of metres in width. If the transition zone is very narrow (relative to the thickness of the freshwater-filled portion of the aquifer), it is sometimes referred to as the *interface*. In this book, this term is reserved, though, for the sharp boundary that is assumed to exist for the purpose of simplifying the mathematical treatment of seawater intrusion problems. Seawater and freshwater are sometimes separated by zones of brackish water that extend over many kilometres that form when the coastline shifts. In that case, it becomes hard to identify the limits of the transition zone, and the concept loses its meaning. The processes that lead to the different kinds of salinity distributions are discussed in Chapters 6 and 8.

1.3.1 Sedimentary Aquifer Systems

Aquifer systems consisting of alternations of various types of sediments form complex hydrogeological settings because the permeability of the geological units is highly variable (e.g. Jiao et al., 2010; Saha et al., 2014). Clay layers can be virtually impermeable, whereas coarse-grained strata can conduct groundwater with greater ease. The permeability is also determined by the degree of consolidation and cementation of the sediments. Coastal plain aquifer systems are typically composed of several aquitards and aquifers, and the sequence generally dips and thickens towards the sea. Owing to past sea level variations, the sediment facies shift between marine and terrestrial types. When sea level was high, deposition of finer-grained sediments dominated where full marine conditions existed, forming strata of clay-rich deposits. When sea level was low, such marine deposits became weathered and eroded, and at the same time, fluvial, aeolian and lacustrine sedimentation resumed, which, depending on the flow conditions, resulted in the deposition of sediments with a range of grain sizes.

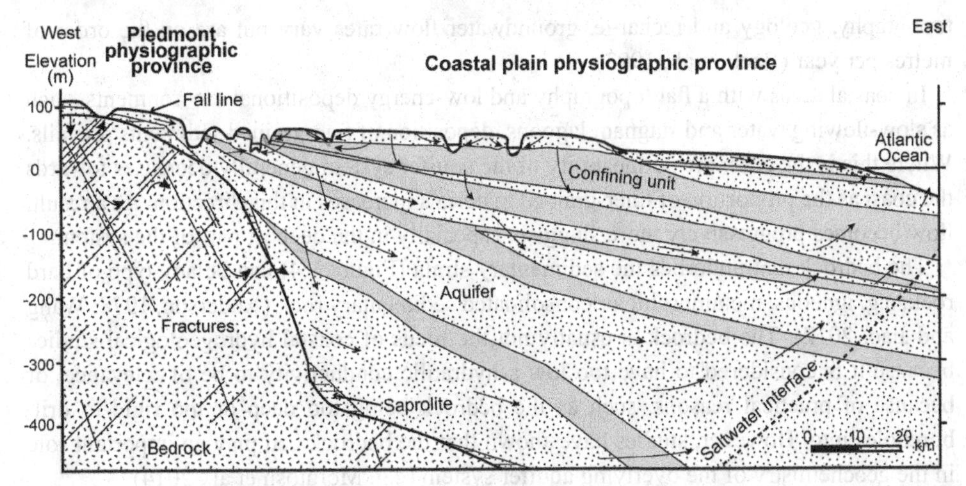

Figure 1.4 Hydrogeological cross section of the Virginia Coastal Plain Province as an example of a sedimentary coastal plain aquifer system. Modified from Nelms et al. (2003).

Alternations of these depositional environments in time resulted in complex, inter-layered fine- and coarse-grained sediment sequences. Individual aquifers, usually consisting of sand and gravel, are separated by aquitards dominated by clay and silt. The degree of aquifer connectivity and inter-aquifer groundwater fluxes are determined by variations of the extent, thickness and lithology of the aquitards (Cherry et al., 2004). In the presence of aquitards that prevent the downward flow of seawater, fresh groundwater flow originating on land may extend offshore, or have done so in the geological past (Post et al., 2013). Conversely, seawater can become trapped inland when low-permeability strata prevent it from migrating (Miller, 2000). The effect of changing flow patterns over (geological) time in combination with the lithological variability gives rise to complex and sometimes counter-intuitive groundwater salinity patterns.

Some coastal plain aquifer systems are volumetrically dominated by clay-rich layers. When aquifers in such systems are extensively exploited, a large part of the abstracted water derives from the aquitards, and the pressure decrease causes land subsidence as the sediments compact (Chen et al., 2003a; Phien-wej et al., 2006; Guo et al., 2015). The processes within aquitards can also exert a control on the chemistry of the adjacent aquifers by releasing salts and ammonium (Mastrocicco et al., 2013; Wang et al., 2013; Colombani and Mastrocicco, 2016) or arsenic and arsenic-mobilising solutes by natural consolidation or in response to pumping (Erban et al., 2013).

An example of a flow regime in a sedimentary coastal aquifer system is depicted in Figure 1.4. Overall, the flow is approximately horizontal on a regional scale. In the hinterland recharge areas, where the aquifer system is dominated by permeable materials with active groundwater flow, the vertical component of flow is downward. Near the shoreline in the coastal plain, the vertical component of the flow can be upward, and springs and seepage zones may be formed in topographic depressions and near the shoreline. Depending on the topography, geology and recharge, groundwater flow rates vary but are on the order of metres per year (Back et al., 1993).

In coastal areas with a flat topography and low-energy depositional environments, such as slow-flowing water and stagnant lagoons, deposition of fine-grained sediments prevails. Where this is the case, the permeability of the aquifer system generally decreases towards the coast as the proportion of finer-grained materials increases in this direction. As a result, flow becomes progressively more sluggish, especially in the deepest part of the system.

The bedrock that underlies the sedimentary strata (Figure 1.4) can be any type of hard rock, e.g. igneous, metamorphic or consolidated sedimentary rock (Anderson, 1978; Wang and Jiao, 2012). The bedrock is usually not the focus of coastal hydrogeological studies because it is considered to have too low a hydraulic conductivity to be of relevance, or because of practical reasons, such as the bedrock being too deep or too rigid to drill boreholes into. However, studies have shown that the bedrock can play an important role in the geochemistry of the overlying aquifer system (e.g. McIntosh et al., 2014).

1.3.2 Limestone Aquifers

Limestone aquifers consist of carbonate minerals. There is a wide variety of limestone types, and their textures and grain sizes are therefore highly variable. Fine-grained marine limestone can have a very low permeability, whereas bioclastic limestone with lots of shell or coral fragments can have many voids that make it permeable. Limestone aquifers are special in the sense that the carbonates dissolve relatively easily as groundwater flows through them. As such, they develop porosity, and permeability increases with time. Since initially, a pristine limestone has a relatively low permeability, groundwater flow will be preferentially along faults and fracture zones, which means that dissolution features will be aligned with such zones of preferential flow. The enhanced permeability due to the dissolution of the rock reinforces this, and as a result, conduits of various size develop. Carbonate rock aquifers that have developed extensive dissolution networks are called *karst aquifers*.

Water flow velocities and volumes inside these features can become high, and flow often resembles that of rivers rather than the slow flow of groundwater through the fine pores of a sedimentary aquifer. Discharge is often localised and in the form of springs, which are sometimes located offshore (Stringfield and LeGrand, 1971). The salinity distribution tends to be highly irregular, as some parts of the subsurface are subject to rapid flow, whereas groundwater can be quasi-stagnant in the less-weathered, less-permeable parts. The seasonal variability of the groundwater salinity can be very high in the permeable parts of the system. Well-studied coastal limestone aquifer systems include those in Florida, USA and Yucatan, Mexico (Barlow, 2003; Bauer-Gottwein et al., 2011), the karst regions of the Mediterranean (Fleury et al., 2007) and carbonate islands, including atolls (Vacher, 2004).

Karstification can be more significant near the water table, where flow is dynamic and groundwater is rich in carbonic acid. The water table elevation in coastal aquifers is controlled by sea level, which has fluctuated over geological time (Section 8.2). Some studies have demonstrated that the position of the zone of karstification in coastal limestone is strongly related to the past sea levels (Audra et al., 2004; Fleury et al., 2007).

1.3.3 Hard Rock Aquifers

Hard rock aquifers are made up by igneous rocks and metamorphic rocks. Just like in limestone aquifers, the flow in hard rock aquifers is mainly localised around faults and joints (Park et al., 2012a), but because the chemical weathering processes in hard rocks are much more sluggish than they are in limestone, the porosity and permeability development is not as pronounced. Instead, permeability distributions in hard rock systems are largely controlled by tectonic processes that result in mechanical weathering by fracturing. Hard rock aquifers are among the least productive aquifer systems, and their strong geological heterogeneity makes them difficult to investigate.

In rocks of volcanic origin, fissures sometimes give the material a high porosity and permeability. Weathering and soil formation make the permeability generally higher near the surface. Flow directions are highly dependent on the orientation of the fissures and the occurrence of impermeable structures such as dykes (Custodio and Bruggeman, 1987). The high degree of structural control on the flow patterns makes it hard to draw generalisations about the hydrogeological behaviour of these systems, and the characterisation of the interaction of freshwater and seawater within these systems remains a challenge.

1.4 Coastal Groundwater Chemistry

1.4.1 Salinity

Salinity is a measure of the total amount of solutes, or more exactly, the sum of the masses of all the individual ions in the water. Absolute salinity is defined as the ratio of the mass of dissolved material in water to the total mass of water (Millero et al., 2008) and is often also referred to as the salt mass fraction. The absolute salinity of standard seawater, which is a carefully defined reference water composition in oceanography, is exactly 35.16504 g kg^{-1}. Since the absolute salinity is tedious to measure due to analytical difficulties, a so-called practical salinity has been defined, which is based on measurement of the conductivity, temperature and pressure of seawater. It is not often used by hydrogeologists, but it is important that they understand its definition so as to avoid confusion when interacting with oceanographers. The practical salinity, which is a dimensionless quantity, of standard seawater is 35.

In hydrogeology, the amount of dissolved ions is most commonly expressed by the mass of dissolved material per litre of water, which is referred to as the total dissolved solids (TDS) concentration, and the unit is mg l^{-1} or g l^{-1}. At 25°C, standard seawater has a density $\rho = 1.02334$ kg l^{-1}, so its TDS is $35.2 \times 1.02334 = 36.0$ g l^{-1}. As the density is a function of temperature, a TDS expressed per unit of solution volume can change as the temperature changes. The absolute salinity is invariant with temperature. *Salinity* and *TDS*

Table 1.1 *Salinity Classes Based on Total Dissolved Solids (TDS) Concentrations*

	TDS (mg l^{-1})	Description
Fresh	0–1000	Sufficiently dilute to be potable; includes rain and is found in rivers, lakes and groundwater
Brackish	1000–10 000	Too saline to be potable; found in rivers, lakes and groundwater subject to evaporation or mixing with seawater
Saline	10 000–36 000	Less than or equal to seawater; found in rivers, lakes and groundwater subject to strong evaporation or intensive mixing with seawater
Hyper-saline	36 000–100 000	Significantly more saline than seawater; found in evaporated seawater or groundwater in discharge zones in arid regions
Brine	>100 000	Seawater or groundwater having undergone intense evaporation or dissolution of rock salt

Note. Descriptions after Post et al. (2018b).

are often used interchangeably, but as the above discussion shows, this is strictly not correct.

Groundwater is often divided into salinity classes (Table 1.1), mainly as an aid in interpreting hydrochemical data or simply for the purpose of discussion. The TDS concentration is often used to define salinity boundaries, but chloride concentrations are sometimes also used (Stuyfzand, 1989). Based on the suitability for human consumption, a TDS value of 1000 mg l^{-1} is usually regarded as the upper limit of freshwater (Alley, 2003). The upper limit for saline water in Table 1.1 is standard seawater, but depending on the salinity of local seawater, it may be more meaningful to set the boundary at a different value.

1.4.2 Sources of Salinity

Besides seawater intrusion, different sources of salt can lead to freshwater salinisation in coastal aquifers. Van Weert et al. (2009) distinguished between three different origins of saline groundwater: (1) marine, (2) natural terrestrial and (3) anthropogenic terrestrial (Table 1.2). Groundwater of multiple origins may form mixtures in coastal aquifers.

Within the class of marine sources of groundwater, connate saline water is the seawater that was trapped in the rock that formed underwater. These rocks may be sedimentary but can also be oceanic basalts, for example. The connate seawater thus has the same age as the rock. Marine transgression groundwater formed when the coastline was further inland and the sea level was relatively high compared to the land surface in the geological history (Chapter 8). It is thus younger than the rock of which it occupies the pore space. The length during which connate or transgression waters can be preserved depends mainly on the permeability of the host rock and the

Table 1.2 *Genetic Categories of Saline Groundwater*

Main class of origin	Genetic category by salinisation mechanism	Typical environment at the time of origin
(i) Marine origin	Connate seawater	Coastal zone (onshore or offshore)
	Marine transgression water	Coastal zone (onshore or offshore)
	Short-lived, episodic flooding	Coastal zone (onshore)
	Lateral seawater intrusion	Coastal zone (onshore)
	Washed-down sea spray	Coastal zone (onshore)
(ii) Terrestrial origin – natural	Evaporative concentration	Shallow water table zones or discharge zones in arid climates
	Dissolution of rock salt	Signficant occurrences of subsurface halite or other soluble rocks
	Salt filtering by membrane effects	Deep sedimentary basins with abundant clay
	Igneous activity	Regions of igneous activity
	Cryogenic concentration	Permafrost regions
(iii) Terrestrial origin – anthropogenic	Irrigation return flow	Agricultural regions in arid and semi-arid zones
	Anthropogenic pollution	Built environments

Note. Largely after Van Weert et al. (2009).

groundwater flow rates, but in some cases, the water can remain preserved for millions of years (Sanford et al., 2013; Ferguson et al., 2018).

Saline groundwater may also form by intrusion following incidental seawater flooding of low-lying coastal areas (Section 6.2.5.2), or it intrudes laterally (Section 6.2.1). The latter process is mostly associated with groundwater abstraction close to the shore, but it may also occur when groundwater recharge rates decrease, or sea level rises (relative to the water table). Such processes are extensively discussed in Chapter 6. Sometimes groundwater has a naturally high salinity because of the effect of sea spray, which is the inland deposition of marine-derived salt particles that are released into and transported via the atmosphere. This process has a strong effect on the salinity of shallow groundwater in areas with frequent strong winds (Van Sambeek et al., 2000), especially when dense vegetation is effective in capturing atmospheric aerosols (Stuyfzand, 1993; Bresciani et al., 2014).

Van Weert et al. (2009) distinguished between several types of natural terrestrial saline groundwater (Table 1.2). The salinity may increase due to evapotranspiration at the land surface (evaporative concentration), or by dissolution of highly soluble rocks such as halite or gypsum. In deep sedimentary basins, clay or shale strata may act as semi-permeable membranes that can filter charged solutes, leading to their enrichment (Section 5.7.7). Juvenile groundwater, formed as a result of igneous activity, is enriched in elements that do not fit in the crystal structure of the minerals that form during the cooling of magma. These incompatible elements, such as for example chlorine, are sequestered into the water that forms during the crystallisation process, which renders them saline. A similar effect occurs during the freezing of seawater as the ions do not fit in the crystal structure of ice, which renders the remaining fraction of water highly saline. This process has been invoked to explain the occurrence of some highly saline groundwaters at high latitudes.

There are two types of terrestrial groundwater with an elevated salinity due to anthropogenic influence. The first is formed by evapotranspiration of water used for irrigation by which dissolved salts get concentrated. Sometimes the irrigation water itself has a higher salinity than the groundwater, for example when wastewater is used. The irrigation water that did not evaporate reaches the water table is called irrigation return flow and is a major contributor to groundwater salinisation in semi-arid areas (Stigter et al., 1998). The second is groundwater contaminated by salt derived from road salt, agricultural and domestic effluents, fertilisers and spills from gas and oil fields or desalinisation plants. Usually these pollutants only cause localised groundwater salinisation.

1.4.3 Problems with High Salinity

Although water with a TDS below 1000 mg l^{-1} is classified as freshwater (Table 1.1), for the water to be pleasant to drink, its salinity should be much lower. Most people find water with a TDS less than 600 mg l^{-1} still acceptable (Van Weert et al., 2009). Taste thresholds for chloride depend somewhat on the associated cations and range from 200 to 300 mg l^{-1} for potassium, sodium and calcium. Chloride concentrations of over 250 mg l^{-1} are easily detected by modern consumers as they are used to the taste of water with lower-level chloride (World Health Organization, 2006). From a historical perspective it is interesting to note that the landmark papers by Drabbe and Badon Ghijben (1889) and Herzberg (1901) (Section 3.2) both contained a treatise about the taste limits for salinity (cf., Houben, 2018; Post, 2018). Salinity threshold concentrations are primarily based on aesthetic grounds, and no salinity or chloride guideline value based on human health is offered for drinking water (World Health Organization, 2006). There are studies suggesting that sustained consumption of water with a relatively high salinity may cause hypertension (Khan et al., 2008).

A high salinity in the soil weakens a plant's ability to extract water from the soil. Excess salt can thus reduce the availability of water to the plant and make it hard for the plant to absorb the nutrients it requires. Very high salinities can have a toxic effect on plants because of the high concentration of certain ions, especially chloride, sodium and boron (Van Weert et al., 2009). Water with salinity lower than 3000 mg l^{-1} can be drunk by livestock, but higher levels can cause physiological problems or even death.

Water of high salinity can be corrosive and a damage infrastructure consisting of bricks, concrete and metals. When in contact with saline groundwater, efflorescence may form on concrete and bricks, which causes scaling and peeling of the surfaces. Salts can then enter further into the inner part of concrete and cause steel bars become rusty. Other problems related to salinity include the elevated hardness of water, which may require to employ more detergents and soap (Van Weert et al., 2009) and, in hydrogeology, the corrosion of measurement equipment and clogging of laboratory equipment by salt residues from saline samples.

1.4.4 Units of Concentration Measurement

The chemical composition of seawater and coastal groundwater, as well as the processes that influence it, is discussed in detail in Chapter 5. Across this book, several units of concentration measurement can be encountered, which are briefly presented here. The SI unit for chemical concentrations in water is the mole m^{-3}, with one mole being equal to 6.022 10^{23} entities such as an element or a molecule. In groundwater studies concentrations are often expressed as mol (l of solution)$^{-1}$ (called molarity), mol (kg of solution)$^{-1}$, or mol (kg of H_2O)$^{-1}$ (called molality). When concentrations are low, a prefix m (mmol = 10^{-3} × mol) is used for millimole, or μ (μmol = 10^{-6} × mol) for micromoles. Since the density of samples of coastal groundwater can be significant, it is important to avoid ambiguity by clearly specifying if a concentration is expressed per litre of solution, per kg of solution or per kg of H_2O. The latter unit is commonly used in geochemistry but less often by hydrogeologists. A concentration expressed per kg of solution is easily converted to per litre of solution by multiplying it by a sample's density (Table 1.3).

Despite the mole being the SI unit, laboratory analysis are more often than not reported as mg l^{-1} or as parts per million. The latter is a mass based concentration ratio so 1 part per million is equivalent to 1 mg per kg of solution. Thus for freshwater with density ρ = 1 kg l^{-1}, a concentration in mg l^{-1} is numerically the same as in ppm, but of course deviations will occur at higher salinities.

Another unit of measure is the equivalent (eq), or milliequivalent (meq), which represents the number of moles of a molecule multiplied by its charge (Table 1.3). Just like a mole represents the number of elemental or molecular entities, an equivalent represents the number of charges. It is frequently used for reactions involving charge effects, such as cation exchange (Section 5.3.3). It is also helpful for quality control purposes as it can be used to calculate the electrical charge balance of a water sample (Table 1.3).

1.5 Hydraulic Heads in Coastal Aquifers

In terrestrial settings with groundwater of uniform density, groundwater flow can be determined using the classical formulation of Darcy's law:

$$q = -K\frac{dh}{dl} \tag{1.1}$$

where q [L T^{-1}] is the specific discharge, or Darcy velocity, K [L T^{-1}] is the hydraulic conductivity and the dimensionless term dh/dl represents the change of the hydraulic head

Table 1.3 *Concentrations of major ions in standard seawater expressed in various units*

	A	B	C	D	E	F	G	H
				Concentration unit				
Solute	Molar mass ($g\ mol^{-1}$)	Charge	ppm	mmol (kg solution)$^{-1}$	mmol (kg H_2O)$^{-1}$	mg l^{-1}	mmol l^{-1}	meq l^{-1}
Na^+	22.990	1	10 781	469	486	11 033	480	480
Mg^{2+}	24.305	2	1283.7	52.8	54.7	1314	54.1	108
Ca^{2+}	40.078	2	412.08	10.3	10.7	422	10.5	21.0
K^+	39.098	1	399.10	10.2	10.6	408	10.5	10.4
							Sum cations	619.5
Cl^-	35.453	−1	19 353	546	566	19 804	559	−559
SO_4^{2-}	96.063	−2	2712.4	28.2	29.3	2776	28.9	−57.8
HCO_3^-	61.017	−1	104.81	1.72	1.78	107	1.76	−1.76
CO_3^{2-}	60.009	−2	14.340	0.24	0.25	15	0.24	−0.49
							Sum anions	−618.6

Note. Data from Millero et al. (2008). One kilogram of seawater contains 0.964835 kg H_2O. The density of standard seawater at 25°C is ρ_s = 1.02334 kg l^{-1}. Relation between columns: D = C / A; E = D / 0.964835; F = C × ρ_s; G = D × ρ_s = F/A; H = G × B. The charge balance of this sample is (Sum cations + Sum anions)/(Sum cations − Sum anions) = (619.5 − 618.6)/(619.5 + 618.6) = 0.07%.

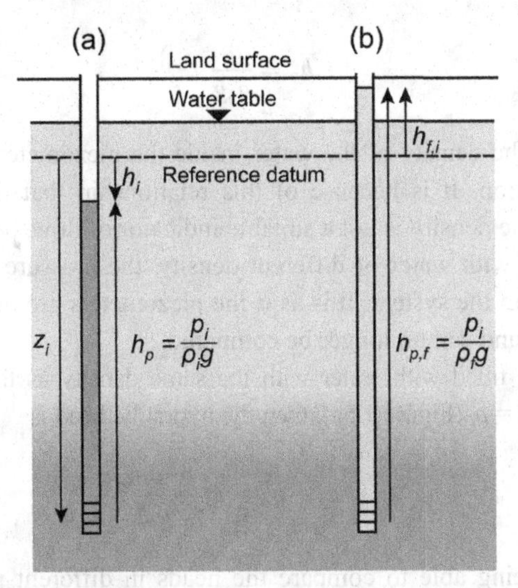

Figure 1.5 Piezometer filled with (a) water of the same density as the groundwater at the well screen, indicating point-water head and (b) freshwater, indicating freshwater head.

h [L] per unit distance l [L] in the direction of the flow. The head is a measure of the energy per unit weight of groundwater (Hubbert, 1940; Freeze and Cherry, 1979), and it can be measured directly in the field with a piezometer (Figure 1.5). A well-constructed piezometer is open at one point in an aquifer only. In practice this means that the slotted section, or the screen, of the piezometer is on the order of a metre because water has to be able to exchange fast enough between the aquifer and the piezometer so that it always indicates the groundwater pressure at the screen. But in principle the measurement represents a point measurement at a specific position and depth, and therefore the hydraulic head has also been termed point-water head (Lusczynski, 1961). In the following sections, the notation h_i will be used for the hydraulic head to emphasise that it represents the value at point i in the system.

Groundwater is able to do work, i.e. overcome its internal friction and the friction by the pore walls during flow, because it contains two main types of energy: potential energy and pressure energy (kinetic energy is small enough to be neglected, which may be an invalid assumption in karst aquifers). The hydraulic head h_i therefore consists of two terms, the elevation head, z_i [L] and the pressure head h_p [L]:

$$h_i = z_i + h_p \tag{1.2}$$

In the field, the elevation head z_i is the vertical distance of the piezometer's screen from a reference datum, typically sea level, and h_p is the length of the water column within the piezometer (Figure 1.5a). h_p is a measure of the pressure p_i [M L^{-1} T^{-2}] in the aquifer because of the relationship:

$$h_p = \frac{p_i}{\rho_a g} \tag{1.3}$$

where ρ_a [M L^{-3}] is the density of the water inside the piezometer and g [L T^{-2}] is the gravitational acceleration. It is because of this relationship that the hydraulic head in groundwater of variable density is not a suitable indicator of flow, because when different piezometers are filled with water of different density, the pressure head h_p is no longer uniquely related to p_i in the system. It is as if the piezometers are instruments that are all calibrated differently, and can no longer be compared.

If the piezometer is filled with water with the same density as the groundwater at the measuring point, i.e. $\rho_a = \rho_i$ (Figure 1.5a), then the hydraulic head is

$$h_i = z_i + h_p = z_i + \frac{p_i}{\rho_i g} \tag{1.4}$$

For the purpose of being able to compare the heads in different piezometers, a single reference density can be introduced that replaces ρ_i in Eqn (1.4). It has become customary to use the density of freshwater ρ_f for this, which leads to the definition of the equivalent freshwater head (or freshwater head in short):

$$h_f = z_i + h_{p,f} = z_i + \frac{p_i}{\rho_f g} \tag{1.5}$$

where $h_{p,f}$ is the length of a column of freshwater of the same weight as the groundwater pressure p_i. Since for saltwater $\rho_f \leq \rho_i$ it follows that $h_{p,f} \geq h_{p,i}$ and $h_f \geq h_i$, so the water level would rise higher in a piezometer if the more saline water inside were replaced by freshwater.

By equating p_i in Eqns (1.4) and (1.5), it follows that h_f is related to the hydraulic head h_i according to

$$h_f = z_i + (h_i - z_i)\frac{\rho_i}{\rho_f} \tag{1.6}$$

Even when all heads are expressed using the same reference density, they can still not be used in the same way to calculate the specific discharge using Darcy's law (Eqn (1.7)) as in single-density systems. The reason is that the freshwater head is a function of z_i and thus increases with depth in saltwater. Therefore, two piezometers with their screens at different depths located within a stagnant body of saltwater will indicate a different freshwater head even if the saltwater body is at rest. The implication is that freshwater head differences can only be indicative of flow along a horizontal surface at elevation z_i.

One can, of course, also use the density of saltwater (ρ_s) to define a saltwater head analogous to Eqn (1.5). In that case, flow can be inferred from the saltwater head gradient within the saltwater body. In other words, as long as the density within a groundwater body is uniform, and all heads are formulated based on that same density, the gradient of that head is in the direction of the flow (unless the aquifer is anisotropic). In theoretical analyses of flow problems, the freshwater and saltwater domains (separated by an interface) are sometimes considered separately, using a freshwater head and saltwater head, respectively, for either of the domains. This type of treatment is not appropriate though when gradual changes in salinity exist, such as in the case of a broad transition zone.

1.6 Darcy's Law in Variable-Density Systems

Equation (1.1) is a simplified form of the more general form of Darcy's law for variable-density fluids (Bear, 1972):

$$\vec{q} = -\frac{\kappa}{\mu}(\nabla p - \rho \vec{g}) \tag{1.7}$$

where κ [L^2] is the intrinsic permeability, and μ [$M\,L^{-1}\,T^{-1}$] the dynamic viscosity of the groundwater. The horizontal components of the specific discharge vector $\vec{q}$ [$L\,T^{-1}$] are

$$q_x = -\frac{\kappa}{\mu}\frac{\partial p}{\partial x} \tag{1.8}$$

$$q_y = -\frac{\kappa}{\mu}\frac{\partial p}{\partial y} \tag{1.9}$$

which can be re-written in terms of freshwater head h_f by rearranging and differentiating Eqn (1.5) with respect to x and y:

$$q_x = -\frac{\kappa \rho_f g}{\mu_f}\frac{\mu_f}{\mu}\frac{\partial h_f}{\partial x} = -K_f\frac{\partial h_f}{\partial x} \tag{1.10}$$

$$q_y = -\frac{\kappa \rho_f g}{\mu_f}\frac{\mu_f}{\mu}\frac{\partial h_f}{\partial y} = -K_f\frac{\partial h_f}{\partial y} \tag{1.11}$$

where it was assumed that $\mu_f/\mu = 1$, and $K_f = \kappa\rho_f g/\mu_f$. Equations (1.10) and (1.11) show that the horizontal component of the specific discharge can be calculated based on the gradient of the freshwater head in the x and y direction. It must be reiterated though that because of the dependency of the freshwater head on depth, the values of h_f to evaluate the gradients in the x and y direction must be taken along the same elevation z.

The component of the specific discharge vector in the vertical direction is given by

$$q_z = -\frac{\kappa}{\mu}\left(\frac{\partial p}{\partial z} + \rho g\right) \tag{1.12}$$

which, again by rearranging Eqn (1.5) and this time differentiating with respect to z, can be converted to

$$q_z = -\frac{\kappa\rho_f g}{\mu_f}\frac{\mu_f}{\mu}\left[\frac{\partial h_f}{\partial z} + \left(\frac{\rho - \rho_f}{\rho_f}\right)\right] = -K_f\left[\frac{\partial h_f}{\partial z} + \left(\frac{\rho - \rho_f}{\rho_f}\right)\right] \tag{1.13}$$

Equation (1.13) shows that the vertical component of the specific discharge can be evaluated using the vertical gradient of the freshwater head if, and only if, the buoyancy term $(\rho - \rho_f)/\rho_f$ is considered. Worked examples of the application of Eqns (1.10), (1.11) and (1.13) were provided by Post et al. (2007) and an application using field data was demonstrated by Post et al. (2018a). While Eqns (1.10), (1.11) and (1.13), in principle, provide a convenient framework to calculate the horizontal and vertical flow components in coastal aquifers, in practice, measurement errors can cause a considerable uncertainty of the flow estimates. Moreover, effects of anisotropy and dipping aquifers further complicate the analysis. More detailed discussions of the latter subject can be found in Bachu (1995) and Bachu and Michael (2002).

Finally, Lusczynski (1961) introduced a quantity known as the environmental water head (h_e) with which q_z can be calculated using the familiar form of Darcy's law:

$$q_z = -\frac{\kappa\rho_f g}{\mu}\frac{\partial h_e}{\partial z} = -K_f\frac{\partial h_e}{\partial z} \tag{1.14}$$

The environmental head is the water level a piezometer would indicate if it were filled with water of the same density stratification as in the adjacent aquifer. The advantages of this ingenious concept are unfortunately somewhat negated by the non-intuitive assumptions that must be made when the water level in the well does not coincide with the water table. This might be the reason that the analysis of groundwater flow based on environmental heads has seen little uptake in coastal hydrogeology. The concept is mentioned here nevertheless because it is sometimes encountered in the literature.

2
Governing Equations for Variable-Density Flow

The simultaneous occurrence of freshwater and seawater in coastal aquifers makes the simulation of groundwater flow and solute transport complex because the governing equations are coupled and nonlinear due to the difference in density of the fluids. In this chapter, the governing equations of mathematical models of flow and transport under variable-density conditions will be derived. Before doing so, the change of the groundwater density as a function of salinity, temperature and pressure will be presented. The full set of equations can only be solved using numerical codes, which are introduced in the last part of this chapter. Later chapters in this book (Chapters 3, 4, 9 and 10) will explore the use of simplified forms of the equations presented here to calculate the water levels and interface position in coastal aquifers using analytical solutions.

This single chapter can only provide an overview of the most important aspects of numerical simulation of coastal aquifer groundwater processes. Much more information can be found in monographs like Pinder and Gray (1977), Bear and Verruijt (1987), Holzbecher (1998) and Zheng and Bennett (2002). A good overview at the introductory level can be found in Bear and Cheng (2010). Here the focus is on the development of the governing equations as well as some important aspects of numerical techniques and practical considerations.

2.1 Governing Equations

To model the flow of groundwater in a coastal aquifer system where both saltwater and freshwater are present, two mass conservation expressions are required: one for groundwater flow and one for solute transport. By using Darcy's law to describe the flow of water and invoking Fick's law for the dispersive flux of solutes (Kolditz et al., 1998), the mass conservation expressions can be written as partial differential equations. Since the density appears in Darcy's law and is dependent on the solute concentration (as well as the temperature and to a lesser extent the fluid pressure), changes in the solute distribution affect the groundwater flow. Since the groundwater flow determines the solute concentrations, the two governing equations are interdependent, which means that the flow and transport equations are coupled and cannot be solved separately (Holzbecher, 1998; Zheng and Bennett, 2002). Before proceeding with the derivation of the differential equations that govern the flow of groundwater and the transport of solutes, the concepts of compressibility and stress are briefly treated, as these are prerequisite for the derivations that follow.

2.1.1 Equations of State

Water has a low compressibility, yet a change in pressure still leads to a change in volume. The strain is the change in volume dV [L^3] per unit of total volume V [L^3] and the compressibility β_w [L T^2 M^{-1}] is defined as the change in strain per unit change in pressure p [M L^{-1} T^{-2}], or stress. Expressed mathematically,

$$\beta_w = -\frac{dV/V}{dp} \tag{2.1}$$

(see also Section 4.4, Eqn (4.13)). For a fixed mass of water m and by making use of $\rho = m/V$ [M L^{-3}], Eqn (2.1) can also be written in terms of the change in density:

$$\beta_w = -\frac{d\rho/\rho}{dp} \tag{2.2}$$

or

$$\frac{d\rho}{dp} = -\rho\beta_w \tag{2.3}$$

In addition to the pressure, the solute mass fraction ω [M M^{-3}] and temperature T [K, or alternatively degrees Celsius] have an important effect on the density, in fact, much more so than pressure. Thus, since $\rho = f(p, \omega, T)$ (Diersch and Kolditz, 2002),

$$d\rho = \frac{\partial\rho}{\partial p}dp + \frac{\partial\rho}{\partial\omega}d\omega + \frac{\partial\rho}{\partial T}dT = \rho\beta_w dp + \rho\beta_\omega d\omega - \rho\beta_T dT \tag{2.4}$$

where β_ω [M M^{-1}] and β_T [K^{-1}] are the solutal and thermal expansion coefficient, respectively defined as

$$\beta_\omega = \frac{1}{\rho}\frac{\partial\rho}{\partial\omega}\bigg|_{T,p} \tag{2.5}$$

$$\beta_T = \frac{1}{\rho}\frac{\partial\rho}{\partial T}\bigg|_{\omega,p} \tag{2.6}$$

Now that $\rho = f(p, \omega, T)$, β_w should formally be re-defined as (compare to Eqn (2.2))

$$\beta_w = \frac{1}{\rho}\frac{\partial\rho}{\partial p}\bigg|_{T,\omega} \tag{2.7}$$

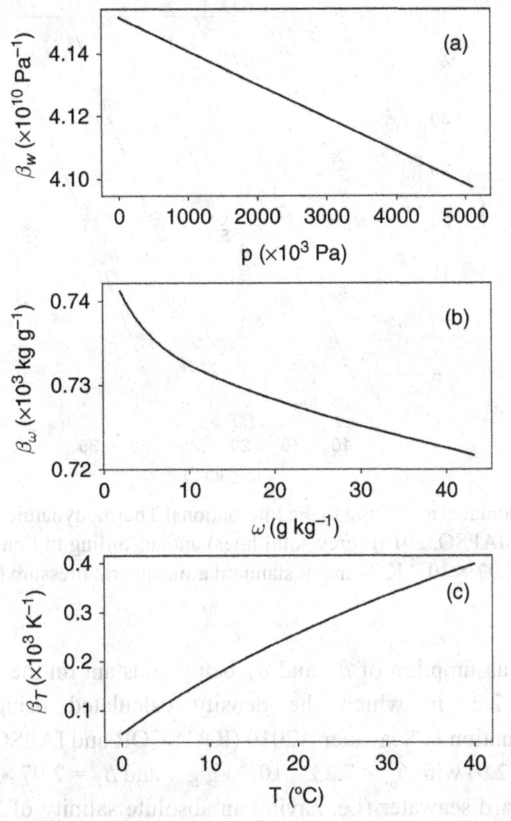

Figure 2.1 Dependence of (a) β_w on pressure p, (b) β_ω on salt mass faction ω and (c) β_T on temperature T. Values were calculated according to the International Thermodynamic Equation of Seawater – 2010 (IOC SCOR and IAPSO, 2010).

Integration of Eqn (2.4) gives

$$\rho = \rho_0 e^{\beta_w(p-p_0)+\beta_\omega(\omega-\omega_0)-\beta_T(T-T_0)} \tag{2.8}$$

which in many numerical codes is approximated by

$$\rho = \rho_0(1 + \beta_w(p - p_0) + \beta_\omega(\omega - \omega_0) - \beta_T(T - T_0)) \tag{2.9}$$

The integration of Eqn (2.4) was done assuming that the β values were not a function of p, T or ω themselves. The graphs in Figures 2.1a and 2.1b show that the variation of β_w and β_ω over common ranges of pressures and salinities observed in coastal aquifers is indeed small (circa 1% and 3%, for β_w and β_ω, respectively, over the range considered). However, β_T depends strongly on temperature (Figure 2.1c) and therefore the assumption of constancy holds only over a small temperature interval. Figure 2.1 further shows that the compressibility β_w is much smaller than either β_ω or β_T.

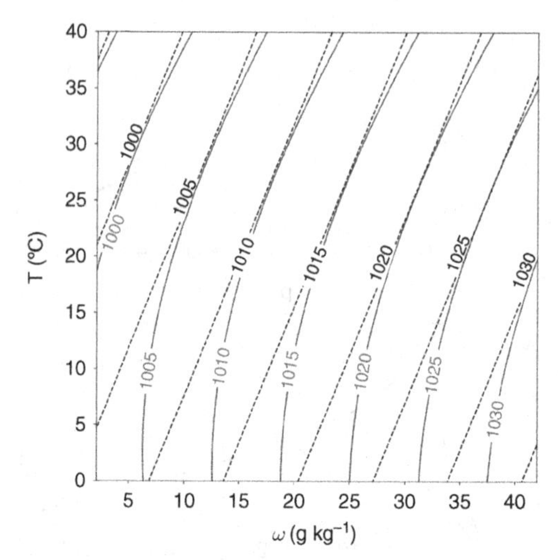

Figure 2.2 Density calculated according to the International Thermodynamic Equation of Seawater – 2010 (IOC SCOR and IAPSO, 2010) (grey solid lines) and according to Eqn (2.9) with $\beta_\omega = 7.23 \times 10^{-4}$ kg g^{-1} and $\beta_T = 2.97 \times 10^{-4}$ K^{-1}, and at standard atmospheric pressure (black dashed lines).

The effect of the assumption of β_ω and β_T being constant on the calculated density is shown in Figure 2.2, in which the density calculated using the International Thermodynamic Equation of Seawater – 2010 (IOC SCOR and IAPSO, 2010) is compared to the result of Eqn (2.9) with $\beta_\omega = 7.23 \times 10^{-4}$ kg g^{-1} and $\beta_T = 2.97 \times 10^{-4}$ K^{-1}, which are the values for standard seawater (i.e. having an absolute salinity of 35.16504 g kg^{-1}), at temperature $T = 25$°C and standard atmospheric pressure ($p = 101.325$ kPa). The pressure was assumed constant. The International Thermodynamic Equation of Seawater – 2010 (or TEOS-10) is a complex algorithm that is based on the Gibbs thermodynamic state function that relates a system's potential energy to its enthalpy and entropy per unit of mass. The Gibbs function depends on absolute salinity, temperature and pressure, and the density can be obtained from its derivative with respect to pressure (IOC SCOR and IAPSO, 2010). The results in Figure 2.2 clearly shows that the accurate densities provided by the TEOS-10 can deviate significantly from those calculated using the approximate expression provided by Eqn (2.9) when the temperatures depart from the reference value of 25°C. Whether or not such deviations are critical has to be decided on a case by case basis.

The most common application of Eqn (2.9) is to use the TDS or the chloride concentration to determine the effect of salinity on the density, which is usually adequate for problems involving variable-density effects in coastal aquifers. The solute species in natural waters each contribute differently to the density though, and for flow problems that are highly density-dependent, and in which the relative proportions of the ions change due to chemical reactions, more sophisticated algorithms based on the concentrations of multiple solutes are more appropriate (Hughes and Sanford, 2004; Post and Prommer, 2007).

2.2 Groundwater Flow

Deriving the mathematical equations for transient groundwater flow in a system where the rocks behave elastically is rather complicated (Bear, 1972) and the way to arrive at the correct formulation has been the topic of some debate in the literature since Jacob (1950) presented his derivation (Freeze and Cherry, 1979; Domenico and Schwartz, 1990). The crucial point is that in a deforming porous medium, the velocity of the groundwater is not relative to a fixed point in space, but relative to the grains or the rock. The movement of the latter may be imperceptibly slow but must be considered when deriving the fluid mass conservation equation in order to arrive at the correct expression for the specific storage coefficient (defined further down) (Domenico and Schwartz, 1990). This is particularly relevant in coastal areas where loading effects due to tidal variations can influence groundwater flow (Reeves et al., 2000; Wilson and Gardner, 2006). This will be discussed in more detail in Section 4.4.

Two types of equations are required to derive the governing equation for groundwater flow. The first is Darcy's law which relates the volumetric flow rate of groundwater per unit surface area (i.e. the flux) to the gradient of the potential. The second is a mass conservation expression, or continuity equation. The equations that are derived are macroscopic, which means that they capture the average behaviour of the flowing groundwater and not the processes at the microscopic scale of the pores. This approach assumes that representative elementary volume exists for which average properties can be defined to describe the state of the system. A detailed discussion is beyond the scope of this book and can be found, for example, in Bear (1972) or Freeze and Cherry (1979).

2.2.1 Darcy's Law

The formulation of Darcy's law that relates the specific discharge flux to the gradient of the hydraulic head, which applies to single-density systems, cannot be used in variable-density flow problems (Section 1.6). The reason is that the hydraulic head is only suitable as a potential for fluid flow when the density is constant everywhere. An alternative formulation of Darcy's law, which can be derived from the consideration of a momentum balance (Bear, 1972), must be used. This form is based on the gradient of the fluid pressure and includes a gravitational force term:

$$\vec{q} = -\frac{\overline{\kappa}}{\mu}(\nabla P - \rho\vec{g}) \tag{2.10}$$

where $\vec{q}$ [L T^{-1}] is the specific discharge vector, $\overline{\kappa}$ [L^2] the intrinsic permeability tensor, μ [M L^{-1} T^{-1}] the dynamic viscosity and $\vec{g}$ [L T^{-2}] the gravitational acceleration vector:

$$\vec{g} = \begin{bmatrix} 0 \\ 0 \\ -g \end{bmatrix} \tag{2.11}$$

The ∇ operator is defined as

$$\mathbf{i}\frac{\partial}{\partial x}+\mathbf{j}\frac{\partial}{\partial y}+\mathbf{k}\frac{\partial}{\partial z} \tag{2.12}$$

and yields the gradient of a scalar quantity.

The components of the permeability tensor $\overline{\kappa}$ are (Bear, 1972)

$$\overline{\kappa}=\begin{bmatrix} K_x & K_{xy} & K_{xz} \\ K_{yx} & K_y & K_{yz} \\ K_{zx} & K_{zy} & K_z \end{bmatrix} \tag{2.13}$$

When a Cartesian coordinate system is used with the axes aligned with the principal directions of permeability, the off-diagonal components of $\overline{\kappa}$ are zero. Hence the components of $\vec{q}$ in the x, y and z direction are

$$q_x=-\frac{\kappa_x}{\mu}\frac{\partial p}{\partial x} \tag{2.14}$$

$$q_y=-\frac{\kappa_y}{\mu}\frac{\partial p}{\partial y} \tag{2.15}$$

$$q_z=-\frac{\kappa_z}{\mu}\left(\frac{\partial p}{\partial z}+\rho g\right) \tag{2.16}$$

where κ_x, κ_y and κ_z are the permeabilities in the x, y and z direction, respectively. Restricting the analysis to three principal directions is not in any way an obstacle for practical applications as almost no detailed information about anisotropy is usually available anyway. Aligning the coordinate system with the three principal directions of anisotropy may not be possible in complex geological environments (Freeze and Cherry, 1979).

It is possible to express Darcy's law in terms of the freshwater head h_f [L] by virtue of the relationship (Eqn (1.5))

$$p=\rho_f g(h_f-z) \tag{2.17}$$

It is recalled that h_f in this equation represents the head with respect to a reference elevation (typically sea level) that would be measured in a piezometer were it filled completely with freshwater. Care must be taken in the application of Darcy's law in terms of freshwater head because freshwater head differences in the vertical direction are not indicative for flow, and horizontal flow directions can only inferred along a plane of constant elevation (Post et al., 2007). Nonetheless the concept has proven extremely useful for the development of numerical codes, because the resulting set of equations is mathematically equivalent to those based on a water pressure formulation. The great advantage of this approach is that model input and output can be specified using heads instead of pressures, which is more intuitive for hydrogeologists.

Differentiating Eqn (2.17) and inserting the result into Eqns (2.14)–(2.16) yields

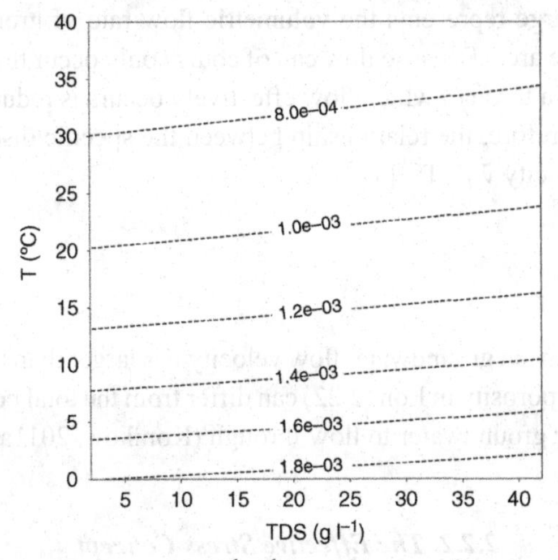

Figure 2.3 Lines of equal viscosity μ (kg m^{-1} s^{-1}) as a function of TDS and temperature. Viscosity values were calculated using the equations and constants in Sharqawy et al. (2012).

$$q_x = -\frac{\rho_f g \kappa_x}{\mu}\frac{\partial h_f}{\partial x} = -\frac{\mu_f}{\mu}K_{f,x}\frac{\partial h_f}{\partial x} \tag{2.18}$$

$$q_y = -\frac{\rho_f g \kappa_y}{\mu}\frac{\partial h_f}{\partial y} = -\frac{\mu_f}{\mu}K_{f,y}\frac{\partial h_f}{\partial y} \tag{2.19}$$

$$q_z = -\frac{\kappa_z}{\mu}\left(\rho_f g\frac{\partial h_f}{\partial z} - \rho_f g + \rho g\right) = -\frac{\mu_f}{\mu}K_{f,z}\left(\frac{\partial h_f}{\partial z} + \frac{\rho - \rho_f}{\rho_f}\right) \tag{2.20}$$

in which the freshwater hydraulic conductivity is defined as

$$K_{f,i} = \frac{\rho_f g \kappa_i}{\mu_f} \tag{2.21}$$

with i being either x, y or z.

Note that these are the same equations as Eqns (1.10), (1.11) and (1.13) in Section 1.6, but here the viscosity ratio has been retained. As Figure 2.3 shows, the viscosity μ in Eqns (2.18)–(2.20) is mainly a function of temperature, and to a lesser extent of concentration. Over the range of temperatures relevant to coastal aquifers, there can be significant changes in viscosity, which might impact on the calculated fluxes. It should be noted that in many constant-density groundwater models, the viscosity does not explicitly appear in the expression for Darcy's law. However, it does exert an influence through the dependency of the hydraulic conductivity on μ, and therefore constant-density models that adopt a fixed value of the hydraulic conductivity implicitly contain the assumption that the temperature remains constant. The dependence of the viscosity on the pressure is relatively small. For example the viscosity increases by about 3% due to the pressure equivalent to a water depth of 500 m (McCain, 1991).

The specific discharge represents the volumetric flow rate of groundwater per unit of cross-sectional surface area. Because flow can of course only occur through the pore space, the cross-sectional area through which flow effectively occurs is reduced by a factor equal to the porosity n. Therefore, the relationship between the specific discharge and the mean groundwater flow velocity $\vec{v}$ [L T^{-1}] is

$$\vec{v} = \frac{\vec{q}}{n} \tag{2.22}$$

In other words, the mean groundwater flow velocity $\vec{v}$ is larger than the specific discharge $\vec{q}$ by a factor of $\frac{1}{n}$. The porosity in Eqn (2.22) can differ from the total porosity as not all pore space is accessible for groundwater to flow through (Konikow, 2011a).

2.2.2 The Effective Stress Concept

Changes in storage of groundwater can be caused by the deformation of rocks (Section 4.4). The rocks themselves may deform but generally the greatest change in bulk rock volume (and thereby porosity) is brought about by rearrangement of the rock assembly (Bear and Verruijt, 1987). For example, grains in an unconsolidated deposit may slip and roll when the forces acting on the contact points between the grains change.

Terzaghi (1925) introduced the concept of effective stress, which is very useful in describing the flow of groundwater in deforming media. A customary and valid assumption for regional groundwater flow problems is that the deformation occurs in the vertical direction only. The total stress σ_T [M L^{-1} T^{-2}] is the force per unit area acting on a horizontal plane cutting across the subsurface, which is caused by the load of the solid material, water and atmosphere above that plane. The effective stress p_s [M L^{-1} T^{-2}] is the difference between the total stress and the groundwater pressure:

$$p_s = \sigma_T - p \tag{2.23}$$

The effective stress is the average of the contact forces between the grain that causes the deformation of the rock skeleton. A detailed discussion of the deformation of rocks is beyond the scope of this book and the reader may want to refer to Verruijt (2013) for in-depth discussions of this topic. This brief discussion and the expression for p_s are provided here because they are needed for the development of the groundwater mass conservation in the next section.

2.2.3 Groundwater Mass Conservation

The derivation that will be presented here largely follows Bear (1972). The starting point is a cubic control volume (Figure 2.4), which has a fixed volume and position in space. The sides of the control volume are aligned with the Cartesian coordinates x, y and z, which

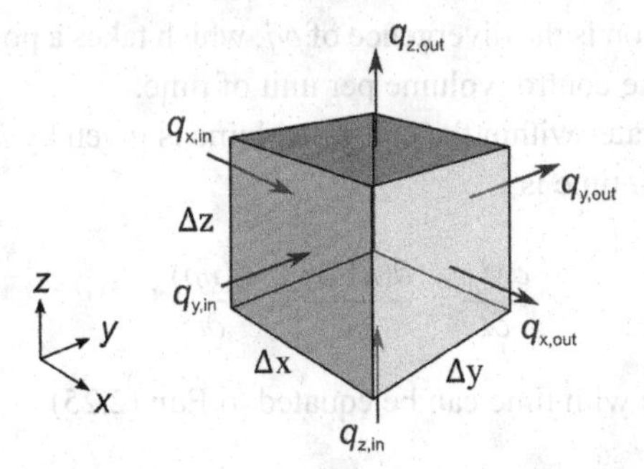

Figure 2.4 Cubic control volume in a Cartesian coordinate system. Arrows indicate the *x*, *y* and *z* components of the specific discharge vector.

are also the principal directions of the permeability. The dimensions of the control volume are Δx, Δy and Δz and its volume is thus $V = \Delta x \Delta y \Delta z$.

The mass conservation principle states that, over a given time interval, any difference in mass flow rate caused by flow across the sides of the cubes must be balanced by a change of mass within the control volume. This can refer to the mass of groundwater or the rock mass. Since in a deforming medium, both the groundwater and the rock are in motion, a mass conservation expression is required for both. Starting with the former, the mass flow rate of water in the *x* direction, i.e. crossing the cube's plane $\Delta y \Delta z$ is

$$\rho Q_x = \rho q_x A = \rho q_x \Delta y \Delta z \tag{2.24}$$

where Q_x [$L^3 T^{-1}$] is the volumetric groundwater flow rate and $A = \Delta y \Delta z$ is the surface area of the face of the cube normal to the *x* axis. The difference of the mass flow rate in the *x* direction over the interval Δx is

$$\rho Q_{x,\text{out}} - \rho Q_{x,\text{in}} = \frac{\partial(\rho Q_x)}{\partial x}\Delta x = \frac{\partial(\rho q_x)}{\partial x}\Delta y \Delta z \Delta x = \frac{\partial(\rho q_x)}{\partial x} V \tag{2.25}$$

Using similar expressions for the mass flow rate in the *y* and *z* directions and summing to get the net mass flow rate for the entire control volume yields

$$\left(\frac{\partial(\rho q_x)}{\partial x} + \frac{\partial(\rho q_y)}{\partial y} + \frac{\partial(\rho q_z)}{\partial z}\right) V = \nabla \cdot (\rho \vec{q}) V \tag{2.26}$$

The $\nabla \cdot$ operator operates on a vector (which is why it is followed by a dot to distinguish it from the gradient operation on a scalar also denoted by ∇, Section 2.2.1) and is defined as

$$\frac{\partial}{\partial x} + \frac{\partial}{\partial y} + \frac{\partial}{\partial z} \tag{2.27}$$

The result of this operation is the divergence of $\rho\vec{q}$, which takes a positive value when there is a loss of mass from the control volume per unit of time.

The mass of groundwater within the control volume is given by $M_w = nV\rho$ [M] and the rate of mass change with time is

$$\frac{\partial M_w}{\partial t} = \frac{\partial(nV\rho)}{\partial t} = \frac{\partial(n\rho)}{\partial t}V \tag{2.28}$$

The rate of mass change with time can be equated to Eqn (2.25)

$$\frac{\partial(n\rho)}{\partial t} = -\nabla \cdot (\rho\vec{q}) \tag{2.29}$$

The minus sign is included because a positive value of the divergence is associated with a decrease of mass with time. This assumes no internal sources or sinks are present within the control volume but the required terms can easily be added later on (Bear and Verruijt, 1987). By recalling the relationship between $\vec{q}$ and $\vec{v}$ (Eqn (2.22)) and remembering that the groundwater velocity must be taken relative to the rock that is itself in motion, the following expression for the specific discharge relative to the moving grains $\vec{q}_r$ [L T^{-1}] holds (Bear, 1972; Verruijt, 2013):

$$\vec{q}_r = n(\vec{v}_w - \vec{v}_s) = \vec{q} - n\vec{v}_s \tag{2.30}$$

where $\vec{v}_w$, $\vec{v}_s$ and $\vec{q}$ are the velocity of the groundwater, the velocity of the grains and the specific discharge relative to a fixed point, respectively (all having dimensions L T^{-1}). Equation (2.29) thus becomes

$$\frac{\partial(n\rho)}{\partial t} = -\nabla \cdot (\rho(\vec{q}_r + n\vec{v}_s)) \tag{2.31}$$

The left-hand side of the Eqn (2.31) can be expanded using the chain rule

$$n\frac{\partial\rho}{\partial t} + \rho\frac{\partial n}{\partial t} \tag{2.32}$$

And subsequently, since ρ is a function of P and C (and temperature as well, but for this derivation isothermal conditions will be assumed),

$$\frac{\partial\rho}{\partial t} = \frac{\partial\rho}{\partial p}\frac{\partial p}{\partial t} + \frac{\partial\rho}{\partial C}\frac{\partial C}{\partial t} = \rho\beta_w\frac{\partial p}{\partial t} + \frac{\partial\rho}{\partial C}\frac{\partial C}{\partial t} \tag{2.33}$$

which makes use of Eqn (2.7). The right-hand side of Eqn (2.31) can be expanded to

$$-\nabla \cdot (\rho\vec{q}_r) - \rho n\nabla \cdot \vec{v}_s - n\vec{v}_s\nabla\rho - \rho\vec{v}_s\nabla n \tag{2.34}$$

so that Eqn (2.31) becomes

$$\rho\frac{\partial n}{\partial t}+n\rho\beta_w\frac{\partial p}{\partial t}+n\frac{\partial \rho}{\partial C}\frac{\partial C}{\partial t}=-\nabla\cdot(\rho\vec{q}_r)-\rho n\nabla\cdot\vec{v}_s-n\vec{v}_s\nabla\rho-\rho\vec{v}_s\nabla n \qquad (2.35)$$

Similar to the groundwater, the mass flow rate of the moving grains can be expressed by (Bear, 1972; Verruijt, 2013)

$$\frac{\partial[(1-n)\rho_r]}{\partial t}=-\nabla\cdot[(1-n)\rho_r\vec{v}_s] \qquad (2.36)$$

where ρ_r [M L^{-3}] is the density of the rock. Expanding the left-hand side of Eqn (2.36) gives

$$\rho_r\frac{\partial(1-n)}{\partial t}+(1-n)\frac{\partial \rho_r}{\partial t}=-\rho_r\frac{\partial n}{\partial t}+(1-n)\frac{\partial \rho_r}{\partial t} \qquad (2.37)$$

and the right-hand side of Eqn (2.36)

$$-\rho_r\nabla\cdot\vec{v}_s-\vec{v}_s\nabla\rho_r+\rho_r n\nabla\cdot\vec{v}_s+n\vec{v}_s\nabla\rho_r+\rho_r\vec{v}_s\nabla n \qquad (2.38)$$

When it is assumed that the rock itself is incompressible (which is reasonable for rocks made up by minerals like quartz or calcite but this assumption may break down for clay minerals or organic material), the derivatives of ρ_r become zero and ρ_r itself cancels on both sides of the equation so that, after rearranging, Eqn (2.36) simplifies to

$$\frac{\partial n}{\partial t}=(1-n)\nabla\cdot\vec{v}_s-\vec{v}_s\nabla n \qquad (2.39)$$

Substituting Eqn (2.39) into Eqn (2.35) gives

$$\rho\nabla\cdot\vec{v}_s+n\rho\beta_w\frac{\partial p}{\partial t}+n\frac{\partial \rho}{\partial C}\frac{\partial C}{\partial t}=-\nabla\cdot(\rho\vec{q}_r)-n\vec{v}_s\nabla\rho \qquad (2.40)$$

Since $\rho=f(p,C)$, where C [M L^{-3}] is the solute concentration (and not mass fraction ω), the second term on the right-hand side can be expanded as

$$n\vec{v}_s\nabla\rho=n\frac{\partial \rho}{\partial p}\vec{v}_s\nabla p+n\frac{\partial \rho}{\partial C}\vec{v}_s\nabla C=n\rho\beta_w\vec{v}_s\nabla P+n\rho\beta_C\vec{v}_s\nabla C \qquad (2.41)$$

in which $\beta_C=\frac{1}{\rho}\frac{\partial \rho}{\partial C}\big|_{T,\varphi}$ [L^3 M^{-1}] is the solutal expansion coefficient in terms of concentration C, with which Eqn (2.40) becomes

$$\rho\nabla\cdot\vec{v}_s+n\rho\beta_w\frac{\partial p}{\partial t}+n\rho\beta_w\vec{v}_s\nabla p+n\rho\beta_C\frac{\partial C}{\partial t}+n\rho\beta_C\vec{v}_s\nabla C=-\nabla\cdot(\rho\vec{q}_r) \qquad (2.42)$$

which can be written in terms of the so-called total (or material) derivative:

$$\rho\nabla\cdot\vec{v}_s + n\rho\beta_w\frac{d^s p}{dt} + n\rho\beta_C\frac{d^s C}{dt} = -\nabla\cdot(\rho\vec{q}_r) \tag{2.43}$$

where $\frac{d^s()}{dt} = \frac{\partial()}{\partial t} + \vec{v}_s\nabla()$ is the total derivative with respect to the moving solids.

For conditions of vertical deformation only, which are valid for practically all regional groundwater problems, Bear (1972) defined the divergence of $\vec{v}_s$ in terms of the total derivative of the effective stress $p_s = \sigma_T - p$ (Eqn (2.23)):

$$\nabla\cdot\vec{v}_s = -\beta_s\frac{d^s(\sigma_T - p)}{dt} \tag{2.44}$$

where β_s is the vertical compressibility [L T^2 M^{-1}] of a fixed mass of solids in motion. Inserting this relationship into Eqn (2.43) the mass conservation expression for groundwater flow in a deforming porous medium finally becomes

$$-\rho\beta_s\frac{d^s(\sigma_T - p)}{dt} + n\rho\beta_w\frac{d^s p}{dt} + n\rho\beta_C\frac{d^s C}{dt} = -\nabla\cdot(\rho\vec{q}_r) \tag{2.45}$$

or

$$-\rho\beta_s\frac{d^s\sigma_T}{dt} + \rho(\beta_s + n\beta_w)\frac{d^s p}{dt} + n\rho\beta_C\frac{d^s C}{dt} = -\nabla\cdot(\rho\vec{q}_r) \tag{2.46}$$

when $\vec{v}_s \to 0$ the total derivatives can be substituted by the partial derivatives and $\vec{q}_r \to \vec{q}$. When the total stress σ_T is assumed to be constant in time as well, the continuity equation finally becomes

$$\rho S_s\frac{\partial p}{\partial t} + n\rho\beta_C\frac{\partial C}{\partial t} = -\nabla\cdot(\rho\vec{q}) \tag{2.47}$$

in which

$$S_s = \beta_s + n\beta_w \tag{2.48}$$

is the specific storage coefficient [L T^2 M^{-1}] in terms of pressure. Physically it represents the change in groundwater volume per unit of (bulk) volume per unit of pressure change (Freeze and Cherry, 1979; Domenico and Schwartz, 1990).

Equation (2.47) is implemented in popular numerical codes like SEAWAT, SUTRA or FEFLOW. A point of confusion though is in the definition of the storage coefficient as in codes like SEAWAT and SUTRA it is defined as $S_s = (1 - n)\beta_s + n\beta_w$. The reason for this is that one can also arrive at Eqn (2.47) by considering a mass balance for a fixed control volume for the fluid only. However, this leads to the introduction of the $(1 - n)$ term in the definition of S_s and the definition of β_s is for a stationary bulk volume, and not for a fixed mass of moving solids (for details, see Bear, 1972). In many cases in practice, the difference is of little significance when assigning a numerical value for S_s in a model as it may be known a priori (e.g. from a pumping test) or gets adjusted during the model calibration. However, if it is to be calculated from the compressibilities of water and the rock, which in coastal aquifers is sometimes possible when the loading efficiency can be determined (Section 4.4), the difference in definitions of S_s does become relevant (Reeves et al., 2000).

With the assumption that the principal axes of permeability are aligned with the axes of the coordinate system, so that Eqns (2.14)–(2.16) can be used for $\vec{q}$, and allowing for external sources and sinks of water through an additional term q_s, Eqn (2.47) turns into the governing equation for groundwater flow in terms of groundwater pressure p:

$$\rho S_s \frac{\partial p}{\partial t} + n\rho\beta_C \frac{\partial C}{\partial t} = \frac{\partial}{\partial x}\left(\frac{\rho\kappa_x}{\mu}\frac{\partial p}{\partial x}\right) + \frac{\partial}{\partial y}\left(\frac{\rho\kappa_y}{\mu}\frac{\partial p}{\partial y}\right) + \frac{\partial}{\partial z}\left[\frac{\rho\kappa_z}{\mu}\left(\frac{\partial p}{\partial z} + \rho g\right)\right] + \rho_q q_s \quad (2.49)$$

where ρ_q [M L^{-3}] is the density associated with the source/sink term and q_s [T^{-1}] is the volumetric flow rate per unit aquifer volume of the source/sink. Differentiating Eqn (1.5) with respect to time gives $\frac{\partial p}{\partial t} = \rho_f g \frac{\partial h_f}{\partial t}$, which together with Eqns (2.18)–(2.20), can be inserted into Eqn (2.47) to express the governing equation for flow in terms of freshwater head h_f:

$$\rho S_{s,f} \frac{\partial h_f}{\partial t} + n\rho\beta_C \frac{\partial C}{\partial t}$$
$$= \frac{\partial}{\partial x}\left(\rho \frac{u_f}{\mu} K_{f,x} \frac{\partial h_f}{\partial x}\right) + \frac{\partial}{\partial y}\left(\rho \frac{u_f}{\mu} K_{f,y} \frac{\partial h_f}{\partial y}\right) + \frac{\partial}{\partial z}\left[\rho \frac{u_f}{\mu} K_{f,z}\left(\frac{\partial h_f}{\partial z} + \frac{\rho - \rho_f}{\rho_f}\right)\right] + \rho_q q_s \quad (2.50)$$

in which $S_{s,f}$ [L^{-1}] is the specific storage coefficient in terms of freshwater head:

$$S_{s,f} = \rho_f g S_s = \rho_f g(\beta_s + n\beta_w) \quad (2.51)$$

which represents the volume of groundwater change per unit bulk rock volume per unit freshwater head change. The term $\rho_q q_s$ was added in Eqn (2.49) to account for the mass balance contribution of sources and sinks. These could be internal sources like abstraction or injection wells, or external sources like groundwater inflow from outside the model domain, or rainfall recharge across the top model boundary.

Either one of these latter two equations is solved using computer codes for variable-density flow, depending on whether it is based on a pressure formulation or a freshwater head formulation. Slightly different and much simplified versions of Eqn (2.50) will be used in Chapters 3, 4, 9 and 10 to develop analytical solutions for the water level in coastal aquifers.

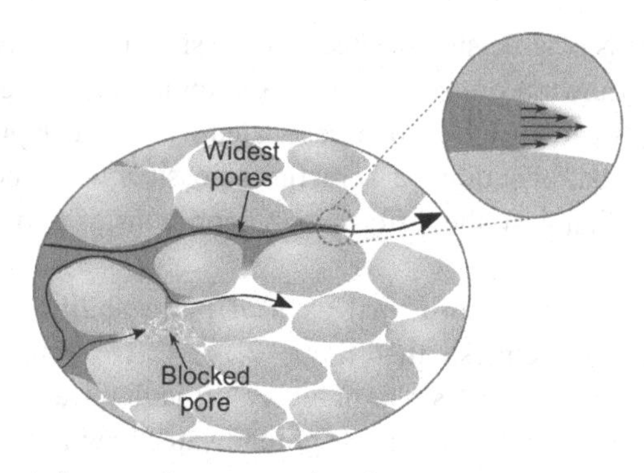

Figure 2.5 Velocity variations at the pore scale of a coastal aquifer that cause hydrodynamic dispersion. Groundwater velocities are highest in the widest pores, which causes the interface between high-salinity water (indicated by darker shading) and low-salinity water (in white) to travel further than in the narrower pores. In blocked pores, flow may be negligible and solute spreading becomes dominated by diffusion. Molecular diffusion causes spreading of solutes at the concentration boundaries. The inset shows the microscopic velocity variations within a pore.

2.3 Solute Transport

The derivation of the governing equation for solutes proceeds in a similar way as that for groundwater. An important aspect for solutes is the dilution and irregular spreading of concentration boundaries, which is caused by diffusion and the effects of hydrodynamic dispersion. The latter refers to the spreading of solutes at a macroscopic level caused by random deviations from the mean groundwater flow velocity due to the irregularity of the pore network and velocity gradients between the pore walls and pore centres (Figure 2.5).

The process of diffusion can be quantitatively described using Fick's law, but effects of hydrodynamic dispersion are not easy to capture in models. Based on the analogy of concentration patterns that develop due to hydrodynamic dispersion and diffusion in homogeneous media, Fick's law has also been used with moderate success to simulate solute transport in groundwater (Konikow, 2011a).

2.3.1 Fick's Law

All solutes dissolved in water are subject to random movements, called Brownian motion, and in the presence of concentration (or rather, chemical activity) gradients, these movements result in a net displacement of the molecules from the region of high concentration towards the lower concentrations. The flux due to diffusion is described using Fick's law, which, for one-dimensional diffusive transport in the x direction of a Cartesian coordinate system, can be written as

$$J_x = -D\frac{\partial C}{\partial x} \tag{2.52}$$

in which J_x [M L^{-2} T^{-1}] is the diffusive mass flux and D [L^2 T^{-1}] is the molecular diffusion coefficient. The negative sign stems from the convention that J_x be a positive quantity in the direction of decreasing concentrations. In a porous medium the pathways taken by the solutes are more tortuous because of the presence of the solids and their irregular geometry. Therefore the diffusion coefficient used in Eqn (2.52) must be the effective diffusion coefficient D_e and not the diffusion coefficient in free water.

For variable-density flow systems with high concentration gradients, such as in the case of dense brines, Hassanizadeh and Leijnse (1995) proposed the following form of Fick's law, written in terms of the solute mass fraction ω rather than the concentration C:

$$J_x = -\rho D \frac{\partial \omega}{\partial x} \tag{2.53}$$

2.3.2 Hydrodynamic Dispersion

Diffusion causes initially sharp concentration gradients to become more gradual with time (or, in a closed system, disappear altogether). At the macroscopic scale spreading by hydrodynamic dispersion also causes sharp concentration boundaries to become more disperse, which has led to the application of Fick's law to transport by hydrodynamic dispersion, albeit with two important modifications. The first is that the tendency for spreading in the direction of the flow, the longitudinal dispersion, can differ in magnitude from that in the direction normal to the flow, the transverse dispersion. The second is that, because the deviations from the mean velocity become larger with increasing velocity, the spreading of solutes is also greater at higher velocity (Zheng and Bennett, 2002). It is therefore assumed that there exists a simple linear relationship between the hydrodynamic dispersion coefficient and the groundwater flow velocity (Bear, 1972). For example, with these assumptions, for the case of a uniform flow field aligned with the axes of a Cartesian coordinate system, and with flow being in the x direction only, the hydrodynamic dispersion coefficients in the x, y and z directions become

$$D_{xx} = \alpha_L v \tag{2.54}$$

$$D_{yy} = \alpha_{TH} v \tag{2.55}$$

$$D_{zz} = \alpha_{TV} v \tag{2.56}$$

where α_L, α_{TH} and α_{TV} are the longitudinal, transverse horizontal and transverse vertical dispersivity, respectively, which all have the dimension of length [L]. The introduction of a horizontal as well as a vertical transverse dispersivity makes allowance for the fact that the spreading normal to the flow may also be directionally dependent. This example shows that even for the simplest case of one-dimensional flow, spreading of the solutes occurs in three directions, provided that a concentration gradient exists in each of these directions. When the flow is not aligned with a particular axis of the coordinate system, solutes will be spreading in the off-axis directions as well. Therefore, to make use of Eqn (2.52) to describe the mass flux due to hydrodynamic dispersion in three dimensions, the coefficient D must be replaced by a tensor and the equation takes the form

$$\vec{J} = -\overline{D} \cdot \nabla C \tag{2.57}$$

or, expanded into matrix form,

$$\begin{bmatrix} J_x \\ J_y \\ J_z \end{bmatrix} = - \begin{bmatrix} D_{xx} & D_{xy} & D_{xz} \\ D_{yx} & D_{yy} & D_{yz} \\ D_{zx} & D_{zy} & D_{zz} \end{bmatrix} \begin{bmatrix} \frac{\partial C}{\partial x} \\ \frac{\partial C}{\partial y} \\ \frac{\partial C}{\partial z} \end{bmatrix} \tag{2.58}$$

The derivation of the components of the hydrodynamic dispersion tensor is a rather involved and lengthy procedure. An excellent treatment is provided by Zheng and Bennett (2002), which shall not be repeated here. For the model in which spreading in the flow direction is controlled by α_L, and that transverse to the flow by α_{TH} and α_{TV} and the effective molecular diffusion coefficient D_e is isotropic, the components of $\overline{D}$ become (Zheng and Bennett, 2002)

$$D_{xx} = \alpha_L \frac{v_x^2}{|\vec{v}|} + \alpha_{TH} \frac{v_y^2}{|\vec{v}|} + \alpha_{TV} \frac{v_z^2}{|\vec{v}|} + D_e \tag{2.59}$$

$$D_{yy} = \alpha_L \frac{v_y^2}{|\vec{v}|} + \alpha_{TH} \frac{v_x^2}{|\vec{v}|} + \alpha_{TV} \frac{v_z^2}{|\vec{v}|} + D_e \tag{2.60}$$

$$D_{zz} = \alpha_L \frac{v_z^2}{|\vec{v}|} + \alpha_{TH} \frac{v_x^2}{|\vec{v}|} + \alpha_{TV} \frac{v_y^2}{|\vec{v}|} + D_e \tag{2.61}$$

$$D_{xy} = D_{yx} = (\alpha_L - \alpha_{TH}) \frac{v_x v_y}{|\vec{v}|} \tag{2.62}$$

$$D_{xz} = D_{zx} = (\alpha_L - \alpha_{TV}) \frac{v_x v_z}{|\vec{v}|} \tag{2.63}$$

$$D_{yz} = D_{zy} = (\alpha_L - \alpha_{TV}) \frac{v_y v_z}{|\vec{v}|} \tag{2.64}$$

Alternative formulations have been proposed (e.g. Bear and Verruijt, 1987), and different numerical models may be based on different dispersion models. A modeller thus has to be aware of this and investigate the potential implications of the choice for a particular implementation.

2.3.3 Solute Mass Conservation

The derivation of the mass conservation expression for solutes follows a similar logic as that of the groundwater, except that it involves consideration of the mass fluxes across the sides of a control volume (Figure 2.4) of the solutes in the water instead of the water itself. Just like before when dealing with a deforming porous medium, the analysis should consider the fact that the deforming rock is moving across the control volume's boundaries as well, but by making the assumption upfront that the rate of movement of the grains is negligible (i.e. $\vec{v}_s = 0$), the derivation can be simplified considerably.

Starting with advective transport, the mass flow rate of a solute with concentration C by advection in the x direction, i.e. crossing the cube's plane $\Delta y \Delta z$ is

$$Q_x C = q_x A C = q_x \Delta y \Delta z C \tag{2.65}$$

The difference of the advective mass flow rate in the x direction over the interval Δx is

$$Q_{x,out} C - Q_{x,in} C = \frac{\partial (Q_x C)}{\partial x} \Delta x = \frac{\partial (q_x C)}{\partial x} \Delta y \Delta z \Delta x = \frac{\partial (q_x C)}{\partial x} V \tag{2.66}$$

Using similar expressions for the solute mass flow rate in the y and z directions and summing to get the net solute mass flow rate for the entire control volume yields

$$\left(\frac{\partial (q_x C)}{\partial x} + \frac{\partial (q_y C)}{\partial y} + \frac{\partial (q_z C)}{\partial z} \right) V = \nabla \cdot (\vec{q} C) V \tag{2.67}$$

The mass of the solute within the control volume is given by $M_C = nVC$ [M] and the rate of solute mass change with time is

$$\frac{\partial M_C}{\partial t} = \frac{\partial (nVC)}{\partial t} = \frac{\partial (nC)}{\partial t} V \tag{2.68}$$

The rate of solute mass change with time is positive when the mass inside the control volume increases, and equal to the negative divergence of $\vec{q}C$ (recalling that the latter is positive for mass outflow):

$$\frac{\partial (nC)}{\partial t} V = -\nabla \cdot (\vec{q} C) V \tag{2.69}$$

Additional terms, however, are required because of the diffusive and dispersive fluxes across the sides of the control volume. Invoking the three-dimensional formulation of Fick's law (Section 2.3.2) the dispersive mass flow rate in the x direction can be written as (Zheng and Bennett, 2002)

$$J_x = -\left(D_{xx} \frac{\partial C}{\partial x} + D_{xy} \frac{\partial C}{\partial y} + D_{xz} \frac{\partial C}{\partial z} \right) n \Delta y \Delta z \tag{2.70}$$

It is important to note that the flux is multiplied by the cross-sectional area of the pore space ($n\Delta y \Delta z$), and not by the total area of the control volume's face, to obtain the dispersive mass flow rate. The difference between the outflow and the inflow in the x direction is

$$J_{x,\mathrm{out}} - J_{x,\mathrm{in}} = -\frac{\partial}{\partial x}\left(nD_{xx}\frac{\partial C}{\partial x} + nD_{xy}\frac{\partial C}{\partial y} + nD_{xz}\frac{\partial C}{\partial z}\right)\Delta x\Delta y\Delta z \tag{2.71}$$

and by analogy in the y and z directions, respectively,

$$J_{y,\mathrm{out}} - J_{y,\mathrm{in}} = -\frac{\partial}{\partial y}\left(nD_{yx}\frac{\partial C}{\partial x} + nD_{yy}\frac{\partial C}{\partial y} + nD_{yz}\frac{\partial C}{\partial z}\right)\Delta x\Delta y\Delta z \tag{2.72}$$

$$J_{z,\mathrm{out}} - J_{z,\mathrm{in}} = -\frac{\partial}{\partial z}\left(nD_{zx}\frac{\partial C}{\partial x} + nD_{zy}\frac{\partial C}{\partial y} + nD_{zz}\frac{\partial C}{\partial z}\right)\Delta x\Delta y\Delta z \tag{2.73}$$

which can be written in vector form:

$$-\nabla \cdot (n\overline{D} \cdot \nabla C)V \tag{2.74}$$

Adding the negative (due to the sign convention) of this term to Eqn (2.69) and dividing by V then gives the following governing equation for solute transport:

$$\frac{\partial(nC)}{\partial t} = -\nabla \cdot (\vec{q}C) + \nabla \cdot (n\overline{D} \cdot \nabla C) \tag{2.75}$$

Some numerical simulators require concentrations to be specified as mass fraction ω rather than concentration C. Equation (2.75) can be written in terms of the solute mass fraction as well based on $C = \rho\omega$ and when also introducing the high-density formulation of Fick's law (Eqn (2.53)), it becomes

$$\frac{\partial(n\rho\omega)}{\partial t} = -\nabla \cdot (\vec{q}\rho\omega) + \nabla \cdot (n\rho\overline{D} \cdot \nabla\omega) \tag{2.76}$$

2.4 Boundary and Initial Conditions

A mathematical model is formed by the combination of the governing equations with the boundary and initial conditions. Boundary conditions are applied at the edges of the model domain in such a way that they are representative for the interactions of the modelled system with the environments that surround it. These environments can be other aquifers or confining units, surface water (such as the sea, rivers or lakes), the unsaturated zone or the atmosphere. Boundary conditions must be chosen both for the governing equation of groundwater flow and solute transport.

2.4.1 Boundary Condition Types

The first of the three commonly used boundary types is to specify the value of the modelled variable (head, pressure, concentration or mass fraction) at the boundary. This is known as a Dirichlet boundary condition. In the case of groundwater flow, the freshwater head or pressure can be specified, for example along the bottom of the sea, where a river is in direct contact with an aquifer or where the groundwater head is known. The latter may be the case along the model's inland boundary when water level measurements are available. For nodes with a Dirichlet boundary condition, a model will calculate the flux of water or solute across the boundary for the corresponding area of influence of the node. A specified head or concentration may be variable in time (hence it is better to avoid the use of the terms constant pressure or constant concentration as they imply that values remains constant in time).

The second type of boundary condition is a specified gradient, also called a Neumann boundary condition. This condition is appropriate for boundaries across which the flux is known, like the rainfall recharge across the land surface, or the inflow of freshwater across a model's inland boundary. A no-flow or a zero-dispersive flux boundary is a special case of a Neumann-type boundary condition as the gradient is zero.

The third type of boundary condition depends on both the model variable itself and its derivatives, and is called a mixed or a Cauchy (or Robin) boundary condition. An example is the (vertical) flux of water across a layer of low-permeability (i.e. not within the model domain but just outside it), in which case the flux is proportional to the difference between the head at the boundary and the water level on the other side of the low-permeability layer (h_s):

$$q_z = -K_z \frac{\partial h}{\partial z} = \frac{(h_s - h)}{c} \tag{2.77}$$

in which c [T] is the hydraulic resistance of the low-permeability unit. By collecting the known terms on the right-hand side of the equation, it becomes clear how the boundary condition involves both the head gradient and the head itself (Bear and Verruijt, 1987):

$$-K_z \frac{\partial h}{\partial z} + \frac{h}{c} = \frac{h_s}{c} \tag{2.78}$$

2.4.2 Choosing Appropriate Boundary Conditions in Coastal Aquifer Models

Care must be taken when assigning specified concentration boundaries along the coastline or seafloor in seawater intrusion models. One issue with this resides in the fact that the outflow zone of freshwater may not be known a priori. Assigning a fixed seawater concentration where there is outflow of fresher water may then overestimate the concentration at the boundary. A second issue that follows from this is that the dispersive flux of solutes across the boundary will be unrealistically high, as the concentration gradient normal to the boundary is overestimated (Smith, 2004). A common choice in seawater intrusion models is to neglect the dispersive flux across the boundary (i.e. a Neumann boundary condition for solutes) and assume that the advective flux dominates the total solute transport flux. The choice between these two boundary conceptualisations can have important effects on submarine groundwater estimates (Walther et al., 2017). Further discussion of the conceptualisation of boundaries for solute transport models when both advective and dispersive fluxes are relevant is provided by Bear and Verruijt (1987).

Two representations of the boundary condition on the inland boundary of coastal aquifers have been used (Werner and Simmons, 2009; Werner et al., 2012): specified flux and specified head, with the choice for either leading to very different conclusions. With a specified head boundary, the system beyond the model boundary can provide or absorb unlimited amounts of water to close the water budget for the modelled part of the system (Jiao et al., 2005). Any analysis adopting this boundary condition should therefore always check if the boundary flux is realistic.

A specified head boundary is realistic only if a surface water feature like a major river, which is connected to the aquifer, exerts an overriding control on the head in the aquifer. A specified head may be appropriate in coastal zones where water levels are maintained artificially (Oude Essink et al., 2010), but the presence of a hydraulic resistance between the surface water and the groundwater usually means that the head the aquifer is not equal to the surface water level. In natural coastal areas, the river course is mostly directed towards the sea near the shoreline. A river flowing parallel to the shoreline is less common. The application of a specified head at the inland boundary is therefore problematic when considering shore-perpendicular cross-sectional models (which run more or less parallel to the rivers).

A fixed-flux boundary, however, is more reasonable for most coastal aquifers under natural conditions. A special case is that the inland boundary is chosen at the groundwater divide where horizontal flux equals zero. If the model boundary cannot extend that far, the flow across the model boundary can be taken as the recharge integrated between the groundwater divide and the model boundary. In this book, the specified flux boundary condition at the inland boundary has been chosen when deriving analytical solutions.

2.4.3 Initial Conditions

Variable-density models require initial conditions, i.e. the values of pressure (or freshwater head) and concentration at the start of the simulation. These must be chosen such that they are consistent with the boundary conditions, and the pressure (or freshwater head) field must be consistent with the concentration field, as the concentrations determine the density and thus the pressure (or freshwater head). As variable-density problems are nonlinear, the initial conditions can have a strong influence on the numerical solution of the problem during later time steps, and therefore it must always be verified to what extent the solution is dependent on the initial conditions, for example by starting the simulation using a different concentration field. Since solute concentration distributions can take a very long time to reach equilibrium, especially in regional problems and when zones of very low advective flow rates are present, the choice of initial conditions is important for an efficient modelling workflow.

2.5 Numerical Methods

Numerical models are a particular class of mathematical models in which, unlike with the analytical equations in Chapters 3, 4, 9 and 10, the variables of interest (e.g. pressure, freshwater head, concentration) are evaluated at discrete intervals in space and time. They have become a commonplace tool in the analysis of coastal groundwater systems. Their popularity stems from their versatility, as numerical techniques allow for consideration of heterogeneity of hydrogeological properties, complex system geometries and temporal variability of system stresses. Moreover, to simulate the coupling between groundwater flow and solute transport that exists in variable-density problems, numerical techniques are indispensable.

A combination of Eqn (2.49) or (2.50) with either Eqn (2.75) or (2.76), coupled by an equation of state like Eqn (2.9), forms the set of governing equations for groundwater flow and solute transport solved by numerical codes. Numerical methods replace the continuous forms of the governing equations by approximate forms and a grid (or mesh) is laid over the model domain and the variables of interest are only calculated at the nodes of the grid.

The earliest studies of numerical models for the simulation of coastal groundwater systems were published in the early 1970s. Pinder and Cooper (1970) solved the coupled groundwater flow and solute transport equations for a problem with a moving interface. Lee and Cheng (1974) reproduced the shape of the wedge of intruded seawater in the Cutler area in Florida (Section 6.2.1.1). Segol et al. (1975) pioneered the now well-established finite element technique for seawater intrusion modelling. A myriad of numerical modelling studies has been published since then, and several codes are available, some of them at no cost, which can be used to construct site-specific models of variable-density flow systems.

Numerical models of real-world coastal aquifer systems require substantial amounts of input data, to the extent that their application sometimes becomes intractable. The runtimes can become very long, especially when long simulation timescales are considered, and there are difficulties related to the numerical solution of the governing equations. Thus, while a powerful tool, the construction of numerical models requires specific expertise, and their meaningful use in aquifer management relies on a concomitant data collection programme.

To save computational time, three-dimensional coastal aquifer models are often reduced to two-dimensional cross-sectional models. This is possible when groundwater flow is towards the shore so that the cross section can be aligned with the main flow direction. It is necessary to subdivide aquifers into multiple model layers to be able to resolve the vertical variations of the groundwater salinity. Practical guidelines for constructing numerical groundwater flow models can be found in Anderson et al. (2015), and for solute transport and variable-density flow models in Oude Essink (2000, 2001b) and Zheng and Bennett (2002).

2.5.1 Solution Techniques

A well-known class of numerical solution techniques is the finite-differences method. The basic principle behind the method is that the derivatives of pressure (or head) and concentration are approximated by numerical equivalents, which are evaluated using the values of the variables at each of the nodes, and for a transient model, for each time step. Since, of course, the variable values themselves are unknown as they are the ones being sought, special techniques are required to arrive at the solution of the mathematical problem. This includes iteration, a process in which the equations are first solved using an initial guess of the variable values, which are then updated in a stepwise fashion until the difference between each subsequent step (iteration) becomes smaller than a pre-defined criterion. Finite difference techniques employ a structured, orthogonal grid, which can be a drawback when simulating irregular natural features like rivers, coastlines or complex geological structures. Moreover, a local refinement of the grid is difficult. The finite difference method is the most intuitive of all numerical techniques and is relatively easy to understand and thus implement in computer codes.

The finite element method has also been widely applied in several computer codes for variable-density flow and transport. Mathematically it is not as intuitive as the finite difference method, but it has the practical advantages that it offers versatility in terms of mesh design (Figure 2.6) and handling of anisotropy. In short, the method works by subdividing the model domain into smaller domains (finite elements), within each of which the unknown variables (pressure, freshwater head, concentration or salt mass fraction, depending on which governing equation is solved) is approximated by a basis function. The basis function expresses the value of the variable within an element in terms of its values at the nodes. This can be a simple linear function, but more complicated functions may be used as well. The principle behind the finite element method is that the governing differential equation can be expressed in terms of a residual, and that the weighted sum of residuals across the entire model domain tends to zero. The basis functions multiplied by the weighting functions are integrated for each element, and a system of algebraic equations can be developed which can be solved for the variables at the nodal points. There are different variants of the finite element method, with the most common in hydrogeological applications being the Galerkin method (in which the basis and weighting functions are the same). Common element shapes can be triangles (which can be tetrahedra or prisms in three-dimensional models) or rectangles (which can be cubes or hexahedrons in three-dimensional models) and the method allows for easy local grid refinement.

In the finite volume method the model domain is sub-divided into control volumes. The basic principle behind the method is that the governing differential equation is integrated over the control volume, and the volume integrals of the advective and dispersive/diffusive flux are replaced by surface integrals across the cell boundaries. The resulting equation is a conservation expression for the control volume, for example of fluid or solute mass, and because the flux across the cell faces from one volume into the next must be equal, the finite volume method is conservative across the entire model domain. The discretised form of the integral equations is used to develop a system of equations that can be solved for the unknown variables. The shape of the control volume can be arbitrary, which makes that the finite volume method is suited for irregular, non-orthogonal grids. The finite element and finite volume methods are both integral methods (unlike finite

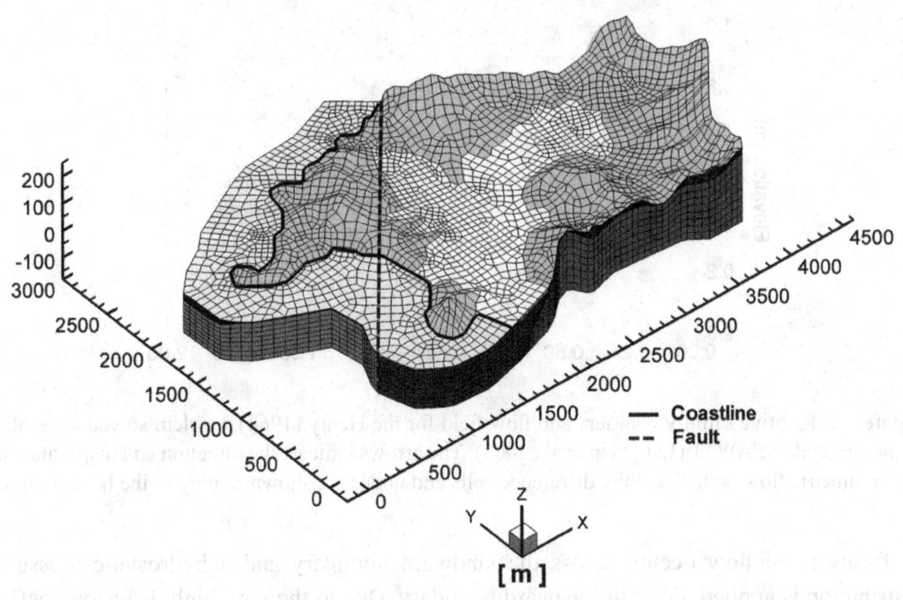

Figure 2.6 Example of a finite element mesh of a coastal aquifer system model (Park et al., 2007).

difference methods, which is a differential method), and the finite volume method is in fact a special case of the finite element method with the basis and weighting functions being 1.

A different category of solution techniques is formed by lattice Boltzmann methods, which have seen some application in the solution of variable-density models. The method finds its origin in the kinetic theory of gases and has the advantage that it is suited for problems where high flow velocities make that inertial effects must be taken into account, such as in fractured rock or karstified limestone aquifers. The method is based on distribution functions for the movement and interaction of hypothetical particles, of which the moments provide the properties like groundwater density and velocity (Sukop and Thorne, 2006). The method has seen some uptake in seawater intrusion modelling (Servan-Camas and Tsai, 2009; Servan-Camas and Tsai, 2010), but has thus far not gained the same popularity as finite difference of finite element solution techniques.

Analytical solutions that can be used to verify the correctness of numerical codes for variable-density flow problems are only available for a very limited number of cases, and moreover these contain stringent assumptions (Chapter 3). Therefore the only way to assess the correct performance of a numerical code is by comparing the model outcomes to the results of a physical experiment or that of other numerical codes. This is known as benchmarking and a considerable effort has been made over the past few decades to develop suitable benchmark problems and procedures. A review of this subject was published by Diersch and Kolditz (2002). The best-known benchmark for seawater intrusion is the Henry problem (Henry, 1964). The Henry problem is a highly schematised representation of a coastal aquifer in the form of a two-dimensional cross-sectional model. A constant rate

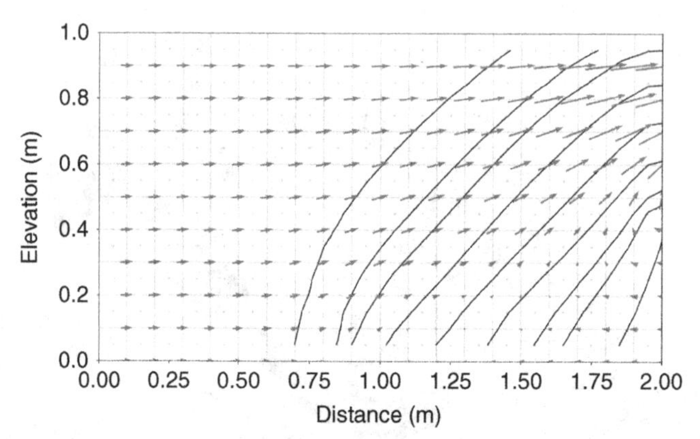

Figure 2.7 Relative salinity contours and flow field for the Henry (1964) problem solved using the numerical code SEAWAT (Langevin et al., 2008). The arrows indicate the direction and magnitude of the volumetric flow rate. The finite difference cells and nodes are shown in grey in the background.

of freshwater inflow occurs across the landward boundary and a hydrostatic pressure distribution is applied along the seaward boundary. Due to the very high diffusion coefficient a very broad transition zone develops (Figure 2.7). Simpson and Clement (2004) found that the choice of boundary conditions in the Henry problem exerts such a strong influence on the simulated concentration distribution that the density only has a moderate effect on the model outcomes. The development of benchmarks for seawater intrusion problems remains an area of ongoing research. A recent example is the definition of a new benchmark by Stoeckl et al. (2016), which is based on a physical laboratory experiment of a freshwater lens and considers transient aspects related to the lens development, and its decay when recharge ceases.

2.5.2 Some Issues with Numerical Techniques

The solute transport equation has proven very difficult to solve for groundwater systems because numerical schemes have difficulty with the advective term in Eqns (2.75) and (2.76), which is often more dominant than the diffusive/dispersive term. Taking the finite difference method as an example, the solute concentration gradient that is required for the advective term is determined from the concentrations at the grid nodes. One can either use the concentration at the upstream nodes (upstream weighting) or at the neighbouring nodes (central weighting). The latter is more accurate but it can lead to artificial oscillations (Figure 2.8), which means that calculated concentrations overshoot the maximum or undershoot the minimum values dictated by the boundary conditions and source terms. The upstream-weighting scheme does not have this problem but it is less accurate and overestimates the spreading of solutes by dispersion (numerical dispersion).

Numerical schemes in which the concentrations are evaluated at fixed spatial coordinates are called Eulerian methods and are prone to the problems with solving the advective term

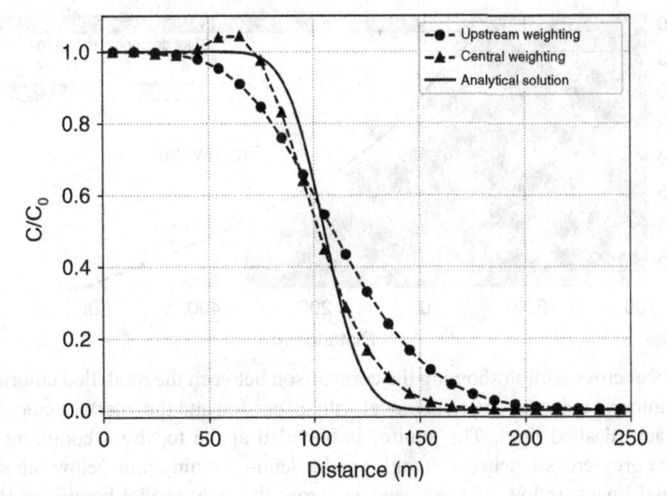

Figure 2.8 Effect of finite difference solution schemes for a one-dimensional solute transport model with advection and dispersion (Zheng and Bennett, 2002). With the upstream-weighting scheme, numerical dispersion is clearly recognisable, as the solute front is broader than for the analytical solution. Numerical dispersion is less for the central-weighting scheme, but the calculated concentrations overshoot the input concentration.

as described above. The solution of the dispersive/diffusive term is less problematic with Eulerian methods. The degree to which the advective term dominates over the dispersive/diffusive term can be expressed by the grid Peclet number (Pe_{gr}):

$$Pe_{gr} = \frac{v_i \Delta x_i}{|D_i|} \tag{2.79}$$

where Δx_i represents the length of a model grid cell and $|D_i|$ is the magnitude of the hydrodynamic dispersion coefficient. The subscript i indicates the direction. Since v, D and cell size can vary across the model domain and, in transient models, with time, the grid Peclet number is not usually a constant value. The performance of numerical schemes has been shown to depend on the Peclet number and from it grid discretisation criteria have been derived (Oude Essink, 1996). For example, for finite difference models with a central in space weighting scheme or a finite element model with linear basic functions, Pe_{gr} should be kept below 2. For finite element models with quadratic basic functions, it should be below 4.

The Courant criterion (Co) is another metric that ensures the stability of numerical solution schemes. It follows from the constraint that a particle cannot travel over a distance greater than the size of a cell during a model time step. It is given by

$$Co = \frac{v_i \Delta t}{\Delta x_i} < 1 \tag{2.80}$$

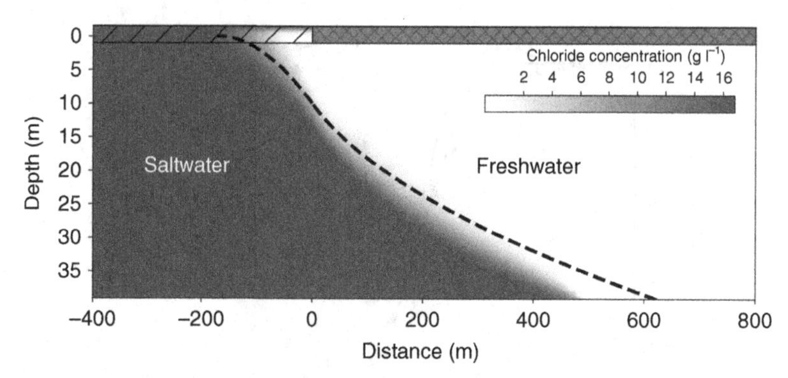

Figure 2.9 Vertical cross section showing the comparison between the modelled chloride concentrations (shaded contours) using a numerical model with dispersion and the interface calculated using an interface approach (dashed line). The aquifer is bounded at the top by a confining unit on land (indicated by the grey cross-hatched rectangle) and a semi-confining unit below the seafloor (indicated by diagonal lines). Inflow of freshwater is across the right model boundary. Based on data provided by Ms Amy Roach.

where Δt is the length of a time step. Because the flow velocities and the dispersion coefficients are given for a hydrogeological problem and therefore cannot be changed, the grid discretisation must be chosen fine enough to meet the Pe_{gr} criterion. A finer grid also necessitates a smaller time step to satisfy the Courant criterion. Often the number of grid cells and time steps can become so large that model run times become impractically long. Because of this, two-dimensional cross-sectional models rather than three-dimensional models are often used to analyse the problem, but this is only possible when a dominant flow direction can be defined, and this approach presents difficulty when the effects of pumping must be analysed. Other authors have even reverted to using coarse grids that necessitate the use of large dispersivities, but the reliability of such models then becomes questionable (Diersch and Kolditz, 2002).

Some authors have reverted to special numerical techniques such as the method of characteristics (Pinder and Cooper, 1970; Konikow and Bredehoeft, 1978; Lebbe, 1983; Oude Essink, 1996). With this method, the advective component of the solute transport equation is solved first by tracing a large number of particles that all have a concentration associated with them. Thus rather than trying to solve the advective term using a Eulerian technique based on a fixed network of nodes, a Lagrangian technique of a moving reference frame is adopted. In the second step, the dispersive and other terms of the solute transport equation are solved in a grid relative to the advective movement (Oude Essink, 1996). This mixed Eulerian–Lagrangian approach thereby lifts the strict constraints of the Peclet criterion. Disadvantages of the method are that it is not mass conservative, and models require large amounts of computer memory, but with modern computers the latter drawback has become less of an issue.

When the thickness of the transition zone is small relative to the aquifer thickness, an interface approach can sometimes substitute for a fully coupled variable-density model

(Essaid, 1990), which offers significant computational efficiency. A comparison between an interface model and a numerical model with dispersion is shown in Figure 2.9. More information about this topic can be found in for example Llopis-Albert and Pulido-Velazquez (2014) and Bakker et al. (2013).

The analysis of the variable-density flow problem can be simplified if the spatial and temporal dependency of the density is neglected in the mass conservation expressions for water and the solute, assuming that only the buoyancy term in Darcy's law is required to capture the effects of density. This is the so-called Oberbeck–Boussinesq approximation (Holzbecher, 1998; Diersch and Kolditz, 2002). Guevara Morel et al. (2015) found that models with this approximation gave similar results as fully coupled models when density contrasts remained below 50 kg m^{-3} for problems with unstable vertical density gradients which lead to the development of descending salt fingers.

2.6 Data Needs

Numerical models of real-world coastal aquifer systems require a lot of input data, especially three-dimensional, transient models. These include

- basic topographical, geological and hydrological data
- the parameters that appear in the governing equations, like hydraulic conductivity, storage coefficients, porosity, dispersivity and effective molecular diffusion coefficients
- boundary conditions, like water levels and recharge rates
- pumping rates
- water level and concentration measurements
- parameters to relate solute concentrations to the density

The degree to which these data are available determines the accuracy with which a system can be modelled. In flow diagrams of the groundwater modelling process, the construction of a numerical model is often placed after the collection of data but more realistically, numerical modelling is part of an ongoing process of planning and groundwater management (Maimone, 2002). The timelines for developing regional, three-dimensional, transient numerical models are in the order of months to years and usually the numerical model is used to identify gaps in conceptual system understanding and additional data needs. This is reflected by the schematic diagram of Figure 2.10, which indicates that a numerical model can be at the heart of the management process, with new information being fed into the model when it becomes available. At the same time the model can be used to prioritise data collection efforts depending on where the greatest uncertainty exists in the model. Modelling in the context of groundwater management is not a static, one-off activity, but an ever-evolving process, which also means that the financial and personnel resources required are substantial.

Data collection is discussed in Section 12.2. Specific information needs for numerical models are related to the parameters of the governing equations. The hydraulic conductivity is one of the most important parameters but often also highly uncertain. Usually the model

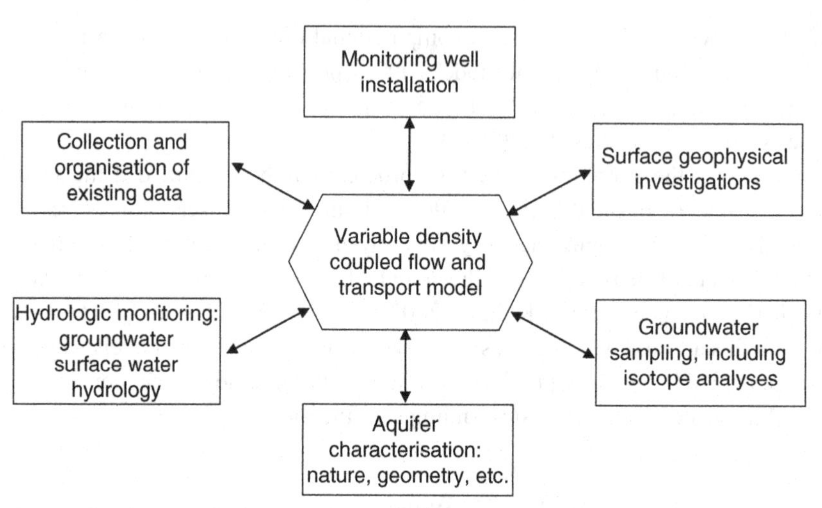

Figure 2.10 Role of a numerical model in coastal aquifer management (Maimone, 2002).

is populated using initial estimates based on aquifer tests, if available, and then the values are adjusted during the model calibration stage. It is common for there to be no data about the dispersivity and its heterogeneity and anisotropy. The rule of thumb that the dispersivity can be approximated by 10% of the solute transport length scale (Gelhar et al., 1992) does probably not apply to regional groundwater models of coastal aquifer systems, as longitudinal dispersivity values on the order of only a few metres were found to yield more realistic salinity distributions than dispersivities of 10 m or more for a circa 51 by 65 km wide model of an area in the western part of the Netherlands (Oude Essink, 2001a).

Automated parameter calibration should be preferred over manual calibration as it provides uncertainty bounds of the optimised parameter values and the model outcomes, but given the long runtimes of variable-density models, a trial and error calibration is often still opted for instead, or only a subset of model parameters is optimised (Sanford and Pope, 2009). Even though calibration is an essential element of the modelling procedure, it is extremely dangerous to regard a calibrated model as a tool with which the future behaviour of a groundwater system can be predicted. Much more can be gained by using a model to understand how the system functions (Voss, 1999) and to test if the conceptual model of the system is actually plausible (Maimone, 2002).

3

Analytical Solutions for a Steady Freshwater–Saltwater Interface

3.1 Introduction

Estimation of the location of the transition zone between fresh- and saltwater is an important topic in coastal aquifer studies. Various processes such as diffusion, dispersion, convection driven by temperature and salinity differences, and the perturbation due to wave action and tidal fluctuations, affect the mixing between saltwater and the freshwater (Section 6.2). If the transition zone between fresh and saline groundwater is narrow, it may be approximated by a sharp interface (called saltwater interface or simply interface hereafter). An example of a sharp interface is a transition zone of a few decimetres to a few metres where the overlying body of freshwater is tens of metres thick. Such interfaces are found, for example, below the freshwater lenses beneath some coastal dunes of the Netherlands (e.g. Stuyfzand, 2017). A wide transition zone can be tens of metres thick with an overlying freshwater body as small as just a few metres in thickness. These are common below oceanic islands (Werner et al., 2017). The definition of sharp is thus also related to the thickness of the freshwater body in the aquifer.

Numerical modelling is required to deal with problems where the transition zone is wide, which is discussed in Chapter 2. This chapter presents some analytical solutions to calculate the position of the interface, assuming that it is sharp and steady. Various coastal aquifer settings are considered, including continental coastal aquifers and circular or strip islands, as well as different hydrological situations such as a seepage zone, precipitation recharge and pumping.

Two common analytical approaches are used to derive equations for the location of the interface. The first one assumes that the flow satisfies the Dupuit assumption (Section 3.3.1) and Ghijben–Herzberg principle (Section 3.3.2), so it is also called Dupuit-Ghijben–Herzberg analysis (Vacher, 1988; Oberdorfer et al., 1990). The second approach is based on potential theory which uses a single potential function for both confined flow and unconfined flow (Strack, 1976). This approach also uses the Ghijben–Herzberg principle but not necessarily the Dupuit assumption. Solutions based on the second approach involve fewer assumptions than the first one. When a solution can be obtained by the first approach, it usually can also be obtained by the second approach (Kashef, 1983; Vacher, 1988; Todd and Mays, 2005). For many practical purposes, the approximate solution using the first approach is indistinguishable from more exact solutions by the potential theory. The solutions in Sections 3.4.6 and 3.5 are based on potential theory.

3.2 Historical Note on Freshwater–Saltwater Relations

Over a century ago, Herzberg (1888) and Drabbe and Badon Ghijben (1889) presented a mathematical relation between the depth of the interface and the elevation of the water table in a coastal aquifer. Although they were not the first to formulate the hydrostatic equilibrium between fresh and saline groundwater, their works can be considered to be the cornerstones of the conceptual model of a freshwater–saltwater interface in a coastal aquifer (Post et al., 2018d).

Several authors (Brown, 1922; Palmer, 1957; Carlston, 1963; Davis, 1978; Reilly and Goodman, 1985; Holzbecher, 2005; Post et al., 2018d) have provided information on the history of the studies of the interface. The oldest known documented observations of fresh-saltwater relationships go back to Greek philosophers. A written document about the relation between freshwater and saltwater was presented in the chapter on 'wonders of fountains and Rivers' in *Natural History* by the Roman natural philosopher Pliny the Elder (AD 77–79): 'It is very remarkable that freshwater should burst out close to the sea, as from pipes. But there is no end to the wonders that are connected with the nature of waters. Freshwater floats on seawater, no doubt from its being lighter; and therefore sea water, which is heavier in nature, supports better what floats upon it' (Bostock and Riley, 1890).

However, the modern scientific theories did not become established until the eighteenth and mostly the nineteenth century (Post et al., 2018d). Joseph Du Commun, a teacher of French at West Point Military Academy has been credited (Carlston, 1963) for being the first to formulate a hydrostatic equilibrium relation between fresh- and saltwater (Du Commun, 1828). He published his paper after questions were raised in a newspaper article about a well that was drilled at a distillery in New Brunswick, New Jersey. It was found that this freshwater well had a water level that rose about 2.4 m to 4.3 m above the local river and the flow rate of the well varied in phase with the tide.

Using the analogy of a U-tube Du Commun (1828) argued in an equilibrium situation, a denser fluid like seawater sitting in arm A of the tube, will rise to a lesser height than the less dense fluid like freshwater sitting in arm B (Figure 3.1). He clearly stated that the density ratio between the fluids controlled the water level difference, and also understood that such a relationship could be used to infer the depth to the saltwater if the height of the freshwater column above the ocean was known (see also Carlston, 1963).

Du Commun (1828) was not the first to use the analogy of a U-tube though, as Inglis (1817) had already used the example of an inverted siphon to explain similar observations in the harbour of Bridlington, along the Yorkshire coast of England (Post et al., 2018d). While Inglis (1817) also spoke of the effect that the density difference between seawater and freshwater must have, he was not as clear in his formulation as Du Commun (1828). However Du Commun (1828) did seem to reverse cause and effect and argued that the depth

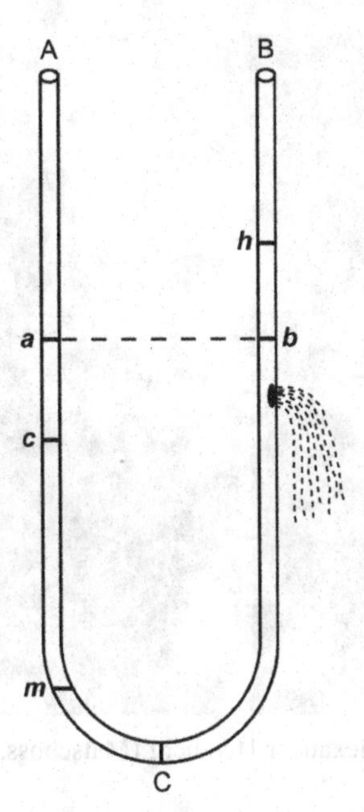

Figure 3.1 U-tube used by Du Commun (1828) to demonstrate the balance of fluids of different densities. The level of freshwater in arm B (right) is indicated by b (or h at high tide). The level a in arm A (left) is the level to which water of the same density as in arm B would rise. The level c is the height to which saltwater in arm A would rise, while m indicates the level for mercury. The hole in arm B was used by Du Commun (1828) to illustrate the analogy with coastal springs.

to the interface controlled the height of the water table and that this was the reason for the occurrence of springs at the top of mountains.

Despite these early nineteenth-century publications, the hydrostatic equilibrium relationship that relates the water table elevation to the depth of the interface has become known as the Ghijben–Herzberg principle. Seemingly unaware of earlier works, Alexander Herzberg (1841–1912) in Germany and Willem Badon Ghijben (1845–1907) in the Netherlands, also formulated the principle about 60 years later. Alexander Herzberg (Figure 3.2, left) was a civil engineer who had investigated the seawater-freshwater relations in the island of Norderney. He wrote down the relation, along with a sketch of a freshwater lens below the island (Figure 3.3) in a technical report that was never publicly released (Herzberg, 1888). His theory did therefore not gain any broad attention until he presented it at a symposium in Vienna, the transcript of which was published in 1901 (Herzberg, 1901). At the time people tended to believe that the groundwater beneath islands must be salt because they are surrounded by the sea. Herzberg's relation had enormous practical significance because it demonstrated that freshwater can extend much deeper than sea level, so that a large resource of freshwater can be found below islands (Holzbecher, 2005).

The figure drawn by Herzberg (1901) clearly shows the shape of the freshwater lens in an island (Figure 3.3): the water table has a bulge and the bottom boundary of the

Figure 3.2 Photographs of (left) Alexander Herzberg (Matschoss, 1921) and (right) Willem Badon Ghijben (De Vries, 1994).

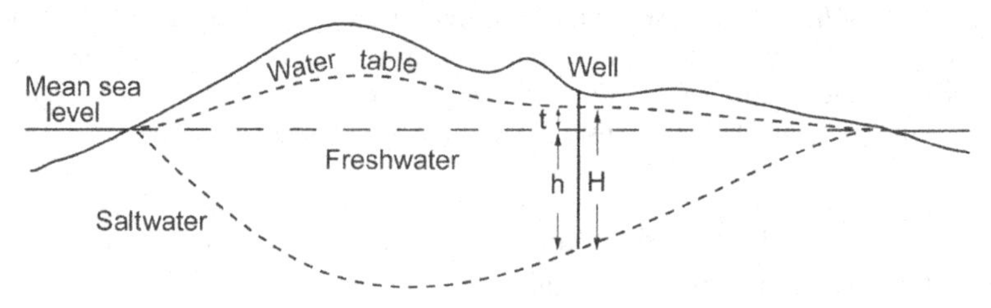

Figure 3.3 Cross section of Norderney Island, showing the application of Herzberg's theory, originally from (Herzberg, 1888), reproduced in English by Brown (1922). H is the total thickness of the freshwater, t is the height of freshwater above mean sea surface and h is the depth of freshwater below mean sea surface. (Note the symbols used here are from the original literature and not consistent with those in this book.)

freshwater is lens-shaped. He seems to be the first scientist who correctly depicted the shape of the freshwater body graphically. Another interesting point to note is the water table, which generally follows the shape of the ground surface but in a subdued manner. Obviously Herzberg understood this over 100 years ago as he drew the water table the highest below the most elevated part of the island.

Willem Badon Ghijben (Figure 3.2, right), a Captain in the Engineering Corps of the Netherlands Army, studied the groundwater salinity distribution around Amsterdam in the Netherlands in order to locate an emergency water supply during times of war (Drabbe and Badon Ghijben, 1889). Based on observations made in locks in surface water, where a wedge of saltwater forms below the outflowing freshwater, he realised that the same situation must exist in groundwater beneath the country's coastal dunes. Using a specific gravity of North Sea water of 1.0238 and a freshwater density of 1.000 he estimated that the interface must be around 42 times the water table height above sea level. The note with the relationship was co-authored by J. Drabbe who was the senior-ranking officer, but did not contribute to the relation scientifically (Carlston, 1963).

There exists some confusion about the spelling of the name of Willem Badon Ghijben. In Dutch, the letter combination 'ij' can be replaced with 'y', but the name under the original note is spelled with 'ij', this is the spelling that must be adhered to (Post, 2018). Also, the note was dated 1887, even though it was not published until 1889, and Herzberg submitted his unpublished report in 1888. So both worked on the subject around almost the same time, and there seems to be no point to try and establish who was actually the first.

Herzberg presented his theory with much more detail than Drabbe and Badon Ghijben (1889), and it did not take long for his work to be noted by scholars outside Germany (e.g. d'Andrimont, 1903; Imbeaux, 1906). The turn from the eighteenth to the nineteenth century was the time when there was great interest in coastal water supply in Europe (Brown, 1922), which is the most likely reason why Du Commun (1828) never received any credit for stating the hydrostatic equilibrium relationship. Davis (1978) argued that this was unfair, but because Drabbe and Badon Ghijben (1889) and Herzberg (1888, 1901) did combine quantitative insight with a sound conceptual model, it seems justified after all to name the relationship in their honour (Post et al., 2018d).

3.3 Mathematical Relations between Freshwater and Saltwater

Various analytical solutions are available to predict the location of the interface. These typically ignore the vertical component of groundwater flow (Dupuit assumption) and adopt the Ghijben–Herzberg principle. Both are discussed in more detail in this section.

3.3.1 Dupuit Assumption

The Dupuit assumption was named after the French hydraulic engineer and economist A. J. Dupuit, who first used it in 1863 (Dupuit, 1863). Figure 3.4 shows two cross sections depicting flow in a confined and unconfined aquifer, respectively. The flow in the confined aquifer is strictly horizontal, which renders the problem one-dimensional. This means that $v_z = 0$, i.e. the equipotential lines are vertical (it is assumed the aquifer is isotropic). The flow lines in the unconfined aquifer, however, bend downward, and the flow along the cross section is two-dimensional. In the presence of vertical flow, the equipotential lines are not vertical. The hydraulic gradient between the points *a* and *b* in Figure 3.4 does not equal the slope of the water table because the vertical lines starting from these points are not coincident with the corresponding equipotential lines.

According to the Dupuit assumption the vertical flow is negligible and the head changes only with horizontal distance *x*. The hydraulic gradient in an unconfined aquifer then equals

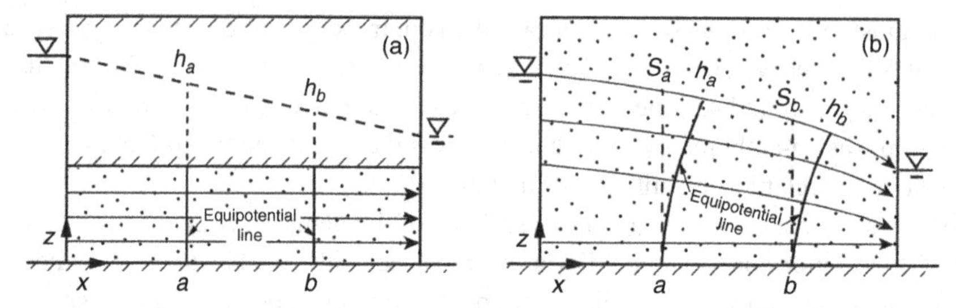

Figure 3.4 (a) Strictly horizontal flow in a confined aquifer and (b) two-dimensional flow in an unconfined aquifer. The curved equipotential lines in (b) are for the true two-dimensional flow conditions; the vertical ones are those for the Dupuit assumption.

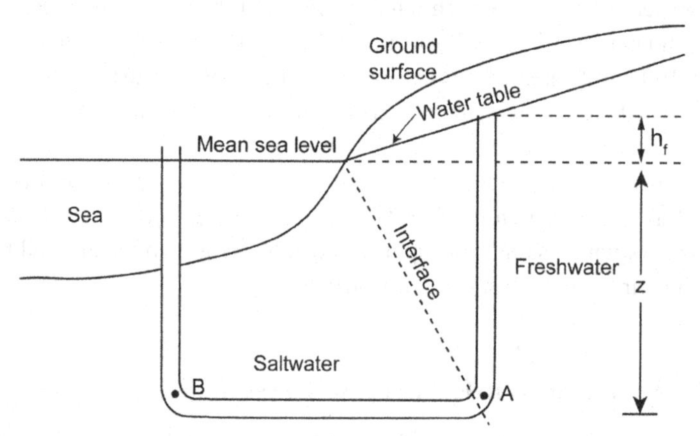

Figure 3.5 Idealised interface between freshwater and saltwater in an isotropic and homogeneous unconfined aquifer under hydrostatic conditions.

the slope of the water table and the flow lines are horizontal and the equipotential lines vertical. At point a for example, the assumption of vertical equipotential lines implies that the head at any elevation equals s_a. In reality the head at point a equals h_a and the head above point a must be monotonically increasing along the line $a - s_a$ because there is vertically downward flow.

The practical purpose of this assumption is that the two-dimensional flow problem can be approximated by a one-dimensional equation. The error generated by using the Dupuit assumption obviously depends on how significant the vertical flow is. For more detailed discussions readers are referred to the papers by Kirkham (1967) and Hager (2004).

3.3.2 The Ghijben–Herzberg Principle

The Ghijben–Herzberg principle can be derived by considering the situation in Figure 3.5, where freshwater and saltwater are separated by an interface. The mean sea level equals zero, and is set to be the datum. The elevation with respect to sea level of an arbitrary point at the interface is z, which is negative downward. If it is assumed that there is no flow in the saltwater part, the saltwater head below the interface is the same everywhere and equal to the sea level and the pressure in point B is equal to that in point A. When the position of the interface is constant in time, the water pressures of the freshwater and the saltwater must be equal where they meet. By expressing the pressures as a function of the lengths of the columns of salt and freshwater respectively, the pressure equality is written as

$$\rho_s g(0 - z) = \rho_f g[h_f + (0 - z)] \tag{3.1}$$

which gives

$$z = -\frac{\rho_f}{\rho_s - \rho_f} h_f \tag{3.2}$$

where h_f [L] is the height of the freshwater column above the sea surface, g [L T^{-2}] is the gravity acceleration constant, and ρ_s and ρ_f [M L^{-3}] are the density of saltwater and freshwater, respectively. As per the Dupuit assumption, the vertical flow in the freshwater part is negligible and the head does not vary with depth, so the height of the freshwater column above sea level is equal to the water table elevation. The head in the freshwater part is then also equal to the freshwater head h_f defined in Section 1.5 (Eqn (1.5)), and hence the same symbol has been used here.

To make some equations more concise, it is common to define

$$\delta = \frac{\rho_f}{\rho_s - \rho_f} \tag{3.3}$$

Some researchers define a parameter called the excess of the specific gravity of seawater over freshwater (Kuan et al., 2012), which is reciprocal of δ. This should not be confused with the specific gravity of seawater, which is defined as ρ_s/ρ_f.

Equation (3.2) is the Ghijben–Herzberg principle. Unlike some texts, it contains a minus sign because z represents an elevation (negative downwards) rather than a depth (positive downwards). If the seawater and freshwater densities are taken as 1025 kg m^{-3} and 1000 kg m^{-3}, respectively, then

$$z = -40h_f \tag{3.4}$$

This relation provides a very simple approach to estimate the depth of the interface. It shows that, if the water table at a point is 1 m above sea level, the interface should be 40 m below the sea surface, and the thickness of the freshwater body is 41 m at that point. The Ghijben–Herzberg principle also provides a way to estimate the change of an interface due to a change in the water table position. For example, a fall of the water table by 1 m will eventually lead to a rise of the interface by 40 m if a new equilibrium situation establishes, demonstrating that small changes of the water table position can have serious effects on the thickness of a freshwater lens. It should be borne in mind though that the rise of the interface will not be instantaneous and that the new position prescribed by the Ghijben–Herzberg principle may take years to attain.

Equation (3.4) implies that the slope of the interface is 40 times of that of the water table (Zhou and Wang, 2009). The latter is generally predominantly controlled by the topography in natural environments where recharge rates are high. In a low-lying coastal plain with a flat ground surface, the water table cannot rise high above sea level, and the interface depth will be restricted. Where the land rises high above the sea and recharge rates are high, the slope of the water table will be steep and the corresponding interface can be nearly vertical.

The density of fresh groundwater varies only minimally around the world, but the density of seawater can deviate significantly from the value of $\rho_s = 1025$ kg m^{-3} used in the example above. For example, for the density of the Dead Sea is $\rho_s = 1230$ kg m^{-3}, so according to Eqn (3.2), the interface is only 4.35 m deep for a water table of 1 m above the sea level (Yechieli, 2000). In a well-mixed estuary, the density of the seawater may be much lower than $\rho_s = 1025$ kg m^{-3}, the depth of the interface may be much greater than that estimated from Eqn (3.4). The presence of aquitards in the aquifer may also significantly modify the depth of interface (Stuyfzand, 2017).

In the absence of chemical or geophysical data for inferring the depth of the interface, the Ghijben–Herzberg principle can provide a first assessment of the depth of the interface when h_f can be determined. When water level observation wells in the freshwater zone are available, a spatial distribution of the interface can be inferred. Extreme care must be taken though in interpreting these values because of the many assumptions underlying the Ghijben–Herzberg principle. For instance, because vertical flow in the freshwater lens is neglected, the interface depth predicted by the Ghijben–Herzberg principle may differ from the actual depth (Custodio et al., 1987; Izuka and Gingerich, 1998). In addition to the assumptions already discussed, the Ghijben–Herzberg principle does not take into account the effect of aquifer geometry. A freshwater lens may be truncated at the bottom by a confining unit for example. Other effects like offshore extensions of the freshwater body and seepage zones, as well as the lens not having reached its equilibrium position yet, make that the Ghijben–Herzberg principle can only be used to obtain an indication of the interface depth.

3.3.3 Hubbert's Formula

The Ghijben–Herzberg principle ignores head losses in saltwater (Lusczynski and Swarzenski, 1966). Hubbert (1940) analysed a more general case in which the hydraulic head does not have to be the same everywhere. In his analysis, the interface may be in motion, and the Dupuit assumption is not required. The heads in the saltwater and freshwater can be defined respectively as (Hubbert, 1940)

$$h_s = \frac{p}{\rho_s g} + z \tag{3.5}$$

$$h_f = \frac{p}{\rho_f g} + z \tag{3.6}$$

where h_s and h_f [L] are heads in saltwater and freshwater respectively at point z and p [M L^{-1} T^{-2}] is the water pressure at the measurement point.

At the interface, equating p from the two expressions leads to

$$z = \frac{\rho_s}{\rho_s - \rho_f} h_s - \frac{\rho_f}{\rho_s - \rho_f} h_f \tag{3.7}$$

If $h_s = 0$, i.e. the head in saltwater is everywhere the same as the sea level datum, Eqn (3.7) reduces to Eqn (3.2). A comparison of the two equations shows that the depth estimated from Eqn (3.7) can be greater or smaller than that based on the Ghijben–Herzberg principle, depending on the head h_s being higher or lower than sea level. The latter situation will be the more common in most coastal aquifer systems as the flow of the saltwater causes heads to fall below sea level (Section 6.2.1.1).

Conceptually, Hubbert's formula is an advancement over the Ghijben–Herzberg principle because the position of the interface is related to the heads in fresh- and saltwater at the interface, while in the Ghijben–Herzberg principle it is related to the water table elevation (Reilly and Goodman, 1985). The interface location z can only be calculated though when the head in both the freshwater and saltwater are known at z, which is impossible in a real-world aquifer. In many aquifers, the difference between Eqns (3.2) and (3.7) will be small because the flow in the saltwater part tends to be slow and therefore deviations of h_s from sea level will be small.

3.3.4 Extension of Hubbert's Formula for Field Applications

Many researchers have discussed the possibility of using water levels measured in wells to identify the location of the interface in a real aquifer, starting with Lusczynski (1961), Perlmutter and Geraghty (1963) and De Wiest (1965). For example, Lusczynski (1961) studied a coastal aquifer with freshwater, brackish water and saltwater, and with a piezometer screen in both the freshwater and the saltwater along a vertical profile, and derived a simple equation to estimate the elevation of the contact between the brackish water and the freshwater.

The early work was reiterated in later publications. For example, Kim et al. (2007) introduced an equation to estimate the interface location using the pressure data from a pair of pressure loggers located below and above the interface in the same vertical profile and employed this equation to study the interface in Jeju Island, Korea. Zhou et al. (2008) expressed the equation in terms of head for a system that is assumed to be in an equilibrium

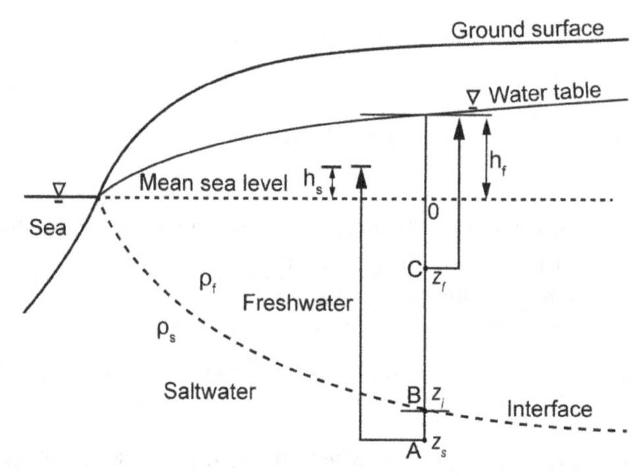

Figure 3.6 Water table and interface in a coastal unconfined aquifer with piezometers in freshwater and saltwater zones (modified from Zhou et al., 2008).

state and where the Dupuit assumption applies. If two piezometers are open at point A and C in saltwater and freshwater zones, respectively (Figure 3.6), and the interface is located at B, the following relation can be written for the pressure equilibrium at point A (Zhou et al., 2008):

$$\rho_s g[h_s + (0 - z_s)] = \rho_s g(z_i - z_s) + \rho_f g[(0 - z_i) + h_f] \tag{3.8}$$

where h_s and h_f are saltwater and freshwater heads at points A and C, respectively and z_s and z_i are the elevation of point A and the interface, respectively (Figure 3.6). Equation (3.8) is a simplified version of Lusczynski's (1961) equation after ignoring the brackish water.

Equation (3.8) can be written as

$$z_i = \frac{\rho_s}{\rho_s - \rho_f} h_s - \frac{\rho_f}{\rho_s - \rho_f} h_f \tag{3.9}$$

The above equation has the same form as Hubbert's formula (Eqn (3.7)). However, in Eqn (3.7), h_s and h_f are heads in saltwater and freshwater at the same point (i.e. B) at the interface, but h_s and h_f in Eqn (3.9) are heads at points A and C measured in piezometers in saltwater and freshwater (Figure 3.6). This highlights that Hubbert's equation can only be applied in practice by assuming hydrostatic conditions between the points of measurements (i.e. the piezometer screens) and the position of the interface.

Because the Dupuit assumption is used, all the heads in the freshwater zone equal h_f and those in the saltwater zone equal h_s along the vertical profile in Figure 3.6. The interface location is therefore independent of the depths of the piezometers at points A and C, but obviously the estimate of z_i will be more in error if the vertical separation becomes greater and conditions are not strictly hydrostatic.

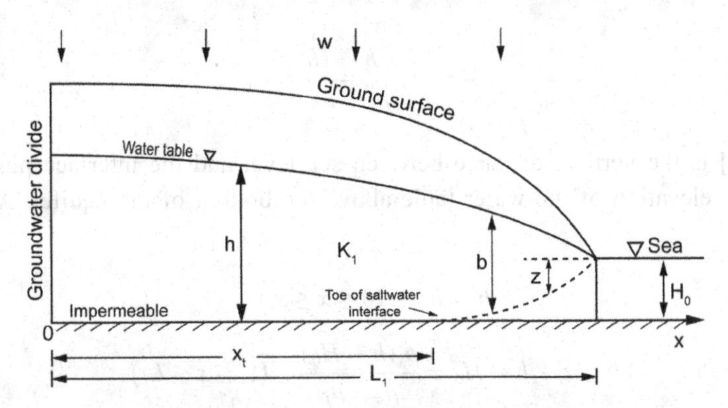

Figure 3.7 Water table and interface in an unconfined continental coastal aquifer.

3.4 Analytical Solutions for the Water Level and Interface in Coastal Aquifers

Equations for the water level and interface position in unconfined and confined aquifers based on the Dupuit assumption and the Ghijben–Herzberg principle will be presented in the following sections. The aquifer configurations considered are (i) continental coastal aquifers, and (ii) circular island or (iii) strip island aquifers. Continental coastal aquifers are bounded on one side by the coast and on the inland side by a specified flow boundary. This can either be a no flow boundary representing the groundwater divide, or an inflow boundary for groundwater from the upstream part of the aquifer (Section 2.4.2). For configurations (i) and (iii) the model cross sections are perpendicular to the coastline and the solutions are expressed in Cartesian coordinates. For configuration (ii) radial coordinates are adopted. In all cases the aquifer is homogeneous and isotropic.

3.4.1 Unconfined Continental Coastal Aquifer

A schematic outline of the conceptual model of this problem is shown in Figure 3.7. The aquifer is assumed to extend infinitely along a straight shoreline so that the flow field and the interface position are invariant along the shore-parallel direction. The domain includes the entire aquifer from the groundwater divide to the shoreline.

The distance from the groundwater divide to the shoreline is L_1 [L]. The system has a uniform hydraulic conductivity of K_1 [L T^{-1}] and rainfall recharge occurs along the top of the domain, with w [L T^{-1}] being the volumetric recharge rate per unit area. The origin of the distance axis x is at the location of the groundwater divide; x_t [L] is the toe location of the interface. The datum of the system is set at the bottom of the aquifer, and the height of the sea level is denoted by H_0 [L].

The interface can be regarded as a no-flow boundary located at a distance below sea level that is equal to (Guo and Jiao, 2007)

$$z = \frac{\rho_f(h - H_0)}{\rho_s - \rho_f} \tag{3.10}$$

where z [L] is the vertical distance between sea level and the interface position and h [L] is the elevation of the water table above the bottom of the aquifer. As seen in Figure 3.7,

$$b = h \qquad (0 \leq x \leq x_t) \tag{3.11}$$

$$b = z + h - H_0 = \frac{\rho_s(h - H_0)}{\rho_s - \rho_f} \qquad (x_t \leq x \leq L_l) \tag{3.12}$$

because $z = H_0$ at $x = x_t$:

$$h = \frac{\rho_s}{\rho_f} H_0 \ (x = x_t) \tag{3.13}$$

With the Dupuit assumption the heads vary only in the x direction, so that the continuity equation for flow in the unconfined aquifer can be written as

$$\frac{d}{dx}\left(\mathrm{K}_1 b \frac{dh}{dx}\right) + w = 0 \tag{3.14}$$

where b is given by Eqns (3.11) and (3.12). Integrating Eqn (3.14) from $x = x_t$ to $x = L_l$ and considering the conditions $h = H_0$ and $-K_1 b \frac{dh}{dx} = wL_1$ (outflow = total recharge) at $x = L_1$, leads to

$$h = \sqrt{\frac{w(\rho_s - \rho_f)}{K_1 \rho_s}(L_1^2 - x^2) + H_0} \qquad (x_t \leq x \leq L_1) \tag{3.15}$$

The toe location of the interface is obtained by letting h equal $\rho_s H_0 / \rho_f$ (Eqn (3.13)) in Eqn (3.15). Solving the equation for x leads to

$$x_t = \sqrt{L_1^2 - \frac{K_1(\rho_s^2 - \rho_s \rho_f)}{w \rho_f^2} H_0^2} \tag{3.16}$$

This equation shows that a higher recharge rate w will cause the interface to move seawards, while increasing hydraulic conductivity K_1 will cause the interface to move further inland. It also shows that the interface position depends on the ratio K_l/w, so in model calibration, either one parameter cannot be determined without knowing the other.

Integrating Eqn (3.14) within the range of $0 \leq x \leq x_t$ and considering the conditions

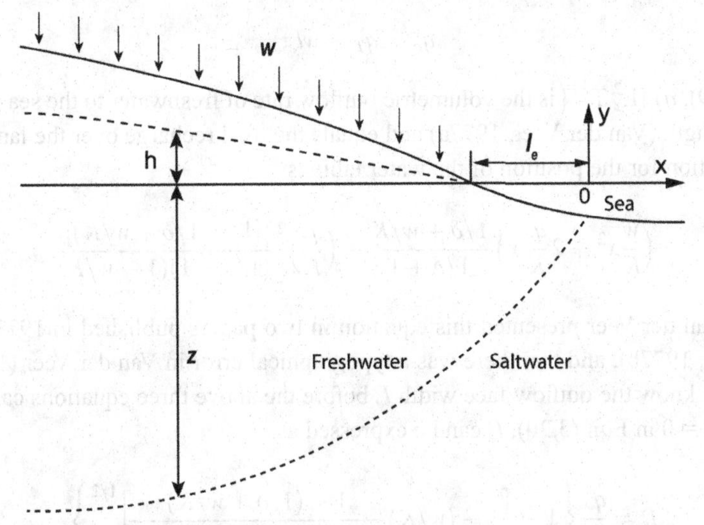

Figure 3.8 Water table and interface in an unconfined continental coastal aquifer with a seepage zone (modified from Van der Veer, 1977a).

$$-K_1 b \frac{dh}{dx} = 0 \quad \text{at } x = 0 \text{ and}$$

$$h = \frac{\rho_s}{\rho_f} H_0 \quad \text{at } x = x_t$$

leads to

$$h = \sqrt{\frac{w}{K_1}(L_1^2 - x^2) + \frac{\rho_s}{\rho_f} H_0^2} \qquad (0 \le x \le x_t) \tag{3.17}$$

If there is no density difference, Eqn (3.17) is reduced to Eqn (9.4) in Chapter 9.

3.4.2 Unconfined Continental Coastal Aquifer with a Seepage Zone

The interface in an unconfined aquifer with rainfall recharge and a seepage zone was studied by Van der Veer (1977a). The conceptual model is shown in Figure 3.8. The width of the outflow zone is l_e. The origin of x is set at the outer edge of the outflow zone.

The solution for the interface position is (Van der Veer, 1977a)

$$z = \left[\frac{-(wx^2/K + 2xq_e/K)}{(1/\delta + w/K)(1/\delta + 1)} \right]^{1/2} \tag{3.18}$$

where q_e [L^2 T^{-1}] is defined as

$$q_e = q_T + wl_e \tag{3.19}$$

In Eqn (3.19), q_T [$L^2\ T^{-1}$] is the volumetric outflow rate of freshwater to the sea per unit of coastline length (Van der Veer, 1977a) and equals the total recharge over the land surface.

The equation for the position of the water table is

$$h^2 = -\left(\frac{w}{K}x^2 + 2\frac{q_e}{K}x\right)\frac{1/\delta + w/K}{1/\delta + 1} - \left(\frac{q_e}{K}\right)^2 \frac{[1-(1/\delta + w/K)]}{(1/\delta + 1)(1 - w/K)} \tag{3.20}$$

Note that Van der Veer presented this equation in two papers published in 1977 (Van der Veer, 1977a, 1977b), and that there was a typographical error in Van der Veer (1977b).

One must know the outflow face width l_e before the above three equations can be used. By setting $h = 0$ in Eqn (3.20), l_e can be expressed as

$$l_e = \frac{q_e}{w}\left\{1 - \left[1 - (w/K)\frac{1-(1/\delta + w/K)}{1-(w/K)(1/\delta + w/K)}\right]^{1/2}\right\} \tag{3.21}$$

The value of l_e can be obtained using Eqns (3.19) and (3.21). An example of how to use the above equations was given by Van der Veer (1977b).

The physical meaning of q_e defined in Eqn (3.19) is not very clear. Vacher (1988) interpreted it as 'a fictitious quantity, equal to the discharge at the shoreline plus the accumulated recharge over the outflow face'. Because q_T is the total inland recharge, it can be assumed that $q_e \approx q_T$ if the outflow face is very small compared to the total inland extent of the recharge area.

3.4.3 Unconfined Circular Island Aquifer

Figure 3.9 schematically shows a freshwater lens beneath a circular island bounded by the interface at the bottom. Recharge is entirely from rainfall infiltration, which discharges to the sea. Studies of this problem were published by Fetter (1972) and Todd (1980).

Continuity requires that the total recharge within any radial distance from the centre of the island r [L] equals the lateral flow at r. This can be expressed by the following continuity equation, which adopts the Ghijben–Herzberg principle to relate the h_f to the interface depth z:

$$\pi r^2 w = -2\pi r K_1 (1+\delta) h_f \frac{dh_f}{dr} \tag{3.22}$$

or

$$w\,dr^2 = -2K_1(1+\delta)\,dh_f^2 \tag{3.23}$$

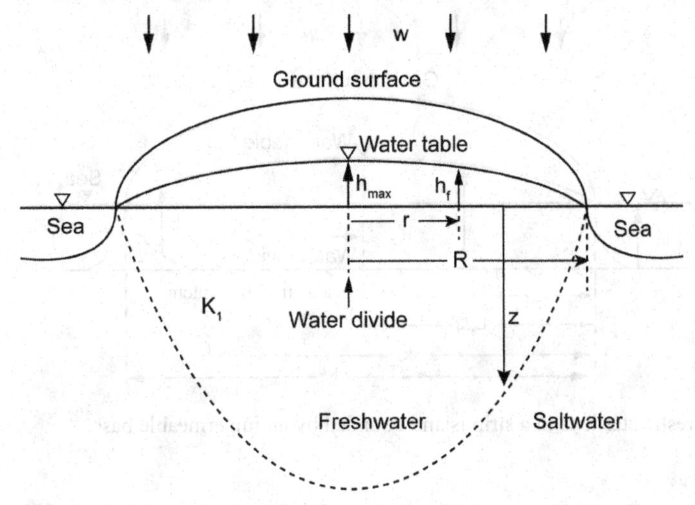

Figure 3.9 Freshwater lens in a circular island bounded at the base by the interface.

Integrating the above equation, letting r change from 0 to R and h_f from $h_{\max}$ to 0 gives

$$h^2_{\max} = \frac{wR^2}{2K_1(1+\delta)} \tag{3.24}$$

where R [L] is the radius of the island. Alternatively, integrating Eqn (3.23) and letting r change from 0 to r and h_f from h_{max} to h_f gives

$$h_f^2 = h^2_{\max} - \frac{w}{2K_1(1+\delta)}r^2 = \frac{w}{2K_1(1+\delta)}(R^2 - r^2) \tag{3.25}$$

From the Ghijben–Herzberg principle, at distance r, the interface depth is

$$z = \delta h_f = \sqrt{\frac{w\delta^2}{2K_1(1+\delta)}(R^2 - r^2)} \tag{3.26}$$

Equations (3.25) and (3.26) show that water table elevation and the interface depth are proportional to the precipitation recharge and the width of the island, but inversely proportional to hydraulic conductivity. An island consisting of highly permeable coral debris with a size on the order of 10–100 m, for example, may not have any freshwater lens of significance, especially since in natural systems tides enhance the mixing between freshwater and seawater.

3.4.4 Unconfined Strip Island Aquifer

When the ratio between an island's length and its width is greater than 2.9 (Vacher, 1988), the flow can be approximated by that below an infinite strip of land, that is, the coastlines are

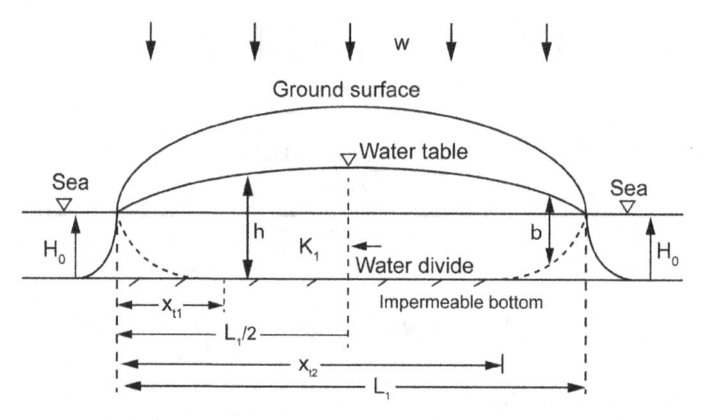

Figure 3.10 Freshwater lens in a strip island bounded by an impermeable base.

parallel and extend infinitely in the direction perpendicular to the groundwater flow direction. Two cases are considered here: either the base of the freshwater flow system is formed by an impermeable layer or by the interface.

Note that in this section the origin of the system is set at the shoreline on the left. For the island situation, usually it is convenient to choose the origin to be at the centre of the island, as in Figure 3.9. However some equations to be derived here will be used also in Chapter 9 about land reclamation where reclamation is assumed to occur along the shoreline on the right. For that reason, it is more convenient to choose the origin at the left shoreline, which is fixed but the centre of an island changes after reclamation.

3.4.4.1 Freshwater Lens Bounded by an Impermeable Base

The conceptual model for an unconfined flow system beneath an infinite strip of land with an impermeable base and uniform rainfall recharge is shown in Figure 3.10. The width of the island is L_1. The distances from the shoreline on the left to the toes of saltwater interface on either side, are denoted as x_{t1} and x_{t2}, respectively. Due to symmetry of the system, the water divide is located in the centre of the island and therefore $L_1 - x_{t2} = x_{t1}$.

Equation (3.15) remains applicable in the region from the toe of the interface to the corresponding shoreline, and Eqn (3.17) is still correct in the region from the water divide to the interface toe. However, according to the coordinate system set-up for the island situation, L_1 in Eqns (3.15) and (3.17) needs to be substituted by $L_1/2$ and x by $x - L_1/2$ leading to the following equations:

$$h = \sqrt{\frac{w(\rho_s - \rho_f)}{K_1 \rho_s}(L_1 x - x^2) + H_0} \qquad (0 \le x \le x_{t1} \text{ or } x_{t2} \le x \le L_1) \tag{3.27}$$

$$h = \sqrt{\frac{w}{K_1}(L_1 x - x^2) + \frac{\rho_s}{\rho_f} H_0^2} \qquad (x_{t1} \le x \le x_{t2}) \tag{3.28}$$

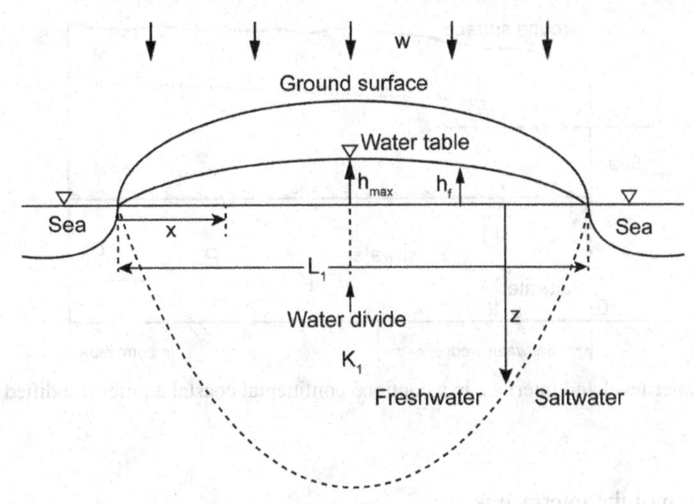

Figure 3.11 Freshwater lens in a strip island bounded at the base by the interface.

Letting h equal $\rho_s H_0/\rho_f$ (Eqn (3.13)) in either Eqn (3.27) or (3.28) leads to

$$x^2 - L_1 x + \frac{K_1(\rho_s{}^2 - \rho_s\rho_f)}{w\rho_f{}^2} H_0{}^2 = 0 \tag{3.29}$$

The locations of the interface toe can be easily obtained by solving the above equation for x_{t1} and x_{t2}. More details on the analytical solutions presented here can be found in Guo and Jiao (2007).

3.4.4.2 Freshwater Lens Bounded by the Interface

Now consider an infinite strip island aquifer system which is the same as the previous one except that the bottom of the system is bounded by the interface. The cross section of the strip island is the same as that of the circular island shown in Figure 3.9 but is repeated in Figure 3.11 with the origin of x located at the left shoreline.

The continuity equation for the flow in the freshwater part can be written as

$$w(x - L_1/2) = -K_1(1+\delta)h_f \frac{dh_f}{dx} \tag{3.30}$$

For the boundary conditions $dh_f/dx = 0$ at $x = L_1/2$ and $h_f = 0$ at $x = 0$ (or at $x = L_1$). Equation (3.30) has the following solution:

$$h_f = \sqrt{\frac{w}{K_1(1+\delta)}(L_1 x - x^2)} \tag{3.31}$$

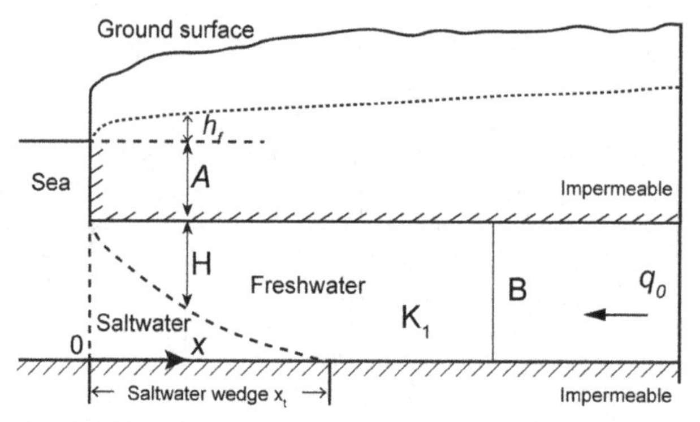

Figure 3.12 Water level and interface in a confined continental coastal aquifer (modified from Oude Essink, 1996).

and the location of the interface is

$$z = \delta h_f = \sqrt{\frac{w\delta^2}{K_1(1+\delta)}(L_1 x - x^2)} \tag{3.32}$$

Equation (3.31) is the same as that obtained by Fetter (1972) and Henry (1964).

3.4.5 *Confined Continental Coastal Aquifer*

A confined aquifer with an interface is shown in Figure 3.12, which is based on Oude Essink (1996). The confining unit terminates at the shoreline. The origin of the horizontal axis is at the shoreline, and x positive landward. The sea level is a height A [L] above the top of the confined aquifer. The height of the freshwater above the interface in the aquifer is H [L], which increases to B [L], i.e. the aquifer thickness inland of the interface. The lateral inflow per unit coastline length is q_0 [L^2 T^{-1}] and there is no recharge through the confining unit. Using the Ghijben–Herzberg principle and Darcy's law, one has for $0 \le x \le x_t$:

$$h_f = (A + H)/\delta \tag{3.33}$$

$$-K_1 H \frac{dh_f}{dx} = -q_0 \tag{3.34}$$

Integrating the second equation using the boundary condition $H = 0$ at $x = 0$ gives after rearranging

$$H^2 = \frac{2q_0\delta}{K_1}x \tag{3.35}$$

At the toe location, $H = B$, so from Eqn (3.35), x_t can be estimated as

$$x_t = -\frac{K_1 B^2}{2q_0\delta} \tag{3.36}$$

Both H and x_t are independent of A.

3.4.6 Confined Continental Coastal Aquifer with a Seepage Zone

The Ghijben–Herzberg principle implicitly assumes that the depth of the interface is zero at the coastline and that the freshwater lens does not extend under the seafloor. In reality, however, there is always some freshwater flowing to the sea. An analytical solution for the interface with a seepage zone was derived by (Glover, 1959). The conceptual model for this problem is shown in Figure 3.13. There is freshwater flow to the sea over a seepage zone of width x_0 away from the coastline. The equipotential lines are not vertical because the flow is not strictly horizontal.

Key assumptions in Glover's model are that the discharge across the land surface above sea level is insignificant in comparison to the discharge across the seafloor and that the aquifer is confined by a confining layer at sea level across which there is no recharge. The latter assumption was not stated explicitly in Glover's original work (Vacher, 1988). Under these assumptions, Glover derived a solution for the head in the freshwater part of the aquifer based on potential theory as

$$h_f = [q_0/(\delta K)]^{1/2}[x + (x^2 + z^2)^{1/2}]^{1/2} \tag{3.37}$$

and the interface depth as

$$z = \left(\frac{2q_0\delta}{K}x + \frac{{q_0}^2\delta^2}{K^2}\right)^{1/2} \tag{3.38}$$

where x [L] is the horizontal distance measured landward from the coastline, z [L] is the interface depth measured downward from the sea level and q_0 [L^2 T^{-1}] is the flow of freshwater per unit length of the coastline. Because the model assumes no local recharge q_0 is a constant and originates from lateral inflow.

The depth of the interface at any point x can be obtained from Eqn (3.38) and at $x = 0$:

$$z_0 = q_0\delta/K \tag{3.39}$$

Also, the head along the line $z = 0$ varies according to

$$h_f = [2q_0x/(\delta K)]^{1/2} \tag{3.40}$$

By setting $z = 0$ in Eqn (3.38), the width of the seepage zone in which the freshwater flows into the sea is

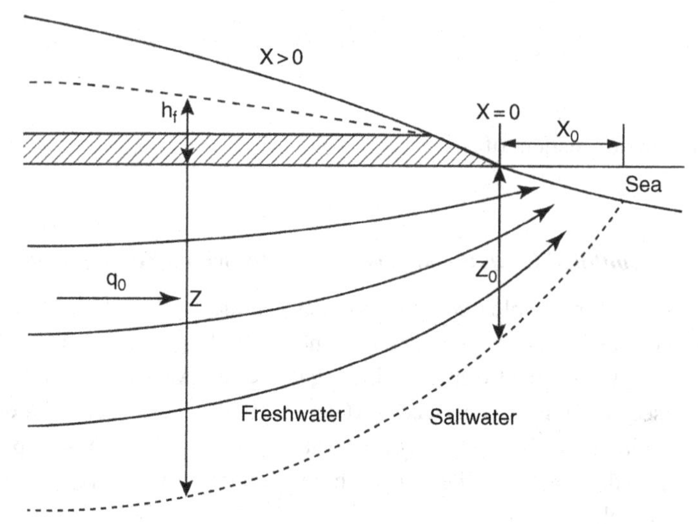

Figure 3.13 Water level and interface in a confined continental coastal aquifer with a seepage zone (modified from Glover, 1959).

$$x_0 = -\frac{q_0 \delta}{2K} \tag{3.41}$$

It should be noted that the seafloor is sloping so the width of the seepage zone from this equation is approximately correct only if the seafloor slope is small. This equation demonstrates that the seepage zone width is proportional to q_0 but inversely proportional to K. Thus, if the regional flow is large, or the coastal aquifer has low hydraulic conductivity, or both, the zone with freshwater submarine discharge may extend far into the sea.

3.5 Pumping-Induced Seawater Intrusion

Analytical solutions for problems involving pumping cannot only consider the head variation along the direction perpendicular to the shore, as the flow field in a horizontal plane is two-dimensional. The solution is thus dependent on x and y, where x is measured along the shore-perpendicular coordinate axis as before, and y is along the shore-parallel coordinate axis.

3.5.1 Unconfined Continental Coastal Aquifer under Pumping Conditions

The aquifer is divided into zones and heads in different zones have different expressions (Figure 3.14). The governing equations in different zones are different and this makes the derivation of the solutions inconvenient. The governing equations for zones 1 and 2 are, respectively,

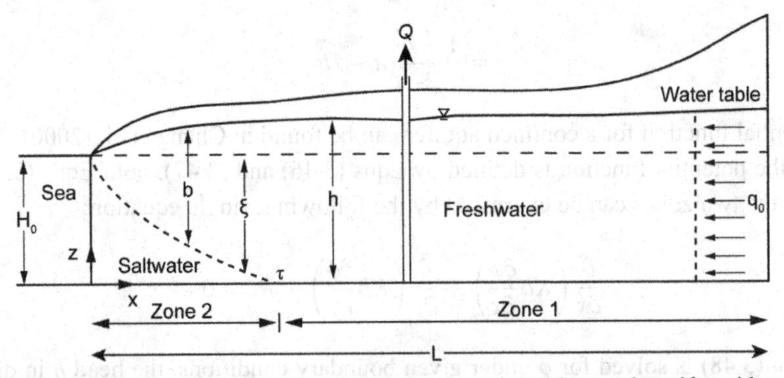

Figure 3.14 Water level and interface in an unconfined continental coastal aquifer with a pumping well (Mantoglou, 2003).

Zone 1:

$$\frac{\partial}{\partial x}\left(Kh\frac{\partial h}{\partial x}\right)+\frac{\partial}{\partial y}\left(Kh\frac{\partial h}{\partial y}\right)+R_s=0 \tag{3.42}$$

Zone 2:

$$\frac{\partial}{\partial x}\left(Kb\frac{\partial h}{\partial x}\right)+\frac{\partial}{\partial y}\left(Kb\frac{\partial h}{\partial y}\right)+R_s=0 \tag{3.43}$$

where R_s [L T^{-1}] is the sink/source term per unit of surface area. Strack (1976) introduced a single potential ϕ for different zones and solutions were derived for this potential. As before, the aquifer is isotropic and homogeneous, the system is in steady state and the Dupuit assumption and Ghijben–Herzberg principle apply. The aquifer terminates at the shoreline but is of infinite extension landward. A single pumping well abstracts groundwater at a constant rate Q [L^3 T^{-1}], and the only source of freshwater inflow is the regional inflow per unit width of coastline q_0 [L^2 T^{-1}].

The value of b in zones 1 and 2 can be expressed respectively as

$$b=h \tag{3.44}$$

$$b=h-H_0+\xi \tag{3.45}$$

where ζ [L] is the depth of the interface below sea level. Under the assumptions stated above and following the concept of potential functions introduced by Strack (1976), Mantoglou (2003) defined a potential function ϕ [L^2] for the unconfined aquifer in Figure 3.14 as

Zone 1:

$$\phi=\frac{1}{2}[h^2-(1+1/\delta){H_0}^2] \tag{3.46}$$

Zone 2:

$$\phi = \frac{1+\delta}{2}(h - H_0)^2 \tag{3.47}$$

The potential function for a confined aquifer can be found in Cheng et al. (2000).

After the potential function is defined by Eqns (3.46) and (3.47), both Eqns (3.42) and (3.43) in the two zones can be expressed by the following, single equation:

$$\frac{\partial}{\partial x}\left(Kb\frac{\partial \phi}{\partial x}\right) + \frac{\partial}{\partial y}\left(Kb\frac{\partial \phi}{\partial y}\right) + R_s = 0 \tag{3.48}$$

After Eqn (3.48) is solved for ϕ under given boundary conditions, the head h in different zones can be obtained from Eqns (3.46) and (3.47).

At the toe, $H_0 = \xi$, which leads to

$$h = (1 + 1/\delta)H_0 \tag{3.49}$$

Then the potential ϕ_t at the toe can be expressed as

$$\phi_t(x_t, y_t) = (1+\delta)H_0^2/(2\delta^2) \tag{3.50}$$

By solving Eqn (3.50) for x_t as a function of y_t, the locus of the toe is obtained (Mantoglou, 2003).

The pumping well location is (x_w, y_w), and the no-flow boundary at the coastline can be incorporated by creating an image well at $(-x_w, y_w)$ assuming there is no interface. The steady state solution for the potential is (Strack, 1976; Cheng et al., 2003)

$$\varphi = \frac{q_0}{K}x + \frac{Q}{4\pi K}\ln\left[\frac{(x - x_w)^2 + (y - y_w)^2}{(x + x_w)^2 + (y - y_w)^2}\right] \tag{3.51}$$

The term $(q_0/K)x$ represents the regional flow and is added using the principle of superposition.

The location of the toe can be obtained by equating Eqns (3.50) and (3.51). Figure 3.15 shows a plan view of the coastal aquifer with the pumping well which intercepts the flow from land to sea. The toe location of the interface is indicated by the dashed line. When there is no pumping, the location of the toe can be obtained as

$$x_t = \frac{(1+\delta)H_0^2 K}{2q_0\delta^2} \tag{3.52}$$

The above discussion is based on the assumption that there is only one pumping well. If there are n pumping wells, Eqn (3.51) can be expanded into on basis of the principle of superposition:

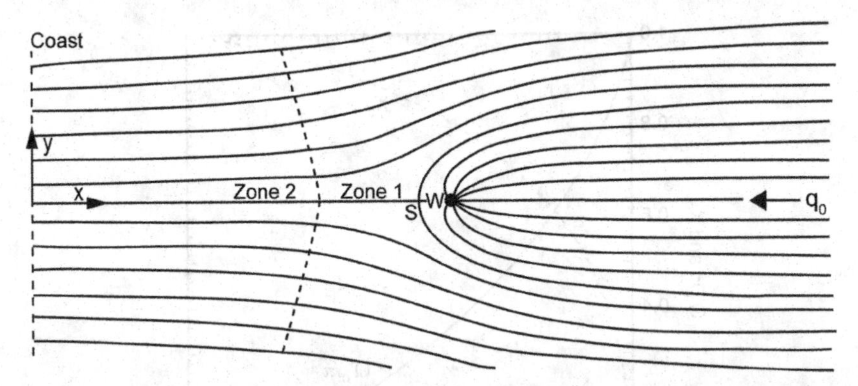

Figure 3.15 Flow field due to regional groundwater inflow and a pumping well near a coast. The broken line indicates the locus on of the interface toe (modified from Strack, 1976).

$$\varphi = \frac{q_0}{K}x + \sum_{i=1}^{n} \frac{Q_i}{4\pi K} \ln\left[\frac{(x-x_i)^2 + (y-y_i)^2}{(x+x_i)^2 + (y-y_i)^2}\right] \tag{3.53}$$

where Q_i and (x_i, y_i) are the pumping rate and location of the ith pumping well. Mantoglou (2003) further discussed the solution for the coastal aquifers which are of finite size using image wells and the superposition method.

3.5.2 Critical Pumping Rate and Critical Intrusion Distance

In coastal groundwater abstraction management, the pumping rate should be controlled in such a way that the interface will never reach the pumping well. The critical pumping rate beyond which the well will pump seawater was discussed by Strack (1976). Cheng et al. (2000) and Mantoglou (2003) used optimisation models to investigate how to maximise the freshwater withdrawal without causing dangerous seawater intrusion. Cheng et al. (2000) also presented a method to devise a design chart (Figure 3.16) for the critical pumping rate and intrusion distance by defining the following dimensionless variables:

$$X = x/x_w \tag{3.54}$$

$$Y = y/y_w \tag{3.55}$$

$$X_t = x_t/x_w \tag{3.56}$$

$$\Phi_t = K\phi_t/(qx_w) \tag{3.57}$$

$$Q' = Q/(qx_w) \tag{3.58}$$

where ϕ_t is given by Eqn (3.50).

The critical pumping rate Q'_c can be obtained by solving the following equation (Cheng et al., 2000):

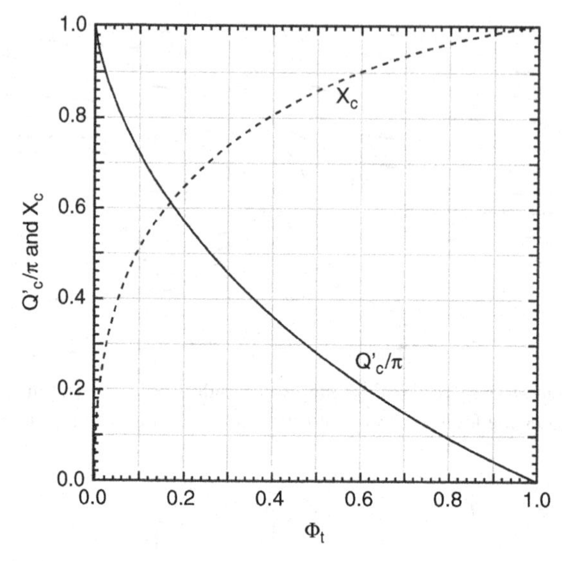

Figure 3.16 Dimensionless plot of maximum pumping rate and intrusion distance versus Φ_t with one pumping well (modified from Cheng et al., 2000).

$$\Phi_t = \sqrt{1 - Q'_c/\pi} + \frac{Q'_c}{2\pi}\ln\left(\frac{1 - \sqrt{1 - Q'_c/\pi}}{1 + \sqrt{1 - Q'_c/\pi}}\right) \tag{3.59}$$

The above equation shows that the dimensionless critical pumping rate is a function of one parameter group (Φ_t) only (Cheng et al., 2000). The critical intrusion distance of the toe along the axis of the well is at

$$X_c = \sqrt{1 - Q'_c/\pi} \tag{3.60}$$

Equations (3.59) and (3.60) have been used to create the design chart in Figure 3.16. When the value of Φ_t is given, the values of Q'_c/π and X_c can be read from the chart.

3.6 Pumping-Induced Saltwater Up-Coning

Pumping in a coastal aquifer may cause not only seawater intrusion laterally but also vertically (Section 6.2.3). When a well pumps freshwater above the interface, the interface moves upwards towards the pumping well, a process referred to as up-coning (Figure 3.17). The up-coning phenomenon was studied mathematically by Tolman (1937), Muskat (1946) and Dagan and Bear (1968). Among them, the work by Dagan and Bear (1968) was the most extensive. The discussion here focuses on up-coning created by vertical wells. For a discussion related to horizontal wells, the reader is advised to consult Hendizadeh et al. (2016).

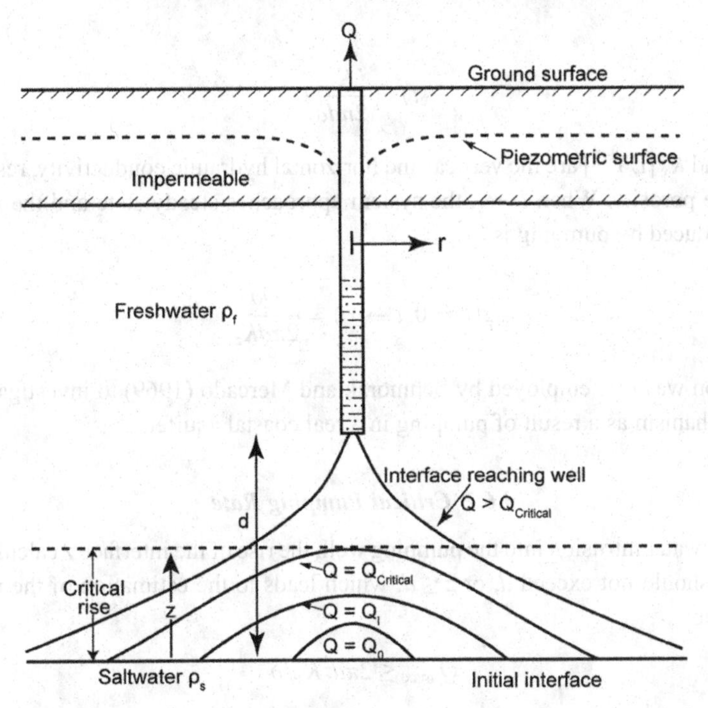

Figure 3.17 Up-coning of the interface induced by a well pumping at a constant rate (modified from Schmorak and Mercado, 1969).

3.6.1 Up-Coning of a Horizontal Interface

To develop an analytical solution for up-coning beneath a vertical well Dagan and Bear (1968), in addition to the assumptions already stated in Section 3.5.1, assumed that the interface is initially horizontal; that the aquifer is areally extensive landward; that the bottom of the aquifer is very deep so both the lateral and bottom boundary effect can be ignored and that the pumping well is initially in freshwater zone. Under these assumptions the position of the interface (relative to its elevation at $t = 0$) can be written as the following function of time t and radial distance r,

$$z(r,t) = \frac{Q\delta}{2\pi K_z d}\left\{\frac{1}{(1+R_d{}^2)^{1/2}} - \frac{1}{[(1+t_d)^2 + R_d{}^2]^{1/2}}\right\} \tag{3.61}$$

where d [L] is the distance between the interface and the well's bottom at $t = 0$ and R_d and t_d are dimensionless distance and time parameters defined as

$$R_d = \frac{r}{d}\left(\frac{K_z}{K_x}\right)^{1/2} \tag{3.62}$$

and

$$t_d = \frac{K_z}{2nd\delta} t \tag{3.63}$$

where K_z and K_x [L T^{-1}] are the vertical and horizontal hydraulic conductivity, respectively, and n is the porosity. When $t \rightarrow \infty$, the system approaches steady state and the rise of the interface induced by pumping is

$$z(r = 0, t \rightarrow \infty) = \frac{Q\delta}{2\pi d K_z} \tag{3.64}$$

This solution was first employed by Schmorak and Mercado (1969) to investigate the up-coning mechanism as a result of pumping in a real coastal aquifer.

3.6.2 Critical Pumping Rate

To avoid seawater intrusion into the pumping well, the rise of the interface z calculated from Eqn (3.64) should not exceed d, or $z \leq d$, which leads to the estimation of the maximum pumping rate:

$$Q_{\max} \leq 2\pi d^2 K_z / \delta \tag{3.65}$$

This equation can be employed to estimate the maximum rate of pumping before seawater intrusion into the pumping well, or safe distance between the bottom of the well and the interface if pumping rate is given. However, experiments carried out by Dagan and Bear (1968) suggested that Eqn (3.65) overestimates the allowable pumping rate and the interface becomes unstable already if the rise exceeds 1/3 of d, the initial distance between the interface and the bottom of the well. On the basis of their finding, the maximum allowable pumping rate is

$$Q_{\max} \leq 0.6\pi d^2 K_z / \delta \tag{3.66}$$

Dagan and Bear (1968) also obtained an analytical solution to calculate the rise of the interface below a drain which discharges groundwater. They obtained their solution using the small perturbation method under the assumptions that the aquifer is uniform and incompressible and contains two immiscible fluids.

The impact of the various parameters and aquifer settings on the up-coning has been widely discussed and new solutions have been developed to relax the assumptions of the above equations. For example, Bear (1979) found that the assumption of an interface is appropriate if the thickness of the aquifer is much greater than the transition zone. Bower et al. (1999) presented an analytical equation to determine the critical conditions for saltwater up-coning in a leaky confined aquifer. Up-coning in two-dimensional sand tank experiments, representative for abstraction from a horizontal well, was visualised using dye tracers in a study by Werner et al. (2009).

4
Groundwater Tidal Dynamics

4.1 Introduction

Groundwater levels in coastal aquifer systems fluctuate in response to the periodic fall and rise of sea level caused by tides. The tidal signal of the sea tides dampens as it penetrates into the aquifers at a rate depending on the hydraulic properties, and with increasing distance inland the signal becomes more attenuated and delayed. Tidal signals can penetrate into coastal aquifers to an inland distance of a few kilometres under favourable conditions.

The measurement of the tidal fluctuation of groundwater heads can be of great help in hydrological aquifer characterisation. Instruments that automatically record and store water level data are nowadays widely available, and high-resolution time series of hydraulic heads can be collected in a cost-effective manner. Analysis of tidal response data can provide information on aquifer parameters, which is useful to supplement other characterisation methods such as pumping tests. In addition, tidal data can provide information on hydrogeological conditions such as the degree of confinement of the aquifer (e.g. Erskine, 1991; Trefry and Johnston, 1998) and the spatial variability of hydraulic properties and connectivity patterns of the aquifer (Alcolea et al., 2007).

The tidal response of coastal aquifers to sea tides can be viewed as a natural hydraulic test at a scale much larger than a conventional pumping test, which impacts the aquifer over a limited radial distance for a finite duration of time. The spatial change of the phase lag and the amplitude attenuation in the aquifer can be used to calculate the hydraulic diffusivity, which is the ratio of hydraulic conductivity and specific storage (or transmissivity and storage coefficient). The tidal approach cannot be used to estimate the hydraulic conductivity and storativity separately, except in the case where the loading efficiency can be estimated (Section 4.4). In other cases, the joint use of a pumping test and tidal fluctuation data is needed to determine these two parameters individually.

Many analytical solutions of tide-induced water level fluctuations are available. This chapter covers only the solutions that were derived for the most common coastal aquifer conditions and are relatively simple, but still instructional to highlight the relevant physical concepts and processes. Following this, a brief review of the analytical solutions in some more specific or more complicated situations is included. Methods for using the equations to estimate aquifer diffusivity are presented, and common issues associated with the propagation of tidal signals in aquifers and parameter estimation are discussed.

In addition to tides, aquifers are also subject to other naturally occurring forces such as Earth tides and atmospheric pressures (i.e. meteorological tides). For the impact of these forces on groundwater system, interested readers are referred to the publications by Hsieh et al. (1987) and Merritt (2004). While the equations and methods presented in this chapter are for analysing cyclic fluctuations of groundwater levels in aquifers near tidal channels or seas, they can also be applied to water level fluctuations induced by other forcings with a periodic nature. Thus, the response of an aquifer to the passage of a flood crest in a stream to which the aquifer is hydraulically connected may also be analysed using the methods here. It may also be possible to relate time-varying recharge to a regional aquifer system using the tidal equation because rainfall recharge and water level fluctuations over multiple years are also approximately periodic (Dickinson et al., 2004).

4.2 Basics of Sea Tides

4.2.1 Periodic Sea Tides

Tides are the falls and rises of the sea surface induced by the resultant effects of the gravitational forces of the moon and the sun and the rotation of the Earth. Because the motions of those astronomical bodies are cyclical, tides are periodic. Ocean tides are extensively discussed in various text books on oceanography (e.g. Affholder and Valiron, 2001; Talley, 2011). The gravitational attraction by the moon is the main force determining the timing and height of tides. This attraction generates two tidal bulges on the Earth's surface. The height of the tidal bulges depends on the gravitational pulling force by the moon and that by the Earth that pulls the water back. As shown in Figure 4.1, on the side of the Earth that is closest to the moon, the moon has the strongest gravitational force to attract the seawater to form a high tide. On the opposite side of the Earth, another high tide is also formed because at this location the gravity force from the moon is the smallest. When seawater is accumulating on the two ends of the Earth, the sides of the Earth that are at right angles to the moon's gravity force must be losing water and thus here the tide is falling. As a result, during every tidal period the seawater at any given point on the surface of the Earth should have two low tides and two high tides.

Every day the Earth rotates around its axis once in about 24 hours, so one tidal cycle would be 24 hours long if the moon did not move. As the Earth rotates, however, the moon moves around the Earth too. As a result, the tidal day is greater than 24 hours and equals 24.83 hours. Every day the high or low tides occur about 50 minutes later than the day before.

The gravity from the sun also influences the tides on Earth, although the influence is smaller than the moon due to greater distance. When the gravity from the moon and the sun attracts the Earth in the same direction (Figure 4.1a), the highest and lowest tides are created and the daily difference between the low and high tides is greatest. These tides are denoted as spring tides. When the gravity force of the moon and that of the sun are perpendicular to each other (Figure 4.1b), the smallest rise and falls in the tides are created and the daily disparity between high and low tides is the smallest. These tides are named neap tides. Both spring and

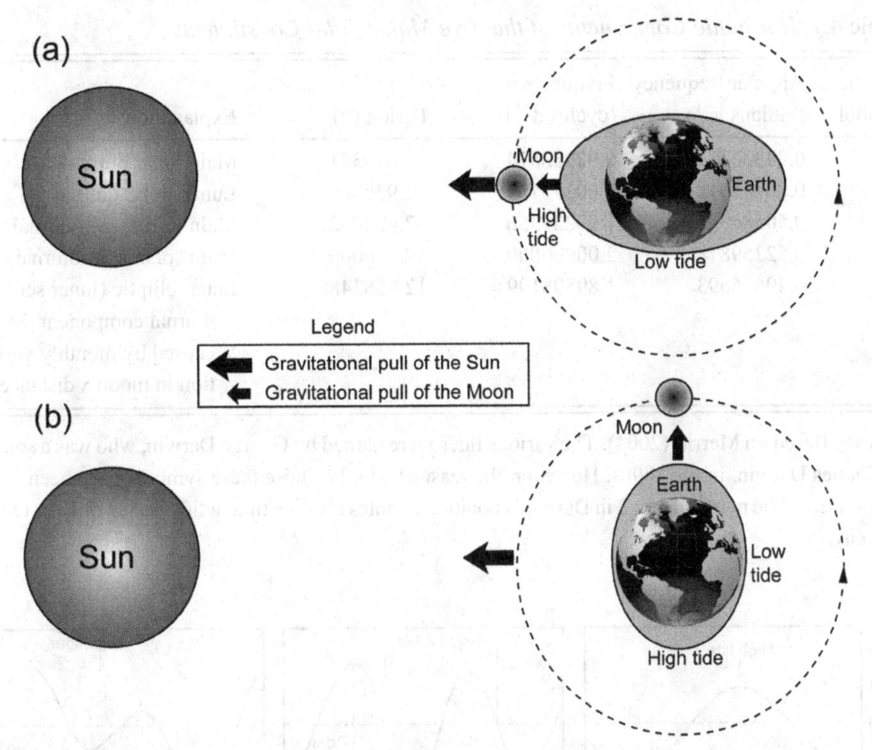

Figure 4.1 Symbolic scheme showing the gravitational forces generating (a) a spring tide and (b) a neap tide (not to scale).

neap tides appear two times a month. The former comes during full and new moons and the latter during the first and last quarter of the moon.

If there were no continents and the Earth were a perfect sphere, all the ocean surfaces should have two equally high and low tides during a lunar day. The existence of landmasses, however, constrains the westward movement of the tidal bulges and blocks the free movement of the tidal waves among different oceans. Each ocean has very irregular coastlines and bathymetry. Consequently, the tides in different oceans and different parts of the same ocean exhibit different patterns.

If in a day the high tide and low tide only occur once, the tidal pattern is called diurnal (Figure 4.2a). If both high tides and low tides occur twice a day, the tide is said to be semi-diurnal if the two high tides and the two low tides are roughly of the same height (Figure 4.2b), or mixed semi-diurnal if the high and low tides differ significantly in height (Figure 4.2c). In most coastal areas, the tide is semi-diurnal.

4.2.2 *Complexity of Tides*

In addition to the moon and the sun, the gravitational pull of other astronomical bodies influences the sea tides. As a result, tides consist of dozens of periodic components (or

Table 4.1 *Harmonic Components of the Five Major Tidal Constituents*

Darwin symbol	Angular frequency (radians h^{-1})	Frequency (cycles d^{-1})	Period (h)	Explanation
O1	0.24335189	0.92953574	25.819341	Main lunar diurnal
K1	0.26251618	1.00273794	23.934469	Lunar–solar diurnal
M2	0.50586804	1.93227356	12.420602	Main lunar semi-diurnal
S2	0.52359878	2.00000000	12.000000	Main solar semi-diurnal
N2	0.49636693	1.89598199	12.658348	Lunar elliptic (lunar semi-diurnal component caused by monthly variation in moon's distance)

Note. Based on Merritt (2004). The various tides were named by George Darwin, who was a son of Charles Darwin, in the 1890s. However, the reasons why he chose these symbols have been forgotten. The numeral 1 or 2 in Darwin's notation denotes if a constituent tide occurs once or twice a day.

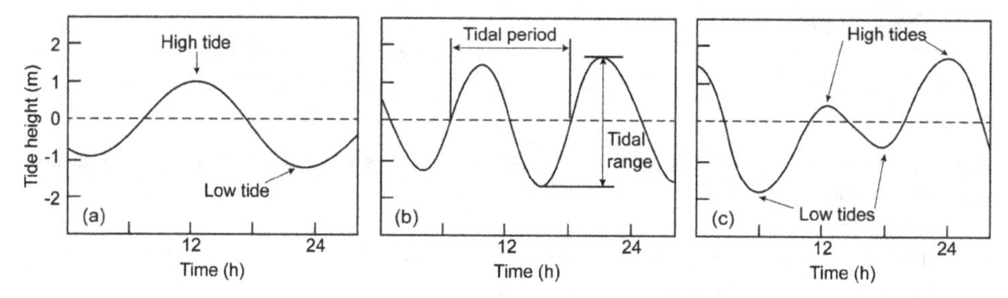

Figure 4.2 Schematic of (a) diurnal, (b) semi-diurnal and (c) mixed tides.

constituents) of different frequencies and amplitudes (Merritt, 2004). Most of them have very small amplitudes. Some components have frequencies very similar to each other so that it is difficult to distinguish them. The amplitude of the tides can be increased if components are in phase but can be dampened if they are out of phase.

Usually Fourier or spectral analysis is used to identify the frequencies of various tides with a record of at least tens of days. However, five constituents (O1, K1, M2, S2, N2) are of principal importance in coastal groundwater studies (e.g. Melchior, 1978; Hsieh et al., 1987; Merritt, 2004) (Table 4.1). These make up approximately 95% of the tidal potential (Merritt, 2004) and their frequencies are well known. Hsieh et al. (1987) presented a detailed method to analyse tides with an application to aquifer parameter estimation. They first removed the low-frequency (< 0.8 cycles d^{-1}) components and only kept the major tides of the semi-diurnal and diurnal constituents. Then they estimated the amplitude and phase shift using Fourier analysis. The constituents with small amplitude were further removed and only the amplitude and phase shift of tide components with large signal-to-noise ratios were finally used for aquifer parameter estimation. For their case study, Hsieh et al. (1987) started with all the five constituents listed in Table 4.1 but eventually only the O1 and M2 constituents were used to estimate aquifer transmissivity.

4.2.3 Periodic Functions

A general form of the sine function with time t as the independent variable that is typically used in tidal analysis is

$$y = h + A\sin(\omega t + c) \tag{4.1}$$

where y is a generic dependent variable, h is the vertical shift, A is the amplitude and the value ω is the angular frequency [T^{-1}], i.e. the number of cycles the function completes in an interval of 0 to 2π radians (or 360° degrees) and c is the shift of the sine curve along the horizontal axis, called phase shift if $\omega = 1$.

As a periodic function, $\sin(t)$ completes one cycle when $t = 2\pi$. So $\sin(\omega t)$ completes one cycle when $\omega t = 2\pi$, or it goes through a complete cycle when $t = 2\pi/\omega$. This is also called the period and is denoted by τ:

$$\tau = 2\pi/\omega \tag{4.2}$$

When t is zero, $\sin(t)$ is zero and crosses the t axis. The first time for the function $\sin(\omega t + c)$ to cross the t axis, or $\sin(\omega t + c) = 0$, is when $\omega t + c = 0$, which leads to

$$t = -c/\omega \tag{4.3}$$

The value c causes the function to shift and the curve becomes delayed (with a time lag) or advanced, depending on c being negative or positive.

Sea tides can be described by combinations of multiple sinusoidal functions. The period of a sea tide is called tidal period, or tidal day, which is the time interval between the occurrences of two consecutive high (or low) tides at the same location (Figure 4.2b). The tidal range is the difference between the lowest and highest tidal height, and the tidal amplitude is half of the tidal range. Instead of the sine function mentioned above, cosine functions are often also used in tide studies.

4.3 Historical Note on Studies of Groundwater Tides

Early documents on tide-induced water changes in wells or springs can be found in the work of ancient scholars in different cultures. A discussion on the relation between sea tide and water wells was presented by Pliny the Elder (AD 77–79). In his encyclopaedia *Natural History*, there is a chapter 'Where the Tides Rise and Fall in an Unusual Manner' (translated by Bostock and Riley, 1890) which states,

At Gades (now Cadiz, Spain), which is very near the temple of Hercules, there is a spring enclosed like a well, which sometimes rises and falls with the ocean, and, at other times, in both respects contrary to it. In the same place there is another well, which always agrees with the ocean. On the shores of the Baetis (now the Guadalquivir river), there is a town where the wells become lower when the tide rises, and fill again when it ebbs; while at other times they remain stationary. The same thing occurs in one well in the town of Hispalis (now Seville), while there is nothing peculiar in the other wells.

Needham and Wang (1959) discussed cases in Chinese literature that mentioned the possible connection between some wells or springs and the sea, where water level was observed to rise and fall with the tides. These wells and springs have different names in Chinese literature such as Hai Yan (the eye of the sea), Chao Jing (tidal well), or Chao Spring (tidal spring). The first record on Hai Yan is found in *Guan shi di li zhi meng* (Mr Guan's Geographical Indicator) written by Guan Lu (AD 209–256).

The well-known British naturalist, Charles Darwin, observed the relation between groundwater level fluctuations and sea tide from small islands located in the Indian Ocean between 3 and 6 April 1832 (Darwin, 1852):

On this island the wells are situated, from which ships obtain water. At first sight it appears not a little remarkable that the freshwater should regularly ebb and flow with the tides; and it has even been imagined, that sand has the power of filtering the salt from the seawater. These ebbing wells are common on some of the low islands in the West Indies. The compressed sand, or porous coral rock, is permeated like a sponge with the salt water, but the rain which falls on the surface must sink to the level of the surrounding sea, and must accumulate there, displacing an equal bulk of the salt water. As the water in the lower part of the great sponge-like coral mass rises and falls with the tides, so will the water near the surface; and this will keep fresh, if the mass be sufficiently compact to prevent much mechanical admixture; but where the land consists of great loose blocks of coral with open interstices, if a well be dug, the water, as I have seen, is brackish.

Dedicated scientific and quantitative studies of tide-groundwater relations, however, only started in the nineteenth century. In Amsterdam, the Netherlands, the relation between the water level in a well near the shores of a sea inlet and the inlet's water level was studied for a period of 2 weeks (Stamkart and Matthes, 1851). Based on detailed observations and calculations they worked out the time lag and amplitude damping, and concluded that there must be a connection between the groundwater and the nearby surface water. In Germany, Olshausen (1904) made an extensive study of the influence of the tides in the river Elbe on the water levels in wells in a confined aquifer. He distinguished between the effect of loading and the direct propagation of a pressure wave. In the USA, Harris (1904) was the first to investigate the tidal loading effect of seawater on groundwater level. He believed that the groundwater level oscillations in wells in Louisiana were caused by loading and unloading of the impermeable formations overlying the confined aquifers as sea level changed due to tides and onshore winds, not by the direct connection between the subsurface water in the confined aquifers and the seawater of the Gulf.

One of the most famous early well-designed field studies on groundwater and tides was conducted by Veatch (1906). Observations were made of the oscillations of the groundwater level in wells driven by various factors (rainfall, temperature, atmospheric pressure, pumping, dams, etc.), including tides. He included an extensive review of the literature and listed 33 publications he could find on the relation between groundwater levels and sea tides in the early days prior to his work, which provides a very valuable source of information for those studying the early history of tidal groundwater analyses. The engineer Johan Heinrich Steggewentz provides an even more exhaustive review of historical sources dealing with tidal effects on groundwater in his PhD thesis (Steggewentz, 1933), albeit that it was published only in Dutch.

Both self-recording and direct-reading gauges were used by Veatch (1906) in various wells near the beaches and bays in Long Island in New York in the summer of 1903. High-resolution water level curves from wells, many of them were flowing wells, together with geological cross sections, were obtained. At one site, they measured water level at almost 1 minute intervals for 30 minutes before and after the times of high and low water for about 3 days in June 1903.

Many important findings on the relation between water level in wells and sea tides were reported by Veatch (1906). He investigated the relation between time lag and inland distance of wells, found that there are different lags between high water and low water, and that there was a difference in the lag in water levels caused by sea tide or the tidal river. He observed that in some wells the obvious evident characteristic of the water level curves driven by tides was 'the greater rapidity and abruptness in the fall of the water than in its rise'. He postulated that the fluctuations in water wells were caused in three ways: transmission of pressure in the open cavities or pathways allowing a free connection between the ocean and the wells; discharge of the groundwater flow by aquifers well connected with the sea; and plastic deformation of the formations extending into the sea due to the alternating loading and unloading of the sea tides. This actually includes all possible ways in which the sea tide can influence groundwater that so far has been identified. Veatch's solid field observation and insightful discussions on tidal-induced water level changes laid down the theoretical foundations of modern studies on groundwater tidal dynamics.

The only thing missing in his work is probably equation development. Forchheimer (1919) has been credited with publishing the first analytical solution that describes the propagation of a periodic wave in groundwater. He developed an equation to calculate the response of the groundwater level to seasonal fluctuations of a river, which he idealised as a sine curve, and was the first to note that such water level fluctuations can be used to infer aquifer properties like transmissivity. Steggewentz (1933) obtained the same solution as Forchheimer (1919) but he specifically developed his equations within the context of tidal analysis. Working for the Dutch National Bureau for Public Water Supply, Steggewentz (1933) sought mathematical expressions for the propagation of tidal waves under a variety of hydrogeological conditions. He took into consideration the effect of the capillary fringe in unconfined aquifers, but he did not consider the elastic storage that plays a major role in confined systems. This was considered later by Jacob (1940), who estimated the coefficient of storage of a confined aquifer on the basis of tide-influenced water level in four wells in a confined aquifer below a barrier beach that is located on Long Island (New York, USA) between Jamaica Bay and the Atlantic Ocean. He found that the ratio between the magnitude of the water level oscillations in the wells on land and the oscillations of the sea tide generating them (the so-called tidal efficiency) was about 42%–44%. Jacob (1940) further derived the equation for calculating the loading efficiency (Section 4.4), although he referred it as tidal efficiency. His equation provides the only way to estimate storativity using tidal data.

In 1949, Jacob presented his now well-known analytical solution to calculate water level oscillations in a confined aquifer influenced by ocean tides at the Fourth Hydraulic Conference at the Iowa Institute of Hydraulic Research. The conference proceedings were published in 1950 (Jacob, 1950) and since then this work has become widely cited. Although the equation is mathematically identical to Forchheimer's (1919) equation (Maas, 1998), the fact that Jacob (1950) applied it to confined aquifers with changes in elastic storage made it a milestone in coastal groundwater studies.

4.4 Tidal Efficiency and Loading Efficiency

Sea tides can influence groundwater level in different ways (Veatch, 1906). When a confined or unconfined aquifer crops out at the coast, sea level fluctuation can directly propagate into the aquifer and cause groundwater level change. When a confined aquifer extends below the seabed, the loading of the aquifer by the overlying seawater also causes a change of the groundwater level in the confined aquifer.

Two dimensionless quantities, tidal efficiency (T_e) and loading efficiency (L_e), are introduced here to describe the relation between sea level change and groundwater level change. The tidal efficiency is defined as the ratio of the amplitude of the water level fluctuation in an inland well A_x (L) and the corresponding amplitude of sea level fluctuation A_s [L] (Van der Kamp, 1972). This is therefore also called amplitude ratio $A_R = A_x/A_s$.

When pressure is applied at the surface by loading, it is carried in part by the aquifer skeleton and in part transmitted to the groundwater, causing a change in the groundwater pressure (Section 2.2.2). The loading efficiency is defined as the ratio of the change in pressure in the aquifer to a spatially uniform change in pressure at the surface (Van der Kamp and Gale, 1983). This definition is general and the pressure change at the surface can be the water or air pressure change, so it applies to sea tides as well as atmospheric tides (i.e. air pressure fluctuations). Obviously it is difficult to measure the loading efficiency in the offshore, because there is usually no well in the sub-sea part of the confined aquifer.

For the sub-sea part of a confined coastal aquifer, the loading efficiency is expressed as the ratio of the change in head Δh within the aquifer and the tidal stage change at the surface Δh_0 (Figure 4.3):

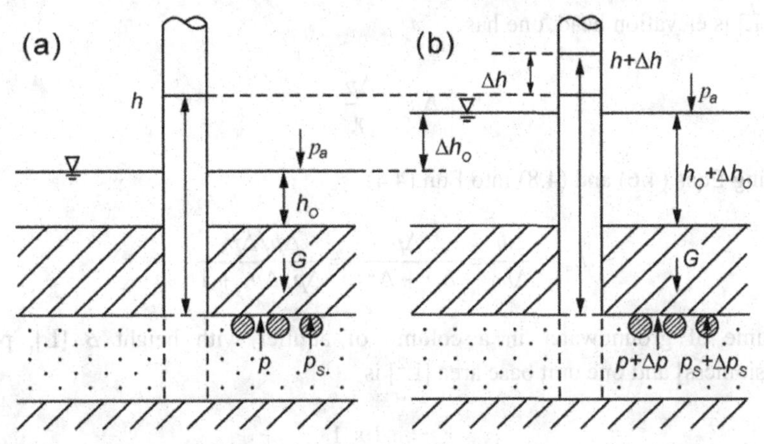

Figure 4.3 Conceptual model used to define loading efficiency. The shaded circles represent exaggerated soil particles (Chen et al., 2011).

$$L_e = \frac{\Delta h}{\Delta h_0} \tag{4.4}$$

The loading efficiency is related to the compressibilities of the aquifer β_s [L T^2 M^{-1}] and the groundwater β_w [L T^2 M^{-1}], and thus the aquifer's elastic storage properties (Section 2.2.3). The following derivation of L_e in terms of β_s and β_w is based on Chen et al. (2011), which is slightly different from that by Jacob (1940) in the sense that Chen et al. (2011) considered the change of the thickness of the aquifer when water volume changes. The final result, however, is identical to that based on the derivation by Jacob (1940).

In the confined aquifer of Figure 4.3, the downward forces per unit surface area at the base of the aquitard created by atmospheric pressure, seawater column height and the weight of the aquitard are balanced by the groundwater pressure and the force exerted by the matrix of the aquifer (cf. Eqn (2.23)):

$$p_a + \gamma h_0 + G = p + p_s \tag{4.5}$$

where p_a [M L^{-1} T^{-2}] is the atmospheric pressure, h_0 [L] is the seawater column height, γ is the specific weight of seawater [M L^{-2} T^{-2}], G is the weight of the aquitard per unit area [M L^{-1} T^{-2}], p is the groundwater pressure [M L^{-1} T^{-2}] and p_s is the effective stress [M L^{-1} T^{-2}].

Assuming that G and p_a are constant, if the sea level increases by Δh_0, then

$$\gamma \Delta h_0 = \Delta p + \Delta p_s \tag{4.6}$$

On the basis of the definition of the (saltwater) head (Section 1.5),

$$h = z + \frac{p}{\gamma} \tag{4.7}$$

where z [L] is elevation head, one has

$$\Delta h = \frac{\Delta p}{\gamma} \tag{4.8}$$

Substituting Eqns (4.6) and (4.8) into Eqn (4.4),

$$L_e = \frac{\Delta h}{\Delta h_0} = \frac{\Delta p}{\Delta p + \Delta p_s} = \frac{\Delta p/\Delta p_s}{\Delta p/\Delta p_s + 1} \tag{4.9}$$

The volume of groundwater in a column of aquifer with height B [L], porosity n [dimensionless] and one unit base area [L^2] is

$$V = nB \times 1 \tag{4.10}$$

Assuming that the solids themselves are incompressible, and neglecting any horizontal expansion or flow, the change of the volume of water ΔV results in a change of the aquifer thickness:

$$\Delta V = \Delta B \tag{4.11}$$

From Eqns (4.10) and (4.11), one obtains

$$\Delta V/V = \Delta B/(nB) \tag{4.12}$$

Based on the definition of β_w (Eqn (2.1)), one has

$$\Delta V/V = -\beta_w \Delta p \tag{4.13}$$

and analogously for the deformation of the aquifer skeleton:

$$\Delta B/B = -\beta_s \Delta p_s \tag{4.14}$$

Therefore, Eqn (4.12) can be written as

$$\beta_w \Delta p = \beta_s \Delta p_s / n \tag{4.15}$$

or

$$\Delta p/\Delta p_s = \beta_s/(n\beta_w) \tag{4.16}$$

Substituting Eqn (4.16) into Eqn (4.9) leads to

$$L_e = \frac{\beta_s/(n\beta_w)}{\beta_s/(n\beta_w) + 1} = \frac{\beta_s}{n\beta_w + \beta_s} \tag{4.17}$$

Equation (4.17) was derived under the assumption that the solid grains that make up the aquifer matrix are incompressible. The compressibility of the aquifer stems from rearrangement of the particles in response to a change in pressure, which causes a small but significant change in the pore space. When the aquifer is made up by compressible materials like clay for example, the compressibility cannot be ignored (Timms and Acworth, 2005). A formula for the loading efficiency considering the compressibility of the matrix part of the aquifer was derived by Van der Kamp and Gale (1983). A discussion of the subject can also be found in Langaas et al. (2006).

Merritt (2004) used Eqn (4.17) to estimate the range of probable values of L_e using the typical range of the values of rock compressibility and the compressibility of water. The water compressibility is $\beta_w = 4.8 \times 10^{-10}$ m^2 N^{-1} (Figure 2.1a). The values for the compressibility of geological materials range from 2×10^{-6} m^2 N^{-1} for plastic clay to 3.3×10^{-10} m^2 N^{-1} for sound rock (Domenico and Schwartz, 1990). Assuming a porosity of 0.35, the loading efficiency was estimated to range from 0.014 to 0.22 by Merritt (2004). This value, however, can be higher if the compressibility of the rock is greater (Rhoads and Robinson, 1979).

The loading efficiency can be used to estimate storage coefficient. Based on the definition of the storage coefficient (e.g. Domenico and Schwartz, 1990) (see also Eqn (2.51))

$$S = B\gamma(n\beta_w + \beta_s) \tag{4.18}$$

one has

$$L_e = \frac{B\gamma\beta_s}{S} \tag{4.19}$$

Equation (4.19) can also be written as

$$L_e = \frac{S - B\gamma n\beta_w}{S} \tag{4.20}$$

or

$$S = \frac{B\gamma n\beta_w}{1 - L_e} \tag{4.21}$$

which is the equation obtained by Jacob (1940, p. 583). This equation can be used to estimate the storativity if the loading efficiency is known.

The term loading efficiency has been often confused with tidal efficiency, starting with Jacob (1940) himself. Van der Kamp (1972) clearly defined the term loading efficiency and pointed out the difference between the two terms (see Eqn (4.50)). The loading efficiency is a constant related to the compressibility of both the water and the aquifer matrix, but the tidal efficiency declines exponentially with inland distance, as will be described later. According to Van der Kamp (1972), what Jacob defined should be the loading efficiency but what he estimated in his case study should be the tidal efficiency. Merritt (2004) also included an extensive discussion on the two terms of tidal efficiency and loading efficiency. In this book, the tidal efficiency and loading efficiency are used in the sense as defined by Van der Kamp (1972). Other terms such as amplitude factor (Jacob, 1950; Ferris, 1951), tidal efficiency factor, true tidal efficiency and apparent tidal efficiency (Carr and Van der Kamp, 1969) are not used in this book.

4.5 Equations for Tide-Induced Water Level Fluctuations

This section will cover equations describing the tide-induced groundwater level fluctuation for four different cases: a fully confined aquifer, a leaky confined aquifer, a confined aquifer under an impermeable unit that cuts off the direct hydraulic connection between seawater and groundwater, and a confined aquifer under an impermeable unit extending offshore for a finite distance. The presented equations are based on papers by Jacob (1950), Jiao and Tang (1999), Van der Kamp (1972) and Li and Jiao (2001a), respectively. For simplicity, only the assumptions, final equations and their implications are presented. The reader should consult the original papers for the derivation of these equations. The work by Merritt (2004) includes a very insightful discussion about some of them.

The general assumptions used in all cases are that the aquifer and aquitard are homogeneous, isotropic and of infinite extent on land. All the units have a constant thickness and are horizontal, and the vertical component of groundwater flow is neglected. The coastline is straight and the x axis is perpendicular to the shoreline, and the x coordinate is positive towards the land, with $x = 0$ at the coast. The difference in density between the groundwater and seawater is ignored because its impact on tidal groundwater level fluctuations is very limited (Li and Chen, 1991; Ataie-Ashtiani et al., 1999; Slooten et al., 2010).

The tidal variations in seas and oceans tend to be complicated. For any coastal area, there are typically dozens of tidal constituents with different amplitudes and phases to be considered. For the confined aquifer cases that will be discussed here, the governing differential equations are linear, so that solutions for individual constituents can be superimposed, or the analysis can be based on the amplitudes and phases of individual constituents. Therefore, in the following discussion, the tidal variation is assumed to be represented by a single sine (or cosine) function.

4.5.1 Tide-Induced Groundwater Level Fluctuation in a Confined Aquifer

In the case considered by Jacob (1950) the aquifer is assumed to be confined and crops out at the coast. Groundwater and the seawater are directly connected. The configuration is shown in Figure 4.4. Due to the direct connection to the open sea, water is forced into and out of the aquifer, causing head $h(x, t)$ in the aquifer to fluctuate.

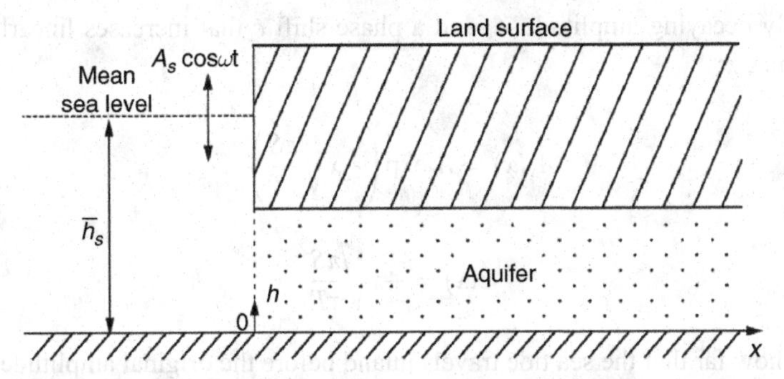

Figure 4.4 Confined coastal aquifer influenced by sea tidal variations.

Under the assumptions presented earlier, the one-dimensional flow equation for the confined aquifer can be expressed as

$$S\frac{\partial h}{\partial t} = T\frac{\partial^2 h}{\partial x^2} \tag{4.22}$$

where T [L^2 T^{-1}] is the transmissivity. At the coast, the head h is subject to the tidal oscillation described by

$$h(0,t) = \overline{h_s} + A_s \cos(\omega t) \tag{4.23}$$

where $h(0, t)$ [L] is the head at the shoreline, $\overline{h_s}$ [L] is the mean sea level and is assumed to be zero in what follows, unless otherwise stated, and ω [T^{-1}] is the tidal speed or angular frequency (cf. Eqn (4.2))

$$\omega = 2\pi/\tau \tag{4.24}$$

where τ [T] is the period of the tide, i.e. time for the tide to complete one cycle. The value of τ is about 24 or 12 hours for diurnal or semi-diurnal tides, respectively (Table 4.1).

On the inland side, the boundary condition for Eqn (4.22) is

$$h(\infty, t) = 0 \tag{4.25}$$

which means that at infinite distance from the coast, the tide has no effect and the head remains constant.

The solution of Eqn (4.22) with these boundary conditions is

$$h(x,t) = A_s e^{-x\sqrt{\frac{\omega S}{2T}}} \cos\left(\omega t - x\sqrt{\frac{\omega S}{2T}}\right) = A_s e^{-x\sqrt{\frac{\pi S}{\tau T}}} \cos\left(\omega t - x\sqrt{\frac{\pi S}{\tau T}}\right) \tag{4.26}$$

Note this equation is identical to Jacob's (1950) solution, except that in his equation a sine function was used. It shows that the sea tide propagates into the aquifer with an exponentially decaying amplitude A_x and a phase shift c that increases linearly with the inland distance x:

$$A_x(x) = A_s \exp\left(-x\sqrt{\frac{\pi S}{\tau T}}\right) \tag{4.27}$$

$$c(x) = -x\sqrt{\frac{\pi S}{\tau T}} \tag{4.28}$$

To know how far that the sea tide travels inland before the original amplitude of the sea tide is damped to a fraction A_x/A_s ($=A_R$, the amplitude ratio, or tidal efficiency), Eqn (4.27) can be re-written to

$$x = -\sqrt{\frac{\tau T}{\pi S}} \ln\left(\frac{A_x}{A_s}\right) \tag{4.29}$$

As can be seen from Eqn (4.29), the magnitude of the hydraulic diffusivity of an aquifer $D_h = T/S$ [L^2 T^{-1}] controls the distance up until which tides are propagated inland. Measurable tidal effects may be observable up to a few hundreds to even a few kilometres inland, as observed by Jiao and Li (2004) and Merritt (2004). High-permeability zones exist in some karst aquifers or in coastal areas reclaimed from the sea. Man-made structures can have a high permeability, such as in Hong Kong, where a buried former sea wall, which consists of huge boulders (Section 9.2.2.1), can be extremely permeable, to the extent that the groundwater in it is essentially part of the seawater. Underneath Hong Kong International Airport, built on an artificial island, the groundwater tidal fluctuation in a piezometer 300 m inland has virtually the same amplitude and phase as the sea tides (Jiao and Li, 2004).

Equation (4.26) can be re-written as

$$h(x,t) = A_s e^{-x\sqrt{\frac{\pi S}{\tau T}}} \cos\left[\omega\left(t - \frac{x}{\omega}\sqrt{\frac{\pi S}{\tau T}}\right)\right] = A_s e^{-x\sqrt{\frac{\pi S}{\tau T}}} \cos[\omega(t - t_{\text{lag}})] \tag{4.30}$$

with t_{lag} [T] defined as

$$t_{\text{lag}} = \frac{x}{\omega}\sqrt{\frac{\pi S}{\tau T}} = x\sqrt{\tau S/4\pi T} \tag{4.31}$$

The time lag t_{lag} represents the time between the peak of the sea high tide (or trough at low tide) and the corresponding peak of the groundwater level (or the trough corresponding to low tide). Both Eqns (4.29) and (4.31) provide an expression to estimate the hydraulic diffusivity D_h.

From Eqn (4.29),

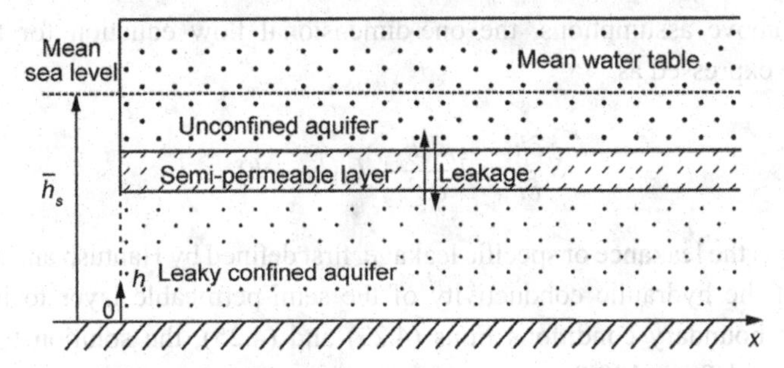

Figure 4.5 Leaky confined coastal aquifer system influenced by sea tidal variations (Jiao and Tang, 1999).

$$D_h = \frac{T}{S} = \frac{x^2\pi}{\tau(\ln A_R)^2} \tag{4.32}$$

where A_R varies with inland distance x. From Eqn (4.31),

$$D_h = \frac{T}{S} = \frac{x^2\tau}{4\pi t_{lag}^2} \tag{4.33}$$

Equations (4.32) and (4.33) are independent equations to estimate D_h using the observed water levels in a well if the amplitude ratio between the sea tide and groundwater level oscillation and the time lag between the oscillations of water level and sea tide are known. Ferris (1951) was the first to apply Eqn (4.26) in this way to calculate aquifer parameters in an actual aquifer. He called the method based on Eqn (4.32) amplitude-ratio method and that based on Eqn (4.33) time lag method. T or S can only be estimated individually if either of the two parameters is known based on other methods such as pumping tests or atmospheric efficiency analyses.

4.5.2 Tide-Induced Groundwater Level Fluctuation in a Leaky Confined Aquifer

A key assumption behind the equation by Jacob (1950) is that the flow in the aquifer is fully confined. Jiao and Tang (1999) considered a multi-layered coastal aquifer system consisting of an unconfined aquifer, semi-permeable layer and a leaky confined aquifer (Figure 4.5). Their solution assumed that the water level in the leaky confined aquifer oscillates with the sea level but that of the shallow unconfined aquifer remains constant. This assumption is reasonable if the specific yield of the shallow aquifer is large enough to damp the tidal effect, so that the tidal oscillations in the unconfined aquifer relative to the confined aquifer become negligibly small (White and Roberts, 1994). Jiao and Tang (1999) further assumed that the semi-permeable layer has negligible storage, and that the leakage rate is proportional to the head difference between the two aquifers.

Under the above assumptions, the one-dimensional flow equation for the confined aquifer can be expressed as

$$S\frac{\partial h}{\partial t} = T\frac{\partial^2 h}{\partial x^2} + L_s(\overline{h_s} - h) \tag{4.34}$$

where L_s [T^{-1}] is the leakance or specific leakage, first defined by Hantush and Jacob (1955) as the ratio of the hydraulic conductivity of the semi-permeable layer to its thickness. Subject to the boundary conditions Eqns (4.23) and (4.25), the solution to Eqn (4.34) becomes (Jiao and Tang, 1999):

$$h(x,t) = A_s e^{-p_l x}\cos\left(\omega t - \frac{\omega S}{2p_l T}x\right) \tag{4.35}$$

where p_l is defined as

$$p_l = \frac{1}{\sqrt{2}}\left\{\left[\left(\frac{L_s}{T}\right)^2 + \left(\frac{\omega S}{T}\right)^2\right]^{\frac{1}{2}} + \frac{L_s}{T}\right\}^{\frac{1}{2}} \tag{4.36}$$

In the case of no leakage ($L_s = 0$), Eqn (4.35) simplifies to Eqn (4.26).

From Eqn (4.35) it follows that the amplitudes of the fluctuation of the sea level and the groundwater level in the leaky confined aquifer are related according to

$$A_x = A_s \exp(-p_l x) \tag{4.37}$$

The time lag t_{lag} between the tidal peak and that of the groundwater level can be written as

$$t_{\text{lag}} = \frac{S}{2p_l T}x \tag{4.38}$$

Using a hypothetical example, Jiao and Tang (1999) showed how the leakage influences the tidal fluctuations. Because of the leakage from the overlying unconfined aquifer with a water table that is assumed to be constant, the amplitude decreases much faster in the semi-confined aquifer than the case of a fully confined aquifer. The leakage also increases the time lag.

The amplitude of water level oscillations and the inland distance to which heads are affected by the tidal fluctuations are traditionally regarded as evidence of the degree of connectivity between the sea and the confined aquifer. A weak tidal influence is thought to be an indicator of a poorly permeable aquifer or a weak aquifer-sea connection (White and Roberts, 1994). The solution here suggests that this can only be concluded with certainty if the hydraulic connection between the leaky and unconfined aquifers is known.

Equations (4.37) and (4.38) can be used to estimate the parameters ratios T/S and T/L_s. First, using the observed amplitudes A_x and A_s of the sea tide and groundwater level, p_l can

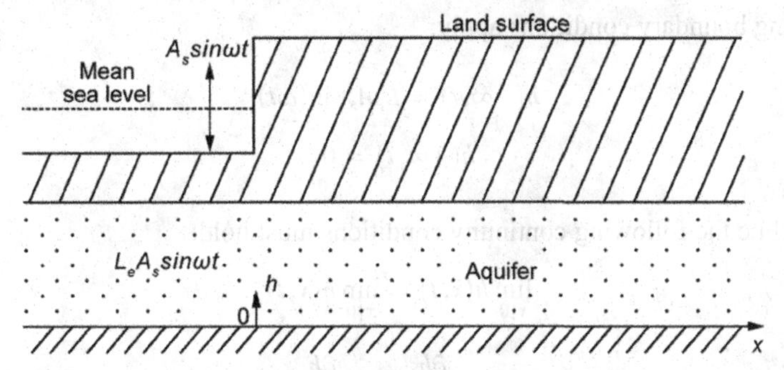

Figure 4.6 Confined coastal aquifer extending under the sea influenced by sea tidal variations.

be calculated from Eqn (4.37). With the estimated p_l and the time lag t_{lag}, taken as the time difference between the peaks of the sea tide and groundwater level, *T/S* can be calculated from Eqn (4.38). Subsequently, T/L_s can be calculated from Eqn (4.36). A case study can be found in Jiao and Tang (1999).

4.5.3 Tide-Induced Groundwater Level Fluctuation in a Fully Confined Aquifer Extending Offshore

Aquifers and their confining units often extend offshore. An analytical solution for the groundwater level change in a fully confined aquifer under this scenario was derived by Van der Kamp (1972). The seawater and the aquifer are not directly connected (Figure 4.6), but due to the loading effect (Section 4.4), the fluctuating sea tide will propagate in the underlying aquifer, albeit with a smaller amplitude. The resulting fluctuation in the aquifer below the seabed is propagated into the onshore part of the aquifer.

The equation for the water level in the aquifer can be expressed as (Van der Kamp, 1972)

$$S\frac{\partial h}{\partial t} = T\frac{\partial^2 h}{\partial x^2} + SL_e\frac{\partial h_s}{\partial t} \tag{4.39}$$

where h_s [L] is the sea level, which fluctuates sinusoidally:

$$h_s = A_s \sin(\omega t) \tag{4.40}$$

Equation (4.39) is solved separately for the two portions of the aquifer:

$$S\frac{\partial h}{\partial t} = T\frac{\partial^2 h}{\partial x^2} + SL_e A_s \omega \cos(\omega t), \quad x<0, \tag{4.41}$$

$$S\frac{\partial h}{\partial t} = T\frac{\partial^2 h}{\partial x^2}, \quad x>0 \tag{4.42}$$

The following boundary conditions apply:

$$h(-\infty, t) = L_e A_s \sin(\omega t) \tag{4.43}$$

$$h(+\infty, t) = 0 \tag{4.44}$$

At the coastline the following continuity conditions must hold:

$$\lim_{x \downarrow 0} h(x, t) = \lim_{x \uparrow 0} h(x, t) \tag{4.45}$$

$$\lim_{x \downarrow 0} \frac{\partial h}{\partial x} = \lim_{x \uparrow 0} \frac{\partial h}{\partial x} \tag{4.46}$$

The solutions for heads in the onshore and offshore regions of the aquifers are then, respectively,

$$h(x, t) = \frac{1}{2} L_e A_s e^{-x\sqrt{\frac{\omega S}{2T}}} \sin\left(\omega t - x\sqrt{\frac{\omega S}{2T}}\right) \quad (x>0) \tag{4.47}$$

$$h(x, t) = L_e A_s \sin \omega t + \frac{1}{2} L_e A_s e^{-x\sqrt{\frac{\omega S}{2T}}} \sin\left(\omega t + x\sqrt{\frac{\omega S}{2T}} + \pi\right) \quad (x<0) \tag{4.48}$$

Equation (4.47) shows that the wave propagates into the inland aquifer with an exponentially damped amplitude. Equation (4.48) shows that the groundwater level change in the offshore region of the aquifer is a combination of a cyclic fluctuation caused by the loading by the tidal seawater and a damped wave travelling seaward from the coastline. Basically, the loading generates a damped wave moving out to the sea, and at the same time, a wave moving inland, with the same damped amplitude but a phase opposite to the seaward wave (Van der Kamp, 1972). Merritt (2004) further pointed out that the first term on the right-hand side of Eqn (4.48) demonstrates that the sea tide is reduced in amplitude by the value of the loading efficiency, which is consistent with the definition of this parameter.

At the coastline ($x = 0$),

$$h(0, t) = \frac{1}{2} L_e A_s \sin(\omega t) \tag{4.49}$$

Using the sea level fluctuation (Eqn (4.40)) and groundwater level fluctuation (Eqn (4.49)), the tidal efficiency at the coastline ($x = 0$) is calculated to be half of the loading efficiency:

$$T_e = \frac{1}{2} L_e \tag{4.50}$$

Recall that Eqn (4.21) can be used to estimate storativity if the loading efficiency is available. However this parameter is hard to estimate because usually there is no well offshore. If the tidal efficiency can be determined using a well at the coastline, the storativity can be obtained from

$$S = \frac{B\gamma n\beta_w}{1 - 2T_e} \tag{4.51}$$

Equations (4.51) and (4.21) are the only equations to estimate storativity using tidal data.

Similar to the way that Eqns (4.32) and (4.33) are obtained from Eqn (4.26), two independent equations estimating the hydraulic diffusivity can be derived from Eqn (4.47):

$$\frac{T}{S} = \frac{x^2 \pi}{\tau[\ln(L_e A_R/2)]^2} \tag{4.52}$$

$$\frac{T}{S} = \frac{x^2 \tau}{4\pi t_{lag}^2} \tag{4.53}$$

Note that Eqn (4.53) is identical to the expression for a confined aquifer (Eqn (4.33)), but that Eqn (4.52) differs from the equivalent Eqn (4.32) as the loading efficiency appears in the equation. If there are groundwater level measurements at the coastline, Eqn (4.50) can be used to determine L_e, and the storativity can be obtained from Eqn (4.51). If not, and there are multiple onshore wells at different distances to the coastline, the measured amplitude of water level fluctuation in each well can be extrapolated to the coastline. Then the tidal efficiency at the coastline can be estimated using Eqn (4.50).

Van der Kamp (1972) discussed the uncertainty of this method. If the aquifer is not fully confined, the leakage may change the tidal efficiency. In the case when the tidal efficiency has a value greater than 0.5, the storativity as per Eqn (4.51) becomes negative, and thus meaningless.

4.5.4 Tide-Induced Water Level Fluctuation in a Leaky Confined Aquifer with Roof Partially Extending Offshore

Li and Jiao (2001a) considered the conceptual model shown in Figure 4.7, which comprises three layers: an unconfined aquifer, a semi-permeable layer and a leaky confined aquifer, with the bottom two layers partially extending into the sea. The distance to which the leaky confined aquifer extends below the sea is called the roof length (L_r). Following Jiao and Tang (1999) the specific yield of the unconfined aquifer is assumed to be so large that the tidal oscillations in the shallow aquifer can be ignored and the water table is fixed at mean sea level. Vertical leakage occurs through the semi-permeable layer that has negligible storage.

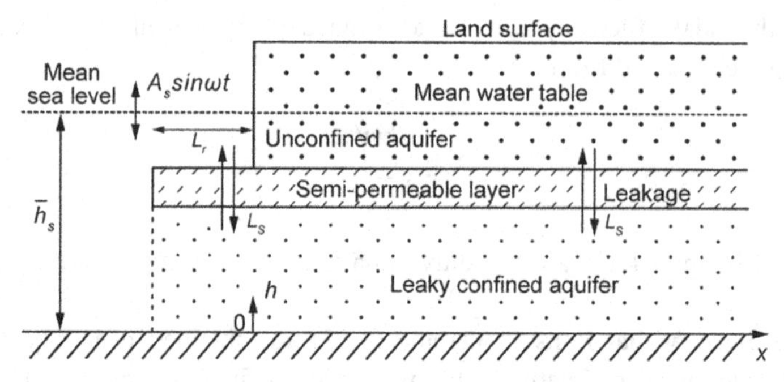

Figure 4.7 Leaky confined coastal aquifer system partially extending offshore influenced by sea tidal variations (Li and Jiao, 2001a).

Under the above assumptions, the one-dimensional flow equation for the onshore part of the confined aquifer is

$$S\frac{\partial h}{\partial t} = T\frac{\partial^2 h}{\partial x^2} - L_s h, \quad x > 0, \tag{4.54}$$

and for the offshore part,

$$S\frac{\partial h}{\partial t} = T\frac{\partial^2 h}{\partial x^2} + SL_e\frac{dh_s}{dt} + L_s(\overline{h_s} - h), \; -L_r < x < 0 \tag{4.55}$$

The boundary conditions are

$$h(-L_r, t) = h_s(t) = A_s\cos(\omega t) \tag{4.56}$$

$$h(+\infty, t) = 0 \tag{4.57}$$

and the continuity conditions at $x = 0$ are

$$\lim_{x\downarrow 0} h(x,t) = \lim_{x\uparrow 0} h(x,t) \tag{4.58}$$

$$\lim_{x\downarrow 0}\frac{\partial h}{\partial x} = \lim_{x\uparrow 0}\frac{\partial h}{\partial x} \tag{4.59}$$

Li and Jiao (2001a) introduced two parameters: the tidal propagation parameter a [L^{-1}] of the confined aquifer,

$$a = \sqrt{\omega S/2T} = \sqrt{\pi S/T\tau} \tag{4.60}$$

and the dimensionless leakage u,

$$u = L_s/\omega S \tag{4.61}$$

and obtained the solutions for the head in both offshore and onshore portions of the aquifer expressed in terms of the above two parameters. Because usually there are wells only in the inland aquifer, only the solution for the inland aquifer is presented here, which is

$$h(x,t) = A_s C_e\, e^{-pax} \cos(\omega t - qax - \varphi), \quad x \geq 0 \tag{4.62}$$

where

$$C_e = \sqrt{(R_1 + \lambda_1/2)^2 + (I_1 - \lambda_2/2)^2} \tag{4.63}$$

$$\lambda_1 = \frac{u^2 + T_e}{u^2 + 1} \quad \lambda_2 = -\frac{(1 - T_e)u}{u^2 + 1} \tag{4.64}$$

$$R_1 = e^{-paL}[(1-\lambda_1)\cos(qaL_r) - \lambda_2 \sin(qaL_r)] + \frac{1}{2}e^{-2paL}[\lambda_1 \cos(2qaL_r) + \lambda_2 \sin(2qaL_r)] \tag{4.65}$$

$$I_1 = e^{-paL}[(1-\lambda_1)\sin(qaL_r) + \lambda_2 \cos(qaL_r)] + \frac{1}{2}e^{-2paL}[\lambda_1 \sin(2qaL_r) - \lambda_2 \cos(2qaL_r)] \tag{4.66}$$

$$p = \sqrt{\sqrt{1+u^2} + u} \quad q = \sqrt{\sqrt{1+u^2} - u} \tag{4.67}$$

$$\phi = \arctan \frac{2I_1 - \lambda_2}{2R_1 + \lambda_1} \tag{4.68}$$

The water level expressed by the water level oscillation amplitude A_x and the time lag t_{lag} of groundwater response to tidal fluctuation can be obtained

$$h(x,t) = A_x \cos(\omega(t - t_{lag})) \tag{4.69}$$

Comparison of Eqns (4.69) and (4.62) yields

$$A_x = A_s C_e e^{-pax} \tag{4.70}$$

$$t_{lag} = \frac{1}{\omega}(qax + \varphi) \tag{4.71}$$

The previous solutions by Jacob (1950), Van der Kamp (1972) and Jiao and Tang (1999) are all special cases of the above solution Li and Jiao (2001a). If there is no leakage and the roof length = 0 ($L_s = 0, L_r = 0$), Eqn (4.62) reduces to Eqn (4.26) if there is no leakage and the roof length extends to infinity ($L_s = 0, L_r = \infty$), it reduces to Eqn (4.47). If the roof terminates at the coastline ($L_r = 0$), this equation is the same as Eqn (4.35). Although as before, Eqns (4.70) and (4.71) can be re-written to obtain equations similar to Eqns (4.52) and (4.53), there is not a simple analytical way to estimate the diffusivity.

Equation (4.54) shows that the groundwater level oscillations in the onshore part of the confined aquifer are modulated by the leakage through the semi-permeable unit, while in the offshore part (Eqn (4.55)) elastic storage changes, caused by the tidal loading rate of the overlying seawater $(SL_e(dh_s/dt))$, also play a role. Li and Jiao (2001a) presented a hypothetical example to show how the water level oscillations in the onshore confined aquifer are controlled by the roof length, leakage and loading efficiency. For a fixed leakage, the fluctuation of water level at a certain inland distance decreases significantly as the length of the roof increases. The amplitude observed at the inland point becomes insensitive to the roof length, however, if the latter exceeds a certain length. The leakage damps the fluctuations of the water level in the confined aquifer but not in a straightforward manner. When leakage is zero or very small, the water level changes are controlled by the sea tide at the boundary at distance L_r and loading rate of the seawater above the roof. A high loading rate over a long roof will lead to significant compression of the aquifer and then water level fluctuations. The loading efficiency has a large influence on water level fluctuation only when the length of the roof is large and the leakage is small. For a fixed leakage and roof length, when the loading efficiency increases, the amplitude increases and the time lag decreases.

When leakage increases, the amplitude of the water level in the inland aquifer decreases due to the leakage from the unconfined aquifer with a head which is assumed to be fixed, but that in the offshore aquifer increases because the leakage from the overlying seawater enhances the fluctuation of the water level in the underlying aquifer, which in turn increases the fluctuation of the water level in the onshore aquifer. However, if leakage continues to increase, the amplitude in the onshore part of the confined aquifer is overwhelmingly determined by the leakage from the fixed-head unconfined aquifer and becomes small.

4.6 A Brief Review of Other Analytical Solutions

Many analytical solutions have been derived in the past 20 years or so, and groundwater tidal dynamics is currently one of the few fields of groundwater research in which quantitative hydrogeologists can still flex their mathematical muscles. As the previous sections have demonstrated, the complexity of the expressions for the head fluctuation depends on the configuration of the aquifer system. However, all the solutions (Eqns (4.26), (4.35), (4.47) and (4.62)) for the fully confined or leaky confined aquifers consist of an exponential function multiplied by a sinusoidal function, so they are periodic with an amplitude that declines exponentially with inland distance. The solution for an unconfined aquifer, however, consists of a series of products of exponential functions and a cosine or sine function (Nielsen, 1990; Li et al., 2000b). Some analytical solutions that are less common than those presented in the above sections but are still important are briefly reviewed and summarised here.

4.6.1 Two-Dimensional Flow

The solutions presented in Section 4.5 assume that the ocean tide induces only one-dimensional flow perpendicular to the coastline. Sun (1997) derived a solution for the two-dimensional groundwater flow induced by tides with an amplitude and phase that change along the coastal boundary. Tang and Jiao (2001) improved his solution by considering a leaky confined aquifer.

Coastlines tend to be irregular and are made up by landforms such as bays, inlets and headlands. A generalised solution can therefore not be derived, but Li et al. (2000a) obtained an approximate non-periodic solution to study tidal water level oscillations in an aquifer bounded by coastlines that form a right angle. An improved solution with a case study was presented by Li et al. (2002).

4.6.2 Leaky Confined Aquifer beneath a Semi-Permeable Layer with Both Leakage and Storage

Li and Jiao (2001b) derived an analytical solution for the head in the semi-confined aquifer considering both leakage across and storage in the semi-permeable layer. This solution shows that while both leakage and storage of the semi-permeable layer are important processes that determine the water level fluctuation in the confined aquifer, leakage is usually the more dominant. The impact of storativity of the semi-permeable layer is negligible only if the storativity is similar or smaller in magnitude than the confined aquifer.

Li and Jiao (2002) presented a complete analytical solution to study tidal water level variations in a two-aquifer system considering the leakage and storage of the semi-permeable layer and the tidal wave interference between the aquifers. The shallow aquifer can be a confined aquifer or an unconfined one with tidal amplitude much smaller than the aquifer depth so that the nonlinear Boussinesq equation can be linearised. The results from the analytical solution show that the storage of the semi-permeable layer between the two aquifers functions as a buffer to the interference of tidal waves of the two aquifers. Quite a few previous solutions such as Jacob (1950), Jiao and Tang (1999), Li and Jiao (2001b), Li et al. (2001) and Jeng et al. (2002) are all special cases of this more complete solution.

4.6.3 Offshore Confined Aquifer with an Outlet

A confined coastal aquifer may extend offshore for a certain distance with its sub-sea outlet covered by sediment with hydraulic properties different from the aquifer. Li et al. (2007) derived an analytical solution for this case. The simultaneous effects of the outlet capping and tidal loading lead to complicated dependency of the water level oscillation on the outlet capping's leakance. Negative phase shift happens if the leakance is low and the offshore confined aquifer is short. The solution of Li et al. (2007) was expanded to multi-layered aquifers by Guo et al. (2007). The influence of leakage at the sub-sea outlet capping was studied by Xia et al. (2007).

4.6.4 Multiple Zones

A coastal aquifer system may have zones of very different aquifer properties. Guo et al. (2010) derived solutions for water level fluctuation due to tidal oscillation in a two-zone aquifer, with the zone nearest to the coast being of finite width and the inland zone of infinite width. The coastal zone decreases the fluctuation in the inland zone by a spatially constant parameter and enhances the phase lag by a spatially constant shift. These two parameters are determined by the contrast of the hydraulic properties of the two zones, and the hydraulic diffusivity and width of the zone adjacent to the coast. Analytical solutions addressing more general spatial heterogeneity in a composite aquifer with multiple aquifer zones were derived by Trefry (1999). Groundwater level oscillations in coastal leaky aquifers with both aquifer and the overlying aquitard consisting of many zones of different hydraulic materials were investigated by Chuang et al. (2010).

4.6.5 Island Aquifers

Island aquifers are subject to periodic forcing along their entire perimeter, creating interference effects of various tidal signals. Townley (1995) studied the response of island aquifers to periodic forcing but assumed a constant head in the centre. He also provided an analytical solution for an aquifer of finite-length subject to simultaneous periodic recharge and tides. Trefry and Bekele (2004) applied this analytical solution to symmetric and synchronous boundary conditions in an aquifer of finite length and with dual-tide signals. Interaction of the tidal fluctuations in the interior of an island was revealed by nonlinear phase lags and hyperbolic, rather than exponential, attenuation of the tidal amplitude. Rotzoll et al. (2008) presented a one-dimensional solution to the groundwater equation in an unconfined aquifer subject to asymmetric and asynchronous fluctuating hydraulic head conditions at opposite boundaries.

4.6.6 Unconfined Aquifers

Ferris (1951) stated that the one-dimensional solution of Eqn (4.26) derived for a confined aquifer also applies to unconfined aquifers if appreciable vertical flows do not occur and the saturated thickness of the aquifer is sufficiently large relative to the amplitude of water table fluctuations. This approach has been applied with some success in certain cases (e.g. Erskine, 1991). Because the storage coefficient in the equation is replaced by specific yield, which may be orders of magnitude greater than the confined storage coefficient, the exponential damping term will quickly reduce the amplitude of tidally forced head oscillations with distance from the ocean in unconfined aquifers (Jiao and Tang, 1999).

An exact analytical solution for tide-induced water table fluctuations in unconfined aquifers is not available due to the nonlinearity of the governing equation. Approximate solutions can be obtained using some kind of linearisation (Kacimov and Abdalla, 2010), or various perturbation methods (Parlange et al., 1984; Nielsen, 1990; Li et al., 2000b; Song et al., 2007). These are complex, and the relation between water table and aquifer parameters cannot be easily understood from these solutions.

4.6.7 Tidal Overheight

An interesting phenomenon emerging from the solutions of the Boussinesq equation is the water table overheight: the mean water level of the unconfined aquifer stands significantly over the mean sea level even when there is no net inland recharge by rainfall or groundwater. This phenomenon has been studied by many researchers (Philip, 1973; Parlange et al., 1984; Nielsen, 1990; Li et al., 2000b; Li and Jiao, 2003a). Philip (1973) concluded that when the amplitude of the tide equals the unconfined aquifer's depth beneath the average sea level, the inland water level stays higher than the mean sea level by about 23% of the tidal amplitude. Laboratory experiments have also confirmed the existence of the overheight (Smiles and Stokes, 1976; Parlange et al., 1984). The overheight will be more significant in the case of a sloping beach (Nielsen, 1990; Li et al., 2000) or decreasing hydraulic conductivity with depth (Li and Jiao, 2003b).

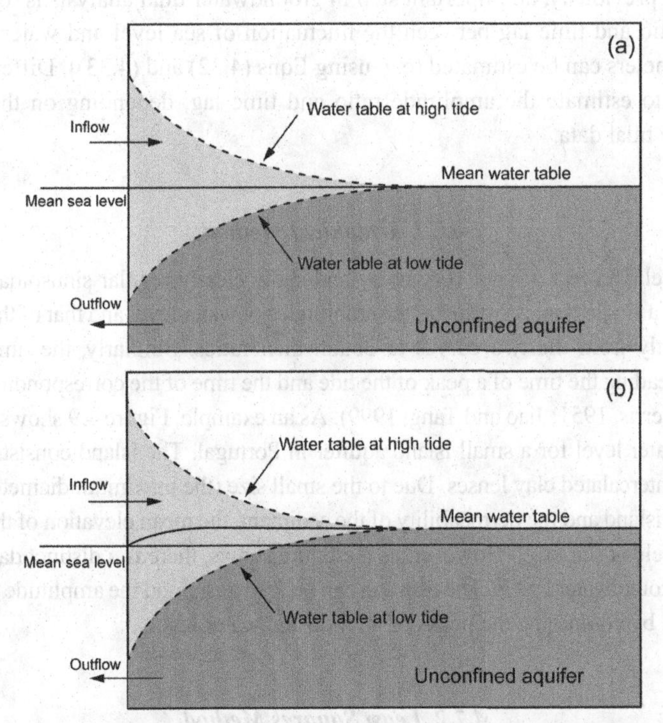

Figure 4.8 Conceptual models to explain why the tide-induced mean water level in an unconfined aquifer should be higher than the mean sea level: (a) hypothetical situation with mean water table equal mean sea level and (b) real situation with mean water table greater than mean sea level (modified from Li and Jiao, 2003a).

Li and Jiao (2003a) conceptually explained why there must exist a water table overheight in a coastal unconfined aquifer. The key reason is that the magnitude of the transmissivity of an unconfined aquifer depends on the aquifer's saturated thickness, i.e. on the water table. If the mean water table and the mean sea level were of the same height, the transmissivity of the aquifer at high tide would be larger than that at low tide (Figure 4.8a). Consequently, the volume of water recharging the aquifer during high tide would be greater than the volume exiting from the aquifer during low tide. If the mean water table is greater than the average sea level, the hydraulic gradient during the low tide will be greater, which compensates for the lower transmissivity and enhances the volume of water flowing back to the sea. Analogously, during high tide, the gradient is smaller and less water flows into the aquifer. In other words, only when the mean water table is higher than the mean sea level can the aquifer water balance be preserved (Figure 4.8b). As will be discussed in Section 7.3.2, the water table overheight in a multi-aquifer system with an unconfined aquifer may have significant consequences in submarine groundwater discharge.

4.7 Estimation of the Amplitude Ratio and Time Lag

As discussed previously, an important step in groundwater tidal analysis is to estimate the amplitude ratio and time lag between the fluctuation of sea level and water level before aquifer parameters can be estimated (e.g. using Eqns (4.32) and (4.33)). Different methods can be used to estimate the amplitude ratio and time lag, depending on the nature and quality of the tidal data.

4.7.1 Graphical Method

For water level data that are not very noisy and show clearly regular sinusoidal fluctuations dominated by a single tidal constituent, the amplitudes of water level and that of the sea tide can be read directly from the two curves to obtain their ratios. Similarly, the time lag can be obtained by reading the time of a peak of the tide and the time of the corresponding peak of the water level (Ferris, 1951; Jiao and Tang, 1999). As an example, Figure 4.9 shows the tide level and groundwater level for a small island aquifer in Portugal. The island consists of sand and gravel, with intercalated clay lenses. Due to the small size (the maximum diameter is less than 100 m) of the island and the permeability of the sediment, the mean elevation of the water table is approximately at sea level. However, as the figure shows, there is a distinct dampening and delay in the groundwater levels. The time lag can be read as t_{lag} and the amplitude ratio $A_R = A_x/A_s$ is obtained by comparing the peaks of the tide and water levels.

4.7.2 Least Squares Method

More often than not, tidal level fluctuations are complicated and the amplitude ratio and time lag cannot be read easily by comparing two curves, especially when the data are noisy.

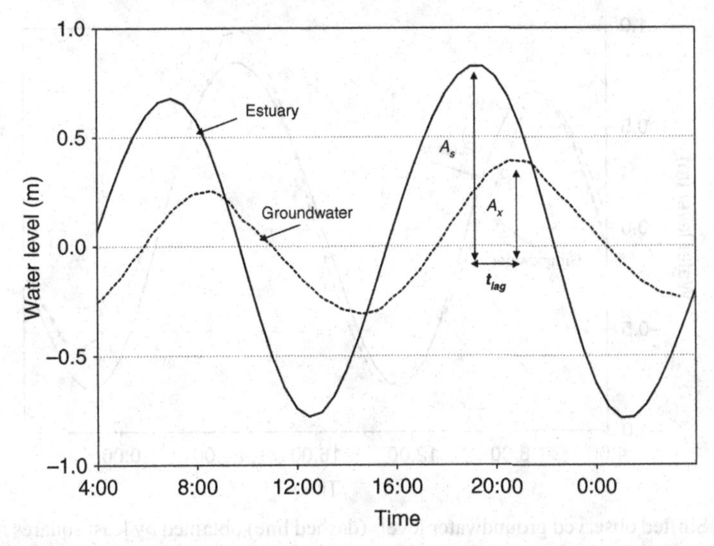

Figure 4.9 Response of the water level in a piezometer (dashed line) located approximately 50 m from the midpoint between the high and low water lines on a small island in Portugal. The solid line shows the water level variation in the estuary in which the island is located.

Erskine (1991) introduced an approach to estimate the time lag and amplitude ratio using multiple water level measurement rather than just the recorded peaks. His method is based on the assumption that if the groundwater level fluctuation is induced by fluctuations of the sea level, the two curves should coincide with each other if the groundwater level is amplified by the amplitude ratio and offset by the time lag.

In the process of matching the two curves to obtain the best fit, there are two parameters which need to be adjusted: the amplitude ratio A_R and the time lag t_{lag}. If the mean levels of the tide and the groundwater level are different, which is usually the case, the groundwater level curve is first shifted in elevation to have the same mean value as the tidal level:

$$h'(t) = \overline{h_s} + [h(t) - \overline{h}]/A_r \tag{4.72}$$

where $h'(t)$ [L] is the transformed groundwater level at time t and $\overline{h}$ [L] is the mean of the groundwater level $h(t)$.

To reduce the effects of errors of individual measurements, Erskine (1991) chose to use the standard deviations of the tidal oscillations to calculate A_R, but this is only possible when the signals are symmetrical around their mean. More generally, A_R and t_{lag} can be determined simultaneously by least squares optimisation to minimise the difference between h_s and h':

$$Obj(A_r, t_{\text{lag}}) = \sum_n [h'(t) - h_s(t - t_{\text{lag}})]^2 \tag{4.73}$$

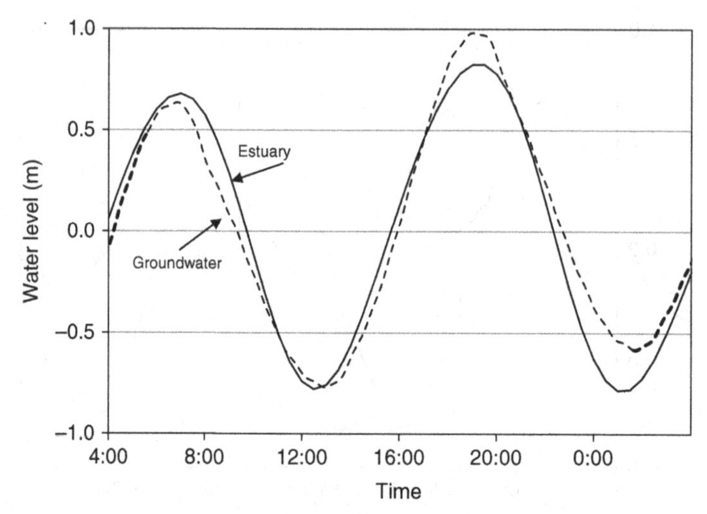

Figure 4.10 Shifted observed groundwater levels (dashed line) obtained by least squares fitting of the observations to the water level variation in the estuary (solid line). The original data are shown in Figure 4.9.

If the water level and the tide time series were measured at different times, the measurements have to be resampled. Graphically the curves of h_s and h' should coincide as in Figure 4.10. They will not be identical because measurement noise and various other factors than the tide also affect the measured water levels. Case studies using this approach can be found in Erskine (1991), Jha et al. (2008) and Trefry and Johnston (1998).

If there is only one well, after the time lag and amplitude of the water level in the well are obtained, both amplitude-ratio method (such as Eqn (4.32)) and time lag method (such as Eqn (4.33)) can be used to estimate the hydraulic diffusivity D_h. If there are a number of wells at different distances from the coast, a line showing the change of time lag versus distance (Eqn (4.31)) can be plotted and the slope of the line can be used to estimate D_h. Theoretically, the line should be a straight line but in practice the data points tend to be scattered and a best-fit line should be used. This approach based on multiple wells should provide a more representative estimation than a single well. Similarly, the amplitude ratio versus distance (Eqn (4.29)) should be a straight line in a semi-log graph, and the slope of the line can be used to estimate the diffusivity.

4.7.3 Processing Water Level Data for Tidal Analysis

The aforementioned methods work only if the water level fluctuations are induced exclusively by tides, so observed water levels should be de-trended to eliminate any non-tidal effects. Fluctuations induced by other factors such as Earth tides, atmospheric pressure, recharge or pumping should be removed. A careful analysis is needed to distinguish the impact of Earth or atmospheric tides on the water level from that of ocean tides (Merritt,

2004). Low-frequency oscillations, such as recharge, aperiodic atmospheric pressure oscillations and long-period sea level fluctuations, can be deleted by subtracting the daily moving average (Rotzoll et al., 2008).

After non-tidal influence is removed from the groundwater level fluctuations, the major tidal constituents contributing to the fluctuations should be identified. In most coastal areas, the tidal signals are dominated by the five major constituents listed in Table 4.1. Two methods are commonly used for analysing the amplitude and phase of periodic components in measured groundwater levels: least squares fitting of the main tidal frequencies (e.g. Hsieh et al., 1987; Merritt, 2004) and spectral analysis based on Fourier transformation (e.g. Trefry and Bekele, 2004; Bye and Narayan, 2009). If the frequencies of the tidal components are known a priori, one can use the least squares regression technique (e.g. Rotzoll et al., 2008). The major tidal constituents and their harmonic components and frequency often are available from nearby tidal gauges. Care must be taken though to ensure that such data are representative for the study location, because the characteristics of the tide can vary significantly along the coast over stretches of a few kilometres.

Some coastal areas may have site-dependent tides different from the major tides listed in Table 4.1. For example, when tidal currents in the sea interfere with each other, or the sea tides collide with currents from river estuaries or semi-enclosed bays, new waves of aberrant frequencies may be generated. In this case, a spectral analysis of the water level in the coastal wells may identify such local effects. Spectral analysis can also be helpful in revealing non-tidal processes such as atmospheric pressure variations and rainfall infiltration fronts (Kuang et al., 2013). The strength of Earth tides should be also checked to assess their contribution to the observed water level fluctuations.

4.8 Notes on the Theory and Practice of Tidal Methods

4.8.1 Tidal Signals in Confined and Unconfined Aquifers

As can be seen from Eqn (4.26) and the other tidal propagation equations in Section 4.5, the attenuation of the tidal signal is a function of the storage coefficient S. Therefore, the damping of the tidal fluctuation with inland distance differs significantly for unconfined and confined aquifers. Due to their low storage coefficient confined aquifers have a weak damping effect and tidal fluctuations may be observed over hundreds of metres away from the sea (White and Roberts, 1994). For example, a borehole at an inland distance of 2.8 km in the well-confined Dridrate aquifer in Morocco still displayed clear sinusoidal fluctuations with a time lag of about 2 hours and amplitude ratio of 14% attributable to the ocean tide (Fakir and Razack, 2003). Tidal signals may also propagate vertically into deep confined aquifers below the seabed. The reservoir pressure in the Ormen Lange gas field offshore Norway reacts with negligible time delay to the tide variation some 2.8 km above (Langaas et al., 2006).

Tidal signals in unconfined aquifers on the other hand are quickly damped. Based on their experience of several case studies in the UK, White and Roberts (1994) summarised that appreciable oscillations are unlikely to exist at inland distances more than 20 or 30 m from the sea. On the basis of tidal study of a coastal aquifer in China, Chen and Jiao (1999) that the unconfined aquifer showed little tidal influence while the underlying confined aquifers indicated a tidal oscillation of 0.1–0.6 m.

Water levels in an unconfined aquifer display tidal signals that vary with depth (Nielsen, 1990; Erskine, 1991; Trefry and Johnston, 1998). Trefry and Johnston (1998) presented a case of an unconfined aquifer with numerous thin peat layers. They speculated that these layers presented a certain confining influence for the unconfined aquifer. Water levels from multi-level observation wells showed that the deeper parts of the aquifer exhibit tidal fluctuations of higher amplitude than the shallow parts, suggesting increasing confinement with depth. Their pumping test data also showed that the effective storativity declined with depth. Erskine (1991) observed the similar phenomenon in a coastal aquifer in the UK: the deeper piezometers had smaller lags and larger amplitude. He concluded that this is a consequence of the pressure waves being damped stronger near the water table where the storage is controlled by the draining and filling of pores, whereas at depth elastic deformation controls the storage capacity.

4.8.2 Composite Signals

Tide propagation into aquifers depends not only the aquifer but also the tide itself. Semi-diurnal frequencies are damped more quickly than diurnal frequencies because the former involves higher internal friction losses (Rotzoll et al., 2008). Sea tides are typically a combination of many harmonic constituents of various frequencies and amplitudes. A minimum of 15 days of continuous groundwater and tidal records are needed to separate the composite waves into their individual harmonic components with a distinct amplitude and periodicity by harmonic analysis (Carr, 1971). For a confined aquifer, which behaves as a linear system with respect to the tidal propagation, the tidal components of the groundwater can be considered as acting entirely independent of each other, so that the total fluctuation at a point is the sum of the fluctuations due to the individual components. Each component can then be analysed separately (Carr, 1971).

Carr (1971) carried out a careful study of tidal fluctuations in a confined aquifer using 16 days of groundwater level records from a tidal gauge and four wells situated along a transect perpendicular to the shore at Borden, Prince Edward Island, Canada. Using harmonic analysis, he identified that there were three major constituents (O_1, K_1 and M_2). Both the tidal and groundwater level records were separated into these three individual components. The time lag and amplitude of each of the constituents were used to estimate the aquifer diffusivity. When the diffusivity was estimated again using the composite wave it was found that the calculated values of aquifer parameters differed very little from those determined by the individual components. Based on this, Carr (1971) concluded that as long as the composite tide is largely diurnal, time series lengths of two to four days are sufficient for the analysis.

4.8.3 Amplitude-Ratio Method versus the Time Lag Method

Ferris (1951) was the first to estimate the hydraulic diffusivity of a real aquifer using both the amplitude-ratio method and the time lag method for the water levels from three observation wells near a river with cyclic fluctuations. He found that the diffusivity estimated from the time lag method was smaller by about 62% than from the amplitude-ratio method and he speculated that this difference was due to the influence of the nearby pumping on the groundwater level and on the timing of the maxima or minima.

Since then many other researchers have found there can be a significant difference in aquifer diffusivity estimated from either two methods. Erskine (1991) found that the diffusivity estimated from the time lag method was about 10 times larger than the amplitude-ratio method. Trefry and Johnston (1998) and Fakir and Razack (2003) found a difference of about 6 to 7 times. The reasons for such major difference seem more fundamental than those postulated by Ferris (1951).

Apparently the analytical expressions used to determine the diffusivity fail to represent key aspects of real coastal aquifers. The discrepancy between the results may be caused by the water table. The water table is likely to have a damping effect on the tidal oscillations. This damping effect has been speculated to have a stronger effect on the amplitude of the tidal oscillations than on the time lag (Erskine, 1991). Other factors such as depth-dependence of the storage coefficient, or spatial heterogeneity may also be responsible for the discrepancy (Trefry, 1999; Jha et al., 2008). Based on numerical simulations, Trefry and Bekele (2004) identified internal layering of an aquifer as the main contributor. Some studies compared tidal approaches with traditional pumping tests and found that the results from the amplitude-ratio method were closer to the pumping test results than the time lag method (Trefry and Johnston, 1998). It therefore seems that the use of time lag method for estimating the aquifer hydraulic diffusivity should be used with caution (Jha et al., 2008).

5

Hydrogeochemistry of Coastal Aquifer Systems

5.1 Introduction

Groundwater in coastal aquifers often is a mixture of seawater and freshwater. Because the salinity of seawater is high relative to groundwater recharge sources of freshwater, a small amount of seawater can dominate the chemical composition. Addition of 1% of seawater to freshwater with 100 mg l^{-1} of chloride nearly triples the chloride concentration; addition of 5% of seawater increases the salinity of the original freshwater to over 1000 mg l^{-1} (Jones et al., 1999). Superimposed on mixing effects are the changes of the groundwater composition in coastal aquifer systems due to chemical processes.

The chapter starts with a discussion of the chemical characteristics of freshwater and seawater, followed by the most common geochemical facies, which are water bodies with a distinct chemical composition (Back, 1966) in coastal aquifers. Some widely used graphical methods in chemical data analyses are also presented. Chemical and isotopic indicators of seawater intrusion or freshwater flushing are illustrated by examples. Various important physiochemical reactions in both natural systems and reclaimed land such as adsorption and ion exchange, decomposition of organic matter, acid sulphate soil formation, membrane filtration and osmosis in clay are discussed. The focus is mainly on coastal plain aquifer systems consisting of unconsolidated sedimentary materials.

5.2 Coastal Water Types

Mixing between seawater and freshwater is one of the most important processes that controls the composition of groundwater in coastal aquifers. The starting point for the interpretation of hydrochemical processes is therefore the identification of the fresh- and seawater end-members that make up the mixture. Because the composition of seawater can vary from one location to the next, especially near the coast and in semi-closed basins like the Baltic, Bohai or Mediterranean seas, an attempt should always be made to identify the solute concentrations, and its variability in space and time, of the local seawater.

5.2.1 Salinity

The total dissolved solids (TDS) concentration of standard seawater is 36 g l^{-1} (Section 1.4.1), but across the globe the salinity of the oceans varies between 33 and 36.5 g l^{-1} and the spatial distribution mirrors the global rainfall minus evaporation pattern. Near the coast water input from major rivers or melting glaciers can cause TDS concentration values to become much lower than 36 g l^{-1}. Because the salinity of terrestrial groundwater is usually much lower than seawater, there can be a drop in the salinity near the sea bottom where significant groundwater input exists (McClatchie et al., 2006). This also occurs around submarine spring vents (Section 7.4). A good example is the water salinity and temperature profiles over the Crescent Beach Spring vent, which show that the spring water is much warmer and fresher than the background seawater, confirming that the spring water has an onshore source (Barlow, 2003).

Inland seas or sea arms which are partially isolated from the open ocean and have little riverine input, may have a salinity higher than standard seawater due to limited circulation and high evaporation. For instance, connected to the Atlantic Ocean only through a very narrow channel, the Mediterranean Sea has salinity of over 39 g l^{-1} and chloride concentration of about 22 g l^{-1} (Table 5.1) because evaporation exceeds precipitation and river runoff (Jones et al., 1999). Conversely, the salinity of the Baltic Sea is markedly lower than standard seawater, because of the high continental freshwater input and low evaporation rates (Meier et al., 2006).

There can be noticeable changes of the seawater composition with time. The change can be seasonal or inter-annual, or over a much longer time scale. For example, in the Dead Sea (a hyper-saline inland lake), the salinity increased from 280 g l^{-1} in 1944 to 340 g l^{-1} in 1996 (Yechieli et al., 1998) due to the diversion of water from the Jordan river for irrigation. On a longer timescale, Zong et al. (2010) reconstructed the salinity history of the Pearl River Estuary in China during the Late Quaternary on the basis of diatom species from sediment cores. The salinity was about 30 g l^{-1} in 9000 yr BP (year before present), became as low as 20 g l^{-1} in the period from 7000 to 6000 yr BP, and has increased gradually to 29 g l^{-1} in the past 1000 years, as the results of the interplay of sea level change and delta progradation. Similar variations during the Holocene have been observed elsewhere as well (Vos, 2015). Moreover, the composition of seawater has not been constant over the past 600 million years due to variations of hydrothermal brine input along mid-ocean ridges (Hardie, 1996). Such historical salinity changes may be relevant when studying long-term groundwater evolution in coastal aquifers (Chapter 8), and care must be taken when selecting the seawater end-member for mixing calculations (Section 5.5.3) or solute transport modelling.

5.2.1.1 Measurement of Salinity

The salinity of water is commonly reported as the TDS concentration (Section 1.4.1). Another common parameter to express salinity is electrical conductivity (EC), which reflects the water's ability to carry an electrical current. The dissolved ions make water more conductive, and thus a higher EC indicates more ions in the water. Since EC is also a function of the temperature (it increases by about 2% per degree Celsius temperature rise), values are reported at a reference temperature (usually 25°C) so that differences between samples of different temperature are attributable to the effect of salinity only. The EC at a reference temperature is called the specific conductance (SC).

The unit for electrical conductivity is Siemens (S), which replaced the 'mho' (the reverse of Ohm) that can still be encountered in some older texts. The electrical conductivity of water is measured as electrical conductivity over a certain distance, thus the SI unit for the electrical conductivity of water is S m^{-1}. In groundwater hydrology it is common practice to use mS cm^{-1} or μS cm^{-1} (1 S = 1000 mS = 1 000 000 μS). Typically, the SC for drinking water is less than 1000 μS cm^{-1}. The SC for standard seawater is 53 mS cm^{-1} at 25°C (47.8 mS cm^{-1} at 20°C). Some researchers define groundwater salinity classes based on SC (Mondal et al., 2008). For example, Park et al. (2012b) divided coastal groundwater into fresh (SC < 1500 μS cm^{-1}), brackish (1500 < SC < 3000 μS cm^{-1}) and saline (SC >3000 μS cm^{-1}) water.

Many handheld instruments to measure SC are available commercially, and it is standard practice to measure it in the field. These instruments usually also report the TDS concentration, which is calculated based on the measured conductivity and temperature. The relationship between EC and TDS depends on the aqueous solutes present, and the reported TDS value may be imprecise when the instrument settings deviate from the site conditions. A simple linear conversion equation between the EC (in μS cm^{-1}) and TDS (in mg l^{-1}) is

$$\mathrm{TDS} = k_e \times \mathrm{SC} \tag{5.1}$$

where the conversion factor k_e depends on the chemical composition of the water sample because different ions in water have a different ability to conduct electricity. The factor k_e can vary between 0.54 to 0.96, but mostly lies between 0.55 and 0.75 (Hem, 1985). For seawater at 25°C k_e = TDS / SC = 36 / 53 = 0.68, which can be used as a first approximation for mixtures of freshwater and seawater.

Measurement of the EC in water with a high turbidity (i.e. high content of suspended particles) can be problematic. Examples include muddy pools, a piezometer or borehole where sludge accumulates, or near the bottom of the sea. As the instrument measures the electrical resistance between a pair of electrodes, the resultant measurement represents the resistance of the water, its solutes and the suspended particles. This may significantly affect the measurement and, for example, a sudden drop of the EC at the seabed should not be mistaken for a dilution effect from fresh groundwater input. An example of such a profile from Hong Kong is shown in Figure 5.1. Repeated measurements of the same seawater samples mixed with mud indicated that the sudden drop of EC near the seabed was a false signal caused by the turbid conditions.

5.2.2 Chemical Composition of Terrestrial Groundwater

Under natural conditions, the chemical composition of terrestrial groundwater is normally dominated by the cations, K^+, Na^+, Ca^{2+}, Mg^{2+} and the anions Cl^-, HCO_3^- and SO_4^{2-}. Each of these ions typically has a concentration > 1 mg l^{-1} and these major ions usually comprise

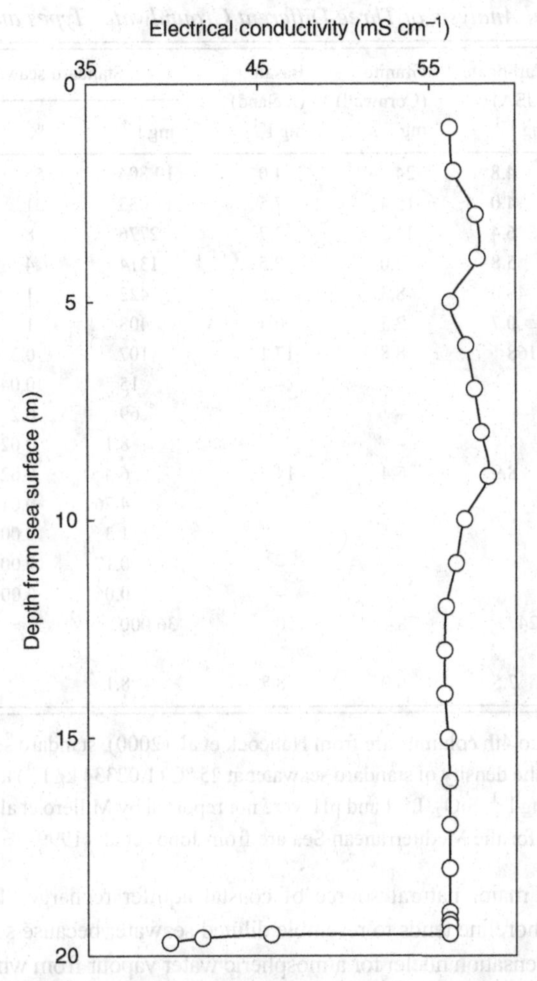

Figure 5.1 EC profile of seawater with a sudden drop above the sea bottom, Tolo Harbour, Hong Kong, in 2009. Initially, this was regarded as indicator of freshwater submarine groundwater discharge. Repeated measurements confirmed, though, that the signal was caused by the mud at the bottom.

over 98% of the total solute content. The chemical composition of fresh, terrestrial groundwater is highly variable though, as it depends on both the composition of the rock through which the water percolates and the composition of recharge sources such as rain and rivers. If not contaminated by anthropogenic activities, fresh groundwater tends to be of the alkaline earth bicarbonate type and often the calcium concentration surpasses the magnesium concentration. The variability between terrestrial groundwater compositions that can be encountered is illustrated by the composition of water from a limestone, granite and basalt aquifer in Table 5.1.

Table 5.1 *Chemical Analysis of Three Different Groundwater Types and Seawater*

Solute	Carbonate (USA) mg l^{-1}	Granite (Cornwall) mg l^{-1}	Basalt (Iceland) mg l^{-1}	Standard seawater mg l^{-1}	Standard seawater %	Mediterranean seawater mg l^{-1}
Chloride, Cl^-	4.8	24	4.0	19 804	55	21 940
Sodium, Na^+	4.0	15.4	7.5	11 033	31	12 500
Sulphate, SO_4^{2-}	6.4	11.5	3.7	2776	8	2700
Magnesium, Mg^{2+}	5.8	2.0	0.5	1314	4	1211
Calcium, Ca^{2+}	48.0	8.3	2.5	422	1	459
Potassium, K^+	0.7	2.3	0.4	408	1	435
Bicarbonate, $HCO_3^=$	168	8.8	17.1	107	0.3	169
Carbonate, CO_3^{2-}	–	–	–	15	0.04	
Bromide, Br^-	–	–		69	0.2	74.1
Strontium, Sr^{2+}	–	–		8.1	0.02	
Silica, SiO_2	8.9	5.4	14.6	6.4	0.02	
Boron, B	–	–		4.76	0.01	
Fluoride, F^-	–	–		1.3	0.004	
Lithium, Li^+	–	–	–	0.17	0.0005	
Iodide, I^-	–	–	–	0.06	0.0002	
Total dissolved solids (TDS)	~247	~78	~50	36 000		39 420
pH	7.5	6.0	8.8	8.1		

Data from the 2nd to 4th columns are from Hancock et al. (2000), standard seawater from Millero et al. (2008) using the density of standard seawater at 25 °C (1.02334 kg l^{-1}) to convert the reported mass fractions to mg l^{-1}. SiO_2, Li, I and pH were not reported by Millero et al. (2008) and are from Hem (1985). Data for the Mediterranean Sea are from Jones et al. (1999).

Rainwater is the major natural source of coastal aquifer recharge. The composition of rainwater near the shoreline tends to resemble diluted seawater because seawater-derived salt particles act as condensation nuclei for atmospheric water vapour from which rain drops form (Deusdará et al., 2017; Vengosh and Rosenthal, 1994; Aswathanarayana, 2001; Silva et al., 2007). Fresh groundwater derived from seaborne precipitation can thus have comparable relative proportion of solute ions as seawater. Fractionation effects and aerosol sources other than seawater, such as soil dust, may lead to significant deviations though. For example, rainwater in northern Portugal was reported to have an annual averaged molar Na/Cl ratio of 1.25 to 1.44, which is much higher than the characteristic value of 0.86 for seawater (Table 5.2) due to the presence of soil dust (Condesso de Melo et al., 2001). The solute content of rainwater decreases with inland distance from the shoreline (Rosenthal, 1987; Stuyfzand, 1993).

5.2.3 Chemical Composition of Seawater

Compared to terrestrial fresh groundwater, seawater has much higher concentrations of major (except for Ca^{2+} and HCO_3^-) and some minor ions, and its chemical composition

Table 5.2 *Molar and Mass Ratios of Concentrations in Standard Seawater as Listed in Table 5.1*

	Molar ratio	Molar to mass ratio conversion factor	Mass ratio
Na^+/Cl^-	0.86	0.65	0.56
Na^+/K^+	45.9	0.59	27.0
SO_4^{2-}/Cl^-	0.05	2.71	0.14
Br/Cl^- (Cl^-/Br^-)	0.0015 (648)	2.25 (0.44)	0.00347 (288)
B/Cl^-	0.00076	0.32	0.00024
K^+/Cl^-	0.019	1.10	0.021
$Ca^{2+}/(HCO_3^- + CO_3^{2-} + SO_4^{2-})$	0.34	0.43	0.15
Cl^-/HCO_3^-	318	0.58	185
Mg^{2+}/Ca^{2+}	5.14	0.61	3.12
I^-/Cl^-	8.5×10^{-7}	3.58	3×10^{-6}

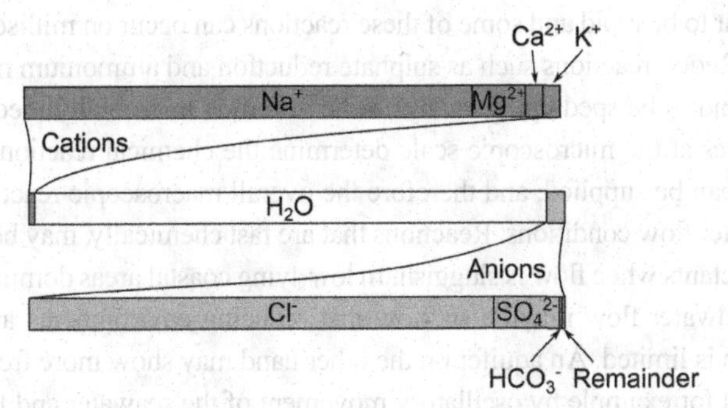

Figure 5.2 Relative proportions of solutes and water in 1 kg of standard seawater based on concentration data from Millero et al. (2008).

shows minor variability within the world's oceans because the long residence time facilitates mixing (Marcet's principle, Millero et al., 2008). Standard seawater is a precisely defined reference standard used in oceanography, the composition of which is based on samples of surface seawater from the North Atlantic Ocean. Its composition is shown in Table 5.1 and Figure 5.2 shows the mass fractions of major ions in 1 kg of standard seawater. Chloride and sodium are the dominant anion and cation, respectively. In terms of mass, the anions contribute to the TDS concentration almost twice as much as the cations and Cl^- dominates over the alkaline (Na^+ and K^+) and earth alkaline (Mg^{2+} and Ca^{2+}) ions (Jones et al., 1999).

The relative proportion of individual ions in seawater tends to be distinctly different from those of terrestrial waters (Richter and Kreitler, 1991), which can be exploited to establish if seawater forms a source of groundwater salinity (Section 5.6). Table 5.2 presents the ratios of some of the ions in seawater. Both molar and mass ratios are commonly encountered in the literature. In this book all ratios are molar ratios unless specified otherwise.

5.3 Chemical Processes in Coastal Aquifers

From the recharge to discharge areas in coastal aquifers, there can be many physical and chemical reactions including acid-base reactions, oxidation-reduction reactions, mineral dissolution and precipitation, sorption and ion exchange, dissolution and exsolution of gases, and biodegradation. Other books (e.g. Stumm and Morgan, 1995; Appelo and Postma, 2005) have covered these topics in much greater detail than possible here, so in this book only the reactions that are the most relevant to coastal aquifers are discussed.

5.3.1 Reaction Rates

The time to reach chemical equilibrium can vary from microseconds for ion association reactions to minutes for some ion exchange and sorption reactions, to many years for certain mineral dissolution and precipitation reactions (Sparks, 1999) (Figure 5.3). Ion exchange can be relatively fast if the exchange occurs at the edges of a mineral grain, such as for kaolinite, but is slower if it involves ions inside the grain, such as the mineral layers of montmorillonite (Merkel et al., 2008). Metal and metalloid sorption reactions on oxides, hydroxides and humic substances appear to be rapid and some of these reactions can occur on millisecond time scales (Sparks, 1999). Redox reactions such as sulphate reduction and ammonium mineralisation are very slow, but tend to be sped up by catalysts and are often microbially mediated.

While processes at the microscopic scale determine the chemical reaction rates, the rate at which reactants can be supplied, and therefore the overall macroscopic reaction rate, depends on the groundwater flow conditions. Reactions that are fast chemically, may become limited by the supply of reactants when flow is sluggish. In low-lying coastal areas dominated by clay-rich aquitards, groundwater flow may be so slow that reducing environments are formed as the supply of oxygen is limited. An aquifer on the other hand may show more frequent changes in chemistry caused for example by oscillatory movement of the seawater and freshwater transition zone due to tides or seasonal changes in recharge (Russak and Sivan, 2010).

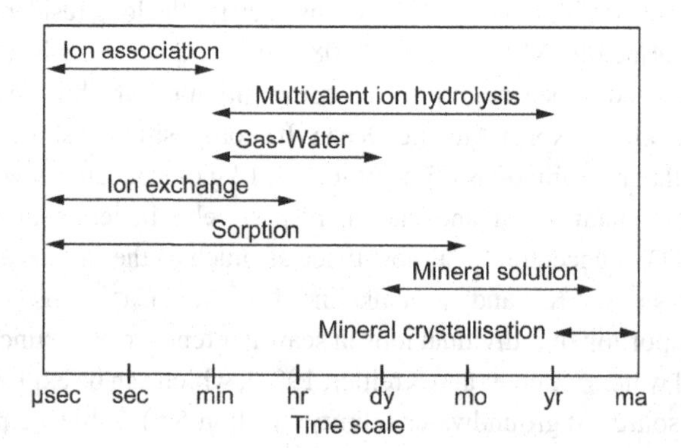

Figure 5.3 Time ranges needed to achieve chemical equilibrium by various reactions in soil environments (Selim and Amacher, 1997).

5.3.2 Redox Conditions

In recharge areas with little organic matter in the soil, water usually has high dissolved oxygen, ranging from a few to about 10 mg l^{-1}, depending on temperature and atmospheric conditions (Abesser et al., 2005). While migrating through the unsaturated zone, oxygen may be lost due to microbiological processes and redox reactions, but by the time the water approaches the water table, there can still be significant dissolved oxygen left. Depending on the geochemical characteristics of the aquifer, the oxygen can continue to be consumed by various biological and chemical reactions, such as oxidation of pyrite or other sulphide minerals and organic matter (Sections 5.7.2 and 5.7.3). A redox boundary forms when all the oxygen has been consumed, which separates the system into aerobic and anaerobic zones (Figure 5.4). Downgradient from the recharge area, especially in deeper confined parts, anaerobic conditions tend to prevail.

Under anaerobic condition, other oxidants are consumed, following a specific sequence which is prescribed by the energy gain of the redox reaction. When present, NO_3 is the first to be reduced once oxygen becomes depleted, followed by manganese and iron oxides, and sulphate. As a result, a geochemical zonation along the flow path develops, which is presented schematically in Figure 5.5. Near the recharge area, where the groundwater is rich in oxygen, the system is dominated by aerobic metabolism (Chapelle, 2001). In the confined part anaerobic processes in the order of nitrate reduction, ferric iron reduction, sulphate reduction and methanogenesis occur along the flow path. This generates dissolved NO_2^-, Fe^{2+}, Mn^{2+} and gases such as N_2, H_2S and CH_4 in the groundwater.

5.3.3 Ion Exchange

It has been well known that clay particles have a net negative electrical charge on the surface since Reuss (1809) discovered that clays suspended in water move towards a positive electrode when a potential is applied. Because the surfaces of clay particles are negative, they attract cations and dipolar water molecules (Figure 5.6). The aqueous phase immediately adjacent to the surface therefore develops a net positive charge. The layer with excess cations bordering the negatively charged surface is called the diffuse double layer (Craig, 2004). As the attractive force decreases with distance from the surface, the excess concentration of cations in the double layer gradually decreases. The thickness of this layer is mainly determined by the charge density: The double layer thickness decreases with an increase in the valency (at equal concentration) or the concentration (for cations of the same valency).

Organic matter, primarily natural humic substances, also has a negative surface charge that attracts positive ions (Figure 5.7). Exposed carboxylic (−COOH) and phenolic (⌬−OH) groups associated with a central organised solid unit are the major source of negative charge on the surface of humus (Deutsch, 1997). Very often clay layers are organic-rich, which adds to the sorption capacity of the clay minerals themselves.

Adsorption is the attraction of ions to the charged surface of soil particles. Among the sorption reactions in aquifers, ion exchange, by which ions adsorbed to solids are

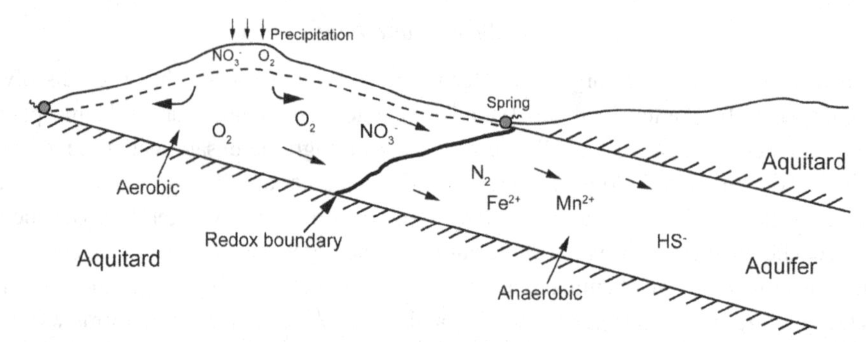

Figure 5.4 Zonation of aerobic and anaerobic conditions separated by a redox boundary (modified from Abesser et al., 2005). Anaerobic conditions preferably develop in the confined part of the aquifer, where the supply of O_2 is limited. Note that in natural systems, NO_3^- is much lower than O_2 in practically all cases, but in areas with intensive agriculture, high NO_3^- concentrations can raise the oxidative capacity to levels far above natural values.

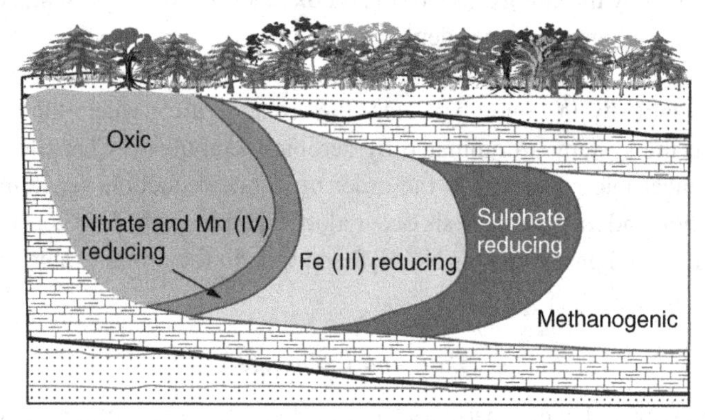

Figure 5.5 Zonation of the redox environments along a groundwater flow path (Chapelle, 2001).

substituted by others in water, is the most important in coastal aquifers. It causes cation concentrations to deviate significantly from those resulting from simple mixing of seawater and freshwater, and can be used as indicator for seawater intrusion or freshwater flushing (Appelo and Postma, 2005).

Clays and soil organic substances are the primary sites for ion exchange, although essentially all solid materials have a certain capacity for ion exchange. Cation exchange is influenced by the concentrations of the cations in the solution and the selectivity of a material for certain cations. For common ions in groundwater, the cation selectivity sequence is: $Na^+ > K^+ > Mg^{2+} > Ca^{2+}$.

Ca^{2+} and HCO_3^- are often the dominant ions in fresh groundwater, so Ca^{2+} is then the main adsorbed cation. Seawater mainly contains Na^+ and Cl^-, so sediments saturated by seawater attract mostly Na^+. When seawater intrudes a freshwater aquifer, cation exchange occurs:

$$Na^+ + 1/2Ca\text{-}X_2 \rightarrow 1/2Ca^{2+} + Na\text{-}X \quad (5.2)$$

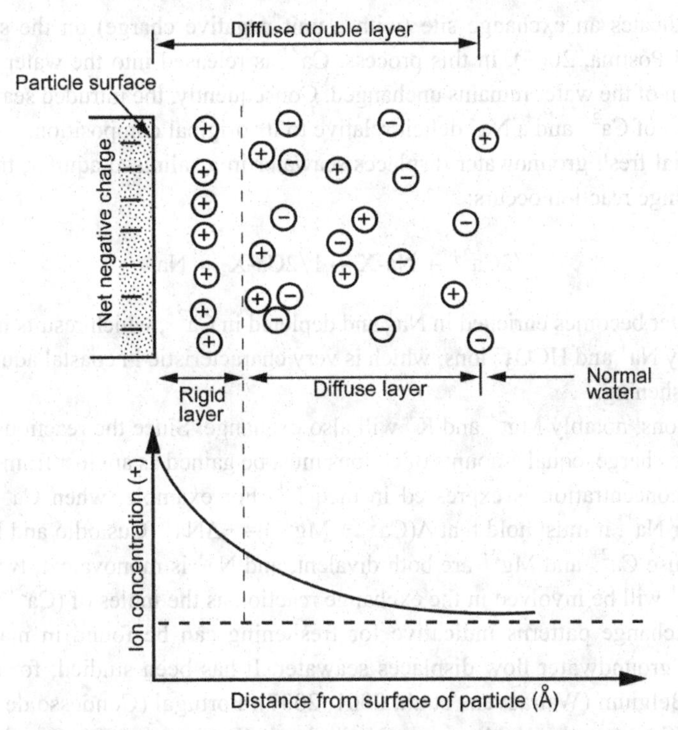

Figure 5.6 Structure of the double layer near the negatively charged surface of a clay mineral (Raj, 1995); 1 Å or angstrom = 10^{-10} m.

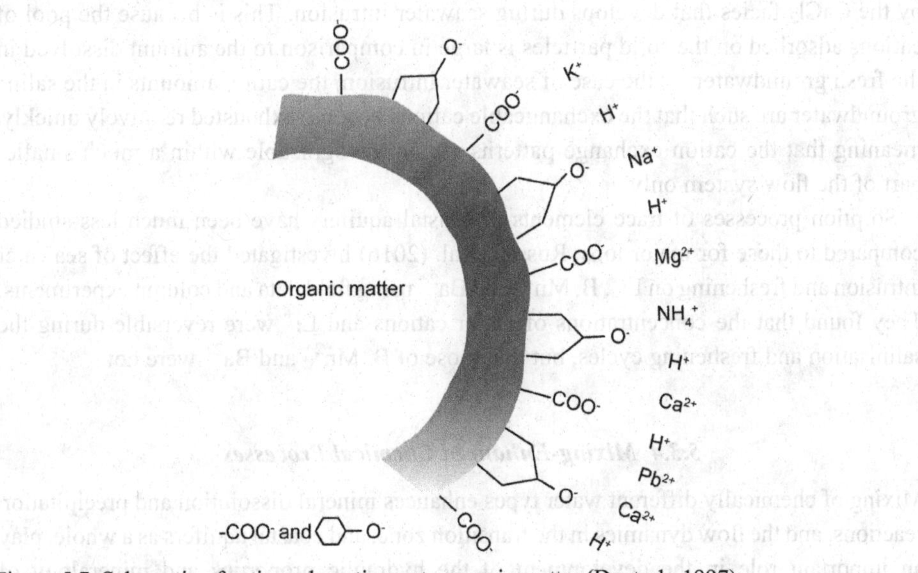

Figure 5.7 Schematic of cation adsorption onto organic matter (Deutsch, 1997).

where X indicates an exchange site (with a unit negative charge) on the soil particles (Appelo and Postma, 2005). In this process, Ca^{2+} is released into the water, but the Cl^- concentration of the water remains unchanged. Consequently, the intruded seawater develops an excess of Ca^{2+} and a Na^+ deficit relative to its original composition.

If terrestrial fresh groundwater displaces seawater in a salinised aquifer, the following cation exchange reaction occurs:

$$1/2Ca^{2+} + Na\text{-}X \rightarrow 1/2Ca\text{-}X_2 + Na^+ \tag{5.3}$$

The freshwater becomes enriched in Na^+ and depleted in Ca^{2+}, which results in freshwater dominated by Na^+ and HCO_3^- ions, which is very characteristic in coastal aquifers subject to active freshening.

Other cations, notably Mg^{2+} and K^+ will also exchange. Since the reactions involve no change in the charge, equal amounts of cations must be gained as are lost from the solution when their concentration is expressed in meq l^{-1}. For example, when Ca^{2+} and Mg^{2+} exchange for Na^+, it must hold that $\Delta(Ca^{2+} + Mg^{2+}) = -\Delta Na^+$ (Custodio and Bruggeman, 1987). Because Ca^{2+} and Mg^{2+} are both divalent, and Na^+ is monovalent, twice as many moles of Na^+ will be involved in the exchange reaction as the moles of $(Ca^{2+} + Mg^{2+})$.

Cation exchange patterns indicative for freshening can be found in many aquifers where fresh groundwater flow displaces seawater. It has been studied, for example, in aquifers in Belgium (Walraevens et al., 2001, 2007), Portugal (Condesso de Melo et al., 2001), the USA (Appelo, 1994) and the Netherlands (Beekman, 1991; Stuyfzand, 1999). The release of sorbed Na^+ in exchange for Ca^{2+} gives rise to the typical $NaHCO_3$ hydrochemical facies across a zone that tends to be much wider than the zone occupied by the $CaCl_2$ facies that develops during seawater intrusion. This is because the pool of cations adsorbed on the solid particles is large in comparison to the amount dissolved in the fresh groundwater. In the case of seawater intrusion, the cation amounts in the saline groundwater are such that the exchangeable cations become exhausted relatively quickly, meaning that the cation exchange patterns remain recognisable within a much smaller part of the flow system only.

Sorption processes of trace elements in coastal aquifers have been much less studied compared to those for major ions. Russak et al. (2016) investigated the effect of seawater intrusion and freshening on Li^+, B, Mn^{2+} and Ba^{2+} using field data and column experiments. They found that the concentrations of major cations and Li^+ were reversible during the salinisation and freshening cycles, but that those of B, Mn^{2+} and Ba^{2+} were not.

5.3.4 Mixing-Enhanced Chemical Processes

Mixing of chemically different water types enhances mineral dissolution and precipitation reactions, and the flow dynamics in the transition zone, and coastal aquifers as a whole, play an important role in the development of the hydraulic properties and mineralogy of carbonate rocks. In limestone aquifers it has been found that when freshwater and seawater mix, depending on the extent of mixing and the precise chemical characteristics of both waters, carbonate minerals may either dissolve or precipitate. Even if both the two waters are in chemical equilibrium with respect to calcite before mixing, their mixture may have the tendency to dissolve or deposit calcite (Wigley and Plummer, 1976; Hanshaw and Back, 1980). The potential to dissolve carbonate minerals is believed to be an important factor that controls cave formation in coastal limestone aquifers (Back et al., 1986; Fratesi, 2013). Modelling studies have shown that porosity development in the transition zone can increase the local permeability, which enhances flow, and this then promotes further dissolution (Sanford and Konikow, 1989; Rezaei et al., 2005).

When calcite dissolves, dolomite may precipitate and the transition zone between freshwater and seawater has been suggested to be an environment conducive to dolomitisation (Hanshaw et al., 1971; Land, 1973). Dolomite formation is believed to occur when the Mg^{2+}/Ca^{2+} ratio is greater than 1, which is met when there is more than 2% of seawater (Magaritz et al., 1980). In the coastal plain of Israel, the Quaternary calcareous sands of the coastal aquifer were transformed in two distinct horizons, with the deeper one interpreted to be associated with a former transition zone and the shallower one with the current transition zone. The occurrence of dolomite below the contemporary transition zone may indicate a palaeo-transition zone that existed when the sea level was low. The continuous formation of dolomite in the transition zone may decrease the hydraulic conductivity of the coastal aquifers (Magaritz et al., 1980) as dolomite forms a void-filling cement (Warren, 2000). Dolomite has not always been encountered in transition zones where calcite was observed to be dissolving, and hence the ubiquitousness of the process has been called into question (Smart et al., 1988). Dolomite formation and the environments where it takes place remain subject to considerable debate (Warren, 2000; Machel, 2004).

When marine limestones become exposed above sea level, the seawater in the pores is displaced by freshwater. This sets into motion a process of recrystallisation of the more soluble carbonate minerals, like aragonite and high-magnesium calcites, to calcite. In the process, ions like Mg^{2+} and Sr^{2+} are released to the groundwater. Studies of this process in Bermuda for example have shown that the degree of recrystallisation correlates with the age of the limestone: the oldest, more inland rocks consist largely of stable calcite forms, whereas the younger rocks near the coast are still rich in aragonite and high-magnesium calcites (Plummer et al., 1976).

5.4 Chemical Evolution in Coastal Plain Aquifer Systems

The general evolutionary trends in a regional flow system have been discussed in many previous publications (e.g. Freeze and Cherry, 1979; Stuyfzand, 1999; Hiscock, 2005), and for coastal aquifers specifically by Custodio and Bruggeman (1987). On the basis of extensive studies of the chemical constituents of groundwater in the sedimentary aquifers of the US Atlantic Coastal Plain, Back (1960, 1966) developed a conceptual model of the chemical changes along a flow path that can be considered exemplary for

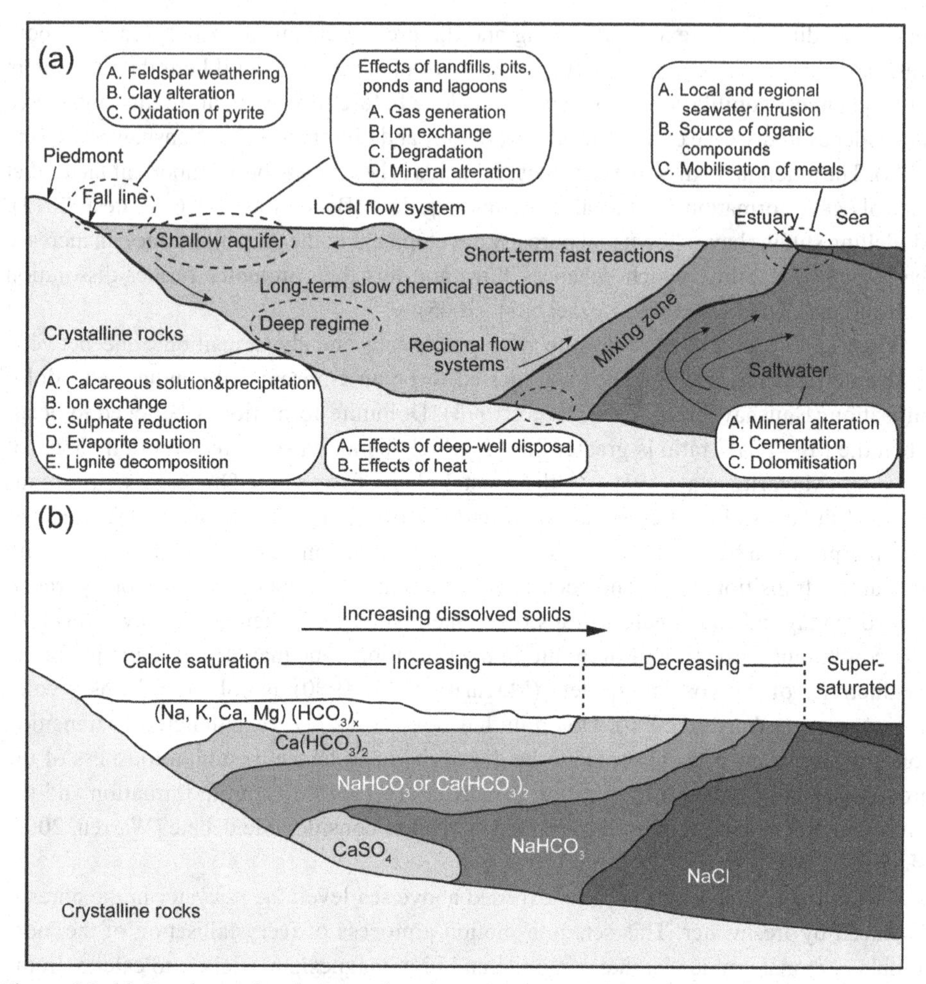

Figure 5.8 (a) Dominant chemical reactions in various parts of a low-lying coastal groundwater system in a temperate humid region. (b) Conceptual model of the chemical evolution of groundwater in a large coastal plain under natural conditions (Back et al., 1993). Graphs are not to scale, but the horizontal distance is about 100 km.

coastal groundwater systems more generally (Custodio and Bruggeman, 1987; Back et al., 1993; Winter, 1998).

The dominant chemical processes in coastal aquifer systems in temperate humid regions at a scale of tens of kilometres is shown in Figure 5.8a. In the recharge area, groundwater derives from rainfall and is therefore fresh. The pH depends strongly on the weatherable minerals in the soil, especially carbonates, but tends to be neutral to slightly acidic. The content of dissolved oxygen, derived from the atmosphere, is high unless reducing substances like organic matter or sulphide minerals are present. In the shallowest part of the system, the chemical composition can change significantly with time in response to

Table 5.3 *Important Sources and Sinks for Some Major Ions in Groundwater in Coastal Plain Aquifers*

Major ion	Primary sources	Primary sinks
Calcium (Ca^{2+})	Dissolution of calcite, dolomite and gypsum; weathering of plagioclase and other feldspar minerals; release by cation exchange during seawater intrusion	Precipitation of calcite; removal by cation exchange during aquifer freshening
Bicarbonate (HCO_3^-)	Biological activity in the soil zone; anaerobic decomposition of organic matter; oxidation of biogenic ethane	Precipitation of carbonate minerals
Sodium (Na^+)	Weathering of silicate minerals; release by cation exchange during freshening	Uptake by cation exchange during salinisation; halite precipitation (at very high concentrations)
Magnesium (Mg^{2+})	Weathering of minerals; alteration of clays; dissolution of dolomite and other Mg-containing carbonates; cation exchange	Dolomitisation; cation exchange
Sulphate (SO_4^{2-})	Oxidation of pyrite and other sulphides; dissolution of gypsum	Sulphate reduction followed by volatilisation or sulphide precipitation

Note. Based on Custodio and Bruggeman (1987) and Back et al. (1993). All solutes occur in low concentrations in rainfall. Seawater is high in Na^+, Mg^{2+} and SO_4^{2-} relative to freshwater and is therefore an important source of these solutes in mixtures. Chloride is not included, as it is chemically, inert unless evaporite minerals, such as halite, form or dissolve, but this occurs only at concentration levels several times that of seawater. Anthropogenic sources are not listed.

transient hydrological processes such as water table changes driven by rainfall recharge. As groundwater flows towards the coast, it interacts with the rock. There can be various chemical reactions that change its composition, the most important one being the dissolution of minerals in the rock. As a result, and depending on the minerals encountered, groundwater tends to become less acidic along its flow path. It is common for dissolved oxygen to become depleted in the presence of reductants (Stuyfzand, 1999).

The above discussion is very generic, and to a large degree, geology and hydrogeochemistry are very site specific (Back et al., 1993). A wide range of different processes can cause the chemical composition of groundwater to change along its flow path. A short summary of the main sinks and sources of major ions and the associated chemical processes is given in Table 5.3. In the parts of the system where the groundwater flow rates are low, such as in the deepest and least permeable parts of the system the groundwater's chemical composition only changes slowly with time as the migration of solutes is limited.

Despite the diversity of natural processes and reactions that influence groundwater chemical composition, they usually yield a foreseeable general pattern of chemical facies (Winter, 1998), as shown in Figure 5.8b. In the recharge areas, groundwater has low concentrations of dissolved solids. The chemical facies is dominated by Ca^{2+}, Mg^{2+}, Na^+ and HCO_3^-, with the relative importance of the cations varying depending on the mineralogy. However, near the coast, mixing of freshwater with seawater causes groundwater to have high concentrations of dissolved solids, and a chemical facies dominated by Na^+ and Cl^-. Details about how chemical facies are defined will be discussed in Section 5.5. For case studies on hydrochemical facies within coastal plain aquifer systems, the readers can read the classical work of Back (1960) and Hanshaw et al. (1967) or later works such as Stuyfzand (1993) or Renken (2005).

5.5 Chemical Data Analysis

There are many ways to present chemical data graphically and visually. The most popular methods are Piper, Stiff and scatter diagrams. It should be noted that one single method alone is often not sufficient to identify temporal and spatial characteristics of groundwater chemistry and multiple methods are recommended to achieve more conclusive results.

5.5.1 Piper Diagrams

The Piper diagram (or trilinear diagram) was first published by Piper (1944). The diagram consists of a central diamond-shaped field and two triangles, the left one for the cations Ca^{2+}, Mg^{2+} and ($Na^+ + K^+$), and the right one for the anions: ($CO_3^{2-} + HCO_3^-$), SO_4^{2-} and Cl^- (Figure 5.9). Other ions are neglected. To draw the data points in the diagram, the ion concentrations are first expressed in meq l^{-1} and the resultant values are added for the cations and the anions separately. These sums are then used to calculate the percentage contribution to the total of each ion. Points are first plotted in the triangular cation and anion diagrams, and then projected to the diamond by finding the intersection of the two lines parallel to the outer sides of the diagram. Figure 5.9 contains three data points for groundwater and one for a seawater sample from the Quaternary basal aquifer in the Pearl River Delta, China (Wang and Jiao, 2012) (Table 5.4).

Because the location of a water sample in the Piper diagram depends only on the relative proportion of the ions, it is possible that two samples with very different TDS plot close together, e.g. when rainwater has similar ionic ratios as seawater. The size or the colour of the data point can be made variable to reflect the TDS concentration of the samples (Domenico and Schwartz, 1990). Since the Piper diagram does not show all major ions it provides incomplete information when other ions are present at significant concentrations, such as for example in sample LL2 (Table 5.4), which has a NH_4^+ concentration higher than SO_4^{2-}.

Waters of similar origin usually plot together in the Piper diagram, so its major function is to graphically identify water types. After categorising samples using the Piper diagram, the

Table 5.4 *Chemical Composition of Water Samples from Boreholes LL2, LL4 and SD6 in the Quaternary Basal Aquifer and Seawater in the Pearl River Delta, China*

Parameter	LL2	LL4	Seawater	SD6
TDS	620	680	34 500	490
Na^+	63	99.3	10 500	24
Mg^{2+}	11.5	9.3	1350	16.2
K^+	6.5	6	390	8.4
Ca^{2+}	63.2	77.8	410	77.7
NH_4^+	16.1	1.2	0	0.8
Cl^-	48	98	19 000	5.8
SO_4^{2-}	9.8	18	2700	4.2
HCO_3^-	372.4	342.3	142	352.8
Charge balance (%)	2.3	3.6	2.9	4.4

Note. Concentrations are in mg l^{-1}.

sample groups can be classified according to location, geologic formation or water origin. Figure 5.10 shows the classification of hydrochemical facies based on a Piper diagram (Back, 1966). The seawater sample from Figure 5.9 is a sodium chloride-type water and the freshwater samples are of the calcium bicarbonate type. The boundaries between the fields shown in Figure 5.10 can be adopted for the specific conditions (Freeze and Cherry, 1979).

Changes in water chemistry in space reflected by samples taken along a flow path show up as a series of points along a trend line in the Piper diagram. Data points near the straight line (F-S) within the diamond (Figure 5.9) represent water samples which are a mixture between the seawater and freshwater (sample SD6) end-members. Usually, however there are additional processes like cation exchange, and this leads to curved paths as shown in Figure 5.9. The upper curved path (F-A-S) tracks the direction along which the water chemistry would evolve as freshwater is replaced by seawater at a particular location. This is because the Na^+ in seawater displaces Ca^{2+} adsorbed to sediment particles (Section 5.3.3), and thus the mixture becomes enriched in Ca^{2+} and depleted in Na^+. The data points cluster around the upward trajectory F-A. But as seawater intrusion progresses, and the exchangeable Ca^{2+} has been depleted, the increase of Na^+ is stronger than Ca^{2+}, and the water samples in the advanced seawater intrusion stage are clustered along or above the downward trajectory A-S. The above evolution path was discussed by Xue et al. (1993) using a case study in Laizhou Bay, China. Other case studies on the evolution path can be found in Jeen et al. (2001), Martens et al. (2011), Park et al. (2012b), and Vandenbohede and Lebbe (2012). The evolution paths will of course be more complicated if chemical reactions other than just ion exchange or multiple end-members are involved.

The cations Na^+ and Mg^{2+} have high concentrations in seawater and therefore their sorbed concentrations are high as well in a seawater-containing aquifer. When such an

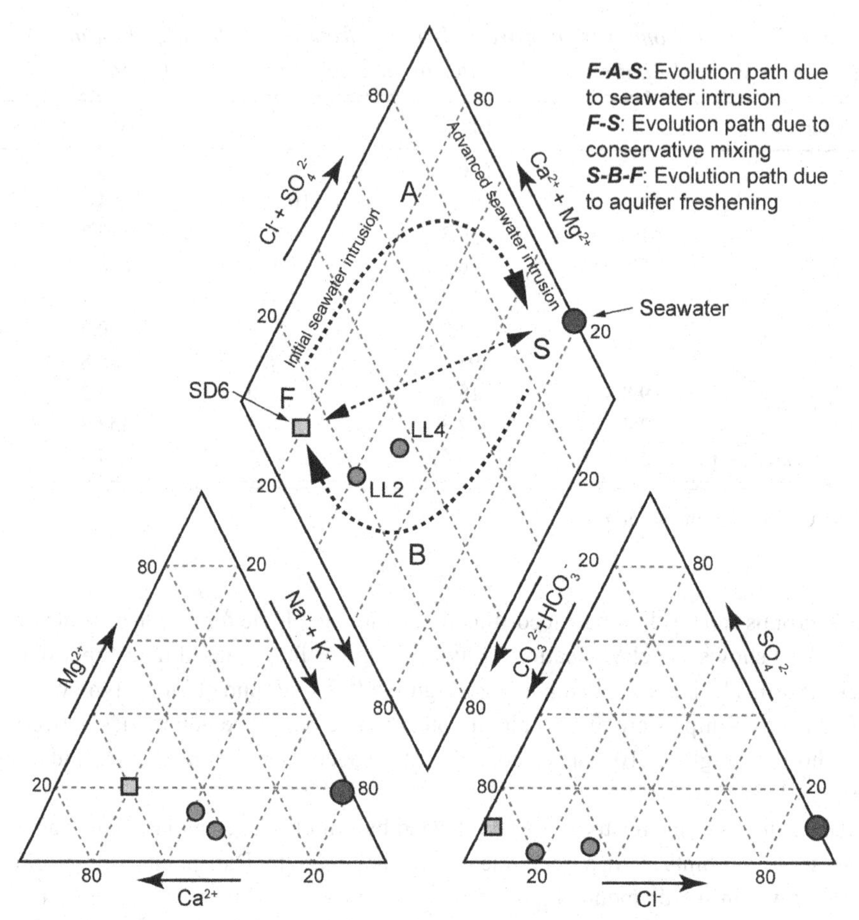

Figure 5.9 Piper diagram with data for fresh groundwater and seawater and the conservative mixing line (straight broken line) of freshwater (F) and seawater (S) end-members. The chemical evolution path for seawater intrusion (upper broken line) and the chemical evolution path for aquifer freshening (lower broken line) are shown. For concentrations, see Table 5.4.

aquifer experiences freshening, the chemical evolution follows the lower path (S-B-F) in Figure 5.9. Freshwater tends to be dominated by Ca^{2+}, which gets exchanged for Na^{+}, followed by Mg^{2+}. The two example data points in Figure 5.9 are near this freshening trajectory and indicate displacement of seawater intruded in the Holocene (Wang and Jiao, 2012).

5.5.2 Stiff Diagrams

A Stiff diagram visualises the chemical analyses and was named after its developer H. A. Stiff (1951). A typical Stiff diagram has three or four horizontal axes showing the major

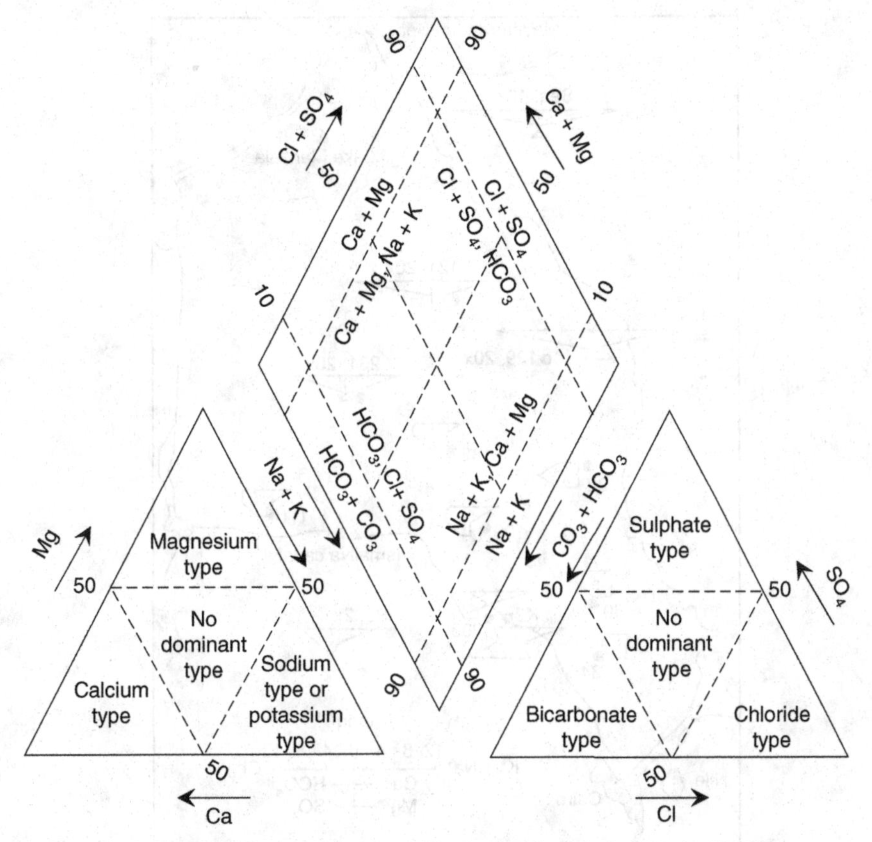

Figure 5.10 Piper diagram with classification of hydrochemical facies (Back, 1966).

ions in meq l^{-1} or sometimes mg l^{-1}, with cations displayed on the left and anions on the right (Figure 5.11). The points are linked to form a polygon, and as such each water type produces a distinct shape.

Appelo and Postma (2005) presented an example of how to use Stiff diagrams to study the spatial change of groundwater chemistry in the Nile delta, Egypt. Figure 5.11 shows the Stiff diagrams of groundwater, seawater (shown immediately below the sea) and fresh Nile water (shown in west of Cairo) samples. Boreholes 121 and 129 have excess of Ca^{2+} relative to seawater, indicating cation exchange as a result of seawater intrusion. Groundwater samples taken along the Ismailya canal that runs from Cairo to the Red Sea, have gradually rising chloride concentrations due to evapotranspiration of Nile water that is employed for irrigation (Appelo and Postma, 2005). A map like in this example can thus provide a good overview of the spatial changes in water composition. If there are too many water samples, it may be hard to show all the Stiff diagrams on a single figure, and if the salinities diverge widely, as typical in coastal areas, scaling is required, as shown by the scaling factor x in Figure 5.11.

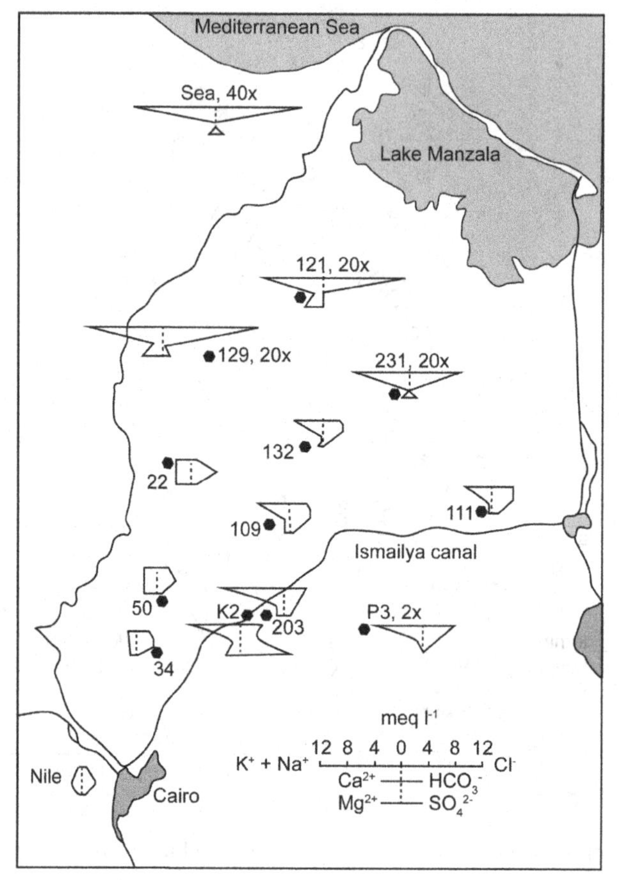

Figure 5.11 Map showing Stiff diagrams of water samples from the Nile Delta, Egypt (Appelo and Postma, 2005).

5.5.3 Scatter Diagrams and Mixing Lines

If seawater and fresh groundwater mix without chemical reactions involved, a theoretical concentration $m_{i,\mathrm{mix}}$ of an ion i in the mixture can be calculated as (Appelo and Postma, 2005):

$$m_{i,\mathrm{mix}} = f_{\mathrm{sea}} \times m_{i,\mathrm{sea}} + (1 - f_{\mathrm{sea}}) \times m_{i,\mathrm{fresh}} \tag{5.4}$$

where $m_{i,\mathrm{sea}}$ and $m_{i,\mathrm{fresh}}$ are the concentrations of ion i in the seawater and the original freshwater, respectively; and f_{sea} is the fraction of the seawater in the mixture. Cl^- is regarded as a conservative ion across a wide range of environments because it does not co-precipitate or exchange with other ions, and is therefore usually chosen as the independent variable to quantify mixing effects. Thus, if the chloride concentrations of the two end-members are known, the fraction of seawater in the sample can be estimated by solving Eqn (5.4) for f_{sea}:

$$f_{sea} = \frac{m_{Cl,sample} - m_{Cl,fresh}}{m_{Cl,sea} - m_{Cl,fresh}} \tag{5.5}$$

where the terms $m_{Cl,fresh}$, $m_{Cl,sea}$ and $m_{Cl,sample}$ represent respectively the chloride concentrations of the freshwater and seawater end-members and the sample, as indicated by the subscripts.

When Cl^- does not behave conservatively, for example where there are evaporite (e.g. halite – a NaCl salt) deposits that dissolve, other tracers may be used to calculate f_{sea}. For example, if stable water isotope data are available, they can perhaps also be used to quantify the mixing process. Sometimes SO_4^{2-} behaves conservatively (Wallis et al., 2011) and can substitute for Cl^- in that case.

Mixing is represented by a straight line in a scatter diagram with linear axes of an ion's concentration versus the Cl^- concentration. Such a line is referred to as the theoretical mixing line between fresh groundwater and seawater. For a linear plot, the mixing line is therefore easily found graphically by connecting the data points for the selected end-members. If most of the water samples have low concentrations, and the range of concentrations is large, log-log or semi-log plots may be better because most of the points may cluster near the origin if shown in a linear graph. When log-log or semi-log plots are used, the conservative mixing line can be obtained by following the three steps below, which are illustrated by using the concentration data from a coastal aquifer system in the Pearl River Delta, China (Wang, 2011):

Step 1: Identify end-members: a freshwater sample with Na^+ 0.9 mmol l^{-1} and Cl^- 0.3 mmol l^{-1} is selected as freshwater end-member. The seawater end-member has Na^+ 456.5 mmol l^{-1} and Cl^- 535.2 mmol l^{-1}.

Step 2: Eqn (5.4) is used to calculate theoretical mixing concentrations ($m_{i,mix}$). Table 5.5 shows the calculated theoretical mixing concentrations of Na^+ or Cl^-.

Step 3: Using the results in Table 5.5, the theoretical mixing lines are drawn in Figure 5.12, together with freshwater and seawater end-members.

After the theoretical mixing line is created, the chemical data of the actual water samples can be plotted on the same graph. Figure 5.13 shows the relation between Cl^- and Na^+ for groundwater samples in the coastal basal aquifer in the Pearl River Delta, China, which experienced palaeo-seawater intrusion (Wang and Jiao, 2012). The Na^+ of the most of the samples is at or close to the theoretical mixing line between fresh groundwater and seawater, suggesting that mixing of the two end-members is the dominant process that determines this ion's concentration.

The concentrations of the end-members, especially fresh groundwater, vary from site to site, and hydrogeological investigation is required to deduce the appropriate

Table 5.5 *Mixing Proportions of Freshwater and Seawater and the Corresponding Theoretical Mixing Concentrations of Na^+ and Cl^-*

Seawater proportion (f_{sea}) (%)	Freshwater proportion ($1 - f_{sea}$) (%)	Theoretical mixing concentration	
		Na^+ (mmol l^{-1})	Cl^- (mmol l^{-1})
0	100	0.9	0.3
20	80	92.0	107.3
40	60	183.1	214.3
60	40	274.3	321.2
80	20	365.4	428.2
100	0	456.5	535.2

Note. The end-member concentrations are $Na^+ = 456.5$ and $Cl^- = 535.2$ mmol l^{-1} for seawater and $Na^+ = 0.9$ and $Cl^- = 0.3$ mmol l^{-1} for freshwater.

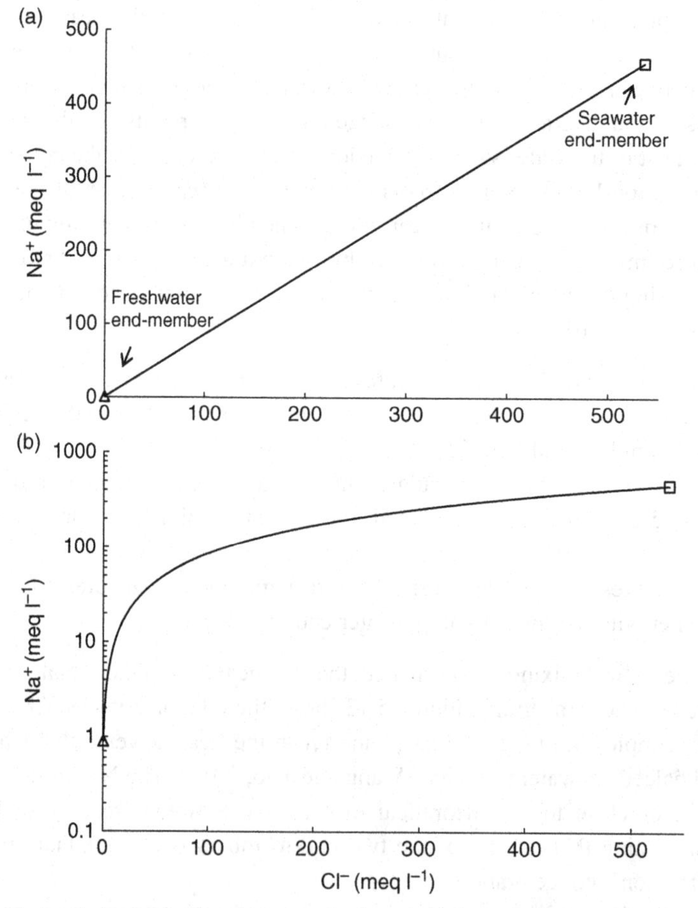

Figure 5.12 Theoretical mixing lines between freshwater and seawater for Na^+ versus Cl^- in a (a) linear and (b) semi-logarithmic coordinate system.

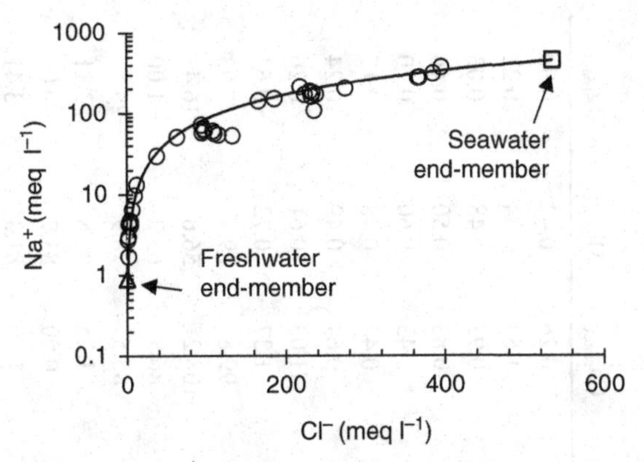

Figure 5.13 Relationship between Na^+ and Cl^- in the coastal aquifer in the Pearl River Delta, China. Seawater and freshwater end-members and their theoretical mixing line are provided for reference (Wang and Jiao, 2012).

end-members for a coastal aquifer system (Mazor, 2004). Usually a water sample in the recharge area of an aquifer system with a salinity negligible compared to the seawater is selected to represent the chemistry of freshwater end-member. In addition to seawater intrusion, saline groundwater can be also formed from return flows from irrigated land, dissolution of evaporite deposits, and evaporative concentration in lakes and lagoons (Daniele et al., 2011). If there are more than two end-members, data points tend to be clustered in an area enclosed by three or more lines in a scatter diagram.

Chemical reactions make that a sample's measured concentration will deviate from $m_{i,\text{mix}}$ as calculated from Eqn (5.4). The magnitude of the deviation is (Appelo and Postma, 2005)

$$m_{i,\text{react}} = m_{i,\text{sample}} - m_{i,\text{mix}} \tag{5.6}$$

where $m_{i,\text{sample}}$ is the measured concentration of the ion i in the water sample and the value of $m_{i,\text{react}}$ is the change in concentration due to chemical reactions. Sometimes $m_{i,\text{react}}$ is denoted as Δm_i and is called ionic delta (Pulido-Leboeuf, 2004).

Table 5.6 shows water samples of an aquifer in the coastal dunes of the Netherlands that was affected by seawater up-coning (Section 6.2.3) caused by over-extraction. Seawater moved up by circa 35 m at this location, displacing the ambient freshwater. The application of Eqns (5.4), (5.5) and (5.6) to infer the deviations from the mixing line is demonstrated in Table 5.6. The samples between 77.5 and 82.5 m below sea level are located with the transition zone. They show an excess of Ca^{2+} and a deficit of Na^+ and Mg^{2+} relative to the mixing line. In fact, when the concentrations are expressed as meq l^{-1}, the excess of Ca^{2+} is balanced by the summed deficit of Na^+ and Mg^{2+} (Figure 5.14), which is a prime example of cation exchange during seawater intrusion.

Table 5.6 *Water Sample Chemistry and the Estimated f_{sea}, $m_{i,mix}$ and $m_{i,react}$ of Na^+, Ca^{2+} and Mg^{2+} of a Multi-Level Observation Well in the Coastal Dunes of the Netherlands Affected by Seawater Up-Coning*

Elevation (amsl)	Na^+	Ca^{2+}	Mg^{2+}	Cl^-	f_{sea}	Na_{mix}	Na_{react}	Ca_{mix}	Ca_{react}	Mg_{mix}	Mg_{react}
−0.1	0.52	4.19	0.43	0.65	0	0.81	−0.29	3.93	0.26	0.43	0.00
−5.5	1.00	5.79	0.76	1.27	0.001	1.35	−0.35	3.95	1.84	0.54	0.22
−10	0.78	4.86	0.51	0.96	0.000	1.08	−0.30	3.94	0.92	0.48	0.03
−20	0.78	4.78	0.96	1.04	0.000	1.16	−0.37	3.94	0.83	0.50	0.46
−31	1.77	5.40	1.30	1.64	0.001	1.67	0.10	3.96	1.43	0.60	0.70
−40	0.70	4.38	0.66	0.96	0.000	1.08	−0.39	3.94	0.44	0.48	0.17
−50	0.65	4.58	0.73	0.99	0.000	1.11	−0.45	3.94	0.64	0.49	0.24
−60	0.58	5.00	0.87	1.72	0.001	1.74	−1.16	3.97	1.03	0.61	0.26
−67	0.81	5.36	1.36	2.31	0.003	2.25	−1.44	3.99	1.37	0.72	0.64
−77.5	36.3	109.7	37.3	169.2	0.347	146.6	−110.3	10.0	99.8	29.4	7.93
−79.5	194.4	123.9	40.1	327.2	0.674	283.2	−88.8	15.6	108.2	56.6	−16.4
−82.5	244.9	104.2	64.2	377.1	0.777	326.4	−81.5	17.4	86.8	65.2	−1.00
−84	331.2	105.7	49.4	456.9	0.942	395.4	−64.2	20.2	85.5	78.9	−29.5
−93	419.3	22.3	85.9	479.5	0.988	415.0	4.4	21.1	1.25	82.8	3.11
−102.5	419.8	21.3	83.7	485.1	1.000	419.8	0.0	21.3	0.00	83.7	0.00
−120	406.7	22.5	84.7	471.0	0.971	407.6	−0.9	20.8	1.70	81.3	3.41
Freshwater end-member	1.13	3.94	0.49	1.02							

Note. Data from Stuyfzand (1988). Concentrations are in meq l^{-1}. The freshwater end-member is in the last row; the saltwater end-member is the most saline sample at 102.5 m depth. amsl = above mean sea level.

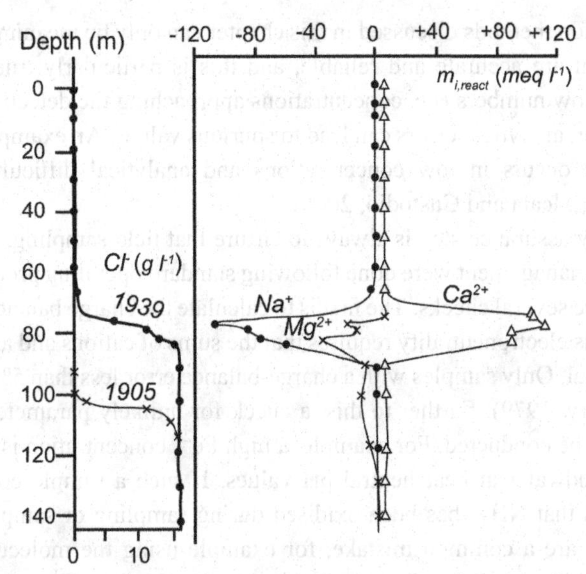

Figure 5.14 Chloride concentration and $m_{i,\mathrm{react}}$ values for Na^+, Ca^{2+} and Mg^{2+} for a borehole in the coastal dunes in the Netherlands (modified from Stuyfzand, 1993).

5.6 Chemical and Isotopic Indicators

The chemical composition of groundwater can be an indicator for seawater intrusion, because, apart from the obvious change in salinity, there are systematic changes in the chemical composition when seawater intrudes into freshwater aquifers. In areas with other potential sources of salt contamination, elevated salinity values in themselves are not conclusive proof of seawater intrusion. Chemical processes and salt origin can be inferred from concentration ratios such as Na^+/Cl^-, Ca^{2+}/Cl^-, Mg^{2+}/Ca^{2+}, Cl^-/Br^-, Ca^{2+}/Mg^{2+} and Cl^-/HCO_3^-. The stable isotopes of water are affected by mixing during seawater intrusion and can also be used to identify salinisation processes other than seawater intrusion.

For the convenience of discussion, the chemical and isotopic indicators will be introduced one by one in this section. However, it is unlikely that a single indicator alone can be used conclusively due to interactions of various chemical processes, site-dependent geochemistry, the range of ratios of different sources, or errors in the chemical analyses. Consequently meaningful conclusions can be made only by combining various indicators. Although chemical indicators can be a first-level screening tool for assessing a possible seawater influence, hypotheses regarding the origin of salinity and the evolution of groundwater need to be complemented by hydrogeological investigations of groundwater flow patterns, including changes thereof with time, to confirm the conceptual ideas (Davis et al., 1998). Reactive transport models in particular are an essential tool with which hypotheses can be tested against the laws of physics and chemistry (Lichtner et al., 1996).

The interpretation methods discussed in this chapter can only be meaningfully used with chemical data that are accurate and reliable, and this is particularly true for ion ratios. Especially when low numbers (i.e. concentrations approaching the detection limit) appear in the denominator, analytical errors can lead to spurious values. An example in this case is Br^-, which often occurs in low concentrations and analytical difficulties can impart significant errors (Alcala and Custodio, 2008).

The first quality assurance step is always to ensure that field sampling, sample storage, analysis and data management were done following standard operating procedures. Quality control can involve several checks. The first is to calculate the charge balance (see Table 1.3 for an example), as electroneutrality requires that the sums of cations and anions expressed in meq l^{-1} are equal. Only samples with a charge-balance error less than 5% are acceptable (Freeze and Cherry, 1979). Further to this, a check for unlikely parameter concentration combinations can be conducted. For example, a high Fe^{2+} concentration is not expected in an oxidised groundwater at near neutral pH values. If such a sample contains NO_3^-, it possibly indicates that NH_4^+ has been oxidised during sampling or sample storage. Unit conversion errors are a common mistake, for example using the molecular mass for N instead of NO_3^- when converting the nitrate concentration from mg l^{-1} to mmol l^{-1}. Finally, field parameters should be compared to those measured in the laboratory, and the theoretical value of the specific conductance can be calculated using geochemical codes like PHREEQC (Parkhurst and Appelo, 2013), which provides yet another quality control check.

5.6.1 Chloride, Salinity and Specific Conductance

Parameters reflecting the salinity such as chloride, TDS or specific conductance are a primary indicator of seawater intrusion. If seawater is the only source of groundwater chloride, this solute is a good indicator of seawater intrusion that can be analysed most easily and cost-effectively. A contour map of the chloride concentration can provide the spatial distribution of seawater intrusion. The baseline or natural background chloride concentration should be established by using historical data or wells not affected by seawater intrusion. If water samples in wells are taken periodically to analyse chloride concentrations, the increase in salinity over time indicates the rate of the seawater intrusion. Complementary analysis of the temporal and spatial trends of falling water levels (Figure 5.15) can provide additional confirmation of seawater intrusion processes (see also Figure 6.19).

For graphical depiction purposes, a chloride value much lower than seawater is used to indicate the advance of the seawater front into an aquifer. The specific value used depends on the ambient freshwater chloride concentration as well as the regulatory guidelines. For example, the chloride concentrations of fresh groundwater along the Atlantic coast of the USA are usually below about 20 mg l^{-1} (Barlow, 2003), so chloride concentrations > 100 mg l^{-1} are considered to be evidence of contamination with seawater (Sonenshein et al., 1997). Most studies, however, have adopted a chloride concentration between 200 and 500 mg l^{-1} for the seawater intrusion front (Xue et al., 1993; Kirsch, 2006). A map showing

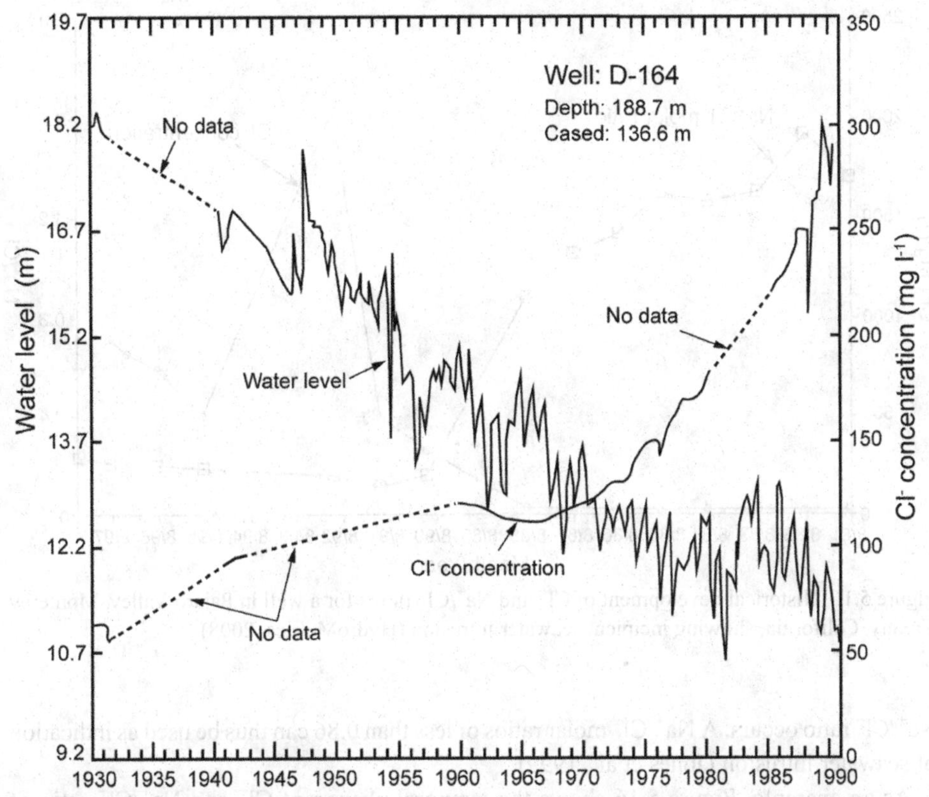

Figure 5.15 Change of the chloride concentration and water level with time in a well in the Floridan aquifer system in north-eastern Florida, indicating seawater intrusion (Spechler, 1994).

the contour lines of the same chloride concentrations at different times can be used to assess the trend of seawater intrusion.

In some studies, it is necessary to define not only the front of the seawater intrusion but also to characterise the transition zone. This zone is depicted most often by measuring the TDS or the groundwater chloride concentration sampled from monitoring wells. So far there is no well-accepted standard of practice regarding the definition of this zone. Some researchers use a TDS range from about 1000 to 35 000 mg l^{-1} or chloride from about 250 to 19 000 mg l^{-1} as indicators of this zone (Barlow, 2003).

5.6.2 Na^+/Cl^- Ratios

Standard seawater has a Na^+/Cl^- molar ratio of 0.86 (or mass ratio 0.56, Table 5.2). Fresh groundwater flowing in an active flow system usually has a Na^+/Cl^- value that is slightly higher (Vengosh and Rosenthal, 1994). During seawater intrusion Na^+ tends to become absorbed, and because Cl^- is not affected by the sorption process, a decrease of the

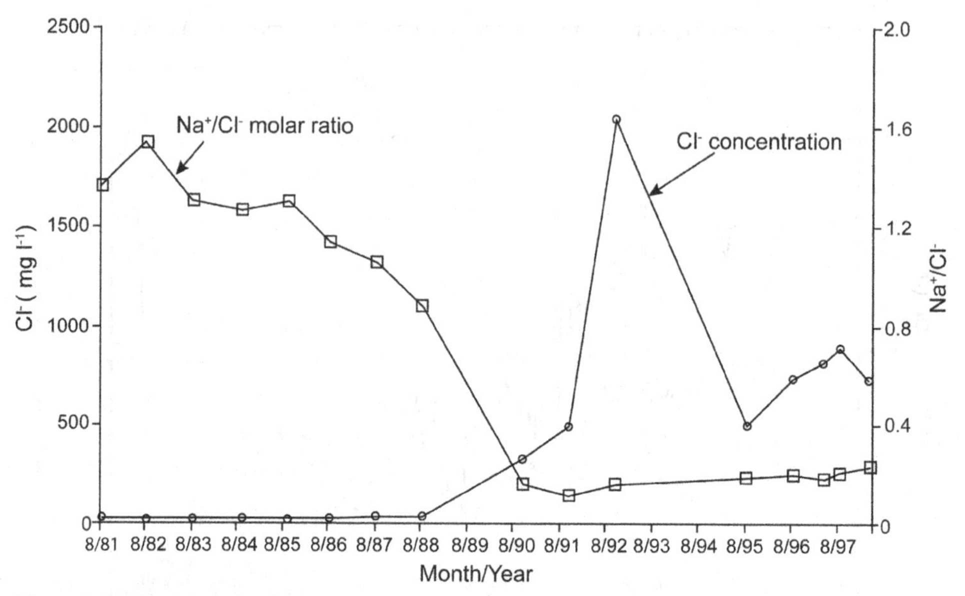

Figure 5.16 Historical development of Cl^- and Na^+/Cl^- ratios for a well in Pajaro Valley, Monterey County, California, showing incipient seawater intrusion (HydroMetrics, 2008).

Na^+/Cl^- ratio occurs. A Na^+/Cl^- molar ratios of less than 0.86 can thus be used as indication of seawater intrusion (Jones et al., 1999).

As an example, Figure 5.16 shows the temporal change of Cl^- and Na^+/Cl^- ratio of water samples from a well in the coastal area in Pajaro Valley, California, USA (HydroMetrics, 2008). Overall the Cl^- concentration and Na^+/Cl^- ratio are inversely correlated: the Cl^- concentration was rising slowly since 1982 and it took more than 6 years for it to exceed 250 mg l^{-1}, the drinking-water standard used in California. Na^+/Cl^- was over 1 before August 1987, but fell to less than 0.4 after August 1990. Despite the increase of the Cl^- concentration before August 1988 being minor, the Na^+/Cl^- ratios already decreased significantly, which shows that the Na^+/Cl^- ratio can be a more sensitive indicator than the Cl^- concentration during the early stage of seawater intrusion when salinity increases almost imperceptibly.

If freshwater intrudes into the aquifer originally filled with saline water, the reverse process occurs (Appelo et al., 1990). That is, the solid phase releases the sorbed Na^+ and takes up Ca^{2+}, which leads to a rise in the Na^+/Cl^- and a decrease in the Ca^{2+}/Cl^- ratio (Custodio and Bruggeman, 1987). The Ca^{2+}/Cl^- ratio may also be affected by the dissolution of calcite, triggered by the depletion of aqueous Ca^{2+} (Appelo and Postma, 2005), which masks the effect of cation exchange. Na^+/Cl^- ratios up to 6.54 caused by cation exchange were reported in a deltaic aquifer in south of Dakar, Senegal (Faye et al., 2005). Mg^{2+} is also displaced from the solid phase by Ca^{2+}, but later than Na^+, which gives rise to characteristic chromatographic patterns along a flowline (Beekman et al., 1991).

The ion exchange process and the consequent changes in the ion ratios are most pronounced in the vicinity of the moving front of the transition zone (cf. Figure 5.14). After one water has been replaced by another water for a long time and the exchange capacity is exhausted, no more exchange occurs, and then other factors and processes control the ionic ratios (Custodio and Bruggeman, 1987). Case studies on using ionic ratios for studying seawater intrusion and aquifer freshening can for example be found in Sanchez-Martos et al. (2002), Faye et al. (2005), Kass et al. (2005) and Coetsiers and Walraevens (2008).

It should be noted that the Na^+/Cl^- ratio in seawater in some coastal areas may be different from 0.86. For example, the Na^+/Cl^- ratio in Dead Seawater it is only 0.25 (Yechieli and Sivan, 2011). Sources of recharge of an aquifer greatly impact on groundwater Na^+/Cl^- ratios. Coastal rainwater has a high Na^+/Cl^- ratio due to the influence of marine aerosols. Rainwater chemistry can be complicated if the rain is mixed with air polluted by soil dust (Section 5.2.2). Groundwater contaminated by anthropogenic sources can have a Na^+/Cl^- ratio greater than 1 (Jones et al., 1999). Elevated Na^+/Cl^- ratios in groundwater can also indicate water-rock interactions. In granitic and alkaline volcanic areas freshwater may attain values from 1.5 to 3 or even greater due to weathering of Na-feldspar (Appelo and Postma, 2005; Leung et al., 2005). In the Millstone Grit aquifer system in UK which consists of sandstones, mudstones and shales, the Na^+/Cl^- ratio was found to be up to 13.7 (Abesser et al., 2005). Much lower values have been found on the other hand in the thermal brines in the coastal aquifer near the Dead Sea in Israel, which have groundwater Na^+/Cl^- ratios between 0.33 and 0.40 (Yechieli and Sivan, 2011).

5.6.3 Mg^{2+}/Ca^{2+} Ratios

In terrestrial fresh groundwater Ca^{2+} usually dominates over Mg^{2+}, so the Mg^{2+}/Ca^{2+} ratio is less than 1. The Mg^{2+}/Ca^{2+} ratio in groundwater in limestone aquifers is usually between 0.5 and 0.7 and in dolomitic aquifers between 0.7 and 0.9 (Rosenthal, 1987). Values of the ratio significantly higher than this range can be an indication of additional sources of Mg^{2+}, including seawater intrusion or weathering of rocks with Mg-containing minerals such as olivine. Dissolution of calcite and gypsum can lead to a Mg^{2+}/Ca^{2+} ratio lower than 0.5. Evaporation of seawater causes the Mg^{2+}/Ca^{2+} ratio to increase due to the removal of Ca^{2+} ions by the precipitation of gypsum. The Mg^{2+}/Ca^{2+} ratios can be lowered in brine involved in dolomitisation whereby Mg^{2+} is trapped in the precipitating dolomite and Ca^{2+} is released by calcite dissolution. The precipitation of gypsum that may accompany this process can limit the Ca^{2+} concentration, thereby modulating the Mg^{2+}/Ca^{2+} ratio.

Seawater is characterised by an excess of Mg^{2+} over Ca^{2+}, with a molar ratio Mg^{2+}/Ca^{2+} ranging from 4.5 to just over 5 (cf. Table 5.2). Seawater intrusion into freshwater aquifers accompanied by ion exchange makes that Na^+ from seawater replaces adsorbed Ca^{2+}. There is also ion exchange between Ca^{2+} and Mg^{2+}, leading to Mg^{2+} being absorbed and Ca^{2+} being released into the water. As a result, in the aquifer intruded by seawater, the Mg^{2+}/Ca^{2+} ratio is smaller than seawater but greater than freshwater (Custodio and Bruggeman, 1987).

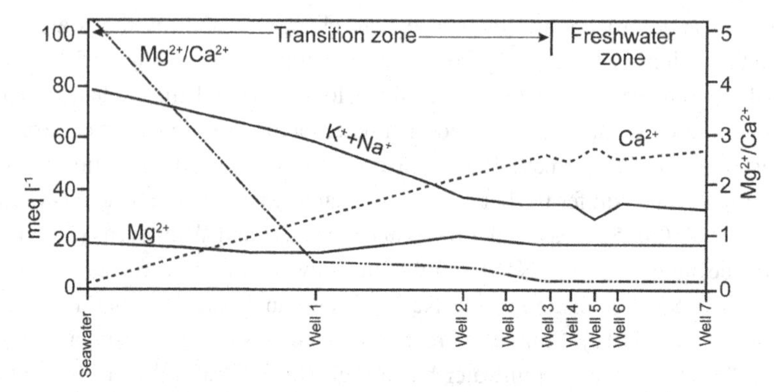

Figure 5.17 Hydrochemical section across the transition zone in a coastal area of Laizhou Bay, China (Xue et al., 1993). The labels on the horizontal axis indicate that one seawater and seven well samples were taken for analysis. The right axis is the molar ratio of Mg^{2+}/Ca^{2+}.

As an example, Figure 5.17 shows the spatial changes of ions along a section passing through the seawater, transition and freshwater zones in a coastal aquifer near Laizhou Bay, China (Xue et al., 1993). In the transition zone, going from seawater to freshwater, Ca^{2+} increases as (Na^{+} + K^{+}) decreases due to ion exchange. Ca^{2+} approaches its maximum roughly near the border of transition and freshwater zones, suggesting that Ca^{2+} was most enriched near the freshwater and saltwater front. Mg^{2+} decreases landward. The ratio of Mg^{2+}/Ca^{2+} was about 5 in seawater and decreases landward, but then became almost stable. Inside the freshwater zone, the spatial variation of the ion concentrations is relatively small.

Figure 5.18 shows Mg^{2+}/Ca^{2+} ratio versus the percentage of seawater in the freshwater–saltwater contact zone of the carbonate aquifer in Castell de Ferro, Spain (Pulido-Leboeuf, 2004). The percentage of seawater is estimated based on the Cl^{-} concentration in the water samples (Eqn (5.5)). The Mg^{2+}/Ca^{2+} ratio increases from about 0.5 in freshwater to about 4 when the percentage of seawater is over 90%. The Mg^{2+}/Ca^{2+} ratio for the seawater at this coast is 4.6. Compared to the conservative theoretical mixing line (not shown here), quite a few samples have excess Ca^{2+} and deficit Mg^{2+}, as calculated from Eqn (5.6). Pulido-Leboeuf (2004) believed that the low Mg^{2+} was partially caused by cation exchange or dolomitisation. However, dolomitisation is a sluggish process, and significant changes due to this process would only be manifest if the residence time of the water in the transition zone is sufficiently long.

Table 5.7 shows the changes in ionic ratios due to ion exchange relative to the position of intruding water (Richter and Kreitler, 1991), assuming there is no more ion exchange in the areas behind the advancing saltwater and freshwater fronts and Cl^{-} is not affected by ion exchange. This shows how Na^{+}/Cl^{-} and Mg^{2+}/Ca^{2+} ratios can be used as potential indicators of seawater or freshwater intrusion.

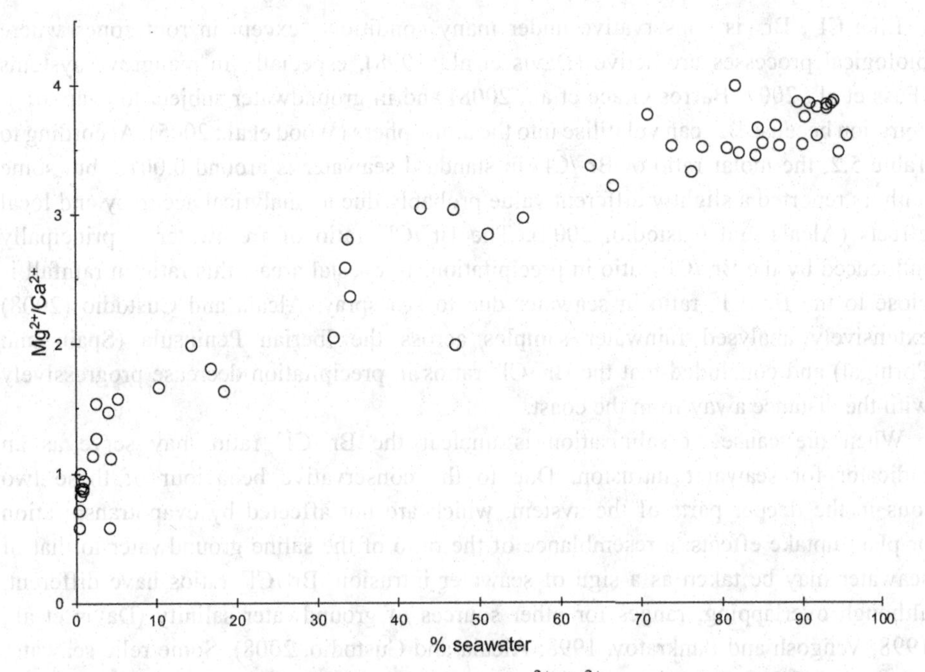

Figure 5.18 Percentage seawater versus the molar Mg^{2+}/Ca^{2+} ratio in the transition zone of a carbonate aquifer (Pulido-Leboeuf, 2004).

Table 5.7 *Changes in Ionic Ratios due to Ion Exchange Relative to the Position of Intruding Water*

	Recent saltwater intrusion		
Ion ratio	Advancing front	Behind advancing front	Freshwater intrusion
Ca^{2+}/Na^{+}	Increase	no change	decrease
$\Delta(Ca^{2+} + Mg^{2+})/\Delta Na^{+}$	Constant	no change	constant
Na^{+}/Cl^{-}	Decrease	no change	increase

Note. Data are from Richter and Kreitler (1991).

5.6.4 Br^{-}/Cl^{-} Ratios

The Br^{-} ion is one of the minor ions and can be difficult to analyse accurately in the presence of chloride. In freshwater the Br^{-} concentration usually is <1 mg l^{-1}, and often <10 µg l^{-1}. The low concentration of Br^{-} in water samples requires that analytical equipment should have high sensitivity (Alcala and Custodio, 2008). In seawater, Br^{-} is usually greater than 55 mg l^{-1} (cf. Table 5.1). When seawater intrudes into an aquifer, dissolved Br^{-} concentrations will increase significantly and this in itself can be an indicator of seawater intrusion, although it is more customary to use Cl^{-}.

Like Cl^-, Br^- is conservative under many conditions, except in root zones where biological processes are active (Davis et al., 1998), especially in mangrove systems (Fass et al., 2007; Barros Grace et al., 2008) and in groundwater subject to evapotranspiration because Br^- can volatilise into the atmosphere (Wood et al., 2005). According to Table 5.2, the molar ratio of Br^-/Cl^- in standard seawater is around 0.0015, but some authors reported a slightly different value probably due to analytical accuracy and local effects (Alcala and Custodio, 2008). The Br^-/Cl^- ratio of freshwater is principally influenced by the Br^-/Cl^- ratio in precipitation. In coastal areas, this ratio in rainfall is close to the Br^-/Cl^- ratio in seawater due to sea spray. Alcala and Custodio (2008) extensively analysed rainwater samples across the Iberian Peninsula (Spain and Portugal) and concluded that the Br^-/Cl^- ratios in precipitation decrease progressively with the distance away from the coast.

When the cause of salinisation is unclear the Br^-/Cl^- ratio may serve as an indicator for seawater intrusion. Due to the conservative behaviour of these two ions in the deeper parts of the system, which are not affected by evapotranspiration or plant uptake effects, a resemblance of the ratio of the saline groundwater to that of seawater may be taken as a sign of seawater intrusion. Br^-/Cl^- ratios have different, although overlapping, ranges for other sources of groundwater salinity (Davis et al., 1998; Vengosh and Pankratov, 1998; Alcala and Custodio, 2008). Some relic seawater has a Br^-/Cl^- ratio higher than the modern seawater Br^-/Cl^- ratio of 0.0015. Anthropogenic sources such as sewage effluents have Br^-/Cl^- ratios as low as 0.00055, while groundwater in which evaporites dissolved have Br^-/Cl^- ratios $<$ 0.00044 (Jones et al., 1999).

5.6.5 Cl^-/HCO_3^- Ratios

Compared to the other major ions, HCO_3^- has a relatively low concentration in seawater and its concentration in groundwater is often higher than in seawater (Table 5.1). The Cl^-/HCO_3^- molar ratio is much less than 2, typically less than 1, in inland freshwater, as opposed to the seawater value of over 300 (Table 5.2). Some researchers have used Cl^-/HCO_3^- ratios to classify groundwater by their degree of contamination by seawater. For example, Gimenez and Morell (1997), Ferrara and Pappalardo (2003) and Jeevanandam et al. (2007) drew contour maps of this ratio to depict the regions with and without marine intrusion. Muhammad and Husam (2011) divided the aquifer into areas strongly affected ($Cl^-/HCO_3^- > 6.6$), slightly and moderately affected (Cl^-/HCO_3^- between 0.5 and 6.6) and not affected ($Cl^-/HCO_3^- < 0.5$) by seawater intrusion. From an overall fall of TDS concentrations and Cl^-/HCO_3^- ratios of samples taken in Potharlanka Island, India between 2001 and 2006, Mondal et al. (2008) inferred that aquifer contamination by seawater had decreased (Figure 5.19). It should be noted though that, apart from seawater intrusion, other reasons such as carbonate mineral dissolution and precipitation or organic matter degradation can also lead to a change in the Cl^-/HCO_3^- ratio (Revelle, 1941). This ratio should therefore be used with extreme caution.

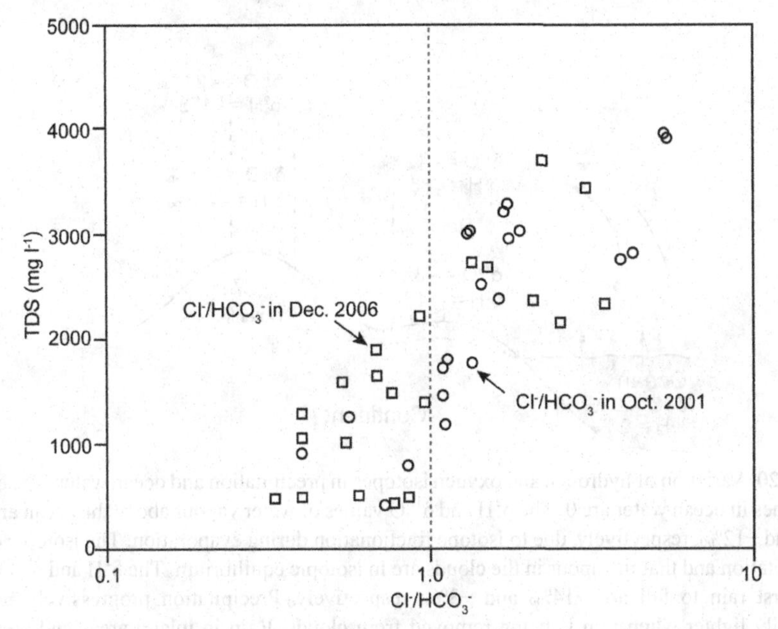

Figure 5.19 TDS versus the molar Cl^-/HCO_3^- ratio for groundwater sampled in Potharlanka Island, India, in 2001 and 2006 (Mondal et al., 2008).

5.6.6 Stable Water Isotopes

Elements with the same number of protons, but different numbers of neutrons in their nucleus are called isotopes. Environmental isotopes are those that are naturally occurring. The water molecule is made up by different combinations of three hydrogen isotopes, ^{1}H, ^{2}H (deuterium) or ^{3}H (tritium, which is unstable), as well as three oxygen isotopes: ^{16}O, ^{17}O or ^{18}O. They can combine into nine different water molecules (Fetter, 2001). The superscript numbers prefixed to the element symbol indicate the sum of neutrons and protons. For hydrogen, ^{1}H is much more abundant than ^{2}H or ^{3}H (the latter being subject to radioactive decay), and for oxygen, ^{16}O is far more abundant than ^{17}O and ^{18}O. The abundance of stable isotopes is measured with measurement techniques such as mass spectrometry and cavity ring-down spectroscopy. The most common way to report the measurement result is by means of the ratio of the rare to the more abundant isotope, normalised by the isotopic ratio of a standard, which for the stable isotopes of water is Vienna standard mean ocean water (V-SMOW). The δ-value of a sample based on this ratio is defined as

$$\delta_{\text{sample}} = \frac{R_{\text{sample}} - R_{\text{standard}}}{R_{\text{standard}}} \times 1000‰ \qquad (5.7)$$

where R_{sample} and R_{standard} are the abundance ratios of the water sample and standard seawater, respectively. Because δ-values are numerically small, they are multiplied by

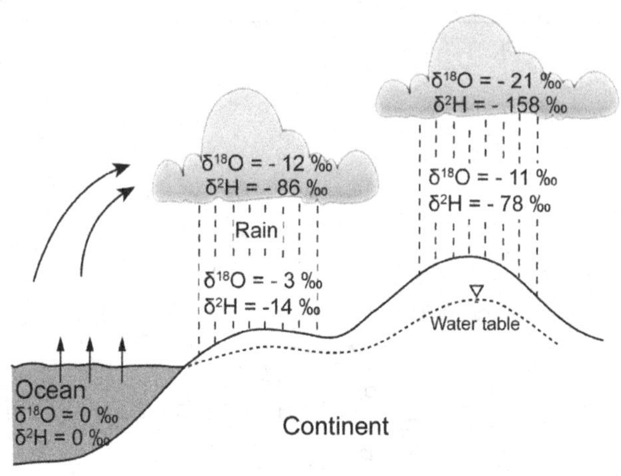

Figure 5.20 Variation of hydrogen and oxygen isotopes in precipitation and ocean water. The δ^2H and $\delta^{18}O$ values in ocean water are 0. The δ^2H and $\delta^{18}O$ values of water vapour above the ocean are about −86‰ and −12‰, respectively, due to isotope fractionation during evaporation. The isotopic content of precipitation and that in vapour in the clouds are in isotopic equilibrium. The δ^2H and $\delta^{18}O$ values of the first rain to fall are −14‰ and −3‰, respectively. Precipitation progressively becomes isotopically lighter when rain is being removed from clouds. Rain in inland areas and areas with high mountains is depleted relative to rain in the coastal plain (Coplen, 1993).

1000, hence the ‰ sign must be included in the notation. δ^2H and $\delta^{18}O$ are widely used to study groundwater salinity origins (Herczeg et al., 2001; Kim et al., 2003). Standard mean ocean water has $\delta^{18}O = \delta^2H = 0$, but in areas where there is significant freshwater runoff, the local seawater may be more depleted (Stuyfzand, 1993; Han et al., 2012). In warm areas with strong seawater evaporation, ^2H and ^{18}O may be enriched relative to mean ocean water (Corlis et al., 2003).

Mass differences between isotopes of the same element make that the light and heavy isotopes behave differently during processes such as evaporation. When water evaporates, it is easier for lighter water molecules than the heavier ones to enter the gas phase because for the latter more energy is required to break the hydrogen bonds in the liquid phase. The vapour that forms is depleted in the heavier isotopes compared to the original water, i.e. the abundance ratio and thus the δ-value decreases. When seawater with δ^2H and $\delta^{18}O = 0$ evaporates, the δ-values of the water vapour are negative (Figure 5.20). A reverse process occurs when clouds form by condensation. The heavier molecules condense first, so the rain is isotopically enriched compared to the water vapour. Because the fractionation effect is temperature dependent, and condensation occurs at lower temperatures than evaporation, the rain that forms remains isotopically lighter than the original seawater (Figure 5.20).

The worldwide correlation between δ^2H and $\delta^{18}O$ in rainwater obeys a linear relationship ($\delta^2H = 8\,\delta^{18}O + 10‰$), which is called the global meteoric water line (Craig, 1961). A local meteoric water line (LMWL) can also be determined for a specific area. If the data points of

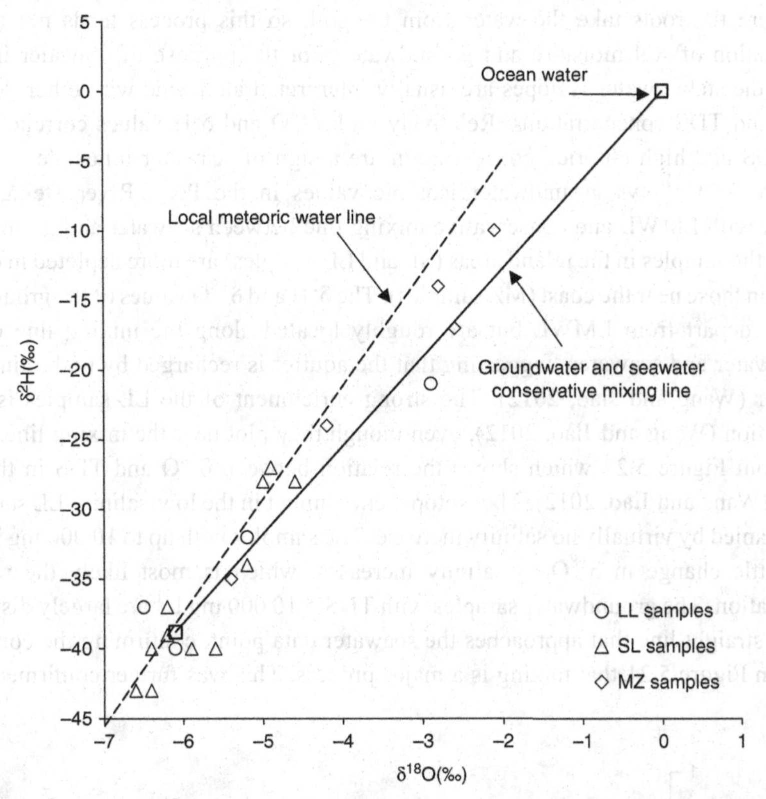

Figure 5.21 δ^2H versus $\delta^{18}O$ values of groundwater samples from the Pearl River Delta, China, compared to the local meteoric water line and conservative mixing line of freshwater and seawater (Wang and Jiao, 2012). The locations of the groundwater samples can be found in Figure 8.17.

groundwater samples are close to the LMWL in a plot of δ^2H versus $\delta^{18}O$ (Figure 5.21), then this forms strong evidence that the source of groundwater is precipitation.

The isotopic composition of the recharge source controls that of groundwater. If the recharge is dominated by precipitation infiltration, fresh groundwater has the isotopic composition very similar to that of precipitation. Seasonal fluctuations in isotopic composition of the rainwater are attenuated during transport in the unsaturated zone, so the isotopic composition of groundwater represents some effective average of the isotopic composition of the local precipitation. Evaporation leads to isotopic enrichment of the remaining water, so when soil water evaporates, the heavier isotopes increase relative to the lighter ones (i.e. increasing δ^2H and $\delta^{18}O$ values) in the soil moisture. Typically, the data points of samples affected by evaporation plot along a line with a slope lower than the meteoric water line in a plot of δ^2H versus $\delta^{18}O$. In the case of coastal aquifers, the sample points may then straddle along the mixing line between freshwater and seawater, which makes it hard to distinguish between the processes. With transpiration by plants, the fractionation occurs in the leaves, and not where the roots take the water from the soil, so this process tends not to cause fractionation of soil moisture and groundwater. For the purpose of seawater intrusion studies the stable water isotopes are usually interpreted alongside with other data such as Cl^- and TDS concentrations. Relatively high $\delta^{18}O$ and δ^2H values correlating with high TDS and high chloride concentration are a sign of seawater influence.

Figure 5.21 shows groundwater isotopic values in the Pearl River Delta, China, together with LMWL and conservative mixing line between seawater and groundwater. Overall the samples in the inland areas (SL and LL samples) are more depleted in δ^2H and $\delta^{18}O$ than those near the coast (MZ samples). The δ^2H and $\delta^{18}O$ values of the groundwater samples depart from LMWL but are roughly located along the mixing line of fresh groundwater and seawater, suggesting that the aquifer is recharged by both rainfall and seawater (Wang and Jiao, 2012). The strong enrichment of the LL samples is due to evaporation (Wang and Jiao, 2012), even though they plot near the mixing line. This is clear from Figure 5.22 which shows the relation between $\delta^{18}O$ and TDS in the same system (Wang and Jiao, 2012). The isotopic enrichment in the low-salinity LL samples is accompanied by virtually no salinity increase. The samples with up to 10 000 mg l^{-1} TDS show little change in $\delta^{18}O$ as salinity increases, which is most likely the result of transpiration. The groundwater samples with TDS > 10 000 mg l^{-1} are largely distributed along a straight line that approaches the seawater data point, confirming the conclusion based on Figure 5.21 that mixing is a major process. This was further confirmed by the

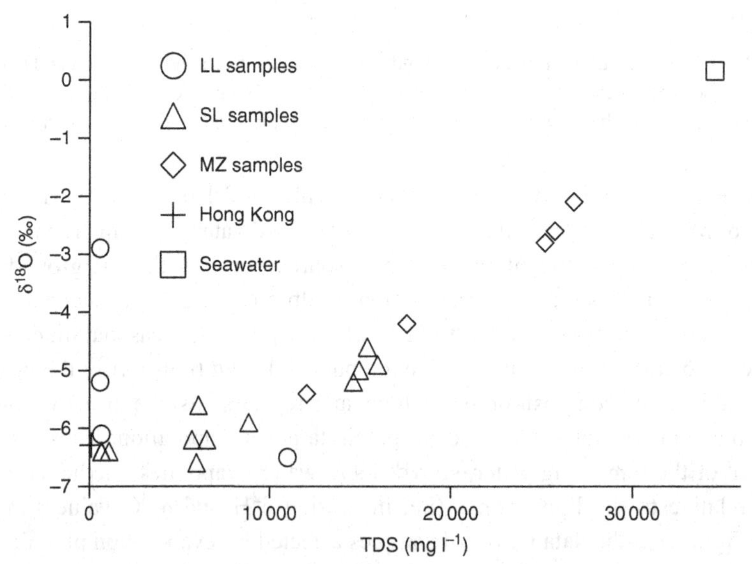

Figure 5.22 $\delta^{18}O$ values versus TDS of samples from the Pearl River Delta aquifer system. The $\delta^{18}O$ values of the seawater in the South China Sea and rainwater of Hong Kong were added as end-members for reference. LL, SL and MZ stand for groundwater samples in different parts of the aquifer system (Wang and Jiao, 2012). The locations of the groundwater samples can be found in Figure 8.17.

geographic location of the groundwater samples as the ones closer to the coast (labelled as MZ samples in the graph) had a comparatively heavier isotope signature as well as greater TDS concentrations.

5.6.7 Age Tracers

Knowing the age of groundwater types is also very important in order to be able to establish the salinity provenance and the timing of seawater intrusion. Radioactive isotopes like tritium (^3H) and radiocarbon (^{14}C) are useful tracers in that regard. Tritium can be used to identify a contribution of young water because it has a half-life of 12.32 years. When its daughter product ^3He is measured at the same time an age can be calculated based on the $^3He/^3H$ ratio (Solomon and Cook, 2000), which also provides constraints on groundwater recharge rates. Tritium is naturally produced in the upper layers of the atmosphere but was also released by nuclear bomb testing in the 1950s and early 1960's (Eyrolle et al., 2018).

Tritium has been detected in intruded seawater in coastal aquifers around the world. Stuyfzand et al. (2012) were able to pinpoint the ^3H values of up to ~30 TU in saline groundwater to intrusion of ^3H-rich seawater. Sivan et al. (2005) detected ^3H in transition zone groundwater in the coastal plain aquifer of Israel and speculated that old, tritium-free seawater could have obtained tritium by mixing with young, tritium-containing freshwater, especially if the freshwater had been impacted by the bomb peak. Han et al. (2012) found ^3H in brackish groundwater in an aquifer with intensive abstraction and explained this by mixing of young groundwater with old, ^3H-free saltwater from greater depth that is being drawn up from greater depth due to the influence of the pumping wells.

Radiocarbon has a half-life of 5730 years and can be used to infer the age of old groundwater up to an age of about 25 000 years. There are many well-known problems with the interpretation of ^{14}C ages, which stem from the reaction of groundwater with old subsurface carbon sources that are radiocarbon free (Sivan et al., 2005; Yechieli et al., 2009a). When carbon sources still containing ^{14}C are adding carbon to the groundwater, like Holocene shell fragments or peat layers, the uncertainties of the calculated ^{14}C ages becomes too large to still be useful (Post, 2004). Another problem in coastal areas is that the mixing of freshwater and seawater complicates the interpretation (Voss and Wood, 1994) and since seawater can have a wide range of ages (Post et al., 2013), there is no single seawater end-member that can be used in calculations of the type presented in Section 5.5.3 to correct for mixing effects.

These limitations leave coastal hydrogeologists with few age tracers for establishing the age of saltwater and determine rates of seawater intrusion. Iodide enrichment has been linked to groundwater age in some settings (Lloyd et al., 1982), but this tracer's behaviour is site specific so that the method only works under certain conditions. There therefore remains a need for new tracers and improved techniques to better constrain ages of coastal groundwater bodies.

5.7 Selected Coastal Aquifer Hydrochemistry Topics

5.7.1 Permeability Changes in Coastal Aquifers

Due to their negative electrical charge, clay particles repel each other and can thus form a face-to-face orientation, leading to a dispersed structure (Figure 5.23a). When cations in the diffuse double layer (Section 5.3.3) balance the negative charge, the repellence decreases and the particles may form a edge-to-face or edge-to-edge orientation, or the clay flocculates (Figure 5.23b) (Craig, 2004). In soils with a high clay content, flocculated clay structures form aggregates that apart from the clay itself may also consist of silt and sand. The hydraulic conductivity of the soil is relatively high because water can easily flow through the large pores between the aggregates. If the repellence between the particles increases, however, the clay particles start to disperse, and the soil's hydraulic conductivity reduces because pore space is reduced. At the same time, clay particles can plug pores between the aggregates when they migrate with the flowing water, thereby adding to the hydraulic conductivity decrease.

One important controlling factor of flocculation behaviour is the total dissolved cation or salt concentration in the soil water. When the amount of dissolved salts in the water increases, soil particles tend to flocculate. This is because the width of the diffuse layer (Figure 5.6) becomes smaller and thus the negative potential reaches less far into the water-filled pore space between the clay particles. But the type of cations present also plays an important role. The major cations in groundwater can be ranked on the basis of their ability to form flocculated structures: $Ca^{2+} > Mg^{2+} > K^{+} > Na^{+}$ (Sumner and Naidu, 1998). Thus, soil particles flocculate if concentrations of ($Ca^{2+} + Mg^{2+}$) increase, or disperse if their concentration decreases relative to the concentration of Na^{+}. This can be expressed by the sodium adsorption ratio (SAR), defined as $SAR = Na^{+}/(Ca^{2+} + Mg^{2+})^{1/2}$ (concentrations expressed in mmol l^{-1}). Figure 5.24 shows the relation among hydraulic conductivity, salt concentration and SAR for a particular soil. Ion exchange in clay-rich materials may significantly change the permeability of the materials (Appelo and Postma, 2005). The

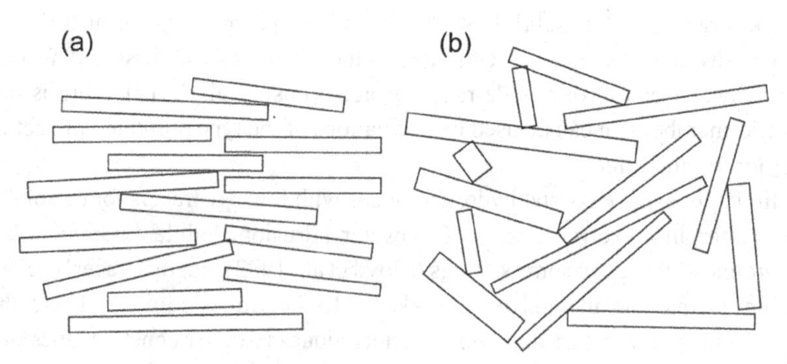

Figure 5.23 (a) Dispersed and (b) flocculated clay structure.

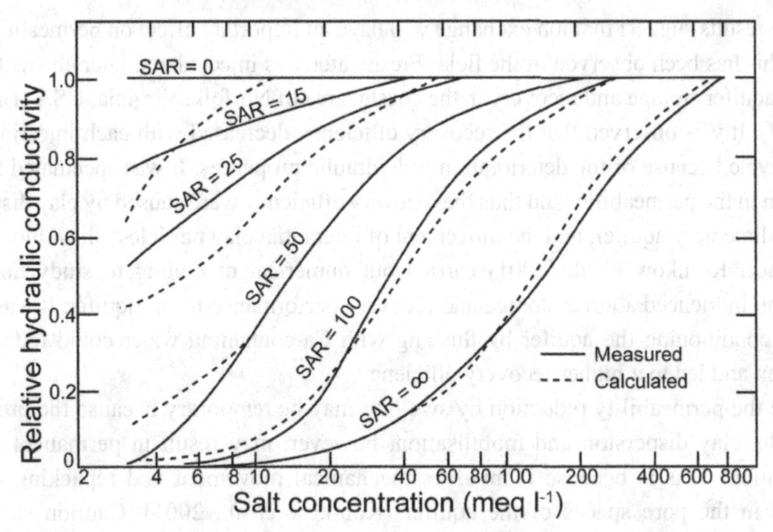

Figure 5.24 Change of relative hydraulic conductivity of a soil versus salt concentration for different sodium adsorption ratios (McNeal, 1968).

replacement of monovalent cations by higher order valency cations can change the particulate arrangement from dispersed to flocculation (Rao and Mathew, 1995).

The above discussion implies that the physical structure of clay-rich layers and consequently their hydraulic conductivity change during the process of seawater intrusion or aquifer freshening. In a soil saturated by seawater, clay particles are flocculated into aggregates, irrespective of the high Na^+ content (Van Weert et al., 2009). This enhances soil aeration, as well as root growth and penetration. If the saline water is displaced by freshwater, the concentration of cations, which bind clay particles together, is significantly reduced. The clay particles are then separated and disperse into the water because of the reduction in the attractive forces between them. This causes the clay particles to swell and the double layer to expand, which eventually decrease porosity and the hydraulic conductivity. Goldenberg (1985) carried out column experiments with dune sand mixed with smectite clay and concluded that the hydraulic conductivity could be decreased by 2 to 3 orders of magnitude when seawater was displaced by freshwater probably due to deflocculation of clay minerals. Rao and Mathew (1995) carried out a laboratory study of the impact of cations on the hydraulic conductivity of a marine clay. Their study demonstrated that the valency and hydration radius of the adsorbed cations have a substantial impact on the clay hydraulic conductivity. They used six cations in their study, and showed that Al-clay was in a flocculated state whereas Na-clay in a dispersed state, with the states for other cations being in between. Sodium saturated clay was approximately six times less permeable than potassium and ammonium clays. Saturation of marine clay with divalent cations increased its hydraulic conductivity nine times in comparison with the sodium clays (Rao and Mathew, 1995).

These results suggest that ion exchange will have an important effect on permeability, and indeed this has been observed in the field. Freshwater was injected and later abstracted as a form of aquifer storage and recovery in the coastal area of Norfolk, Virginia, USA (Brown et al., 1977). It was observed that the recovery efficiency decreased with each injection/withdrawal cycle because of the deterioration of hydraulic properties. It was speculated that the reduction in the permeability and thus the recovery efficiency were caused by clay dispersion in the sedimentary aquifer, and the movement of interstitial clay particles which blocked the pore space. Konikow et al. (2001) carried out numerical modelling to study how clay dispersion influenced aquifer storage and recovery performance in the aquifer. It was found that pre-conditioning the aquifer by flushing with Ca-containing water could reduce clay dispersion and led to a higher recovery efficiency.

While the permeability reduction by swelling may be temporary because the process is reversible, clay dispersion and mobilisation, however, may result in permanent aquifer deterioration. This is because it involves mechanical movement and repacking of clay particles in the pore spaces of the aquifer (Konikow et al., 2001). Caution should be exercised when designing column tests to study the impact of ion exchange or solution electrolytes on hydraulic conductivity. Any factors such as gas bubbles in water, microbial growth, and water evaporation may overshadow the impact of the concentration of ions and their exchange on hydraulic conductivity. Moreover, when conducting permeability tests in the laboratory using field samples, care must be taken that the displacement of the ambient water from the sample by a flushing solution does not lead to changes in the physical structure that is not representative for the in situ conditions in the field.

5.7.2 Decomposition of Organic Matter

Many coastal plain aquifer systems consist of unconsolidated sedimentary materials formed in riverine, estuarine, deltaic or marine environments. Such strata are often dominated by fine materials such as silt and clay. Clay-rich aquitards tend to become thicker and more spatially continuous towards the coast (McFarland et al., 2006; Jiao et al., 2010) and are rich in organic matter, including terrestrial plant fragments, algae and remains of various organisms (Chapelle and McMahon, 1991; Zong et al., 2006).

The sedimentary organic matter in aquifer materials is usually low, ranging from 0.01 to 0.2 wt% (Tambach et al., 2009). For example, McMahon et al. (1990) reported that the organic matter comprised about 1% of fine-grained confining unit materials versus only 0.2% of the sandy aquifer materials in the coastal plain aquifer system in South Carolina. Figure 5.25 shows the changes of total organic carbon (TOC), total nitrogen (TN) and particle size with depth in the aquitard sediments in the Pearl River Delta, China (Jiao et al., 2010). Note that in this figure the grain size is expressed in ϕ scale (Krumbein and Aberdeen, 1937). Both TN and TOC vary with depth and are always greater than 1 g kg^{-1} and 10 g kg^{-1} (or 0.1% and 1%), respectively, at the depth interval between 15 and 30 m. Particle sizes in the same interval are finer than in other parts of the core, indicating that fine-grained

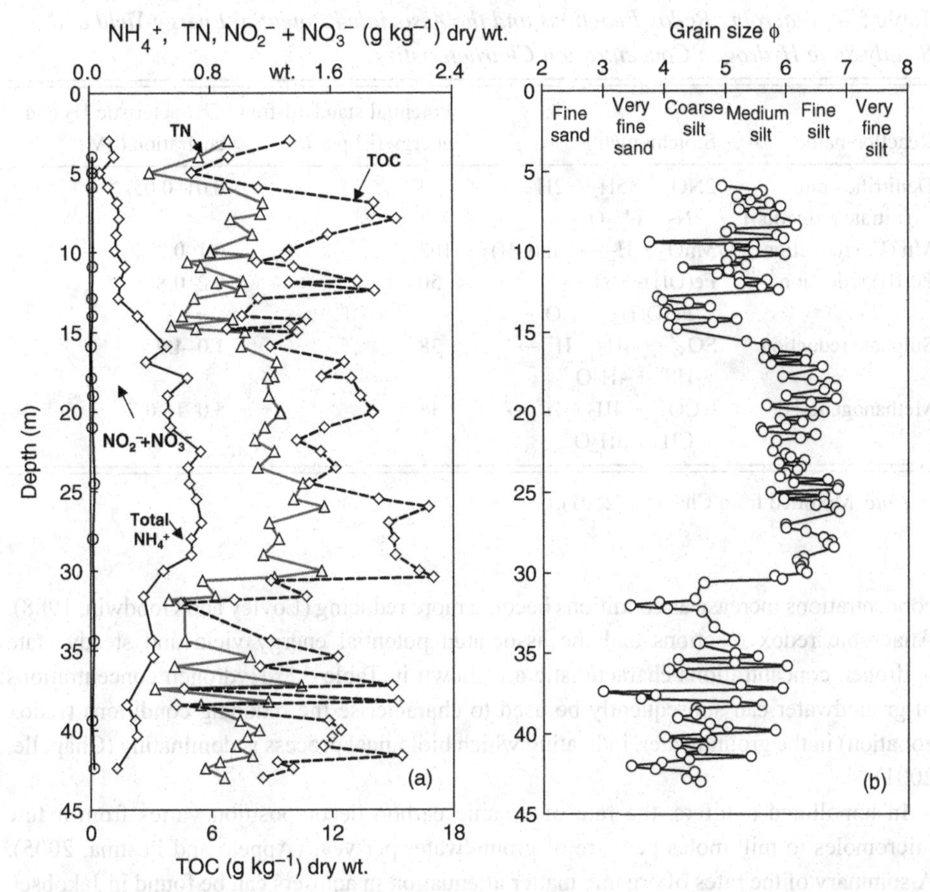

Figure 5.25 (a) Total organic carbon (TOC), $NO_2^- + NO_3^-$, NH_4^+ and total nitrogen (TN) versus depth and (b) particle size versus depth in the aquitard sediments at borehole SD14 in the Pearl River Delta, China.

sediments are associated with more TOC (Wang, 2011). In the zone of high TOC, the total ammonium (water-soluble plus exchangeable) is also high, with concentrations up to ~0.6 g kg^{-1} as N. This indicates that ammonium nitrification is limited by the low permeability of the fine materials that severely restricts oxygen invasion (Jiao et al., 2010).

The decomposition of the organic matter in anaerobic system is mediated by microbes, and proceeds via three steps (Appelo and Postma, 2005). First the large organic particles and molecules are decomposed into smaller ones. Then these smaller fragments are decomposed into compounds like acetic acid (CH_3COOH) and formic acid (HCOOH). Finally these intermediates are oxidised during the so-called terminal electron accepting processes (McCarty, 1997). In each of the steps, hydrogen (H_2) is produced by the fermentative bacteria, which is then used by respiring microorganisms such as nitrate reducers, Fe^{3+} reducers, sulphate reducers and methanogens (Chapelle, 2001). H_2

Table 5.8 *Anaerobic Redox Reactions and the Associated Potential Energy Yield and Steady State Hydrogen Concentration Characteristics*

Reaction name	Stoichiometry	Potential standard-free energy (kJ per H_2)	Characteristic H_2 concentration (nM)
Denitrification (nitrate reduction)	$2NO_3^- + 5H_2 + 2H^+ \rightarrow N_2 + 6H_2O$	224	0.01–0.05
Mn (IV) reduction	$MnO_2 + H_2 \rightarrow Mn(OH)_2$	163	0.1–0.3
Fe(III) reduction	$Fe(OH)_3 + H^+ \rightarrow Fe(OH)_2 + H_2O$	50	0.2–0.8
Sulphate reduction	$SO_4^{2-} + 4H_2 + H^+ \rightarrow HS^- + 4H_2O$	38	1.0–4.0
Methanogenesis	$HCO_3^- + 4H_2 + H^+ \rightarrow CH_4 + 3H_2O$	34	5.0–15.0

Note. Modified from Chapelle (2001).

concentrations increase as conditions become more reducing (Lovley and Goodwin, 1988). Anaerobic redox reactions and the associated potential energy yield and steady state hydrogen concentrations characteristic are shown in Table 5.8. Hydrogen concentrations in groundwater can subsequently be used to characterise the reducing conditions (redox zonation) in the groundwater, indicating which biological process is dominating (Chapelle, 2001).

In unpolluted aquifers, the rate of organic carbon decomposition varies from a few micromoles to millimoles per litre of groundwater per year (Appelo and Postma, 2005). A summary of the rates of organic matter attenuation in aquifers can be found in Jakobsen and Postma (1994). Various reduction processes associated with organic matter decomposition are shown in Figure 5.26. The sequence of these reduction processes is controlled by the efficiency of different microorganisms to use electron donors at decreasing concentrations (Clemmer, 2003).

5.7.3 Sulphate Reduction

Sulphate can be reduced in anoxic environments in the presence of organic matter. Seawater is an important source of sulphate in coastal aquifers and newly reclaimed lands. The oxidation of organic matter by sulphate reduction whereby H_2S is formed can be expressed by the following reaction equation:

$$2CH_2O + SO_4^{2-} \rightarrow 2HCO_3^- + H_2S \tag{5.8}$$

where CH_2O represents organic matter. The sulphate reaction is catalysed by sulphate-reducing bacteria, which rely upon fermentative bacteria to break down the large organic

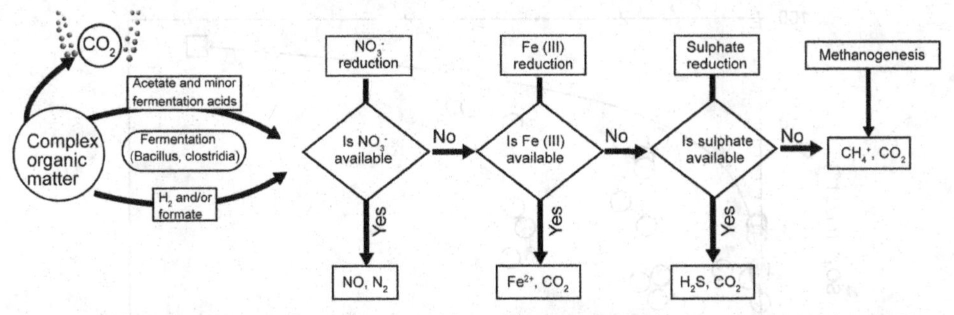

Figure 5.26 Biologically mediated decomposition of complex organic matter and associated reduction processes (Chapelle, 2001).

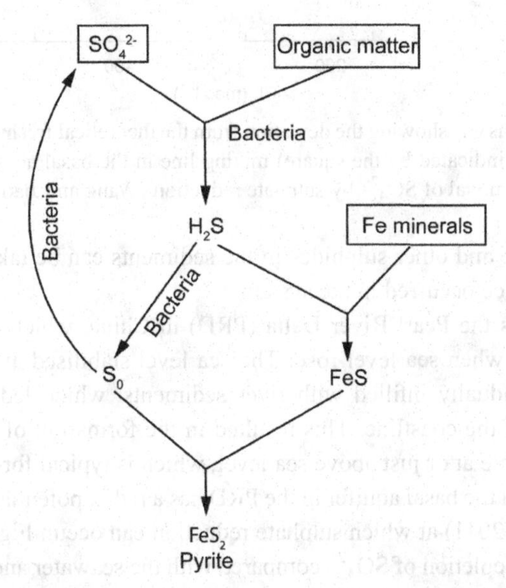

Figure 5.27 Schematic representation of the formation of pyrite (FeS_2) under reducing conditions (Berner, 1984).

molecules into smaller ones first (Chapelle, 2001). As a result of this process, sulphate concentrations fall, whereas HCO_3^- concentrations increase. The H_2S gas that is produced can be easily identified because it has the unmistakable odour of rotten eggs.

In aquifers affected by seawater intrusion, sulphate depletion is widely observed (Chapelle and McMahon, 1991; De Montety et al., 2008; Wang et al., 2013). In an aquifer system with entrapped seawater, sulphate may be consumed to levels below the detection limit (Wang, 2011). Because Cl^- is conservative, the SO_4^{2-}/Cl^- ratio can be used to assess the degree of sulphate reduction by comparing this ratio with that of the seawater of about 0.05 (Table 5.2). If the sediment has Fe oxides, some of the formed H_2S may be consumed (Jakobsen and Postma, 1999) as Fe-sulphide minerals (FeS_2) form (Figure 5.27). So

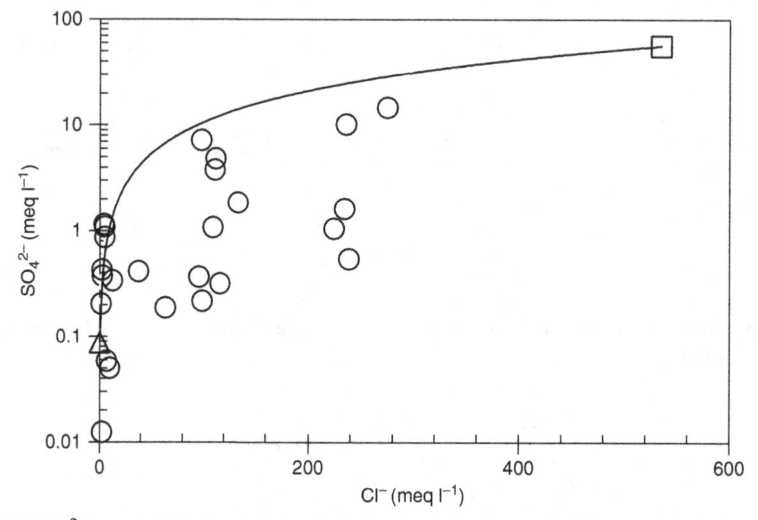

Figure 5.28 SO_4^{2-} versus Cl^- showing the deviation from the theoretical freshwater (indicated by the triangle) and seawater (indicated by the square) mixing line in the basal aquifer in the Pearl River Delta, suggesting the removal of SO_4^{2-} by sulphate reduction (Wang and Jiao, 2012).

the presence of pyrite and other sulphides in the sediments can be taken as evidence that sulphate reduction once occurred in the system.

A good example is the Pearl River Delta (PRD) in China, which was largely formed during the Holocene when sea level rose. The sea level stabilised at about 6 ka BP and depressions were gradually infilled with river sediments, which led to a progressively seaward migration of the coastline. This resulted in the formation of a low-lying coastal zone, with a water table at or just above sea level, which is typical for many deltas.

The groundwater in the basal aquifer in the PRD has a redox potential between −34.1mV and −314mV (Wang, 2011) at which sulphate reduction can occur. Figure 5.28 shows that there is a significant depletion of SO_4^{2-} compared with the seawater and freshwater mixing line, indicating considerable removal of SO_4^{2-} from the groundwater. Another process that can cause sulphate depletion is gypsum precipitation, but the fact that the saturation index for this mineral was negative in most samples indicates that it has no tendency to precipitate (Wang, 2011).

Several studies concluded that sulphate reduction does not occur in aquitards due to a lack of sulphate-reducing bacteria (Jones et al., 1989; Chapelle and McMahon, 1991), but this is not the case in the aquitard in the PRD. Figure 5.29 shows the profiles of Cl^- and SO_4^{2-} from pore water samples collected from a cluster of seven piezometers installed in the town of Minzhong. All the piezometers were installed in the Holocene and Pleistocene marine sediments except the last one which was in the Pleistocene basal terrestrial aquifer. This site is less than 3 km from the shore and was under the sea about 300 years ago, with relic seawater still remaining in the marine mud (Jiao et al., 2010). The SO_4^{2-} concentration is only a few mg l^{-1} in the mud, which is significantly depleted compared to the SO_4^{2-}

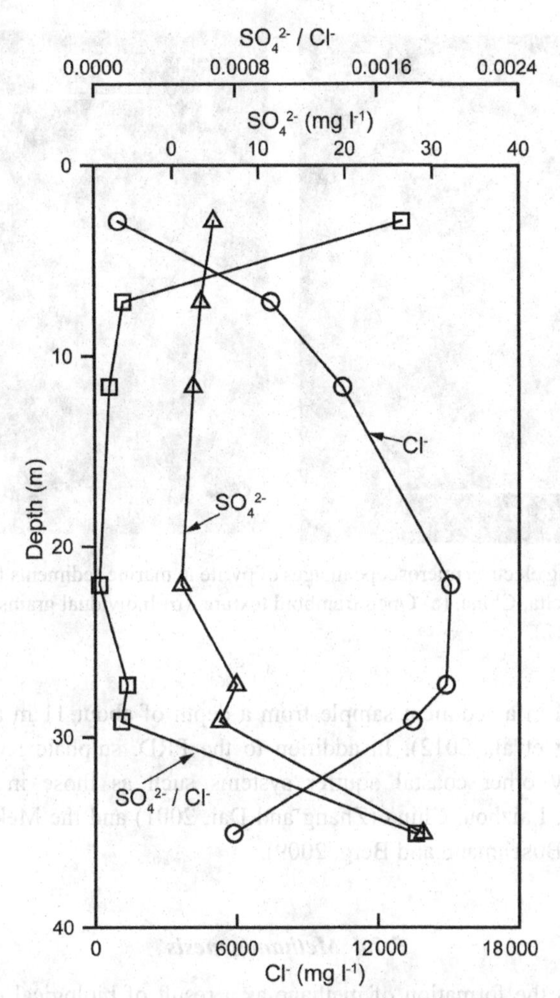

Figure 5.29 Profiles of SO_4^{2-}, Cl^- and the SO_4^{2-}/Cl^- ratio versus depth of pore water samples collected from a cluster of seven piezometers installed in the Holocene and Pleistocene marine aquitards and the basal aquifer (deepest sample) in Minzhong town, Pearl River Delta, China. This site was under the sea about 300 years ago, and relic seawater still exists in the marine mud.

concentration of ~2800 mg l^{-1} in seawater. The SO_4^{2-}/Cl^- molar ratio is less than 0.002 (Figure 5.29), which is well below the seawater ratio (Table 5.2). This SO_4^{2-} depletion is believed to be caused by sulphate reduction, which is substantiated by the smell of H_2S of the water from some of the piezometers.

The sediments at this site are rich in Fe oxides, including hematite, limonite, magnetite and ilmenite (Wang, 2011). With the mediation of microorganisms, the sulphate reduction products H_2S and Fe oxides may combine to form Fe-sulphide minerals (FeS_2) (Figure 5.27). Indeed, pyrite is common in PRD Quaternary sediments. Figure 5.30 show the pyrite

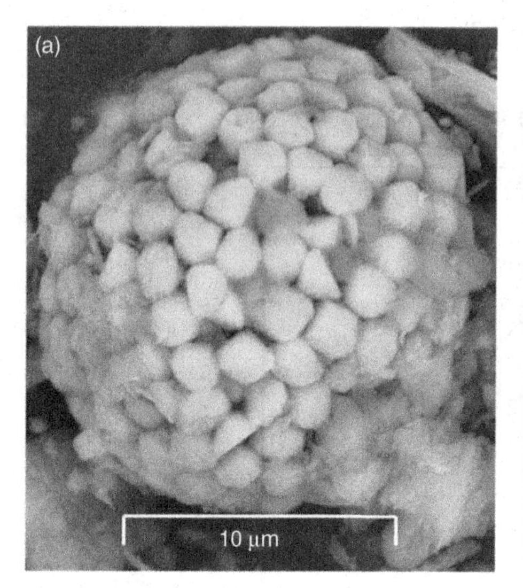

Figure 5.30 Scanning electron microscope images of pyrite in marine sediments from borehole cores in the Pearl River Delta, China. (a) Open framboid texture. (b) Individual grains with diameter >10 μm (Wang et al., 2012).

minerals identified in a sediment sample from a depth of about 11 m at the piezometer cluster site (Wang et al., 2012). In addition to the PRD, sulphate reduction has been observed in many other coastal aquifer systems such as those in the Netherlands (Stuyfzand, 1993), Laizhou, China (Zhang and Dai, 2001) and the Mekong, Bengal and Red River deltas (Buschmann and Berg, 2009).

5.7.4 Methanogenesis

Methanogenesis is the formation of methane as a result of biological decomposition of organic matter in the absence of oxygen and is usually the last stage in the biomass decomposition under reducing conditions. Methane generation is a widespread and important way of microbial metabolism. In addition to biogenic methane, methane can be thermogenic, i.e. related to natural gas petroleum reservoirs (Darling and Gooddy, 2006). Methane has a global warming potential 34 times greater than CO_2 (Stocker, 2014). Consequently methanogenesis in coastal zones is worthy of scientific investigation.

Biogenic CH_4 can be generated by two pathways, by the reduction of CO_2,

$$CO_2 + 4H_2 \rightarrow CH_4 + 2H_2O \tag{5.9}$$

or by acetate fermentation,

$$CH_3COOH \rightarrow CH_4 + CO_2 \tag{5.10}$$

The first pathway produces CH_4 only, whereas the second one produces both CH_4 and CO_2.

The presence of CH_4 in groundwater is indicative of an extremely reducing environment. Methane has no colour or smell, but is highly flammable. Some boreholes installed in the basal aquifer in Pearl River Delta have such high methane contents that the gas emanating from them can be lit. CH_4 concentrations in the air in the wellbore in the PRD can be over 10 times of that in the atmosphere. Gas samples in some boreholes had a methane concentration amounting to about 88% of the total gas volume (Wang, 2011). Darling and Gooddy (2006) studied sedimentary aquifers and aquitards of various redox conditions and found that the highest concentrations of naturally occurring dissolved CH_4 were usually associated with aquitards.

The Cretaceous Middendorf coastal plain aquifer in South Carolina (Figure 5.31a) is a prime example of a system with a well-defined zonation caused by redox processes mediated by microbiologic activities (Lovley and Goodwin, 1988; Chapelle and McMahon, 1991; Chapelle and Lovley, 1992; Park et al., 2006). This aquifer system comprises a wedge-shaped body of clay, silt, sand and limestone strata of Cretaceous age and younger, which dip to the southeast. The Middendorf aquifer is one of the major aquifers in the system. The aquifer in the upper coastal plain comprises sediments of non-marine origin with ample ferric iron but little calcite. Towards the coast, the aquifer material changes gradually into marine sediments with abundant calcite but little ferric iron. It is buried below a depth of over 1 km close to the Atlantic Ocean. Groundwater receives recharge in the outcrop area, flows to the lower coastal plain in the southeast and discharges via the overlying artesian aquifers to the ground and sea.

Lovley and Goodwin (1988) investigated the distributions of sulphate, nitrate, ferrous iron, methane and hydrogen along the flow path (Figure 5.31b). Four well-defined redox zones can be inferred from the figure: denitrification near the outcrop (as demonstrated by the significant decrease in nitrate from the agricultural area), Fe^{3+} reduction (as demonstrated by ferrous iron accumulation) and sulphate reduction (as demonstrated by sulphate depletion) and methanogenesis (as demonstrated by methane accumulation). Each zone is dominated by different groups of microorganisms, and hydrogen changes along the flow path from less than 0.1 nM, a concentration characteristic of nitrate reduction, to 10 nM, which is characteristic of methanogenesis (Chapelle et al., 1993).

5.7.5 Ammonium Generation

Naturally occurring ammonium has been observed in several coastal aquifers. For example, the upper aquifer in the Red River Delta in Vietnam has ammonium concentrations of up to 61.7 mg l^{-1} (Berg et al., 2001). Concentrations of up to 390 mg l^{-1} occur in the basal aquifer in the Pearl River Delta and the pore water in the aquitards above the aquifer has even higher concentrations (Jiao et al., 2010). A typical feature of these coastal aquifer systems is that the groundwater is anaerobic and the systems are rich in

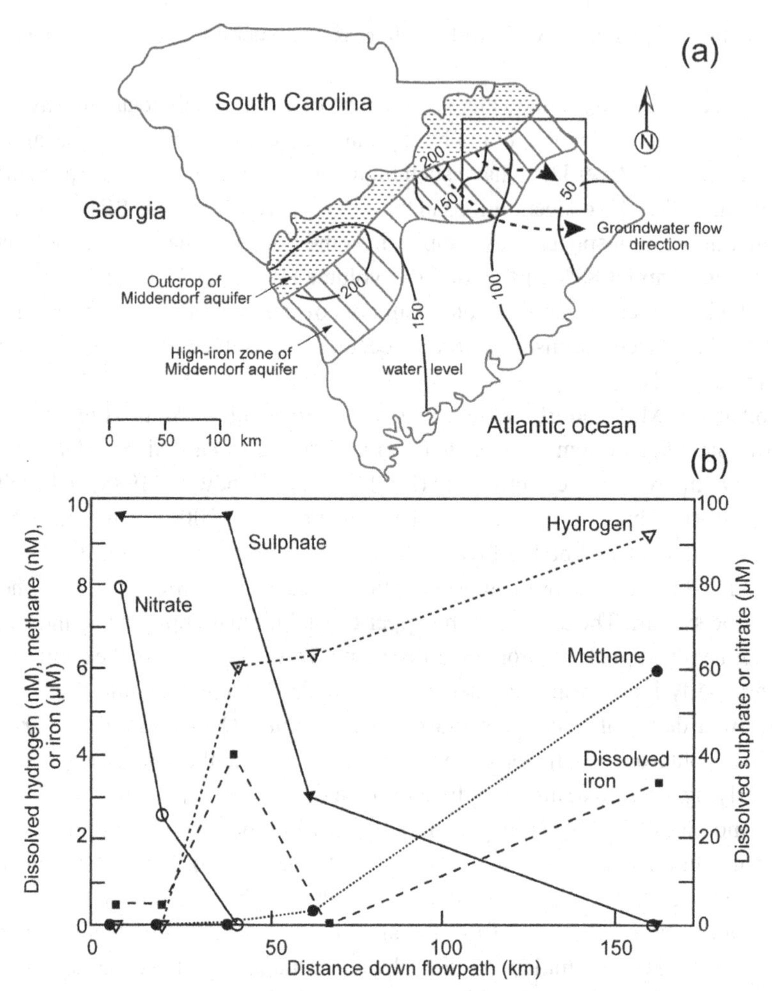

Figure 5.31 (a) Zonation of groundwater in the Cretaceous Middendorf coastal plain aquifer in South Carolina (Chapelle and Lovley, 1992). (b) Hydrogen, sulphate, nitrate, dissolved iron and methane along the flow path of the aquifer (see the box with flow lines in a) (Lovley and Goodwin, 1988).

organic matter (Figure 5.25) due to burial of large quantities of plants, animals and microorganism remains collected on the seafloor or estuaries during the late Holocene regression.

The dissolved ammonium in such systems is believed to be formed by the microbial fermentation of sedimentary organic matter (Berg et al., 2001; McArthur et al., 2001). Ammonium can be generated in two ways, which are controlled by different functional genes of the microorganisms involved: from organic nitrogen sources, which is known as ammonium mineralisation or ammonification, and from nitrite to ammonium, which is called dissimilatory nitrate reduction to ammonium. The process of ammonification by decomposition of organic matter turns organic nitrogen into ammonia (NH_3) or ammonium (NH_4^+).

Ammonium concentrations can be attenuated by conversion into nitrate or nitrite via biological oxidation processes by ammonia-oxidising bacteria (AOB) or aerobic ammonia-oxidising archaea (AOA). Nitrate formed by the above processes may be subsequently reduced into nitrogen gas by denitrification process. Anaerobic ammonium oxidation (anammox) has also been identified (Mulder et al., 1995), where nitrate and nitrite replace O_2 as the electron acceptor for the oxidation of ammonium, with conversion of both species to elemental nitrogen, N_2:

$$NH_4^+ + NO_2^- \rightarrow N_2 + 2H_2O \tag{5.11}$$

These processes decrease the concentration of ammonium in the groundwater and the related ammonium pool. However, if the microbiological processes which generate ammonium overwhelm those which oxidise ammonium, abnormally high ammonium levels in the aquitard-aquifer system will persist.

Stable nitrogen isotopes can be used to detect the origin of inorganic nitrogen because the values of $\delta^{15}N$ of nitrogen of natural and anthropogenic sources is distinctly different. Ranges of the values of $\delta^{15}N$ of various nitrogen sources can be found in Scheible et al. (1993) and Clark and Fritz (1997). A case study demonstrating that the ammonium is originating from natural sedimentary organic matter in a coastal aquifer system based on stable nitrogen analysis can be found in Wang et al. (2013).

NH_4^+ ions formed by mineralisation of organic matter in sediment may be also active in water-rock exchange. Wang and Jiao (2012) studied the ion exchange processes in the Holocene organic- and clay-rich aquitard in the coastal area of the Pearl River Delta. Statistically significant inverse correlations were observed between the reacted amounts of Mg^{2+} and NH_4^+, and at locations where the concentration of NH_4^+ is higher relative to values due to conservative mixing, that of Mg^{2+} is lower. The Mg^{2+} ion usually has a greater affinity to be adsorbed than NH_4^+ ion (Domenico and Schwartz, 1990), so the negative correlation indicates that NH_4^+ may be exchanging for Mg^{2+} (Wang and Jiao, 2012).

The coastal Swan River in Western Australia provides a good example of sediment-derived NH_4^+ displacement by cations in the intruded seawater (Smith and Turner, 2001). The riverbed sediment is rich in organic matter and humic substances, and nutrient concentrations, particularly of NH_4^+, in the riverbed are very high (Figure 5.32). Sediments with freshwater contains more exchangeable NH_4^+ than sediments containing saline water. When seawater intrudes into the river, during what is called a salty tide, ammonium can be desorbed from the clay or organic matter. The density difference between saltwater and the originally freshwater below the river bed forces saltwater to sink into the underlying aquifer but regional lateral groundwater flow forces groundwater below the river bed to flow upward. This process flushes the NH_4^+

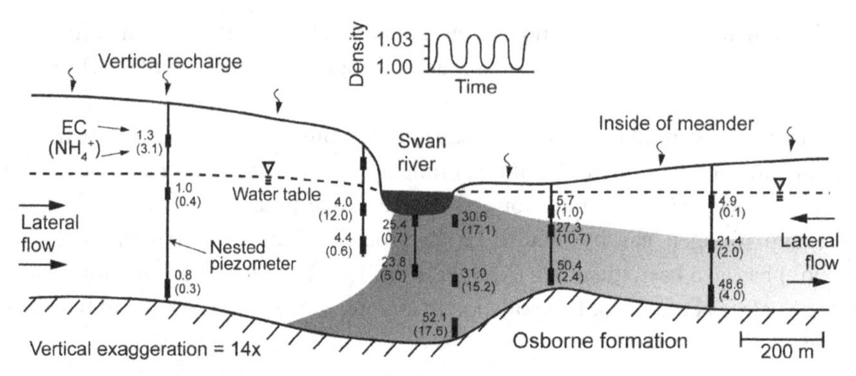

Figure 5.32 Electrical conductivity (mS cm^{-1}) and ammonium concentrations (mg l^{-1}) of groundwater below the Swan River, Australia (Smith and Turner, 2001).

back to the river. Smith and Turner (2001) speculated that because of this a red tide (i.e. algal bloom due to excess nutrients) sometimes may occur soon after a salty tide.

5.7.6 Acid Sulphate Soils in Low-Lying Coastal Areas

During sulphate reduction, sulphide minerals can be formed in waterlogged, anaerobic and organic-rich low-lying tide-dominated estuarine floodplains. The soils containing these sulphides, mostly pyrite (FeS_2) and iron monosulphides (FeS), are called acid sulphate soils (Dent, 1986). The sulphides become oxidised if they are exposed to atmospheric oxygen, caused by a lowering of the water table due to drainage for agriculture activities (Dent, 1986) or groundwater pumping (Fitzpatrick et al., 2009) (Figure 5.33). The mineral oxidation process produces dissolved iron, sulphate and trace metals, and the pH in groundwater can drop to values less than 4 (Dent, 1986). The low pH may lead to the dissolution of aluminium-containing minerals. Acid sulphate soils occur in many coastal areas in the world (Dost, 1973; Grant, 1973; Lin et al., 1995) but have been most extensively studied in Australia (Hicks et al., 2009; Poch et al., 2009), where they are a common occurrence in the absence of buffering minerals like calcium carbonates.

Lin and Melville (1994) studied acid sulphate soils and the physical geography of the Pearl River Delta. Here, acid sulphate soils are usually formed around bedrock islands where sedimentation rates were lowest, which promoted the formation of pyrite. As a result of centuries of land reclamation, and intensive agricultural activities, oxidation of pyrite has occurred and acidic conditions have developed. Water derived from acid sulphate soils causes many environmental and ecological problems. It can corrode concrete, and kill fish and plants. Oysters, which are permanently attached to rock surfaces and cannot migrate, may experience breakdown of their shell and a reduced growth rate. Seagrass species may be lost, or there may be colonisation by acid-tolerant weeds (Crook et al., 2008).

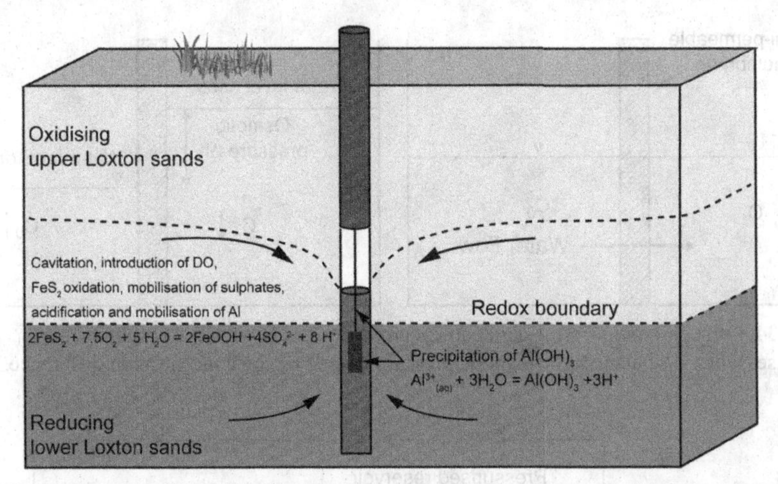

Figure 5.33 Conceptual diagram indicating various geochemical processes in and around a cone of depression induced by pumping. When the water table falls, air will enter the formerly waterlogged soil, which causes pyrite oxidation, acidity generation and dissolution of aluminosilicate minerals such as clays. Leaching due to infiltration of precipitation brings the acid in the unsaturated zone to the saturated zone, causing locally-derived acidic groundwater within the cone of the depression (Fitzpatrick et al., 2009).

5.7.7 Membrane Filtration and Osmosis in Clay

5.7.7.1 Basic Principles

A membrane is a thin layer of selectively permeable materials which acts as a barrier that allows certain particles to pass through but not others. A membrane can filter ions by geometric restrictions that allows only the passage of solutes smaller than the pores, or by processes such as electrostatic repulsion of the ions by the double layers (Section 5.3.3). In the subsurface, clay can behave like a membrane.

For aqueous solutions like groundwater, chemical osmosis can be defined as the net migration of water through a membrane to a reservoir with greater solute concentration driven by the difference of the chemical activities on both sides of the membrane (Figure 5.34a). A concentration gradient across the membrane will cause water to flow from the reservoir with the lower salinity on the right to one with the higher salinity on the left. As a result, the water level in the left reservoir increases while that in the right reservoir decreases. The flow ceases when the head difference is large enough to balance the osmotic driving force (Figure 5.34b). The pressure due to the water column overheight at equilibrium is called the osmotic pressure.

The osmotic pressure is determined by the ion concentration of the solution. If the liquid in the right reservoir is pure water or has a low ion concentration, and the liquid on the left is a concentrated solution such as seawater, then the osmotic pressure of the liquid on the right is negligible, and the osmotic pressure difference is approximately equal to the osmotic pressure of the concentrated solution on the left. For typical seawater, the theoretical

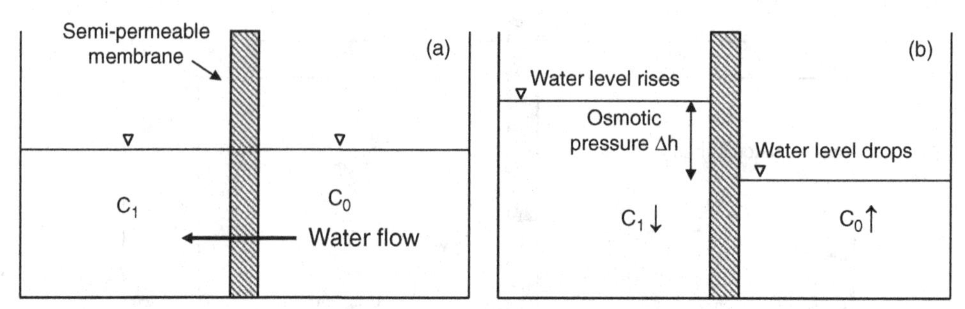

Figure 5.34 Osmosis across a semi-permeable membrane. (a) Osmotic flow occurs when $C_1 > C_0$. (b) Flow ceases when Δh balances the osmotic pressure caused by the concentration difference.

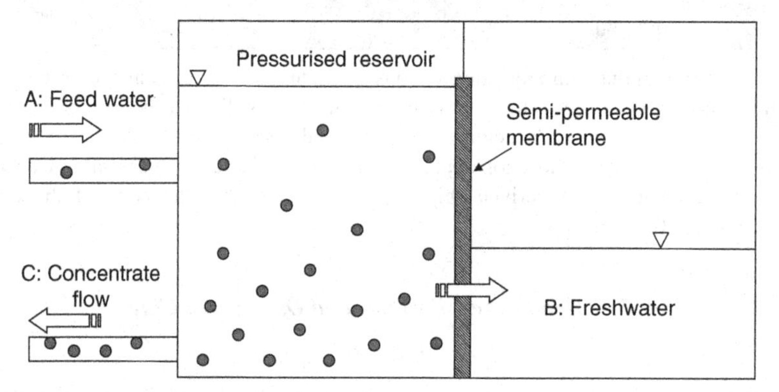

Figure 5.35 Saline feedwater, low-salinity product water and very saline concentrate (brine or reject water) in a reverse osmosis desalination plant. The density of the grey dots indicates the salinity.

maximum osmotic pressure is 2.9×10^6 Pa, equivalent to a water column of almost 300 m (Gerstandt et al., 2008).

If the water level in the left reservoir is increased so that the pressure difference across the membrane exceeds the osmotic pressure, water will flow from the left reservoir with the high concentration to the right reservoir with the low concentration (Figure 5.35). This process is called reverse osmosis, which forms the basic principle of desalinisation to produce potable water from seawater. Three liquids are involved in the process of desalination (Figure 5.35): the saline feedwater (brackish water or seawater); low-salinity produced water; and a highly concentrated residual of the feedwater (brine or reject water) (McGinnis, 2002). In the production process, the feedwater (similar to the water in the left reservoir) is driven into a sealed reservoir to pressurise it against the membrane. After the water penetrates through the membrane, the salinity of the residual water increases, and the resulting brine must be removed.

The major operating cost of a desalination plant is the energy required to pressurise the feedwater (Section 12.5). The water pressure must exceed the osmotic pressure of seawater, which is equivalent to about 300 m of water. The desalination cost can be significantly reduced if brackish water is used. Salts, bacteria and other impurities such as sand, shells and seaweed clog the membranes, and necessitate costly clean-up and membrane replacement, which add to the cost of desalination.

5.7.7.2 Osmotic Processes in Natural Geological Systems

Sedimentary geological systems often consist of interbedded less-permeable layers of clay and shale. Under favourable geological and geochemical conditions, the latter may behave as a semi-permeable membrane, and flow driven by osmosis flow may occur. The membrane behaviour of clay layers has been proposed as an explanation for anomalous pressure and salinity distributions in deep geological basins (Belitz and Bredehoeft, 1988; Ortoleva, 1994). The generation of abnormally low or high pressure in the geological formations can be explained using Figure 5.34. If the reservoirs at both sides are closed, relatively high and low fluid pressure can be generated in the left and right reservoirs, respectively, by osmotic effects. Similarly, anomalous pressures in deep, high salinity groundwater system can be formed by the cross-formational movement of water through semi-permeable shale or clay membranes (Neuzil, 1986; Belitz and Bredehoeft, 1988). When groundwater flows across a stratum that functions as a membrane, the solute concentration on the inflow side of the membrane is elevated, but the concentration of the outflow side falls. This phenomenon is called salt filtering, ultrafiltration or hyper-filtration (e.g. Freeze and Cherry, 1979).

Anomalously high hydraulic heads have sometimes been found in some coastal plain aquifers which are hard to explain by the topographic driving forces of regional flow system (Marine, 1974). Marine (1974) and Marine and Fritz (1981) studied the over-pressurised Triassic sandstone basin in Aiken, South Carolina (Figure 5.36). This basin is overlain by 330 m of coastal plain sediments which contain freshwater with a TDS concentration of 38 mg l^{-1} and has water level of 58 m above sea level. The buried Triassic basin is a graben structure surrounded by low-permeability crystalline metamorphic rock. Sediments within the Triassic basin consist of intercalated lenses of fine sand and clay, which is mainly illite. Two wells, screened only inside the Triassic rocks, yielded water with a TDS concentration of 11 900 and 18 500 mg l^{-1}, and had heads of 192 and 140 m above sea level, respectively. Marine and Fritz (1981) contended that the clayey parts of the Triassic basin behave like membranes, whereas the sandy parts function like the container of the saline solution. Assuming that the overlying coastal plain sediments are the container of the freshwater, the osmotic flux could have created the high pressures inside the basin.

5.7.8 Chemical Processes in Coastal Areas Reclaimed from the Sea

When land is reclaimed using the land filling approach, fill materials are placed on the seafloor (Section 9.2.2.1). Consequently, there will be chemical and physical interactions between the seawater, groundwater, marine sediments and fill materials (Figure 5.37). In

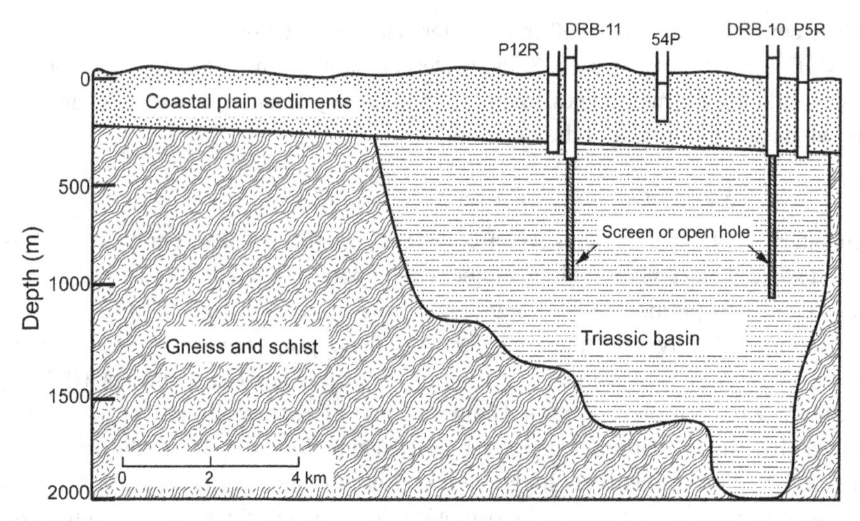

Figure 5.36 Cross section of the Dunbarton Triassic Basin. Wells DRB-11 and DRB-10 penetrate the osmotically pressurised deep part of the basin. The shallow wells show little signs of anomalous head and yield low-salinity water (based on Marine and Fritz, 1981).

the subsurface beneath the original land bordering the reclamation site, there will be gradual freshening due to the seaward migration of the transition zone. This freshening process will eventually also extend into the reclaimed areas, which receive freshwater recharge from rainfall as well (Figure 5.38).

Weathered igneous rock such as granite and volcanic rock is a dominant fill material used in many coastal reclamation areas in southeast China. These materials tend to be oxidised and rich in iron. The chemical characteristics of reclaimed areas can be very complex when fill materials are sourced from different areas, which is very common. The seabed usually has a thick layer of dark marine mud. For some important coastal infrastructures such as the Hong Kong International Airport, an effort was made to dredge the mud away to minimise the ground subsidence after reclamation but still at least one metre of mud remained. The mud is rich in organic compounds, has a high oxygen demand, has high contents of heavy metals due to contamination, and is rich in sulphides. Under undisturbed conditions, alkaline and anoxic conditions prevail in the mud, and organic matter and heavy metals (mainly as metal sulphides) are fairly stable.

At a reclamation site materials and waters of distinctly different chemical characteristics mix (Figure 5.37). As the seawater originally occupying the pore space of the fill material and former marine sediments is displaced by freshwater (Figure 5.38), the freshening can be easily recognised by cation exchange. A case study about the freshening of a coastal aquifer around the reclamation site in Shenzhen Bay can be found in Chen and Jiao (2007).

Various physical and chemical reactions may occur between the original marine sediments and the fill materials in the reclaimed land and these reactions may result in the

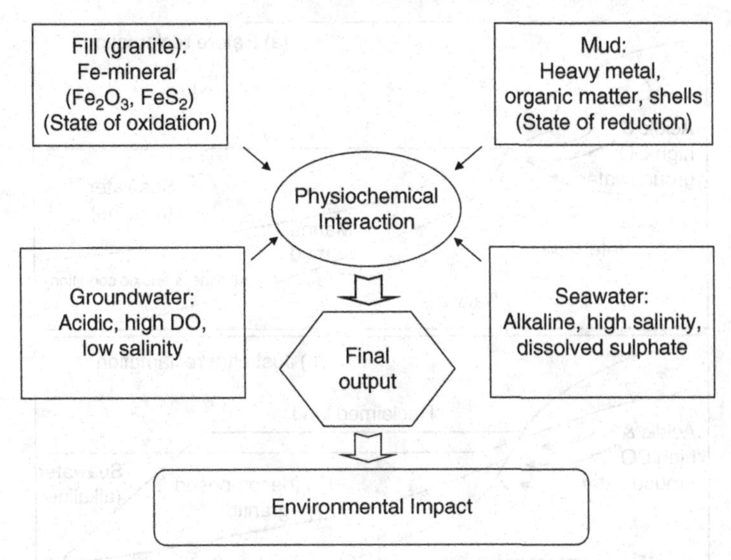

Figure 5.37 Possible physicochemical interaction of mud, fill, groundwater and seawater if weathered granite is used as fill material.

release of chemicals unfavourable to plants and organisms (Jiao et al., 2005). For example, most reclaimed areas will experience extensive construction and deep excavation. The sediments can then be disturbed, be exposed to air and become oxidised, which may result in the release of metals that were stable under reducing conditions. Oxidation of sulphides generates sulphuric acid, leading to acid sulphate soils. This process may be enhanced by acid rainfall recharge. The heavy metals and the organic matter released to groundwater will eventually discharge to the sea.

Marine mud is usually rich in shell fragments (Suh, 2004). For example, the mud at Shekou, Shenzhen, China has a high shell content as a consequence of commercial oyster cultivation (Jiao et al., 2005). The $CaCO_3$ in the fragments may be dissolved by the acidic groundwater. Eventually this may cause ground settlement and land subsidence. Dissolution of shell fragments was observed as an important process in other coastal areas (Stuyfzand, 1984).

Suh (2004) evaluated the geochemical processes governing the distributions of trace metals near Rozelle Bay, Sydney, Australia. This site was infilled between 1972 and 1980 to create an area for recreation. The fill materials included dredged estuarine mud and sand, construction materials, as well as demolition, domestic and industrial waste. Groundwater samples collected in the reclamation site indicated that these fill materials controlled the concentrations of both major ions and trace metals in groundwater. In the areas containing construction waste, the major ions in groundwater could be ranked based on their average concentration according to: $Na \gg Mg > Ca > K$ for the cations and $Cl^- > SO_4^{2-} > HCO_3^-$ for the anions, whereas in the areas with fills of dredged marine sediments the sequence was: $Na^+ > Ca^{2+} > Mg^{2+} > K^+$ for the cations and $HCO_3^- > Cl^- > SO_4^{2-}$ for the anions. Suh (2004) further found that groundwater flow delivered trace metals to the estuary.

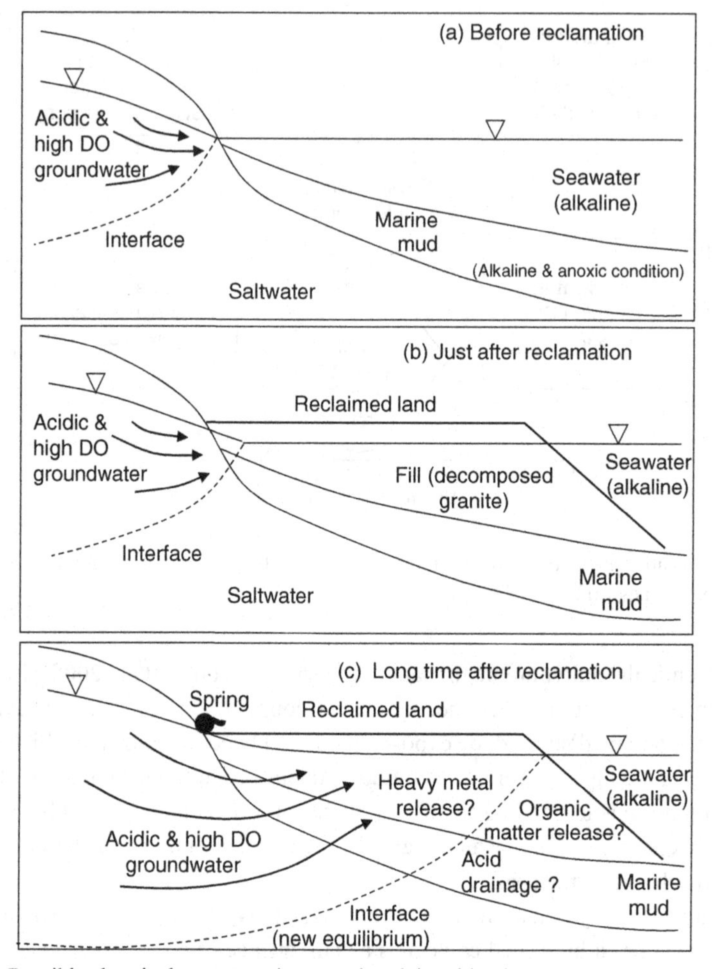

Figure 5.38 Possible chemical processes in coastal reclaimed land.

Chen et al. (2007) investigated the concentrations and mobility of metals in groundwater and marine sediments in the reclaimed land around Shenzhen, China. They found that the longer the sediments had been buried, the lower were the metal concentrations in the sediments, but the higher were the metal concentration in the groundwater inside the reclaimed site relative to that in the original coastal area. The decrease in salinity as the original seawater was flushed from the marine mud by freshwater was accompanied by a decrease of the pH, which caused the mobilisation of metals such as V, Cr, Mn, Ni, Cu and Cd. Similar to land fill operations, Pit et al. (2017) found that beach nourishments (i.e. the replenishment of beaches with sand sourced elsewhere) can lead to heavy metal mobilisation. Before carrying out any large-scale land reclamation project, these potential adverse effects on the environment should be assessed.

6
Seawater Intrusion

Broadly speaking, seawater intrusion is the displacement of fresh groundwater by seawater. The cause of seawater intrusion can be natural or anthropogenic. The most common natural causes include a decrease in groundwater recharge, or an increase of relative sea level. The latter may or may not lead to the inundation of land by seawater. Typical anthropogenic causes include abstraction of groundwater from wells, a decrease in recharge due to land use change, which both lead to a lowering of groundwater levels, or the disposal of seawater used in industrial processes. These changes occur at a range of timescales, typically from months to centuries, or even more. This chapter will focus on the different modes of seawater intrusion, and their causes. Examples are provided based on case studies.

6.1 A Closer Look at Seawater Intrusion

The most common and general definition of seawater intrusion is the displacement of fresh groundwater by seawater (Bruington, 1972). Sometimes saltwater is used instead of seawater, to allow for other sources of salt other than the sea. For example Klassen and Allen (2017) defined saltwater intrusion as the process by which a freshwater aquifer becomes salinised due to natural or anthropogenic stresses. On the other hand, Van Dam (1999) defined it as the inflow of saline water in an aquifer system, while Werner et al. (2013) understood seawater intrusion to be 'the landward incursion of seawater'. The latter definitions differ subtly but importantly from the former, as they do not necessarily imply a loss of freshwater: Indeed there can be landward flow of saltwater (i.e. within the saline groundwater body itself) without a change in the groundwater salinity distribution.

There are other subtleties to note as well. The definition of Klassen and Allen (2017) does not presuppose a flow of water, which means that it captures the possible effects of salinisation by diffusion. Albeit a slow process, diffusion dominates solute transport in aquitards, and plays a role in the salinisation of aquifers from bodies of trapped saltwater (e.g. Dakin et al., 1983). And finally, one aspect to note from the definition by Klassen and Allen (2017) is that they explicitly refer to freshwater, thereby following others (e.g. Bruington, 1972). It is thus restrictive in the sense that it excludes brackish water, which may also become more saline by seawater intrusion. Given the growing importance of brackish groundwater as a resource this is not just a point of pure academic interest.

Therefore, to capture the full range of conditions under which the salinisation of coastal groundwater by seawater can occur, the following comprehensive definition is proposed:

Seawater intrusion is the increase of the salinity of groundwater, which may or may not be in direct hydraulic contact with the sea. It is the result of flow processes by which the ambient groundwater is being displaced by or mixed with seawater, or of the migration of solutes from seawater into fresher groundwater under stagnant conditions.

Not requiring there to be a hydraulic connection with the sea allows for the possibility that connate seawater may form the salt source. This definition also explicitly considers the role of diffusion in systems where stagnant conditions dominate.

Seawater intrusion can have a variety of causes. A decrease of the groundwater pressure will induce flow which will affect the salinity distribution. Such a change in pressure can result from groundwater abstraction, land drainage, a decrease in recharge, or even enhanced evapotranspiration (Comte et al., 2014). The latter may occur when excavation of materials from the ground causes the water table to be closer to the surface. It may also be that the sea level or the water level of saline surface water bodies like estuaries or lagoons rises relative to groundwater levels. If a transition zone exists, it will shift landward, or vertically upward, or both.

Groundwater salinity may also increase when saline surface water shifts landward, or if a surface water body becomes more saline. The landward shift may be rapid and catastrophic such as during a storm surge or tsunami, or it may occur more gradually such as during a transgression. Salinisation of surface water is common in rivers and estuaries, and may be occurring on a seasonal basis (see also Section 5.2.1). For example, the waters of the Gulf St Vincent and Spencer Gulf in South Australia show a marked trend in salinity which is driven by evaporation in summer, which can drive salinity up to 50 g kg^{-1} and cause density-driven circulation (Kämpf, 2014).

Seawater can intrude further upstream in estuaries and rivers as runoff decreases, or if the river is deepened, either by natural erosion or by anthropogenic activities, such as for example when bottom materials are removed during dredging operations (Parker et al., 1955). The connectivity between the aquifer and a saline surface water body thereby will become greater, thus enhancing the exchange of water between the aquifer and the surface water. This increases the susceptibility of the aquifer to seawater intrusion.

Some cases of seawater intrusion involve the deliberate or undeliberate disposal of saline water by humans. Examples of reported cases include the leakage of saltwater used to cool power plants and industrial facilities (Hughes et al., 2009; Brassington and Taylor, 2012) and the disposal of high-salinity production water from petroleum wells (Otton, 2006). Cross-contamination of aquifers may also occur via leaking wells if they have been improperly constructed or damaged (Winslow et al., 1957; Custodio and Bruggeman, 1987). Problems can be particularly severe in irrigation areas, where thousands of wells have sometimes been drilled by private land owners (Barlow, 2003). Observation wells with long well screens are also highly susceptible to borehole flow, and may result in contamination of freshwater reserves especially when there are abstraction wells nearby (Rotzoll, 2010).

6.2 Modes of Seawater Intrusion

Seawater intrusion can occur in many different forms. They have been schematically displayed in Figure 6.1. Each of these will be discussed in detail in this section.

6.2.1 Lateral Intrusion

6.2.1.1 Steady State Interface

Under natural, steady state conditions a wedge of seawater extends inland from the coastline, as discussed in Section 1.3. The saltwater in this wedge is separated from the freshwater above it by the transition zone. By diffusion and hydrodynamic dispersion, there is upward movement of the solutes in the intruded seawater into the freshwater that flows towards the coast (Figure 6.2). The solutes are entrained with the freshwater and ultimately discharge in the ocean. For the seawater wedge to maintain its position, it is therefore necessary that the resultant dissolved salt loss is compensated, which means that there is a continuous, slow landward flow of seawater within the wedge. The earliest studies in which this was recognised were carried out in Japan (Nomitsu et al., 1927), but the most famous description of the circulatory flow pattern within a seawater intrusion wedge became that by Cooper (1959). The image of the saltwater wedge in the Cutler area near Miami, Florida, is now the quintessential model of the distribution of fresh- and saltwater in a coastal aquifer (Figure 6.2).

The entrainment of dissolved salts in the freshwater flow towards the sea constitutes a significant loss of salt mass. Cooper (1959) realised that the action of molecular diffusion alone would not be sufficient at the prevailing concentration gradients across the transition zone to account for the flux required. He correctly inferred that hydrodynamic dispersion (Section 2.3.2) was an important solute transport process, even though at the time, the research on this subject was still in its infancy. Further to this, he contended that the oscillatory tidal displacement of the transition zone should promote hydrodynamic dispersion, an idea that was pioneered by Wentworth (1948), but also by Senio (1951) and Carrier (1958), and that dispersive mixing effects should be greater in heterogeneous aquifers than in homogeneous aquifers. The effects of tidal oscillation and aquifer heterogeneity on the shape of the transition zone are discussed in more detail in Section 6.2.1.4.

Based on field measurements along the Cutler area transect, Kohout (1960) demonstrated the existence of the landward-directed flow of seawater within the wedge. He acknowledged that heads based on water levels measured in the observation wells could not be used directly, as the point-water head (Section 1.5) is not a suitable potential from which groundwater flow can be inferred where differences in density exist between the wells.

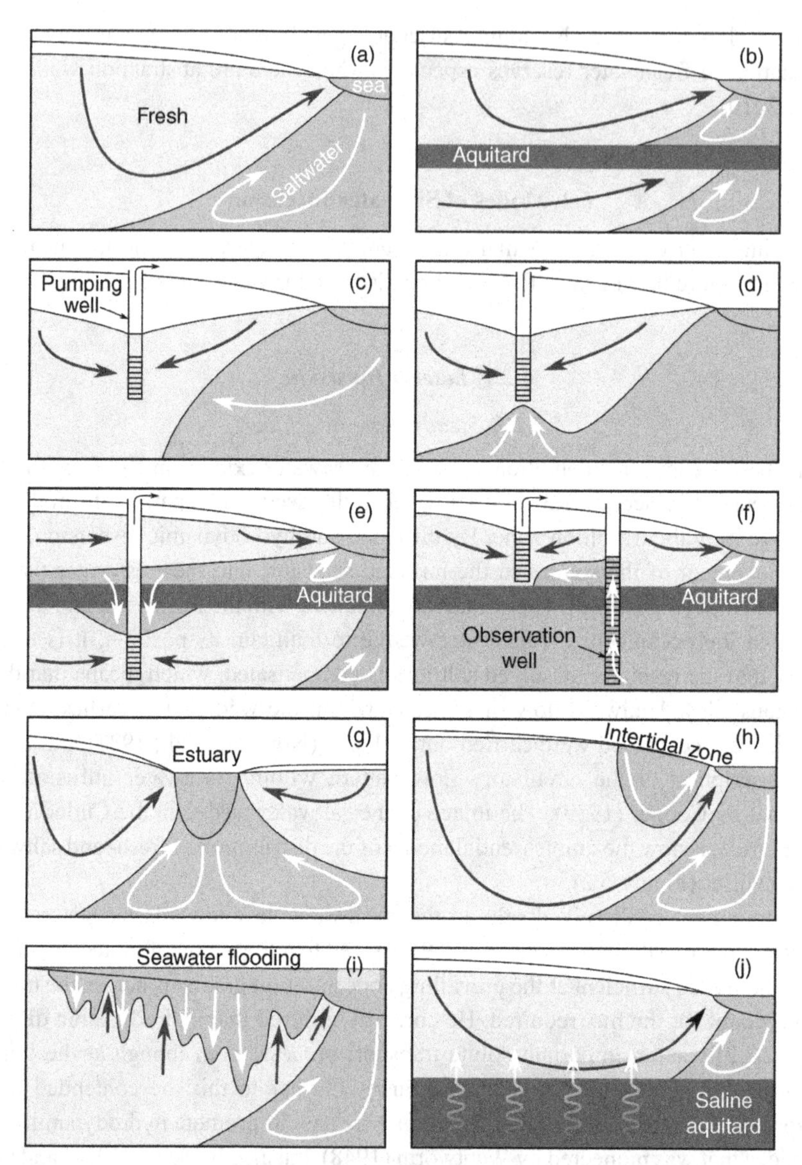

Figure 6.1 Schematic cross sections of a coastal aquifer showing different types of seawater intrusion. (a) Stable wedge in equilibrium with a steady water table. (b) Multiple wedges in a multi-layered aquifer system. (c) Active horizontal seawater intrusion caused by pumping. (d) Up-coning beneath a pumping well. (e) Down-coning caused by abstraction of groundwater beneath a saltwater aquifer. (f) Inter-aquifer leakage of saltwater by borehole flow. (g) Seawater intrusion via estuaries or canals. (h) Tide- and wave-driven seawater circulation between the high- and low water tidal mark. (i) Density-driven downward seawater intrusion during flooding. (j) Diffusion of dissolved salts from a confining unit with remnant seawater.

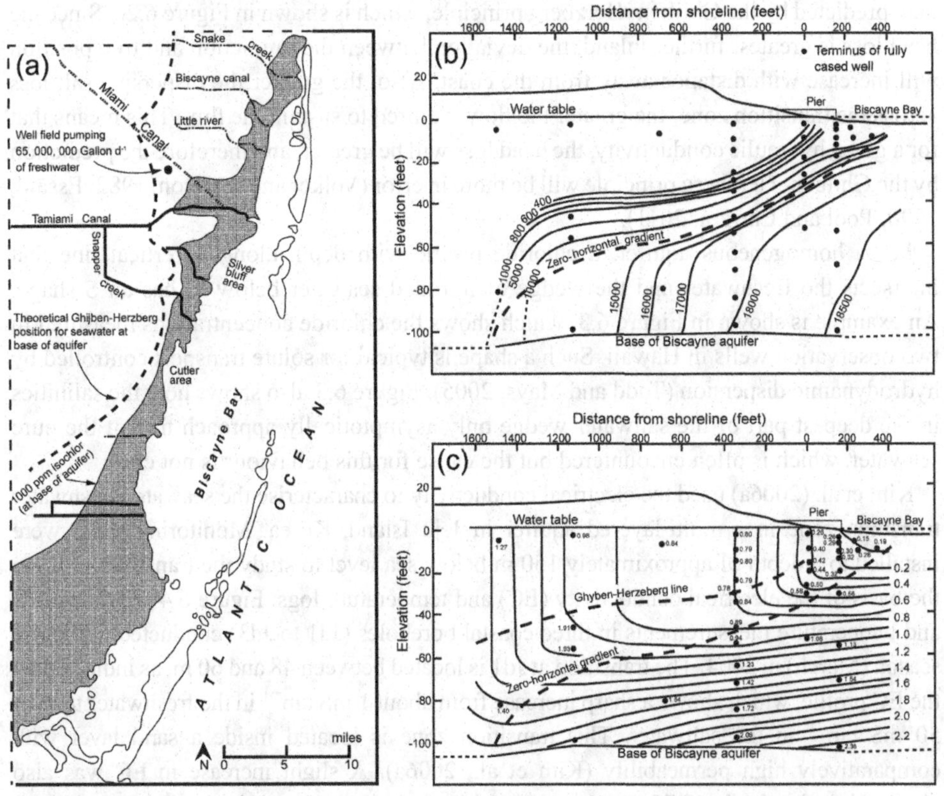

Figure 6.2 (a) Map showing the inland extent of intruded seawater marked by the 1000 ppm (parts per million, approximately equal to mg l^{-1}) isochlor at the base of the aquifer in Dade County, Florida. The theoretical position according to the Ghijben–Herzberg principle is indicated by a thick broken line. (b) Transect perpendicular to the coast showing the isochlors (in ppm) in the Cutler Area. (c) Same transect as in (b), but now showing the freshwater head contour lines. Figures from Kohout (1960).

Figure 6.2c shows the pattern of freshwater heads that was inferred, and the landward decrease (along a horizontal plane) in the lower part of the aquifer is a clear indication of there being flow. Near the coast, groundwater with 16 g l^{-1} of Cl^- (with the Bay having 18.3 to 18.8 g l^{-1} of Cl^-) is flowing seaward, highlighting that a part of the groundwater discharging is strongly mixed.

The head loss in the saltwater wedge has the effect that the position is more seaward than predicted by the Ghijben–Herzberg principle. This follows from the fact that the pressures (at a given elevation) of the water inside the dispersed wedge must be lower than the pressures within a wedge consisting of pure seawater, as otherwise there could not be flow. This means that the saltwater head inside the wedge is below sea level, and from Hubbert's formula (Eqn (3.7)) it then follows that the position of the interface moves to a lower elevation. So for any given horizontal position, the interface will be found at a greater depth than predicted by the Ghijben–Herzberg principle, which is shown in Figure 6.2c. Since the head loss is greatest further inland, the deviation between the prediction and true position will increase with distance away from the coast. Also, the greater the dispersive salt loss across the transition zone, the greater the flow required to sustain the flux. This means that for a given hydraulic conductivity, the head loss will be greater, and therefore the prediction by the Ghijben–Herzberg principle will be more in error (Volker and Rushton, 1982; Essaid, 1990; Pool and Carrera, 2011).

For a homogeneous aquifer, a chloride profile with depth along a vertical line that intersects the freshwater and the wedge of intruded seawater below it, has an S-shape. An example is shown in Figure 6.3, which shows the chloride concentrations measured in two observation wells in Hawaii. Such a shape is typical for solute transport controlled by hydrodynamic dispersion (Todd and Mays, 2005). Figure 6.3 also shows how the salinities in the deepest part of the saltwater wedge only asymptotically approach that of the pure seawater, which is often encountered but the cause for this behaviour is not clear.

Kim et al. (2006a) used the electrical conductivity to characterise the seawater-freshwater transition zone in a multi-layered aquifer in Jeju Island, Korea. Monitoring wells were installed to a depth of approximately 150 m below sea level to study the transition zone on the basis of the electrical conductivity (EC) and temperature logs. Figure 6.4 shows the EC and temperature measurements in three coastal boreholes (Jd1 to Jd3) conducted in the dry season (March) in 2004. The transition at Jd1 is located between 48 and 60 m, as indicated by the EC profile, which shows a sharp increase from about 1 mS cm^{-1} in the freshwater to over 50 mS cm^{-1} in the saltwater. This transition zone is located inside a sand layer with comparatively high permeability (Kim et al., 2006a). A slight increase in EC was also observed at the depth of 75 m where a sand layer is located. A transition zone of about 40 m thick exists from 80 to 120 m in borehole Jd2. However, in Jd3, only a narrow transition

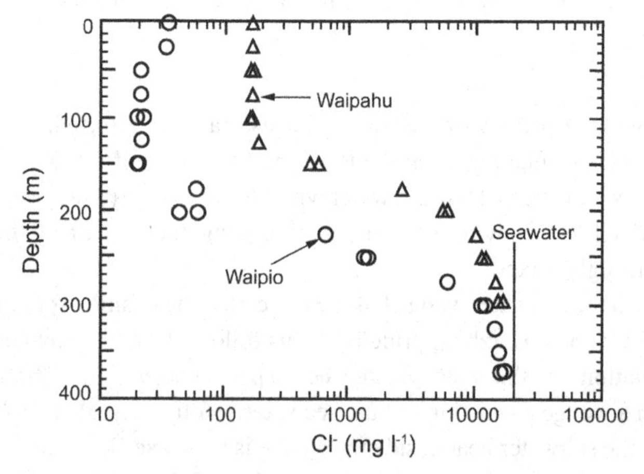

Figure 6.3 Chloride concentration versus depth in two observation wells (Waipio and Waipahu) in Hawaii (Voss and Wood, 1994).

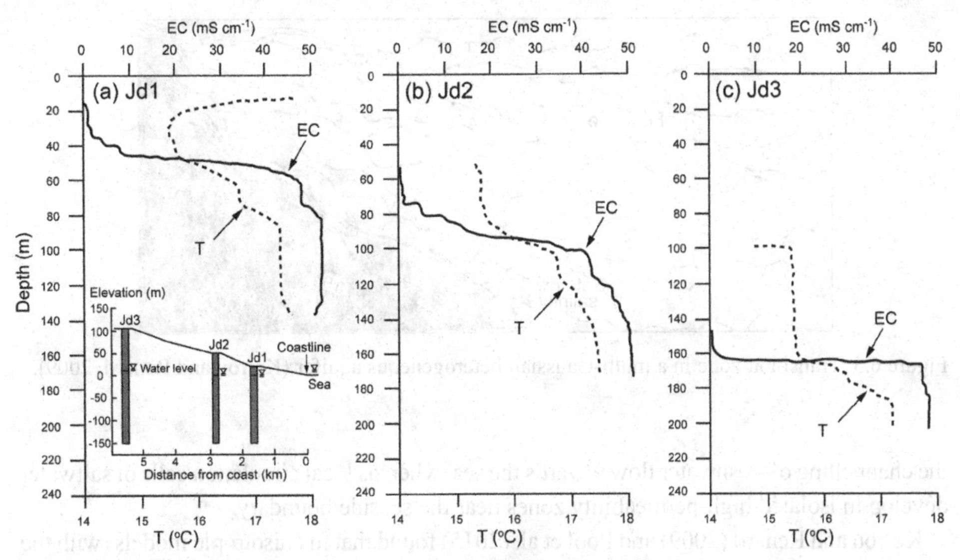

Figure 6.4 EC and temperature versus depth in three wells on Jeju Island, Korea. The inset shows the simplified cross section of the aquifer from Jd1 to Jd3 (modified from Kim et al., 2006a).

zone exists at a depth of about 165 m. The pattern of the temperature profiles in Jd2 and Jd3 is similar to the EC profiles. The temperature difference in the saltwater and freshwater zone is greater than 1°C. In Jd1 the temperature is lowest between 20 and 40 m. This may suggest a very permeable zone (Kim et al., 2006a) where flowing groundwater has a cooling effect. It cannot be excluded though that some of these depth trends are influenced by the borehole itself (Tellam et al., 1986; Shalev et al., 2009; Post et al., 2018a).

6.2.1.2 *Effects of Heterogeneity and Anisotropy*

Since actual aquifers are subject to multiple stresses that change with time (Prieto et al., 2006) and the spatial variability of their permeability is difficult to determine, the effects of lithological heterogeneity and anisotropy have been studied almost exclusively using mathematical models. Pioneering work was done by Dagan and Zeitoun (1998), who developed an analytical solution with which the uncertainty of the position of an interface in a layered confined aquifer can be quantified. Abarca (2006) and Kerrou and Renard (2009) studied an anisotropic dispersive variant of the Henry Problem (Section 2.5.1) and considered two-dimensional models with small-scale heterogeneity (i.e. caused by features much smaller than the dimensions of the wedge) and medium-scale heterogeneity (i.e. caused by features within the same aquifer that are comparable in size to the wedge). In these models the transition zone widens as a result of the irregular flow pattern caused by the log-normally distributed hydraulic conductivity field. At the same time the isocontour lines become more jagged as a result of the variability of the velocity field (Figure 6.5). When high-permeability zones are connected to the freshwater part of the aquifer, these promote

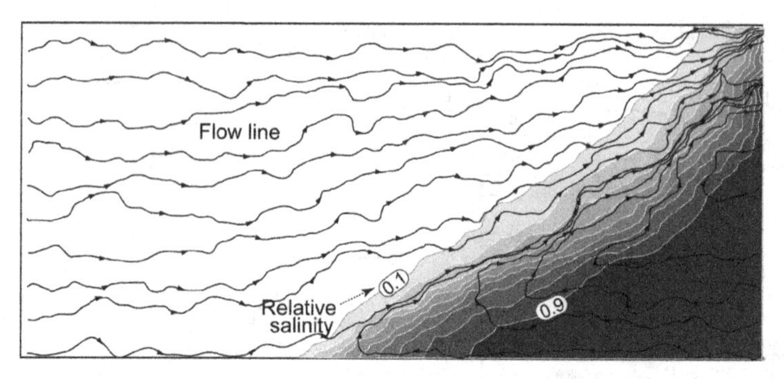

Figure 6.5 Transition zone in a multi-Gaussian heterogeneous aquifer (Kerrou and Renard, 2009).

the channelling of freshwater flow towards the sea, whereas local circulation cells of saltwater develop in isolated high-permeability zones near the seaside boundary.

Kerrou and Renard (2009) and Pool et al. (2015) found that in anisotropic models (with the hydraulic conductivity in the vertical direction being lower than in the horizontal direction), the wedge moves more landward when the degree of heterogeneity increases, and that due to the anisotropy, the transition zone widens in the horizontal direction more so than in the vertical direction. The greater landward movement of the wedge is because the effective horizontal hydraulic conductivity of a log-normally distributed, anisotropic hydraulic conductivity field increases with increasing heterogeneity and thus for a given rate of seaward freshwater flow, a smaller hydraulic head gradient suffices. With the freshwater influx being constant, and the head fixed on the seaward side, the inland heads fall, and as a direct consequence of the Ghijben–Herzberg principle (Section 3.3.2), the wedge moves inland.

It is important to stress though that these outcomes are strongly determined by the statistical properties of the log-normal conductivity fields. For example, in a two-dimensional system, the wedge penetration varies in an opposite manner for an isotropic field than for an anisotropic field, i.e. the transition zone moves seaward when heterogeneity increases (Kerrou and Renard, 2009). Somewhat confusingly, this was also observed for anisotropic fields in the simulations of Abarca (2006), but she scaled the fields between simulations to avoid boundary effects, thereby largely eliminating the increased hydraulic conductivity effect caused by increasing the heterogeneity. The results are therefore more comparable to the isotropic case studied by Kerrou and Renard (2009).

The dimensionality of these models has an important effect as well. Kerrou and Renard (2009) found that the aforementioned effects are more pronounced in two-dimensional models than in three-dimensional models. In addition, for their isotropic models they found that the landward penetration of the wedge decreases with increasing heterogeneity in two-dimensional models, but increases in three-dimensional models. Again, these findings are directly related to the nature of the hydraulic conductivity fields that were adopted because for isotropic conditions the effective hydraulic conductivity only changes with increasing heterogeneity in two-dimensional models but not in three-dimensional

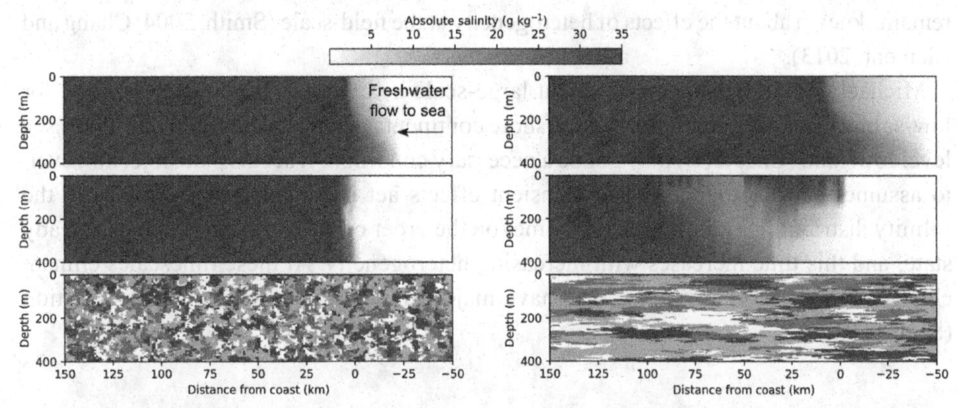

Figure 6.6 Simulated salinity distributions for two different heterogeneous fields of hydraulic conductivity (based on the modelling outcomes by Michael et al., 2016). The middle graphs show the simulated salinity for the heterogeneous field, and the top graphs show the results for a model with a homogenous permeability equal to the effective value of the heterogeneous field. The structure of the hydraulic conductivity fields is shown in the bottom graphs.

models. The dispersion, however, always becomes larger when increasing the heterogeneity, regardless of model dimensionality. The complex behaviours observed in the models are caused by the interplay of the effects caused by a change in effective hydraulic conductivity and the dispersion (Abarca, 2006). Unlike for isotropic aquifers, this behaviour is the same for two- and three-dimensional models alike in anisotropic models, as in anisotropic permeability fields, the effective hydraulic conductivity increases with increasing heterogeneity regardless of the model dimensionality.

Abarca (2006) and Kerrou and Renard (2009) found that the landward flow of seawater within the wedge is much greater in heterogeneous models than in their homogeneous equivalents. Michael et al. (2016) showed that large-scale geological heterogeneity can exert such a strong effect on the groundwater flow pattern that the salinity distribution that develops, no longer bears any resemblance to the classical portrayal of a wedge-shaped saltwater body (Figure 6.6). Using model simulations incorporating lithological data from the Lower Bengal Delta they showed that as horizontal continuity of geological strata increases, preferential flow becomes able to drive freshwater discharge up to tens of kilometres below the seafloor. A very wide transition zone develops beneath the seafloor, and the tortuous flow paths that result from the heterogeneity drive multiple saline circulation cells.

From these modelling studies, it seems that the widely held assumption that the landward saltwater flux is much smaller than that of the seaward flowing freshwater, is not generally valid. This also has important implications for submarine groundwater discharge and geochemical cycles, because the exchange of water between the ocean and offshore aquifers is much more vigorous and widespread in heterogeneous than in homogeneous models. The former are likely to be better representations of real aquifer systems, but little remains known about the effects of heterogeneity at the field-scale (Smith, 2004; Chang and Clement, 2013).

Michael et al. (2016) contended that large-scale heterogeneity is in itself enough for low-salinity groundwater to form in offshore continental shelf aquifers, and thus that a sea level lowstand (Chapter 8) may not be a necessary condition. It seems plausible, however, to assume that heterogeneity and transient effects act in tandem. This is because the salinity distributions take a very long time (on the order of 10^4–10^5 years) to reach steady state, and this time increases with increasing heterogeneity. At these timescales climate change and sea level variations also have major influences on the salinity distribution (Section 8.2).

6.2.1.3 Layered Aquifer Systems

The modelling studies discussed in the previous section are all characterised by permeability fields that vary gradually according to a pre-defined statistical model. These are appropriate to study the effect of heterogeneity within single aquifers, or, at a larger scale, in geological environments with gradually changing lithological properties, such as delta environments. When the lithology changes abruptly though, the hydrogeological properties of the rocks can change in a discrete fashion, and this has important implications for seawater intrusion, of course. Multiple wedges of intruded seawater can form (Figure 6.1b) in systems where more than one aquifer is present (Mualem and Bear, 1974).

An example of a wedge intruded into a multi-layer system is shown in Figure 6.7. It was published by Toyohara (1935), who conducted one of the earliest detailed and targeted field studies of seawater intrusion. Unfortunately this work has received little attention in the international literature. It forms a prime example of the role of aquitards in creating a staggered transition zone that predates the better-known works of Lusczynski and Swarzenski (1966) in Long Island. As Figure 6.7 shows, two pronounced vertically separated wedges exist near the coast, which Toyohara (1935) attributed to a silt layer (that he called the Hedro stratum), at a depth of approximately 10 m. When viewed as graphs of concentration versus depth, the initial increase of the chloride concentration with depth is followed by a decrease and then an increase again. This pattern persists until the borehole at 100 m from the coast. The wedge intruded into the lower aquifer much further than into the upper aquifer. Toyohara (1935) also measured water levels and realised that the tides will have an influence on the salinity distribution.

Oki et al. (1998) explored the effects of large-scale geological heterogeneity using a numerical model of the coastal aquifer system of Oahu, Hawaii, USA, where permeable volcanic rocks are overlain by marine and terrestrial sedimentary strata (Figure 6.8). The top part of the volcanic sequence is strongly weathered and less permeable than the unweathered volcanic rock. Together with the overlying sedimentary units, the weathered volcanic rocks are known as caprock, because it forms a barrier against groundwater discharge to the ocean. The model results showed a strong dependence on the flow field

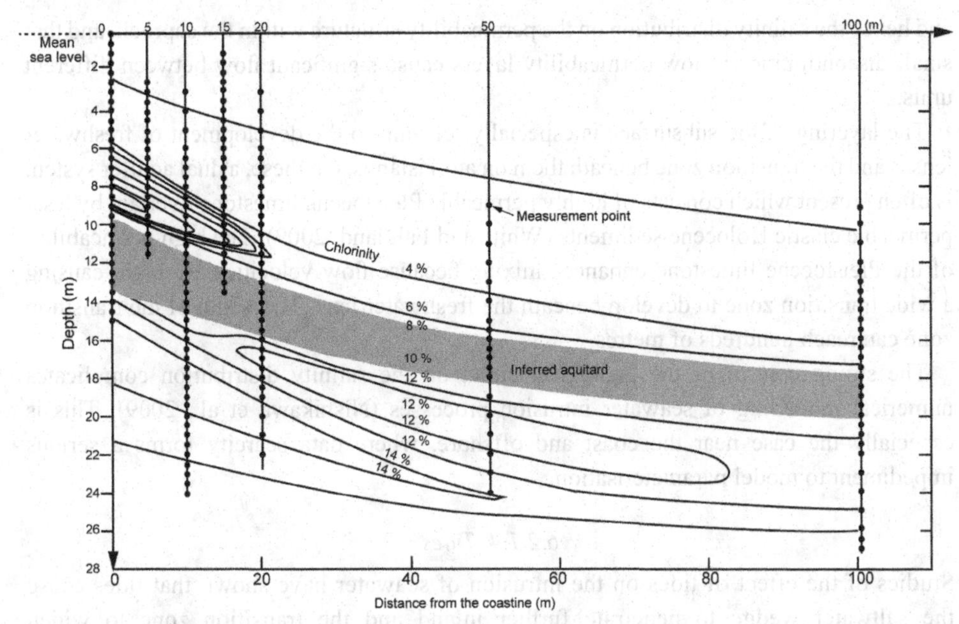

Figure 6.7 Cross section showing the distribution of salinities along a transect in Yumigahama, Japan (Toyohara, 1935). The shaded region shows the inferred position of the silt layer mentioned by Toyohara (1935).

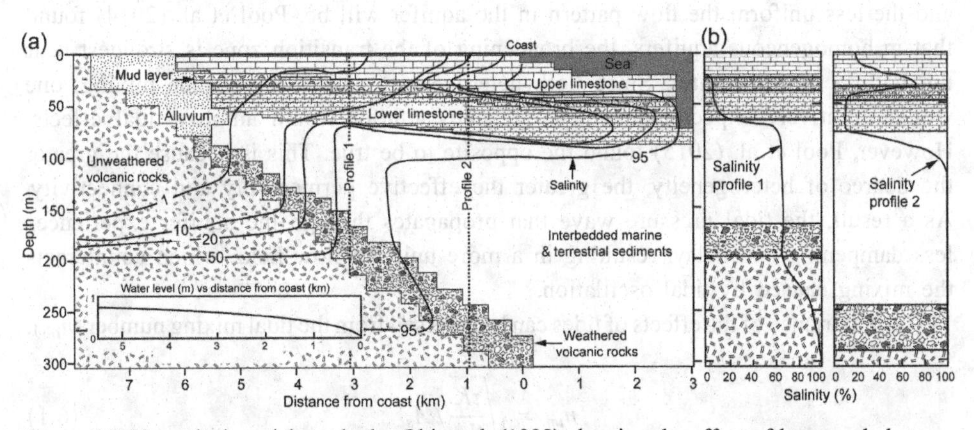

Figure 6.8 Numerical model results by Oki et al. (1998) showing the effect of large-scale heterogeneity on the salinity distribution in the coastal aquifer system of Oahu, Hawaii. (a) Relative salinity contours (values expressed as a percentage of seawater) along a cross section perpendicular to the coast. (b) Two profiles showing relative salinity versus depth. Inset in (a) shows simulated water level as a function of inland distance from the coastline.

and hence the salinity distribution on the permeability structure within the caprock, and that small discontinuities in low-permeability layers cause significant flow between different units.

The layering of the subsurface is especially relevant to the development of freshwater lenses and the transition zone beneath them on atoll islands. On these, a dual aquifer system is often present which consists of highly permeable Pleistocene limestone overlain by less-permeable clastic Holocene sediments (White and Falkland, 2009). The high permeability of the Pleistocene limestone enhances mixing because flow velocities are high, causing a wide transition zone to develop beneath the freshwater lens. The width of the transition zone can reach hundreds of metres.

The strong control of the geological strata on the salinity distribution complicates numerical modelling of seawater intrusion processes (Nishikawa et al., 2009). This is especially the case near the coast and offshore, where data scarcity forms a serious impediment to model parameterisation.

6.2.1.4 Tides

Studies of the effect of tides on the intrusion of seawater have shown that tides cause the saltwater wedge to penetrate further inland and the transition zone to widen (Underwood et al., 1992; Ataie-Ashtiani et al., 1999). Tides enhance mixing because the dampening of the tidal pressure wave with increasing distance from the coastline causes groundwater flow to respond in a non-uniform way, which tends to promote mixing. The effect is greater for greater tidal amplitude, but also increases with the aquifer storativity, because the higher the storativity, the stronger the damping effect, and the less uniform the flow pattern in the aquifer will be. Pool et al. (2014) found that in homogeneous aquifers, the broadening of the transition zone is strongest near the bottom of the aquifer. Since both heterogeneity and tides enhance mixing, one would intuitively expect that a heterogeneous aquifer would amplify tidal effects. However, Pool et al. (2015) found the opposite to be true. This is because the greater the degree of heterogeneity, the greater the effective permeability and connectivity. As a result, the tidal pressure wave that propagates through the aquifer experiences less dampening and delay, resulting in a more uniform flow field, which counteracts the mixing caused by tidal oscillation.

The importance of the effects of tides can be assessed from the tidal mixing number (n_{tm}),

$$n_{tm} = \sqrt{\frac{\tau K}{S_{s,f}}} \Big/ A_s \qquad (6.1)$$

in which τ [T] is the tidal period, K [L T^{-1}] is the hydraulic conductivity, $S_{s,f}$ [L^{-1}] is the specific storage coefficient in terms of head (Eqn (2.51)) and A_s [L] is the tidal amplitude. For heterogeneous aquifers, K must be replaced with an effective value of the hydraulic conductivity. For both homogeneous and heterogeneous aquifers alike, tidal effects on the

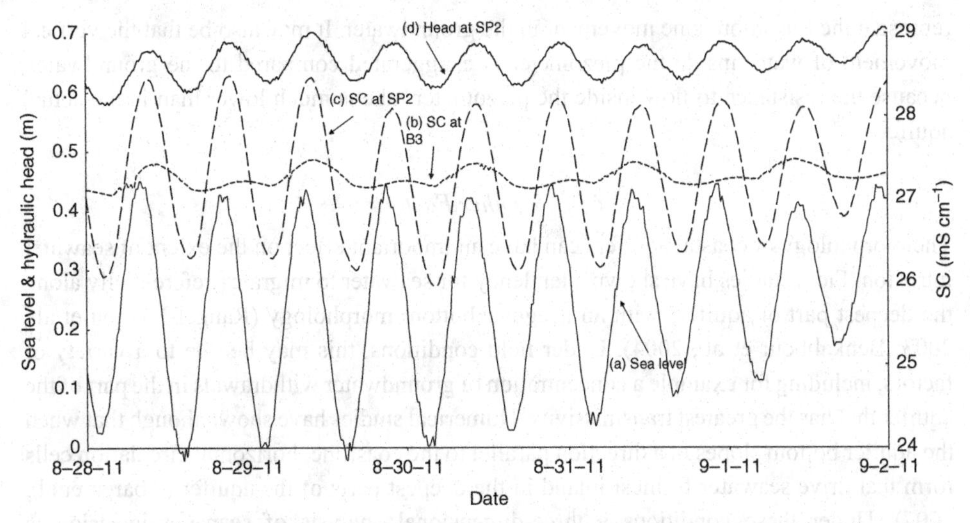

Figure 6.9 Sea level, specific conductance (SC) and hydraulic head versus time in a coastal aquifer in Israel during a five-day measurement period (Levanon et al., 2017). The groundwater measurements were taken at a distance of 70 m from the shoreline using a buried conductivity sensor (B3) as well as a piezometer (SP2). The specific conductance variations in the piezometer overestimate the true fluctuations within the aquifer as recorded by the buried sensor.

location and shape of the transition zone can become significant when $n_{tm} \leq 600$ (Pool et al., 2015).

While the effect of tides on the width and position of the transition zone can be significant, the oscillatory movement of the transition zone is small during a tidal cycle. Salinity fluctuations at a fixed point in the aquifer are therefore small, as shown in Figure 6.9. Using these field measurements, Levanon et al. (2017) showed that in an unconfined aquifer the vertical movement of the transition zone is linked to the movement of the water table (not shown in Figure 6.9). The physical explanation for this is that the propagation of the pressure wave is so fast in the deeper parts of the system that it does not create any significant head gradients that drive flow. Near the water table, however, the movement of (fresh) groundwater into and out of the unsaturated zone in response to the tidal pressure fluctuation causes a noticeable vertical displacement. The transition zone follows this vertical groundwater motion.

When comparing the specific conductance at B3 (a buried sensor) and SP2 (a sensor in a piezometer), it can be seen that when the specific conductance is measured using a buried sensor, the amplitude recorded during a tidal cycle is much smaller than when the sensor is placed inside a piezometer. A piezometer will overestimate the amplitude when the salinity of the water column inside it varies with depth, due to leaking tubes, or the entry of freshwater through the opening at the land surface (Section 12.2). As the water inside the piezometer moves up and down, the sensor records the oscillatory movement of the transition zone inside the piezometer as a salinity variation in time but this does not represent the transition zone movement in the groundwater. It may also be that the vertical movement of water inside the piezometer is exaggerated compared to the groundwater, because the resistance to flow inside the piezometer tube is much lower than in the actual aquifer.

6.2.1.5 Other Factors

The morphology of coastal aquifers can have an important effect on the extent of seawater intrusion. Field studies have shown a tendency for seawater to migrate preferentially along the deepest part of aquifers with an irregular bottom morphology (Rangel-Medina et al., 2003; Benkabbour et al., 2004). Under field conditions, this may be due to a variety of factors, including for example a concentration of groundwater withdrawals in the part of the aquifer that has the greatest transmissivity. Numerical studies have shown though that when the aquifer bottom slopes in a direction parallel to the coastline, horizontal circulation cells form that drive seawater furthest inland in the deepest parts of the aquifer (Abarca et al., 2007). Under these conditions a three-dimensional analysis of seawater intrusion is warranted.

Vertical geological structures such as faults or dykes can further exert an important control on the pathways of intruding seawater, as well as the seaward flowing fresh groundwater (Barker et al., 1998; Houben et al., 2018). Such structures can be relatively permeable or impermeable depending on the host rock and nature of the faults. For example, weathered dykes in crystalline hard rocks tend to form relatively permeable structures, whereas unweathered dykes in more plastic sedimentary strata have a lower permeability than the surrounding units (Comte et al., 2017). Seawater intrusion in the Clarendon Basin in Jamaica was found to be along permeable conduits associated with fault zones (Howard and Mullings, 1996). In addition to faulting, folding can create complex geological structures that lead to compartmentalisation of the groundwater system (Yechieli et al., 2009b).

The effect of fractures has been found to critically depend on their orientation and position within the flow system. Insights obtained from modelling studies have shown that vertical fractures within the wedge can enhance the thickness of the transition zone, whereas channelling of freshwater flow along horizontal fractures can cause a narrowing of the transition zone (Sebben et al., 2015).

6.2.2 Moving Transition Zone

Pumping from coastal wells is the most common cause of seawater intrusion (Barlow, 2003). As the pressure in the aquifer gets lowered, a landward flow of seawater is induced (Figure 6.1c). Some scholars distinguish between passive and active seawater intrusion. With passive seawater intrusion, the flow of freshwater towards the sea is maintained, but with active seawater intrusion, the effect of pumping is so strong that the freshwater outflow ceases and all the water in the aquifer moves landward (Fetter, 2001; Werner, 2017).

Turner and Foster (1934) were among the earliest to relate the shape of the transition zone to the fact that it was moving inland, and argued that its blunt shape may have to do with groundwater withdrawal. In part the blunt shape of the transition zone has to do with the condition that the isocontours have to be perpendicular to an impermeable aquifer bottom across which flow and dispersive transport from underlying layers is negligible (Henry, 1964). Numerical, laboratory and field studies have shown that the landward movement of the transition zone is often accompanied by a steepening of the transition zone (Calvache and Pulido-Bosch, 1991; Andersen et al., 2005; Badaruddin et al., 2015). The greater the rate of pumping, the greater the steepening. This is because the relative importance of the density-dependent driving forces becomes smaller as the inland groundwater flow rate increases (Badaruddin et al., 2017).

The effect of pumping on the transition zone was analysed in great detail in a field experiment of induced seawater intrusion into an unconfined beach aquifer in Denmark (Andersen et al., 2005; Jørgensen et al., 2008). Measurements of the chloride concentration of water samples from a dense array of multi-level observation wells were used to determine the response of the transition zone to pumping. Prior to the experiment, a wedge of intruded seawater was present in the unconfined aquifer with the contour lines of equal chloride concentration dipping landward. The position of the wedge hardly changed during the months before the experiment (Andersen, 2001). Pumping took place between 5 April 2000 and 18 January 2001, and during this period the transition zone moved inland by approximately 90 m. This resulted in a steepening of the transition zone, which is apparent from the near vertical position of the chloride concentration contour lines (Figure 6.10b). Upon termination of pumping, the transition zone tilted back and the intruded seawater body regained a wedge shape, albeit that it remained located far inland of the natural equilibrium position at least up until the last measurements in February 2003 (Figure 6.10c). Strontium isotope data measured on the water samples revealed that the pumping also led to the upward flow of freshwater from an underlying limestone aquifer (Jørgensen et al., 2008), and the experiment thus showed the intricate flow patterns and mixing between water types that can occur under field conditions (Figures 6.10d–6.10f).

In some numerical studies of the motion of the seawater wedge an overshoot phenomenon has been observed (Watson et al., 2010), which has also been confirmed in physical laboratory experiments (Morgan et al., 2013). It has been described under conditions where sea level was changed instantaneously, in which case the transition zone first moved landward of the steady state position associated with the new sea level, before retreating seaward and assume this steady state position. It is presently unclear if the process might actually occur under field conditions, or if it is limited to the special conditions studied in numerical and sand tank experiments. Given the complexity of coastal aquifers, and the multitude of interacting processes, the relevance of overshoot phenomena for management purposes is difficult to assess and thus far has attracted scientific interest only.

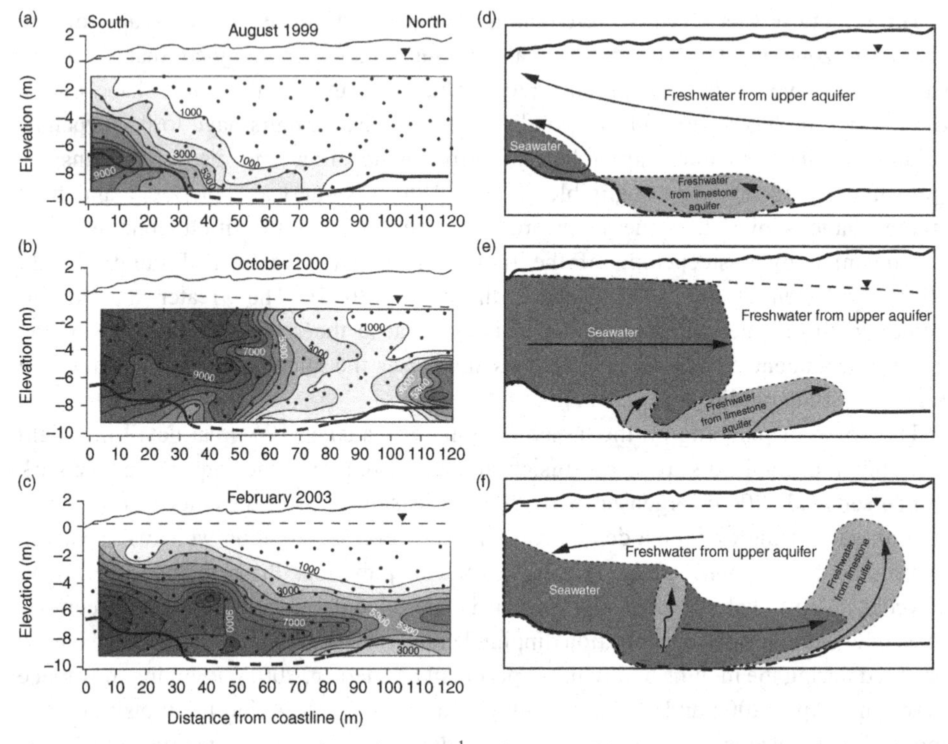

Figure 6.10 Chloride concentrations (in mg l^{-1}, left pane) and water types (right pane) during a controlled seawater intrusion experiment in Denmark (modified from Jørgensen et al., 2008).

6.2.3 Up-Coning

Up-coning is the process where the abstraction of water from a pumping well leads to the upward migration of more saline groundwater (Figure 6.1d). It was first identified in the field by Pennink (1905) and forms a major nuisance in many water supply systems (Houben and Post, 2017). Since coastal aquifers tend to be much thinner than they are wide, the vertical migration distances are much shorter than the horizontal ones, which makes up-coning one of the foremost hazards to drinking-water wells. Because of this, it has been extensively studied (Reilly and Goodman, 1985). The mathematical treatment of up-coning by means of analytical solutions is discussed in Section 3.6.

Theoretical studies have shown that a stable cone of saltwater can develop beneath a pumping well if the pumping rate does not exceed a critical threshold (Dagan and Bear, 1968). The apex of the cone will be located below the pumping well, and freshwater flows along the transition zone, which remains in a stable configuration. Nonetheless, because of dispersion of solutes across the transition zone, the salinity of the pumped water can be elevated relative to the freshwater (Reilly and Goodman, 1987). An asymmetric transition zone tends to develop, which is wider on the freshwater than on the saltwater side of the

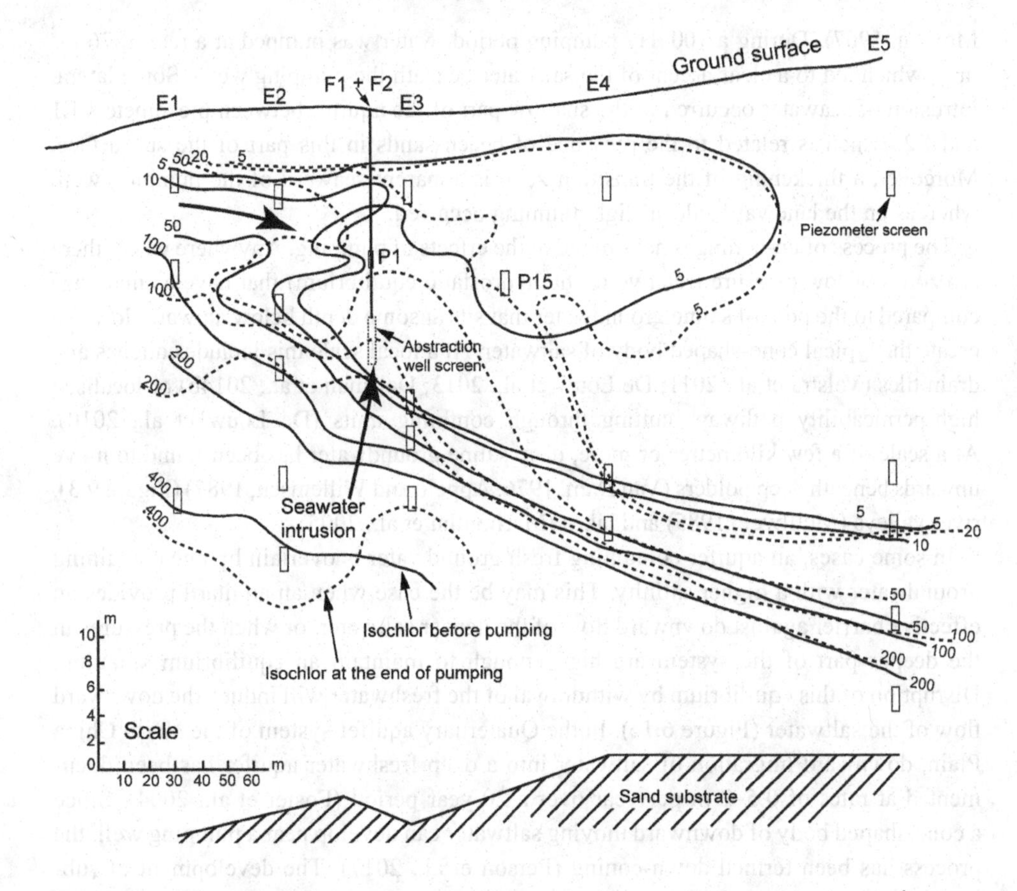

Figure 6.11 Cross sections showing the contour lines of equal chloride concentration (meq l^{-1}) before and after pumping in a coastal aquifer in Senegal (Debuisson and Moussu, 1967).

50% isochlor. This is because flow velocities in the freshwater zone are higher than those in the saltwater zone. During the transient phase of cone development, i.e. the transition from a horizontal boundary between fresh- and saltwater to a cone-shaped transition zone, the longitudinal dispersion is critically important. This is because the motion of the saltwater towards the well promotes mixing in the direction of the flow (Jakovovic et al., 2011). As conditions stabilise, the flow in the saltwater body becomes smaller, and the mixing of solutes across the transition zone is dominated by transverse dispersion (Reilly and Goodman, 1987). The early stages of saltwater up-coning are therefore characterised by a wider transition zone than the final stable stages.

Heterogeneity and transient processes complicate the aforementioned general model under field conditions. A field example of up-coning beneath a pumping well is shown in Figure 6.11. The cross section displays lines of equal chloride concentration near a pumping well in a sand aquifer prior to and after a period of groundwater abstraction (Debuisson and

Moussu, 1967). During a 100 day pumping period, water was pumped at a rate of 26 m^3 hr^{-1}, which led to a clear ascent of the saltwater beneath the pumping wells. Some lateral intrusion of seawater occurred in the shallow part of the aquifer between piezometers E1 and E2, which is related to the presence of beach sands in this part of the subsurface. Moreover, a thickening of the transition zone is apparent seaward of the pumping well, whereas on the landward side, a slight thinning occurred.

The process of up-coning is not limited to the effects of pumping. Anywhere where there are zones of low pressure (relative to the hydrostatic equilibrium) that have a small size compared to the pool of saline groundwater that sits at some depth below, upward flow can create the typical cone-shaped body of saltwater. At a local scale this includes ditches and drain tiles (Velstra et al., 2011; De Louw et al., 2013; Delsman et al., 2014b) or localised high-permeability pathways cutting through confining units (De Louw et al., 2010). At a scale of a few kilometres or more, deep saline groundwater has been found to move upwards beneath deep polders (Van Dam, 1976; Appelo and Willemsen, 1987) (Figure 9.3), river valleys (McElwee, 1985) and lakes (Marimuthu et al., 2005).

In some cases, an aquifer containing fresh groundwater is overlain by one containing groundwater with a higher salinity. This may be the case when an aquitard provides an effective barrier against downward flow of the denser saltwater, or when the pressures in the deeper part of the system are high enough to maintain an equilibrium situation. Disruption of this equilibrium by withdrawal of the freshwater will induce the downward flow of the saltwater (Figure 6.1e). In the Quaternary aquifer system of the North China Plain, downward migration of saltwater into a deep freshwater aquifer has been documented at rates of 0.5–2 m per year over a 20 year period (Foster et al., 2004). Since a cone-shaped body of downward moving saltwater can develop near a pumping well, the process has been termed down-coning (Person et al., 2017). The development of submarine low-salinity groundwater reserves is likely to be negatively impacted by this process.

6.2.4 Downward Seawater Intrusion

Surface waters that are connected to the sea form pathways for seawater intrusion (Figure 6.1g). They include estuaries, rivers, canals and tidal creeks. They may be filled with seawater entirely and all the time, but freshwater may also occur, or seawater is only present part of the time. In mixed seawater-freshwater systems, the seawater may pool on the bottom due to its high density, or when there is vigorous mixing, brackish water may form. Seawater intrusion along surface water is especially prevalent in low-lying coastal regions where the water stage is up to only a few metres above sea level, and where freshwater discharge to the sea is low enough for the seawater to move into the channel against the direction of the freshwater current. As surface waters are often contaminated with nutrients or man-made chemicals, this form of intrusion does not only constitute a source of salt to the aquifer but of other contaminants as well.

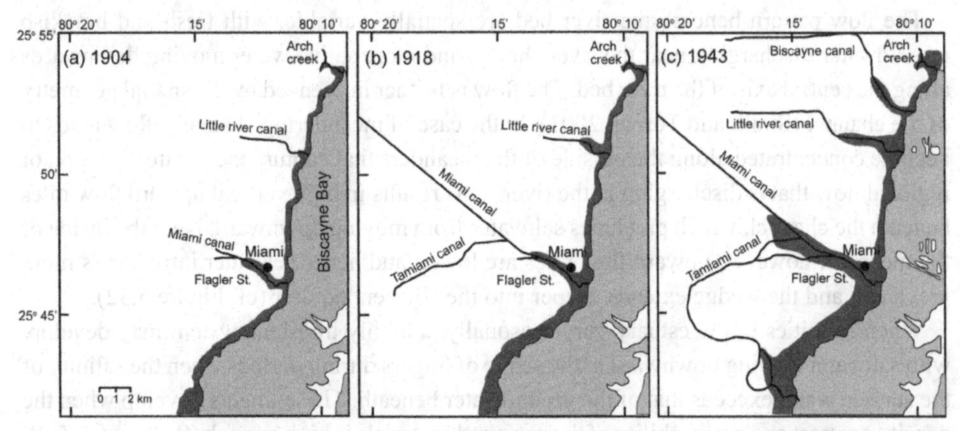

Figure 6.12 Maps showing the progression of seawater intrusion along canals in Miami-Dade County, Florida, between 1904 and 1943. Seawater-affected areas are shown in dark grey, and the front of the saltwater and freshwater is defined at a chloride concentration of 100 mg l^{-1} (Renken, 2005).

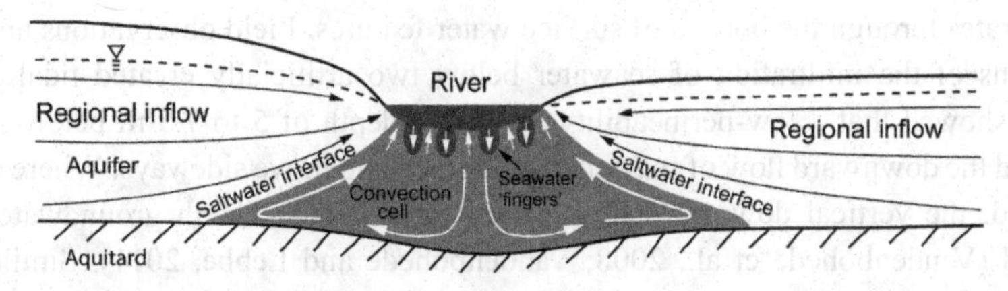

Figure 6.13 Schematic cross section showing vertical seawater intrusion below a saline river (Smith and Turner, 2001).

As an example, Figure 6.12 shows seawater intrusion in the Biscayne aquifer in Miami-Dade County, Florida, USA from 1904 to 1943 (Renken, 2005). The uncontrolled drainage by means of the canal system caused a lowering of the water table by nearly 1.8 m, which allowed seawater to intrude into the superficial aquifer. Dredging of materials from the canal bottoms aggravated the situation, as the removal of fine-grained material increased the hydraulic connectivity between the aquifer and the canal (Parker et al., 1955). Groundwater withdrawal also resulted in high-salinity water from the canals to be drawn into the aquifer.

An idealised example of seawater intrusion below an estuary is shown in Figure 6.13 (Smith and Turner, 2001). Just like intrusion of seawater across the seafloor, a circulatory flow pattern of seawater develops inside a wedge-shaped body of saltwater in the aquifer. Mixing of intruded seawater and discharging freshwater occurs along the transition zone, which may drive additional seawater to intrude into the aquifer to compensate for the salt mass lost as this mixed water discharges into the river.

The flow pattern beneath the river bed are spatially variable, with fresh and brackish groundwater discharging near the river shores, and more saline water moving downwards along the central axis of the river bed. The flow is further influenced by the spatial geometry of the channel (Smith and Turner, 2001). In the case of meandering channels, flow tends to become concentrated along the outside of the meanders that capture the greatest portion of regional flow that is discharging in the river. This results in high vertical upward flow rates beneath the channel, which precludes saltwater from moving downward. Near the inside of the meander, however, upward flow rates are lower, and hence seawater intrusion is more prevalent, and the wedge extends deeper into the adjacent aquifer (cf. Figure 5.32).

When salinities in the estuary vary seasonally, a highly transient system may develop, with saltwater moving downward in the shape of fingers during periods when the salinity of the surface water exceeds that of the groundwater beneath. These fingers develop when the density contrast and permeability of the river bed materials is high enough (Section 6.2.5.2), but the distribution of upward flow plays a key role as well. Finger development is prevented when upward flow rates are high enough, but in zones of low upward flow this stabilising influence is not as strong and focal points of salinisation form here (Smith and Turner, 2001).

Low-permeability units play an important role in controlling the distribution of saltwater that infiltrates through the bottom of surface water features. Field observations and model simulations of the infiltration of seawater below two artificially created tidal inlets in Belgium showed that a low-permeability layer at a depth of 5 to 7.5 m below sea level obstructed the downward flow of seawater, and forced it to move sideways. Where the layer terminated, the vertical downward flow of seawater into the fresh groundwater below continued (Vandenbohede et al., 2008; Vandenbohede and Lebbe, 2011). Similarly, the salinisation of an aquifer connected to an estuary on the Mediterranean coast in Israel extended much further in permeable sand units than in less-permeable silt layers (Shalem et al., 2014).

6.2.5 Flooding

Many low-lying coastal areas experience episodic flooding. The flooding can be frequent and regular in nature, for example with it is related to tidal cycles, or it can be related to catastrophic events like tsunamis or hurricanes. Inundation of the low-lying areas can occur in different ways. It can be direct by a simple landward displacement of the coastline, but it can also occur by the overtopping or breaching of a natural or artificial barrier (Ramsay and Bell, 2008). In the latter case, the retreat of the seawater is difficult and long-lasting seawater intrusion can be the result.

6.2.5.1 Regular Flooding

Beach aquifers and salt marshes are subject to regular flooding driven by tidal cycles (Section 7.3.2). Depending on the environment, the flooding occurs in regular cycles varying in frequency from once or twice per day, to once every month or every few months. The regularity of this cycle gives rise to particular hydrodynamic conditions that have a strong influence on the distribution of solutes in groundwater.

Below beaches, the infiltration of seawater during high tides causes seawater to infiltrate into the often high-permeability sediments of this high-energy depositional environment. As the sea retreats during low tide, the infiltrated seawater flows towards the low water line and discharges there. Even though the flow directions in the aquifer can change to landward during a rising tide, the constant infiltration of seawater drives a net seaward flow when averaged over multiple tidal cycles (Figure 6.1h).

The size of a saline circulation cell that becomes established this way depends on a number of factors, including the slope of the beach, the permeability of the beach sediment and the tidal amplitude. The cell can be as shallow as a metre or less (Ullman et al., 2003; Robinson et al., 2007; Abarca et al., 2013), but can also reach a thickness of over 10 metres (Lebbe, 1981, 1999) (Figure 6.14). The position of the cell is highly dynamic in time as a result of spring-neap cycles (Abarca et al., 2013), seasonal changes of freshwater outflow (Heiss and Michael, 2014), storms (Robinson et al., 2014) and changes to the beach morphology caused by sedimentation and erosion (Huizer et al., 2017).

The presence of the cell means that freshwater flowing towards the sea is forced to flow underneath before discharging, which means that the outflow zone is often seaward of the low water line. This complicates the detection of this water flux, which can be an important source of nutrients and other dissolved substances to the coastal environment.

6.2.5.2 Catastrophic Flooding

Low-lying coastal regions are susceptible to flooding by seawater which makes that fresh groundwater resources are vulnerable to vertical seawater intrusion. Their vulnerability is increasing due to a number of influences. The incidence of minor flooding events has increased in the USA as a result of the accelerated rise of sea level during the twentieth century (Kopp et al., 2016; Sweet et al., 2014). Moreover, rising temperatures lead to a greater risk of cyclone-related storm surges (Grinsted et al., 2013). Apart from climate related factors, the conversion of natural coastal areas like mangroves into fish farming areas or by land reclamation, as is happening at a large scale in Asia (Section 9.2.1), removes the natural buffers that protect coastal plains from the sea. Finally, as population numbers increase, a greater number of people move to previously un- or sparsely inhabited areas closer to the shore.

These concerns over future environmental and demographic change have led to intensified research into the effect of vertical seawater intrusion caused by seawater flooding. Another major driver was the Boxing Day tsunami that hit the countries bordering the Indian Ocean in December 2004. The impact of a tsunami on coastal water resources is two-fold (Villholth and Neupane, 2011): the powerful waves and the enormous amount of water can cause immediate damage to infrastructure such as wells, storage tanks, sewer systems and water pipelines; and the inundation by seawater contaminates the coastal water resources. The inundation of seawater over the tsunami-affected areas may last a few

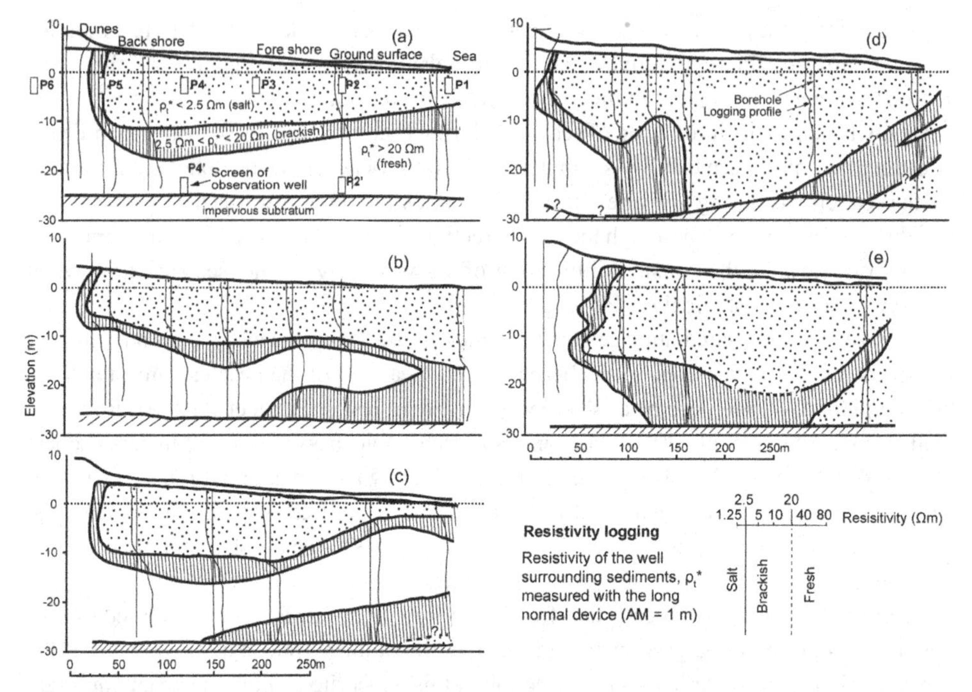

Figure 6.14 Distribution of fresh, brackish and saltwater based on borehole resistivity (ρ_t^*) measurements along transects perpendicular to the coast of the De Panne area, Belgium (modified from Lebbe, 1999). The cross sections are separated by a few hundred metres to just over a kilometre along the beach.

minutes to a few days, but the infiltration of the saltwater through the soil can last for much longer. Seawater trapped in depressions on the land surface continues to contaminate the underlying aquifer after a tsunami has retreated.

Open, large-diameter wells can act as a fast and direct pathway for the seawater into the aquifer (Villholth and Neupane, 2011). Although sealed wells with raised standpipes on raised platforms help to avoid this problem, improper well construction, broken seals or corroded standpipes still form pathways for aquifer salinisation following a flood (Cardenas et al., 2015). In Sri Lanka, seawater filled over 50 000 domestic open dug wells up to 1.5 km distance inland following the 2004 tsunami. The salinity record of three wells before and after the tsunami in Kaggola, Sri Lanka, is shown in Figure 6.15 (Illangasekare et al., 2006). A sharp increase in the salinity, as measured by the specific conductance in X-141 and P-09, is clearly apparent immediately following the tsunami. The salinity in well P-06 showed a slightly more gradual response, which Illangasekare et al. (2006) attributed to the lateral spreading of infiltrated seawater.

Direct groundwater observations that allow studying the salinisation processes in response to flooding are scarce. A notable exception is the study by Andersen et al. (2008) of which the results are shown in Figure 6.16. A dense network of approximately

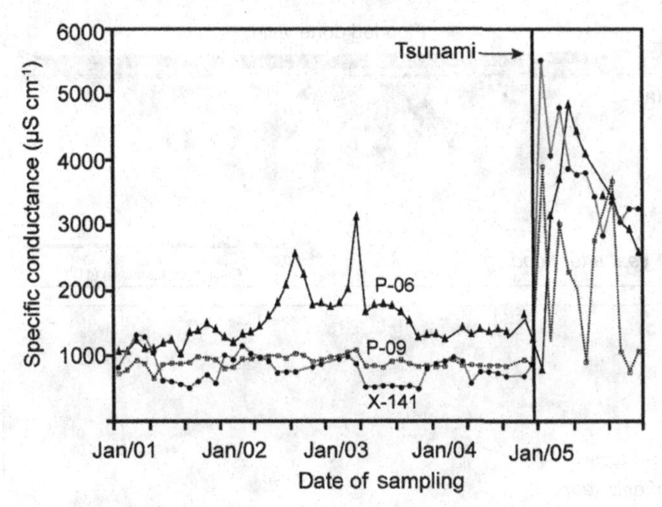

Figure 6.15 Salinity recorded before and after the 2004 Boxing Day tsunami in three tube wells in Kaggola, Sri Lanka (Illangasekare et al., 2006).

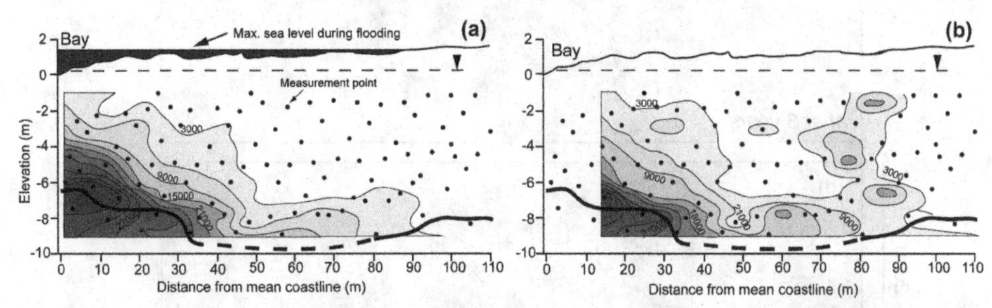

Figure 6.16 Vertical cross sections of the specific conductance (in μS cm^{-1}) of groundwater along a transect of a beach aquifer in Denmark on (a) 28 August 1999 and (b) 27 February 2000 (Andersen et al., 2008). Note that this is the same transect as in Figure 6.10.

100 observation wells were installed along a 120 m transect across a beach that was hit by a storm surge on 30 January 1999. The seawater flooded the beach up to 90 m from the coast, and pooled for some days in the depressions. Electrical conductivity measurements of water samples taken before the surge show a wedge of intruded seawater that is typical given the hydrological conditions at the site (Andersen et al., 2008). The effect of the inundation event is apparent from the measurements taken 28 days after the storm, which show that there is a plume of seawater moving down into the aquifer at roughly 70 to 90 m from the coastline.

Subsequent studies have highlighted the importance of salt finger formation following a flood (Cardenas et al., 2015; Yang et al., 2015; Post and Houben, 2017). The infiltration of seawater directly after the event is driven by the head generated by the water standing on the land surface, as well as the higher density of seawater than that of the fresh groundwater

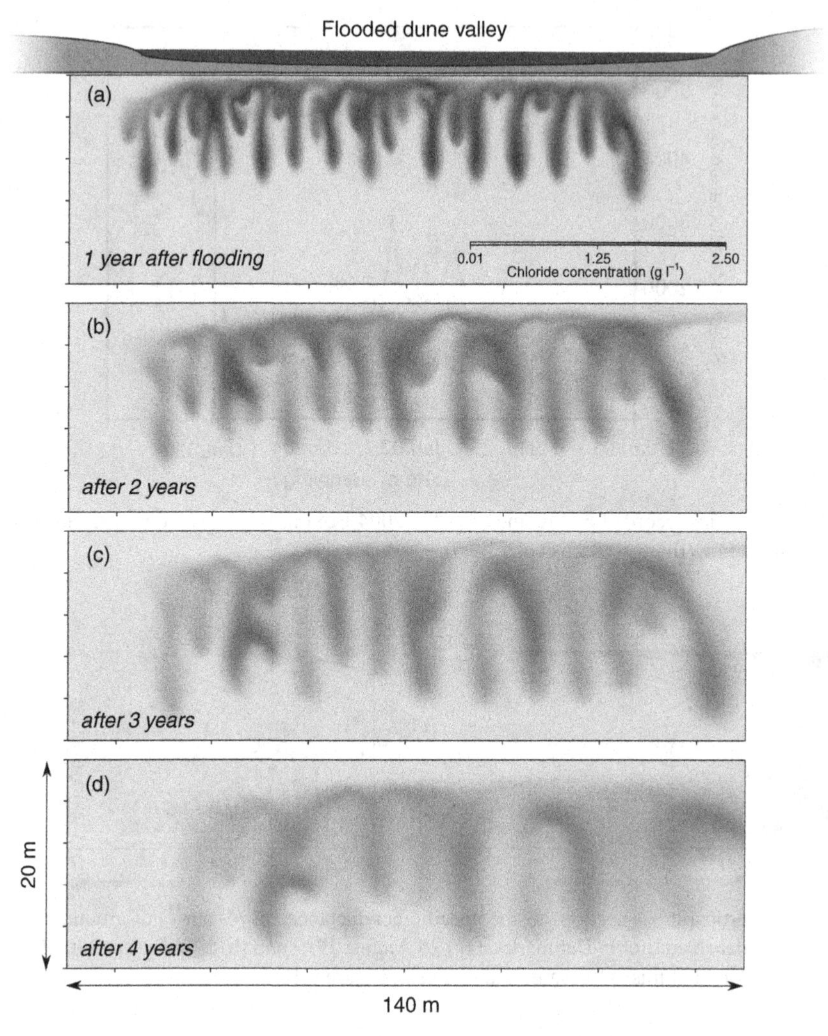

Figure 6.17 Cross sections showing modelled chloride concentrations following the inundation of a dune valley by seawater on the island of Baltrum, Germany, after (a) 1, (b) 2, (c) 3 and (d) 4 years since the flood. Based on modelling results by Post and Houben (2017).

(Illangasekare et al., 2006). Initially, when the density contrast is relatively high, salt fingers may form (Figure 6.1i) that can descend with a velocity of several metres per year (Figure 6.17). This salinisation process is known as convective fingering, convection or free convection. The latter is reserved for cases where flow is driven exclusively by density differences.

Whether or not fingers form is primarily a function of the permeability of the subsurface and the density contrast between the seawater and the groundwater. This is formally expressed by the Rayleigh number

$$Ra = \kappa \Delta \rho g B / (D_e \mu) \tag{6.2}$$

where κ [L^2] is the intrinsic permeability, $\Delta\rho$ [M L^{-3}] the density difference between the seawater and the groundwater, g [L T^{-2}] the gravitational acceleration, B [L] the thickness of the aquifer (L), D_e [L^2 T^{-1}] the effective diffusion coefficient and μ [M L^{-1} T^{-1}] the dynamic viscosity of water. The often quoted critical value of $Ra > 4\pi^2$ as a criterion for the onset of convective finger formation relies on the assumption of an initial linear decrease of $\Delta\rho$ with depth over the interval B. Other complicating factors like heterogeneous permeability and the presence of a background flow field limit the definition of an onset criterion under field conditions (Simmons et al., 2002), but the Rayleigh number is nevertheless useful as it reveals the quantitative relationship between the parameters that influence the formation of salt fingers.

Field observations and model simulations (Terry and Falkland, 2010; Post and Houben, 2017) have shown that vertically intruded seawater slowly disperses by mixing as it moves downward. Due to the corresponding dilution of the seawater the relative importance of density-driven flow decreases, and the downward migration rate becomes restricted by the natural vertical flow rates that are due to rainfall recharge. This means that the downward migration of the salt-contaminated water becomes slower while at the same time a larger volume of the fresh groundwater becomes contaminated.

Remediation by pumping needs to be done with great care, as it can induce up-coning of saltwater (Illangasekare et al., 2006) or draw in the descending brackish water that still has not reached the bottom of the freshwater lens. Vithanage et al. (2009) demonstrated that the salinity is higher in the areas using pumping as a remediation method than in the sites only flushed by rainfall. This is because pumping induced further mixing of the freshwater with the seawater.

In addition to contamination by seawater, there is also a threat of contamination related to the leakage and spread of bacteria and pathogenic contaminants in the floodwaters which mix with sewage from damaged sanitary systems (Shaw et al., 2006). There may be also unique contamination problems related to a specific coastal area. For example, the most serious problem caused by the east Japan earthquake and tsunami in March 2011 was the contamination of both surface and groundwater by damaged nuclear power stations (Chague-Goff et al., 2012; Sanial et al., 2017).

6.2.6 Diffusion

Salinisation of fresh groundwater by seawater can occur in the absence of flow, albeit at a much slower rate than in the case of advective transport by flowing groundwater. Diffusion refers to the migration of solutes from regions of a high chemical concentration to where the concentration is lower, and it results from the random motion of solutes at the microscopic scale. Where a concentration difference exists, and in the absence of other forces, there will be a net flux of solutes in the direction of decreasing concentration

(Figure 6.1j). The higher the concentration gradient, the higher the diffusive flux of solutes (Section 2.3.1).

When flow exists, driven by topographic or density differences, complete salinisation may be accomplished with a few years or less (cf. Figure 6.17). For diffusion to cause the complete salinisation (up to seawater concentration) of a 10 m thick layer of freshwater, thousands of years are required. Nonetheless, diffusion is important to explain seawater intrusion processes in low-permeability units such as clay layers. Examples of systems where diffusion plays a key role in controlling the salinity distribution over timescales of thousands of years are discussed in Section 8.3.1.1.

Diffusion is further important to understand solute transport processes in highly heterogeneous coastal aquifers. While advection by groundwater flow can effectively flush and homogenise the permeable zones of an aquifer, solute concentrations do not change as rapidly in less-permeable zones where transport is dominated by diffusion. Where permeability changes significantly over short distances, this may lead to a dual-domain behaviour in which the kinetic mass transfer between high- and low permeability regions needs to be considered to understand the characteristics of the transition zone. For example, Lu et al. (2009) found that wide transition zones can develop because of kinetic mass transfer effects for transient seawater intrusion. The regular movement of the transition zone, alternating between land and seaward, for example, due to tidal oscillations or seasonal fluctuations of freshwater recharge, sets up temporally variable concentration differences between the mobile- and immobile zones, which act to enhance diffusive spreading of solutes and thus create a wider transition zone.

While the principle of this mechanism has been demonstrated in laboratory tank experiments and numerical models, its importance in the field has not yet been systematically investigated. Observations have shown though that alternations of salinity due to permeability can occur over space intervals of a few tens of metres (Van Geldern et al., 2013). Where low-permeability zones filled with saltwater are enclosed by more permeable materials with fresher water, diffusion of salts from the low-permeability units has in some cases been found to lead to salinisation of the freshwater (Colombani and Mastrocicco, 2016).

6.3 Reversibility of Seawater Intrusion

Seawater intrusion caused by over-abstraction is, in principle, a reversible process as freshwater can flush out the intruded seawater. The rate at which this occurs is primarily dependent on flow rates of the freshwater, but the heterogeneity of the aquifer also plays a role since it determines the amount of mixing between fresh- and saltwater. Its greater density makes seawater pool on the bottom of the aquifer, which hampers it being flushed out by the freshwater, as the freshwater flow will be focused in the upper part of the aquifer above the pooling saltwater. The severity of the seawater intrusion, in other words the

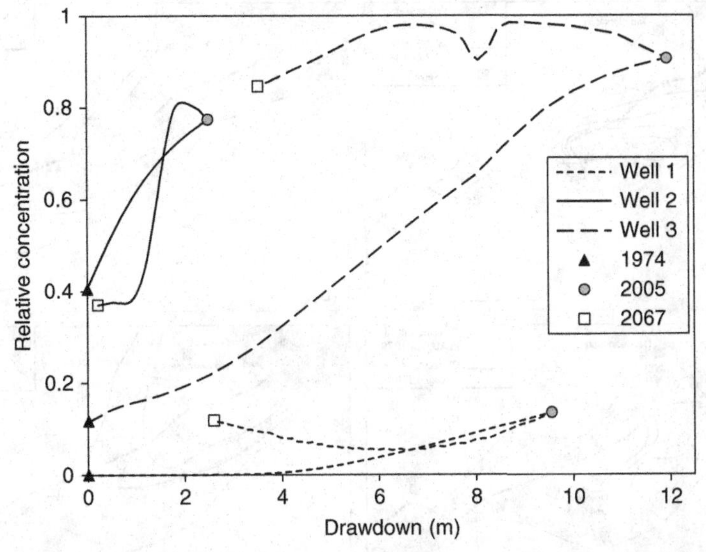

Figure 6.18 Concentration versus drawdown as calculated by a numerical model of a coastal aquifer in Oman (modified from Walther et al., 2014). Each line corresponds to one observation well, and data are shown since 1974 (triangle marker symbol). It can be seen that after pumping is stopped in 2005 (circle marker symbol), the salinities can still increase, even though the drawdown is decreasing (i.e. heads are recovering). In well 2, the increase is relatively short and immediately follows cessation of pumping. The increase is more long-lived in well 3 but is interrupted by a brief period of salinity decrease. In well 1, the salinities first decrease after pumping has ceased, but then slowly start to rise until the final simulation year 2067.

volume of seawater that has entered the aquifer, is another important factor that determines to what extent, and over what period of time, the groundwater salinities can be restored.

When abstraction ceases, a body of intruding seawater can nonetheless maintain its landward motion for a while (Howard, 1987; Walther et al., 2014). This is because, as the water levels in the aquifer recover, the seawater partially compensates for the volume of freshwater lost by pumping. This continues until the heads have adjusted to the post-pumping conditions, which can take many years to decades. This can mean that even though the heads in an aquifer are recovering, the salinity of the groundwater at some locations can continue to increase (Figure 6.18).

The recovery of salinities of pumping wells affected by seawater intrusion has been documented in a number of studies. In some, the aquifer was flushed by water derived from natural recharge processes (Han et al., 2015) (Figure 6.19) and in others from artificial recharge (Rey et al., 2013). In highly permeable systems, the advance and retreat of the transition zone can follow a seasonal pattern driven by the transient dynamics of the pumping and recharge regime. This has been observed in areas with a Mediterranean climate where the dry summers necessitate high pumping rates, which lead to seawater intrusion in summer. The severity of the salinisation is a function of the summer rainfall

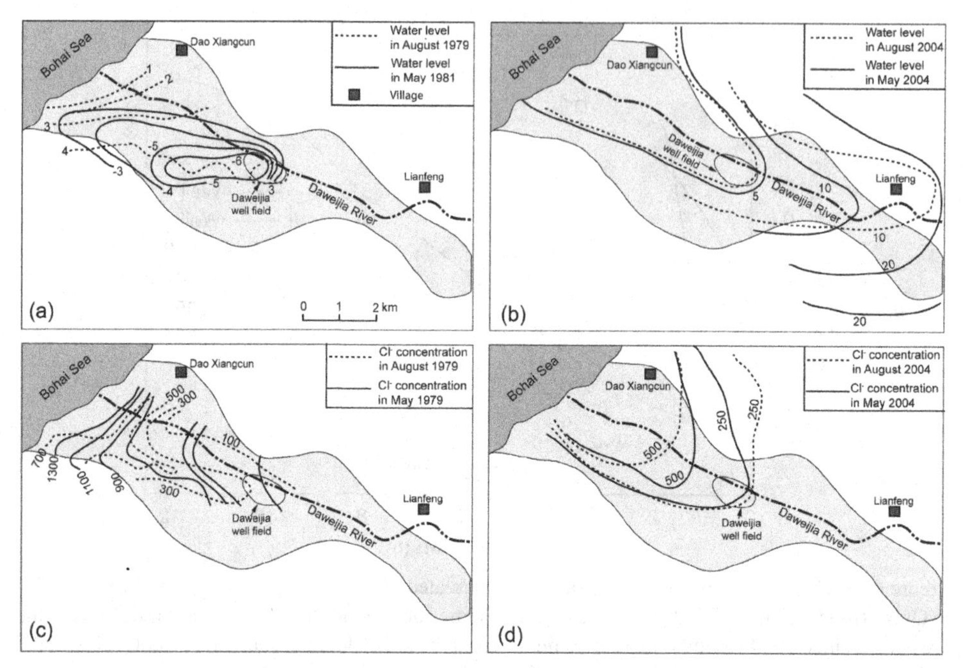

Figure 6.19 Maps of the Daweijia area adjacent to the Bohai Sea in northeast China showing groundwater levels (in metres relative to mean sea level) in (a) 1979 and 1981 and (b) 2004, as well as chloride concentration (in mg l^{-1}) contour lines in (c) 1979 and (d) 2004 for a carbonate aquifer (modified from Han et al., 2015).

amount as it influences both the water table decline and the water demand (Calvache and Pulido-Bosch, 1994). Cessation of pumping and recharge lead to flushing out of the intruded seawater during the subsequent winter. However, incomplete flushing leads to a build-up of salt concentrations in the groundwater over multiple years (Calvache and Pulido-Bosch, 1991).

García-Menéndez et al. (2016) studied the evolution of the groundwater salinity in a coastal aquifer on the eastern shore of Spain where groundwater is used intensively for irrigation over a time span of 42 years. Excessive pumping of groundwater with wells concentrated in a 2 km^2 area since the 1960s had led to lateral seawater intrusion as well as up-coning. The resulting salinisation of irrigation wells by the middle of the 1980s forced the abandoning of many wells, which, combined with more efficient irrigation practice and the use of treated wastewater instead of groundwater, led to a significant decrease in the amount of pumping during the 1990s. This led to a general improvement of the water quality, which was accelerated by the high rainfall amounts in the years 2002–2004. The front of seawater retreated towards the coast, but up-coning still persisted in 2013 and salinities remained high in many wells.

7
Submarine Groundwater Discharge

7.1 Introduction

Submarine groundwater discharge (SGD) is a young subject area within coastal hydrogeology, and has seen an enormous increase in interest over the past two decades because of mounting concerns over its role in delivering contaminants from coastal aquifers to the ocean and its role in providing essential nutrients to coastal waters. This chapter provides an overview of the basic concepts of SGD as well as common estimation approaches. Methods based on radon and radium isotopes are extensively discussed because they are among the most common in estimating SGD fluxes. At the end of the chapter, the limitations, problems and uncertainties in SGD estimation are discussed and the future research directions are suggested.

7.2 Basic Concepts

7.2.1 Definition and Multi-Disciplinarity of SGD

SGD is defined as the seaward flow of water across the sea and land interface. It includes both fresh groundwater discharge originated from terrestrial recharge and circulated saline groundwater originated from the sea (Taniguchi et al., 2002; Burnett et al., 2003). Usually the circulated saline groundwater flux dominates SGD. Here, SGD is referred to as the amount of the flow crossing the sea and aquifer interface, but some researchers also use the term in a more generic sense to indicate the processes of the flow crossing the interface (Burnett et al., 2006).

SGD takes place at the interface between land and sea at a continuum of scales: the nearshore scale, with a range of 0–10 m; the embayment scale, with a range of 10 m to as much as 10 km offshore; and the continental shelf scale (Bratton, 2010; Luo et al., 2014). It thereby spans different scientific disciplines as well as physical environments. Oceanographers and hydrogeologists are literally examining the same problem from different ends (Burnett et al., 2006). Although there are early examples of hydrogeological studies of seepage of land-derived groundwater through the seabed (Glover, 1959; Frind, 1982; Hill, 1988), the traditional task of practising hydrogeologists is to find fresh groundwater, who thus have primarily focused on the terrestrial hydrological cycle. The shoreline has been considered as a boundary, and the scope of coastal hydrogeology long remained limited to subjects such as seawater intrusion (Volker and Rushton, 1982; Andersen et al., 1988) and tide-induced groundwater level fluctuations (Jacob, 1950; Van der Kamp, 1972). It still remains commonplace for regional groundwater flow models to terminate at the shoreline, and the coast is often assumed to be a fixed-head boundary (Walther et al., 2017).

The multi-disciplinary nature of SGD studies requires collaboration among hydrogeologists, marine scientists, geochemists and oceanographers (Burnett et al., 2006; Kazemi, 2008). Distinct cultural and structural differences exist between these groups, and those working in SGD have to overcome any reluctance to work outside their own field of specialisation. There is also the need to develop a common vocabulary. For example, in ocean research any subsurface water is considered to be groundwater, but traditional hydrogeologists reserve the term groundwater for the water that originates from a terrestrial aquifer, which excludes circulated seawater. This difference in terminology has led to confusion when SGD estimates from oceanographic mass balance models and hydrogeological models are compared (Moore, 1999; Taniguchi et al., 2002).

Another example of terminology differences is the term 'subterranean estuary', which was defined by Moore (1999) as 'a coastal aquifer where groundwater derived from land drainage measurably dilutes seawater that has invaded the aquifer through a free connection to the sea'. Or in short, the subterranean estuary is 'a reaction zone of ground water and seawater' (Moore, 1999) or 'the mixing zone of fresh and salty groundwater' (Moore, personal communication, 2011). Because the term subterranean estuary is vivid and appears to be easy to understand, it soon gained popularity in the field of SGD since its introduction. Hydrogeologists, however, would argue that there is no need for this new term because the term transition zone (or mixing zone) has long been used in coastal hydrogeology (e.g. Herzberg, 1901; Meinzer, 1936) to indicate the same thing. Even the general term *coastal aquifer* may be sufficient because in it, there is usually a transition zone at the coast. The term estuary derives from the Latin word *aestuarium*, which denotes a tidal inlet of any size or 'tidal part of a shore' (Oxford Dictionaries, 2018). However, this usage is 'rare in modern use' (Cox and Gordon, 1970; Pierson et al., 2002), as in contemporary English, the word estuary is usually associated with a river or river mouth (Oxford Dictionaries, 2018). By implication the term subterranean estuary literally means subsurface river mouth, so it can be confusing and misleading. In this book, the term transition zone is used, instead of subterranean estuary.

7.2.2 Importance of SGD

It is well established that in the global hydrological cycle, river discharge is the key pathway for terrestrial water to the ocean. Unlike groundwater, rivers are visible, and their flow can be more easily quantified than that of groundwater (Taniguchi et al., 2002). Shiklomanov (1998) estimated the difference between global terrestrial precipitation (119 000 km^3 yr^{-1})

and evapotranspiration (74 200 km^3 yr^{-1}) to be 44 800 km^3 yr^{-1}, which eventually discharges to the oceans in the form of river and groundwater discharge.

SGD has been traditionally overlooked scientifically because it has been difficult to identify and measure SGD (Taniguchi et al., 2002). The subject has been receiving increasing attention since Moore (1996) demonstrated that the chemical mass flux originated from SGD could amount to as much as 40% of the river water chemical mass flux into the ocean based on a radium budget analysis in the seawater of the South Carolina coast. It should be noted that SGD is extremely site dependent. Taniguchi et al. (2002) collected SGD results from eight local case studies and the estimates of SGD range from 3% to 87% of surface runoff.

On a global scale fresh groundwater input to the sea has been estimated to range from 0.01% to 10% of river runoff (Church, 1996; Shiklomanov, 1998; Zektser et al., 2007). This portion of fresh groundwater input to the oceans represents the freshwater component of SGD. The total SGD, which includes both fluxes of freshwater and circulated seawater, tends to be much greater than fresh SGD, and can be comparable to the river flux or even a few times greater (Kim et al., 2005; Moore, 2010a). The high total flux of the SGD has been confusing to some hydrogeologists who believe that groundwater input to the sea is only a small fraction of the river water input (Younger, 1996). However, the large total SGD is not contradictory to the traditional thinking because the SGD is dominated by circulated seawater and does not represent the input of terrestrial fresh groundwater only.

As summarised by Taniguchi et al. (2002), SGD may be expressed in different units: volume per unit time (e.g. m^3 d^{-1}), which is widely used by groundwater hydrologists; volume per unit time per unit length of shoreline (e.g. m^3 d^{-1} m^{-1}); and volume per unit time per unit area (e.g. m^3 d^{-1} m^{-2}). For the second and third cases, the total volumetric rate of SGD is known only when the total shoreline length or the SGD discharge area is known. Even though SGD per unit length of coastline may be low, the total SGD along a coastline can still be significant since often large tracts of coastline are involved. This is distinctly different from rivers, which are only concentrated in river mouths.

The chemical mass input into the oceans from coastal groundwater is relatively important because groundwater in coastal aquifers tends to be rich in various dissolved chemical substances due to natural geological and geochemical processes or anthropogenic pollution. Some chemical elements such as nitrogen, phosphorus, carbon and silica can have concentrations several orders of magnitude higher in groundwater than in surface waters (e.g. Moore, 1996; Slomp and Van Cappellen, 2004; Santos et al., 2008). These nutrients have important implications for marine ecology and biogeochemistry (e.g. Johannes, 1980). Consequently, SGD must be considered in analysing the nutrient budgets of coastal systems.

7.3 Driving Forces of SGD

The process of SGD involves many driving forces such as topography-driven flow, tidal pumping, wave set-up and convection caused by salinity and temperature differences between seawater and groundwater (e.g. Ataie-Ashtiani et al., 1999; Taniguchi et al., 2002; Robinson et al., 2007) (Figure 7.1). For a comprehensive review of the SGD driving forces, the reader is referred to the review papers by Robinson et al. (2018), Moore (2010b) and Santos et al. (2012). The latter discusses other forces such as bioirrigation and bioturbation, gas bubble upwelling and sediment compaction as driving forces of the SGD. This section focuses primarily on topography-driven flow and tidal pumping.

7.3.1 Topography-Driven Flow

When the water table is a subdued replica of the topography of the land surface, this type of seaward groundwater discharge is referred to as topography-driven. Patterns of topography-driven flow depend on many factors, such as the permeability distribution, recharge rates, landforms and drainage network, as discussed in many traditional hydrogeology textbooks (e.g. Freeze and Cherry, 1979; Domenico and Schwartz, 1990; Fetter, 2001; Tóth, 2009). The resistance of the subsurface to flow hinders the dissipation of the groundwater pressure, and as a result, the head above sea level at the coast extends a certain distance into the offshore part of the aquifer, which implies that there must be sub-sea groundwater flow.

For an unconfined coastal aquifer the topography-driven freshwater discharge is greatest immediately near the shore and theoretically decreases exponentially with offshore distance (Fukuo and Kaihotsu, 1988). The intertidal zone is then often a groundwater seepage zone and evidence of groundwater discharge can be found there, e.g. small streams flowing down the beach with a temperature and salinity obviously different from seawater. Sometimes soil pipes and other erosion features are formed by outflowing groundwater. Some researchers stated that the flow occurs mostly within a distance of about 100 m of the coastline, but since the distance can be very site dependent, any definition of SGD that includes a specific distance is unlikely to be generally meaningful (Scientific Committee on Oceanic Research, 2004)

In confined aquifers the hydraulic resistance offered by the aquitard means the dissipation of the groundwater pressure slow, so land-derived flow can persist far away from the coastline (Kooi and Groen, 2001; Bakker, 2006). Faults or fracture zones, conduits in karst aquifers, or coarse deposits within finer sediments such as palaeo-river valleys or channels may form preferential flow paths along which seepage zones or even submarine springs can develop. Using numerical models, Mulligan et al. (2007) found that offshore palaeo-river channels increase the fluid exchange between offshore groundwater and seawater. Michael et al. (2016) simulated more general heterogeneous offshore aquifer systems and concluded that preferential flow can create active flow systems extending from land to sea. Their simulations showed that discharge of fresh groundwater can occur up to tens of kilometres offshore (Figure 6.6) if the overall horizontal hydraulic conductivity is much greater than vertical hydraulic conductivity so that discharge near the coastline is impeded.

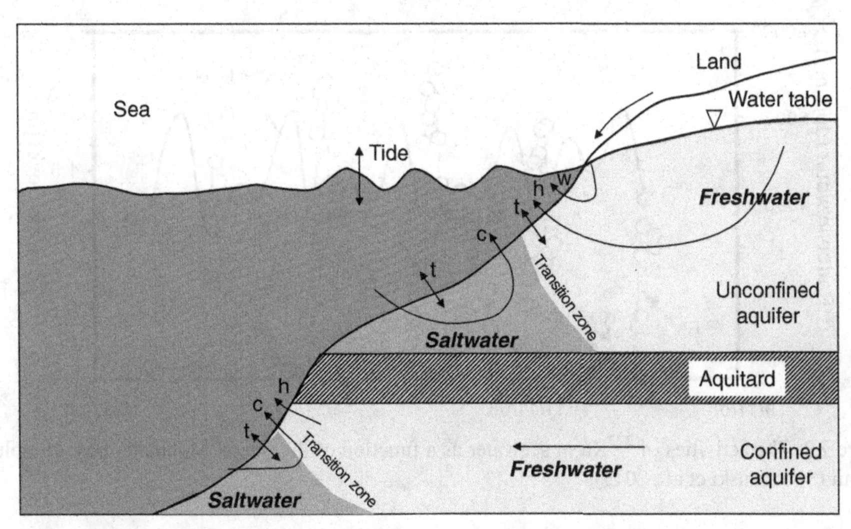

Figure 7.1 Schematic of SGD processes (not to scale). Driving forces are h = hydraulic potential gradient, c = convection, t = tidal pumping, w = wave set-up. Arrows indicate fluid movement. Based on Taniguchi et al. (2002).

7.3.2 Tidal Pumping and Waves

Tidal pumping refers to the process whereby seawater is driven into the aquifer when the tide is high and groundwater is driven into the sea when the tide is low. The effect of tidal pumping has been observed using continuous measurement of the radon isotope ^{222}Rn in coastal seawater. Radon is a natural tracer enriched in groundwater and a high ^{222}Rn activity in seawater can thus be indicative of SGD. As an example, Figure 7.2 shows that there is a close relation between ^{222}Rn in the seawater and tidal level tide at Black Point, Maunalua Bay, Honolulu, USA (Swarzenski et al., 2013): ^{222}Rn increased sharply at the primary and the secondary low tide events and dropped quickly to background values at high tidal level. Such a close relation suggests that groundwater input to the sea is mainly controlled by the difference between groundwater level and the sea level. When sea level is lower, the larger head difference between groundwater and sea level induces larger groundwater discharge; when sea level is higher, the lower head difference leads to a reduction in groundwater input.

The negative correlation between sea level and ^{222}Rn activities has been observed by continuous ^{222}Rn measurements in many other coastal areas (Burnett et al., 2008; Tse and Jiao, 2008; Santos et al., 2010). However, if the seawater depth is great, the response of ^{222}Rn to water level change may not be so quick and a delay may be observed because it takes time for the groundwater input to be mixed with the seawater.

As shown in Figure 7.3, tidal pumping can induce SGD not only laterally from onshore aquifer but also vertically from the offshore aquifer. It is expected that this so-called benthic

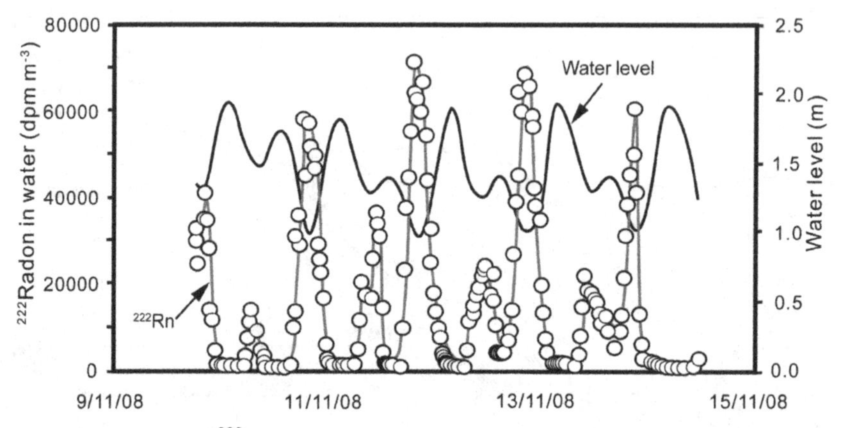

Figure 7.2 The activities of ^{222}Rn in seawater as a function of the tide in Maunalua Bay, Honolulu, Hawaii (Swarzenski et al., 2013).

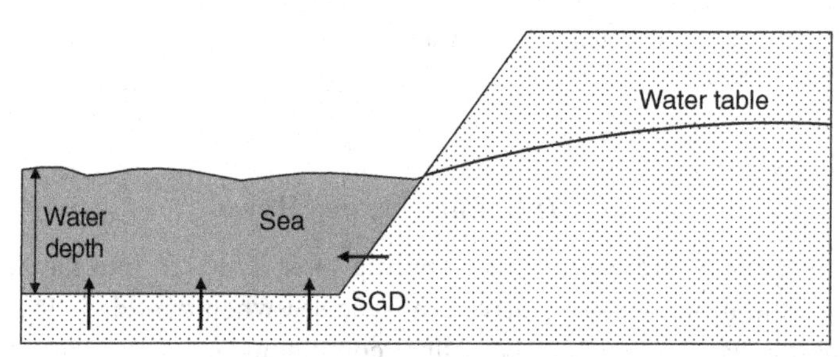

Figure 7.3 Seaward groundwater discharge controlled by the hydraulic gradient between the groundwater and the sea. Note that tide-driven SGD includes groundwater input from both the sloping shoreface and the horizontal seabed.

tidal pumping becomes progressively weaker away from coast where seawater becomes deeper.

The idealised models of the seawater wedge in coastal aquifers discussed in Chapter 3 result in an interface depth of zero at the coastline under the Ghijben–Herzberg principle (Figure 3.5) or greater than zero when freshwater discharge to the sea is accounted for (Figures 3.8 and 3.13). The latter of course is more realistic, but tidal fluctuation and waves will complicate the transition zone behaviour near the shoreline. Periodic flooding of the land surface by tides and waves creates a subsurface seawater circulation cell in the intertidal zone between the high and low tide levels (Figure 7.4). Fresh groundwater driven by the regional hydraulic gradient exits to the sea through the gap between the upper saline plume and the classic saltwater wedge, forming a so-called freshwater discharge tube.

The upper saline plume has been observed in the field (Lebbe, 1983, 1999; Michael et al., 2005; Vandenbohede and Lebbe, 2006) and further studied by numerical models and

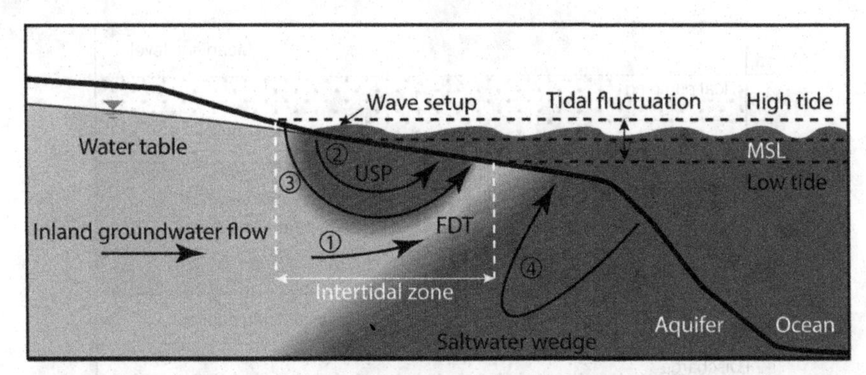

Figure 7.4 Coastal unconfined aquifer with an upper saline plume (USP) and a freshwater discharge tube (FDT). (1) fresh groundwater discharge through the FDT, (2) wave set-up circulation, (3) tide-induced circulation and (4) density-driven circulation. The shading indicates the groundwater salinity distribution. Modified from Robinson et al. (2007).

laboratory experiments (Robinson et al., 2007; Kuan et al., 2012; Greskowiak, 2014). The size of the upper saline plume depends on factors such as the tidal regime, beach slope, waves and fresh groundwater outflow. The upper saline plume, however, is limited to the shallow intertidal zone (Michael et al., 2005). As far as water resource or regional flow in a large-scale coastal aquifer studies is concerned, it is reasonable to ignore the upper saline plume and the classic saltwater wedge remains a good approximation.

The upper saline plume is very dynamic compared to the saltwater within the deeper transition zone. The frequent exchange between fresh groundwater and seawater may have considerable impact on the distribution of nutrients and metals and their subsequent submarine groundwater discharge in the areas around the intertidal zone (Charette et al., 2005; Michael et al., 2005; Robinson et al., 2007; Heiss et al., 2017; Kim et al., 2017; Liu et al., 2018).

Li and Jiao (2003a) investigated the tidally driven SGD processes in an unconfined-confined aquifer system. In the unconfined aquifer, the average water table elevation is higher than the mean sea tidal level (Section 4.6.7). Because the transmissivity varies with the elevation of the water table, the aquifer transmissivity is highest during high tide and lowest during low tide. Consequently, inflow during high tide occurs at a higher rate than outflow at low tide, so to maintain the water balance, i.e. the inflow equals the outflow over a tidal cycle, the hydraulic gradient must be higher during low tide than during high tide, and the end effect is that the time-averaged water table in the unconfined aquifer is higher than sea level (Figure 4.8). This tidal overheight drives a flow across the aquitard into the underlying semi-confined aquifer (Figure 7.5). A sloping land-water surface enhances the circulation because it leads to a higher water table in the shallow unconfined aquifer than for a vertical shoreface as shown in Figure 7.5 (Nielsen, 1990; Li and Jiao, 2003a).

Wave set-up refers to the mean sea level increase as a result of waves running up the shore (Robinson et al., 2018). Similar to tides, the water level profile caused by wave set-up

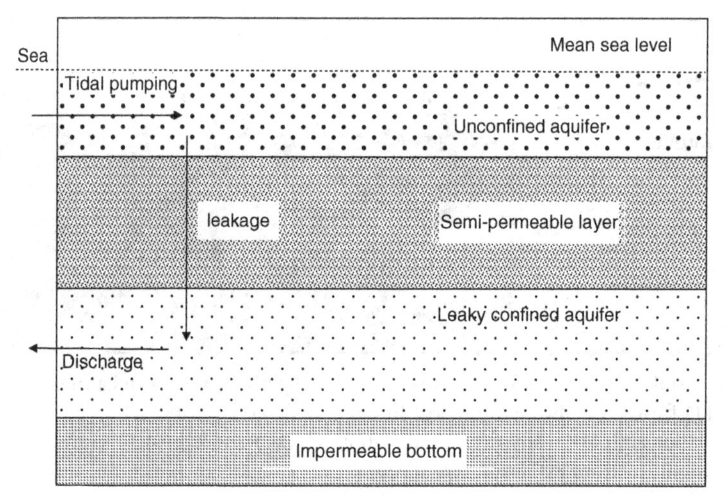

Figure 7.5 Tide-induced seawater-groundwater circulation in a layered coastal leaky aquifer system consisting of an unconfined aquifer, semi-permeable layer and leaky confined aquifer (modified from Li and Jiao, 2003a).

drives seawater circulation through the beach sediments. This process is also called wave pumping. Wave set-up can affect large stretches of coast during swash events (Sous et al., 2016). Individual waves also drive infiltration-exfiltration cycles, but at a very high frequency and this process is therefore challenging to measure. Field studies of this SGD process are very limited (Heiss et al., 2014), but it has been studied numerically and in the laboratory (Bakhtyar et al., 2013; Geng et al., 2014).

The constant water circulation through the land-sea interface by various driving forces is the key reason that SGD can generate extensive chemical mass input to the sea. Even if there is no net SGD water flux, there may be a net influx of solutes from the groundwater into the ocean (Moore, 1996, 2010a). Figure 7.6 demonstrates how some chemicals, for example, radium, can be driven to the sea by tidal pumping. Radium is strongly absorbed by the soil matrix or flocculated particles in a freshwater aquifer and gets desorbed when the salinity rises. This means that when seawater displaces fresher water in the subsurface during high tide (Figure 7.6), Ra is desorbed from the aquifer matrix; during low tide, the Ra-enriched seawater flows back into the sea. Consequently, although there is no net flow from the aquifer to the sea, or even if there is seawater intrusion, there is still a net supply of Ra to the sea (Figure 7.6). Many elements such as manganese, iron, uranium, chromium and vanadium, as well as other species adsorbed on Mn and Fe oxides may have similar geochemical behaviour as Ra (Moore, 1996).

SGD may not always enhance the input of a chemical into the sea. The physiochemical properties and hydrochemical processes in the zone immediately near the shoreline determine the concentrations of the seawater that flows back. Within this zone there may be chemical equilibrium so that solute concentrations are not enhanced, or the chemical

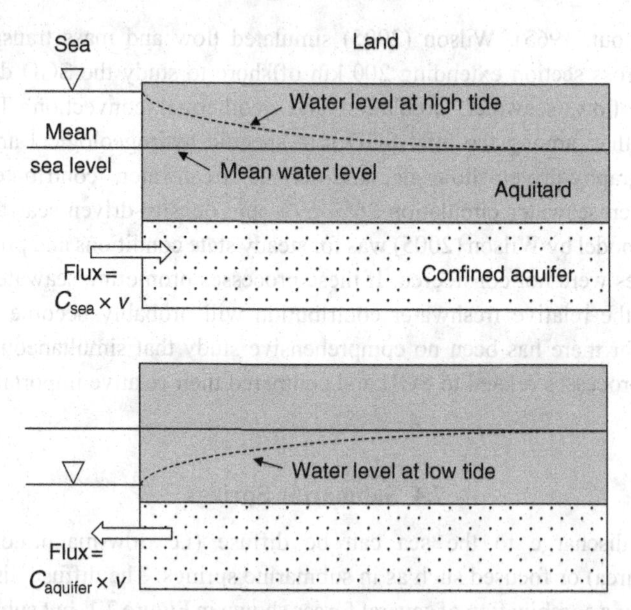

Figure 7.6 Sketch showing supply of chemical mass to the sea [Δflux = $(C_{aquifer} - C_{sea}) \times v$] by tidal pumping in a confined aquifer when there is no net water volume exchange between the sea and the aquifer. During high tide, seawater flows into the aquifer with a low concentration (C_{sea}) of a chemical (such as Ra) and velocity v; during low tide, brackish water leaves the aquifer with the same velocity but with higher concentration of the chemical ($C_{aquifer}$).

process may take a much longer time to generate an increase in solute concentrations than the duration of a tidal cycle. Some chemicals may disappear because of chemical processes induced by tidal fluctuations. For example, in coastal aquifers rich in ammonium due to highly reducing conditions (Jiao et al., 2010; Mastrocicco et al., 2013), the oxygen-rich circulated component of SGD may oxidise the ammonium in the coastal aquifers, reducing the seaward discharge of ammonium.

7.3.3 SGD Components

The relative importance of various components of SGD is site and scale-dependent. Using a small-scale flow and transport model and a specific aquifer configuration and parameters, Li et al. (1999) estimated that seawater circulation caused by tide and wave set-up may contribute 96% of the total SGD compared to 4% of fresh SGD from land-derived groundwater. Xin et al. (2010) also stressed the importance of wave forcing on the circulation patterns and mixing processes. They conducted model simulations of an unconfined aquifer with a sloping shoreface and found that because of the combined effect of tides and waves, 61% of total SGD was due to seawater circulation.

The differences in density and temperature between seawater and the underlying groundwater also causes the mass exchange between seawater and groundwater when convection develops (Kohout, 1965). Wilson (2005) simulated flow and mass transport in a two-dimensional cross section extending 200 km offshore to study the SGD driven by topography-driven flow, seawater circulation and geothermal convection. The simulation demonstrated that, among the total SGD in a specific hydrogeological and topographic setting, topography-driven flow or land-derived freshwater, contributed 37%–72%, thermally driven seawater circulation 26%–57% and density-driven seawater circulation 2%–6%. The model by Wilson (2005) was for steady state conditions and processes such as tides and waves were not considered. If these processes promoting seawater recirculation are included, the relative freshwater contribution will probably become much smaller. However, so far there has been no comprehensive study that simultaneously considered all the major processes related to SGD and compared their relative importance.

7.4 Submarine Springs

Groundwater discharge to the sea can be diffuse (i.e. low-magnitude fluxes over a widespread area) or focused such as in submarine springs. The diffuse discharge can be driven by one or a combination of several forces shown in Figure 7.1, but submarine springs are primarily driven by the hydraulic gradient. Submarine springs are an enigmatic hydrogeological phenomenon, and have been documented in various ancient cultures. Reviews of the historical sources accounting of submarine springs can be found in Clendenon (2009) and Kohout (1966). A review of the use of submarine springs for a range of societal purposes can be found in Moosdorf and Oehler (2017).

As described by the Greek philosopher Aristotle (384–322 BC) in his treatise Meteorology: ‘Many great rivers fall into it [a lake] and it has no visible outlet but issues below the earth off the land of the Coraxi about the so-called ‘deeps of Pontus’ (a deep place in the sea) ... At this spot, about three hundred stadia [an ancient Greek unit of length, 1 stadia = 157–209 m] from land, there comes up sweet water over a large area, not all of it together but in three places’ (Aristotle, 350 BC). Clendenon (2009) interpreted this description as surface water that sinks underground in a limestone karst region, flows to the seabed along subterranean karst channels, and then resurges as three submarine springs in the Black Sea almost 60 km offshore. Strabo (63 BC – AD 21), another Greek geographer and philosopher, accounted of freshwater emerging from a spring a few kilometres offshore from Latakia, Syria that was collected and then transported to the nearby city as a source of freshwater.

Figure 7.7 is a historical sketch showing how sailors collected freshwater from submarine springs more than 160 years ago (Humboldt and Thrasher, 1856). Many geographers and travellers summarised how coastal communities used submarine springs. In the book of the *History of Phoenicia* by Rawlinson (1889), there is a note on the use of offshore springs by the Phoenicians: ‘When this (water) supply was cut off by an enemy Aradus (islet and town of ancient Phoenicia) had still one further resource. Midway in the channel between the island and the continent there burst out at the bottom of the sea a freshwater

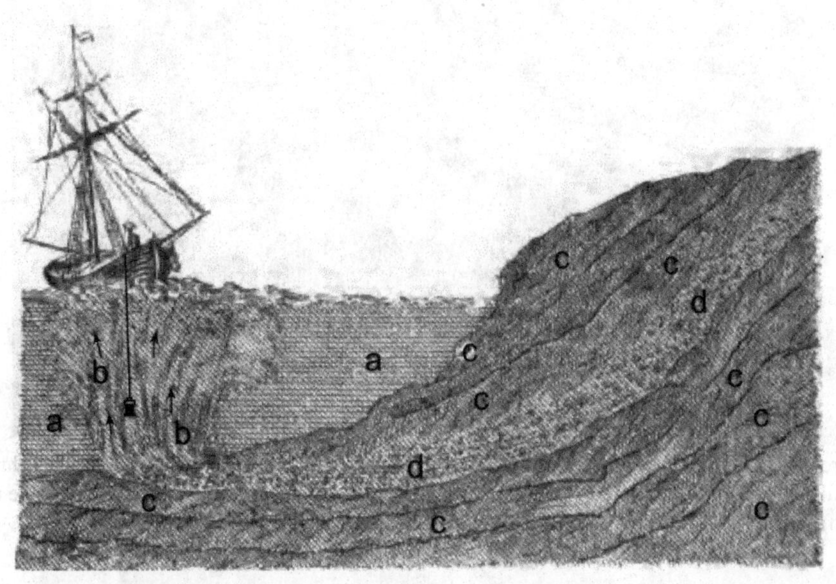

Figure 7.7 An old sketch showing the emergence of freshwater from a submarine spring and how sailors collected freshwater from the spring (Humboldt and Thrasher, 1856) (a: saltwater; b: freshwater; c: impermeable layer; d: permeable layer).

spring of great strength; by confining this spring within a hemisphere of lead to which a leathern pipe was attached the much-needed fluid was raised to the surface and received into a vessel moored upon the spot, whence supplies were carried to the island'. Joseph Du Commun, who was among the first to formulate the hydrostatic equilibrium between freshwater and seawater in a coastal aquifer (Section 3.2), believed that there were subterranean streams in the sea just as numerous as on the Earth surface (Du Commun, 1828). He speculated that the reason that some voyagers found freshwater in 'spaces of several miles in extent in the open sea' was that these were the mouths of huge subterraneous rivers. Zwemer and Zwemer (1911) described submarine springs at the Persian coast that 'find their source on the mainland of Arabia or Persia, and the water not only bubbles out in pools and wells onshore, but below the tide level there are freshwater springs several miles out at sea'. The Arabs 'go out in their boats, place a bamboo over the opening in the rock and then collect freshwater above sea level in their great leather skins' (Zwemer and Zwemer, 1911).

Most of the large-scale submarine springs are associated with karstic conduit systems developed in limestone rock (Kohout, 1966; Swarzenski et al., 2001; Povinec et al., 2006; Clendenon, 2009; Valle-Levinson et al., 2011). A famous example of such a submarine spring is Crescent Beach Spring which is located almost 4 km away from the shoreline of north-eastern Florida, USA (Brooks, 1961; Swarzenski et al., 2001). The spring was surveyed in 1875 and has been well studied since then. Even though the spring vent is 18 m below the sea surface, the strong upward flow of

Figure 7.8 A submarine spring in north-western Lombok where groundwater discharges through a coral reef. The diameter of the plume at the sea surface is about 3 m (the persons provide scale in the photo) (photograph courtesy of Till Oehler).

spring water generates a visible circular plume at the sea surface as the turbulence from the spring water that interrupts the local wave field. The emerging water gives a strong smell of sulphide gas. The spring is sourced by thick limestone in the Floridan aquifer system and has a discharge of about 42.5 $m^3\ s^{-1}$. Note for comparison that the total water consumption of Hong Kong in 2017 with its population of over 7 million is equivalent to about 31 $m^3\ s^{-1}$. Figure 7.8 shows a submarine spring in north-western Lombok, where groundwater discharges through a coral reef. The spring is about 150 metres away from the shore line. The vent of this spring has a water depth of about 4 metres. The plume at the sea surface has a diameter of about 3 m. The main source of water is probably from the volcanic aquifer below the coral reef. The spring water is brackish due to mixing with seawater (Oehler, personal communication, 2018). Another example is the submarine springs around the Jeju Island, Korea, which are widely used for drinking water or bathing. The springs are fed by the very permeable volcanic basalts. The cover image of this book shows one of the spring sites at Iho Tewoo Beach at the north coast of the island. At low tide, water level in the well is at least 1 m above the sea level and groundwater from the well and numerous springs in the intertidal zone fills two ponds with a total area of about 10 000 m^2. At high tide, however, this site is entirely submerged by seawater.

Most submarine springs, however, are much smaller, with much less remarkable features. When not visually apparent, they can be detected by measurements of EC or temperature near the seabed. For example Marui (2003) identified eight submarine springs around Rishri Island in Hokkaido, Japan. One of them has sand particles boiling up around the spring vent as a result of upward flowing groundwater. Figure 7.9 shows a karstic intertidal spring in Gunung Kidul, Indonesia, which generates so-called sand boils (Moosdorf and Oehler, 2017). Because they are localised features, submarine springs can be very difficult

Figure 7.9 Submarine springs in the intertidal area in Gunung Kidul, Indonesia, with sand bubbling up around the spring vents (modified from Moosdorf and Oehler, 2017).

to detect, and when flow rates are low, mixing of the emerging spring water with seawater smears out the salinity or temperature contrast, especially when there are high waves or strong currents.

It should be noted that due to strong upward groundwater flow, coastal areas with submarine groundwater springs are also the areas with quicksand, where the sand behaves almost like a liquid. The effective stress becomes zero (Eqn (2.23)) and the areas cannot support much loading, so people and animals can get stuck.

Evidence of former, inactive submarine springs can sometimes be found during offshore geological investigations. Figure 7.10a shows a sketch of a recrystallised spring vent structure from Smooth Ridge in Monterey Bay (Silvestru, 2001). Figure 7.10b shows a 'water-escape structure' in coarse-grained sediments. This flame-like and cylindrical pillar structure cuts through the laminated fine-grained sandstone (Lowe, 1975). Figure 7.10c shows a feature identified from cores in a borehole drilled into weathered volcanic soil in the coastal area of Tung Chung near Hong Kong International Airport (Fletcher, 2004) (for location, see Figure 9.2). Quite a few boreholes installed there show similar features. Local geotechnical engineers call this a soil pipe. Light brown sandy silt soil pipes penetrate through completely decomposed volcanic rock. The sediments inside the soil pipes consist of transported sands and occasionally show laminated features. Such a feature is believed to be formed by water flowing through the soil mass along relic joints.

It should be noted that not only submarine groundwater springs can form these structures (Figure 7.10). Other pressured fluids such gas and oil may also form them. Some of these

Figure 7.10 (a) Sketch of a suspected spring vent structure from Smooth Ridge in Monterey Bay, California. The sample diameter is about 30 cm (Silvestru, 2001). (b) Photograph of a water escape structure with a diameter of about 2.5 cm in layered sediments (Lowe, 1975) (cited in Judd and Hovland, 2007). (c) Sandy silt soil pipe within completely decomposed volcanic rock formed by water flowing through the soil mass along relic joints (Fletcher, 2004).

structures are macrofossil rich. A more comprehensive discussion of these structures, which can offer insights into past hydrological, climate and biochemical conditions, can be found in Judd and Hovland (2007).

7.5 Methods for Assessing SGD

The basic approaches to estimate SGD can be divided into three categories (Burnett et al., 2001): analytical or numerical modelling; direct measurements; and environmental tracer techniques. Modelling may involve simple water balance models and complex numerical models. Direct measurements are usually carried out by seepage meters, sometimes aided by geophysical or thermal methods. Tracer techniques usually are based on natural environmental tracers such as radon and radium isotopes.

7.5.1 Regional Groundwater Models

Many coastal groundwater models assume that the aquifer terminates at the coastline. At this boundary the (saltwater) head is assumed to be the same as sea level (He et al., 2008; Datta et al., 2009). However, since there still can be significant flow in the offshore part of aquifers, models ideally extend beyond the coastline. For an unconfined aquifer, it is suggested that a boundary should be chosen such that at least 95% of seaward discharge occurs between the coastline and the boundary. A trial and error method can be used to determine the boundary location after an initial model is set up. For a confined aquifer which crops out at the seabed, the model must accommodate this outcrop location. When this location is uncertain or unknown, tidal information can be used to determine the offshore extent of the confined aquifer (Cheng et al., 2004). In this approach tide-induced water level fluctuations in observation wells are analysed to obtain the relation between tidal amplitude and distance. The extrapolated offshore distance at which the tidal amplitude is the same as that of the sea tide is treated as the location of the equivalent model boundary. A brief discussion about how to set up boundaries conditions for confined and unconfined aquifers in a multi-layered aquifer system can be found in Hu and Jiao (2010).

The seaward discharge from these models represents the freshwater input to the sea and is only a small part of the SGD. If the density difference is included in the model, the density-driven circulated seawater component of the SGD can be estimated (Smith, 2004). However, for practical reasons, regional models cannot consider all SGD components such as those driven by short-term tidal pumping and wave set up or geothermal convection. As a result the SGD estimates from such a model are not directly comparable to those from a mass balance calculation based on tracers such as radium and radon that theoretically includes all the SGD components.

7.5.2 Seepage Meters

A seepage meter (Figure 7.11) is the only device that measures groundwater discharge through the seafloor directly (Burnett et al., 2001). The basic principle and design were first explained by Lee (1977). Groundwater that seeps through the sediment covered by a chamber with its open end pushed into the seafloor forces the water already in the chamber to flow through a narrow opening in the top of the chamber into a plastic bag. The water volume increase over a certain time period is taken as the mean groundwater discharge rate during that period (Burnett et al., 2001). This type of seepage meter is straightforward and cost-effective, and can provide a direct measurement of the flux. If the seabed is deep, usually a diver is needed to install the seepage meter, and then collect the bags. The collection and measurement need to be carried out at an interval of 1 to 4 hours to provide sufficient data to resolve SGD changes over a tidal cycle.

In order to solve the labour-intensive sampling problems and also to deliver high-resolution time series data, modern, automated seepage meters have been invented. For example, there are ultrasonic (Paulsen et al., 2001), heat pulse (Taniguchi and Fukuo,

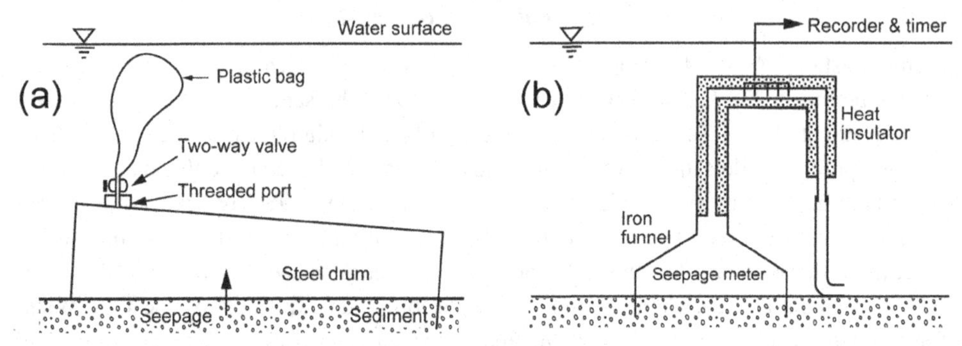

Figure 7.11 (a) A traditional seepage meter and (b) a heat pulse seepage meter (Burnett et al., 2006).

1993) and electro-magnetic (Rosenberry and Morin, 2004) seepage meters. A comparison of the performance of different seepage meters at a field site was carried out by Taniguchi et al. (2003). Automated metres can measure very low SGD rates and the measurement interval can be as short as a few seconds. Some can also measure flow both in and out of the seabed and provide information on the chemistry of the SGD. A main drawback is that they are costly so that it can become expensive to install automated seepage meters in multiple locations. For SGD studies over a fairly large area, which require a large number of seepage meters, the traditional seepage meters continue to be widely used (Michael et al., 2003; Russoniello et al., 2013).

Various factors that impact the performance of the traditional chamber-and-bag seepage meters were examined by Murdoch and Kelly (2003). A main factor is that the flow to the seepage meter will be different from the natural flow through the seabed to a certain degree because the chamber inevitably modifies the flow. There may be mechanical problems related to the plastic bags that lead to anomalous water flow into the bag, which is not due to natural seepage (Burnett et al., 2001). It is difficult to install the seepage meters firmly to the seabed if the seabed has coarse materials such as gravels and shell fragments, which leads to leakage around the base of the seepage meter. Another practical problem is that the seepage meters can be dislodged by fishermen and the action of waves and currents. For this latter reason they have mostly been deployed where ocean conditions are calm and thus SGD under more turbulent conditions is relatively under-investigated.

7.5.3 Natural Tracers

Natural tracers that have been widely employed to quantify SGD include radium quartet (^{223}Ra, ^{224}Ra, ^{226}Ra, ^{228}Ra), ^{222}Rn, CH_4, SF_6, stable isotopes ($^{86}Sr/^{87}Sr$, ^{2}H and ^{18}O) and salinity (Cable et al., 1996; Moore, 1996; Basu et al., 2001; Burnett et al., 2006; Beck et al., 2007; Charette, 2007; Moore et al., 2008; Tse and Jiao, 2008; Lin et al., 2010; Kim et al.,

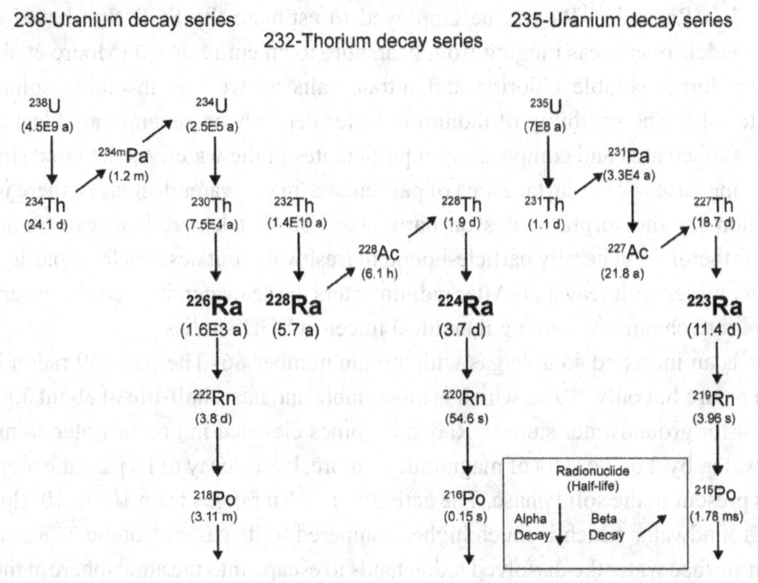

Figure 7.12 Radionuclides of the uranium and thorium decay series (Charette et al., 2012). The number in the bracket is the half-life in milliseconds (ms), seconds (s), minutes (m), days (d) or years (a).

2011; Lee et al., 2012). Among them, radium and radon isotopes are the most popular natural tracers for SGD studies.

Radium and radon are daughter nuclides of uranium or thorium (Figure 7.12), which occur in almost all natural rocks, especially igneous rocks (Tung et al., 2010). Groundwater therefore has elevated radium and radon activities relative to surface waters, especially seawater. An enrichment of radon and radium in seawater or estuaries is thus a potential indicator of groundwater discharge.

7.5.3.1 Basic Characteristics of Radium and Radon

Radium is a radioactive metal with atomic number 88 and has 33 isotopes. Among them, ^{228}Ra, ^{226}Ra, ^{224}Ra and ^{223}Ra, with half-lives of 5.7 years, 1600 years, 3.66 days and 11.4 days, respectively, are widely used as natural tracers in SGD studies. ^{226}Ra is always the most abundant in terms of atoms, but varies in terms of activity with ^{228}Ra. Because of the much lower abundance of ^{235}U compared to ^{232}Th, ^{223}Ra activities are usually lower than ^{224}Ra activities by 1 or 2 orders of magnitude (Porcelli, 2008). Since radium has two electrons in its outermost shell it belongs to the alkaline earth group and thus possesses chemical characteristics similar to elements such as barium and calcium.

The differences in half-lives can be exploited in to study different SGD aspects. The short-lived ^{224}Ra and ^{223}Ra can be used to calculate water mass ages in nearshore areas for offshore distances up to 150 km (Hancock et al., 2006; Dulaiova et al., 2009) while long-lived ^{228}Ra and ^{226}Ra can be employed to estimate the SGD flux based on mass balance models over areas ranging from nearshore to an entire ocean (Moore et al., 2008).

Radium forms soluble chloride and nitrate salts as well as insoluble sulphate and carbonate salts. The solubility of radium in water depends on several variables: the ionic strength, surface area and composition of particulates in the water. As the ionic strength of the water increases or the surface area of particulates in the water decreases, there is greater competition for the sorption sites on particulates so that less radium can be adsorbed. Radium is therefore generally particle-bound in freshwater but desorbs from particles when freshwater mixes with seawater. After radium enters the seawater, it behaves conservatively with regard to chemistry, making it an ideal tracer in SGD studies.

Radon is an inert and soluble gas with atomic number 86. There are 39 radon isotopes found in nature but only ^{222}Rn, which is most stable and has a half-life of about 3.8 days, is of interest for groundwater studies. Radon becomes elevated in groundwater compared to surface water, by 2 or 3 orders of magnitude or more, by α-decay of its parent isotope ^{226}Ra which is present in the solid phase. The activity of ^{222}Rn ranges from 10^3 to 10^5 dpm l^{-1} in natural groundwater which is much higher compared to its parent isotope ^{226}Ra (Porcelli, 2008). In surface water the dissolved radon tends to escape into the atmosphere at the water-air surface.

7.5.3.2 Measurement of Radon and Radium Isotopes

Radium sampling and measurement. Dissolved radium in water can be extracted by allowing the water of a known volume to flow through a cartridge filled with manganese-coated acrylic fibre (referred as Mn fibres hereafter) (Moore, 1976). This fibre can be easily made by soaking an acrylic fibre in a hot $KMnO_4$ solution, or be purchased directly from a commercial vendor. The sampled water is first passed through 0.5 μm filter to remove suspended particulates, and then through a cartridge containing 20–40 g of Mn fibres to which the dissolved radium sorbs. To ensure complete adsorption of radium onto the fibres, the flow rate is kept below 1 l min^{-1}.

The water volume needed for groundwater radium extraction is typically about 2 to 5 l. To extract radium from seawater, a large volume of 20 to 1200 l is needed to meet the detection limit because the radium activities in the seawater are much lower than groundwater. Usually seawater is pumped and stored in large water tanks or buckets. Radium concentrations in river water are usually between groundwater and seawater and about 10–50 l of river water is needed for a measurement.

Ideally, the column with the Mn fibres should extract 97% of the radium. To test the radium sorption efficiency, another column with Mn fibres can be added to absorb the residual radium from the water that has already flowed through the first column. Both columns are then analysed for radium to determine the radium sorption efficiency of the first column. If the efficiency is low, measures such as a lower flow rate, a longer column or more tightly packed fibres should be considered to increase the contact between the water and the fibres. After sampling of high-salinity water, the Mn fibres should be flushed with radium-free freshwater to remove salts before measuring Ra. Purging with compressed air for 2 to 7 minutes should bring the water/fibre weight ratio in the range of 0.6–0.8, which is optimal for the maximum emanation of Rn from the Mn fibres (Sun and Torgersen, 1998; Burnett et al., 2008).

The short-lived ^{223}Ra and ^{224}Ra can be measured directly using a portable Radium Delayed Coincidence Counting (RaDeCC) device, based on an approach developed by Moore and Arnold (1996). ^{224}Ra is measured best within a few days after sampling to avoid significant decay. To avoid interference from ^{224}Ra decay, ^{223}Ra is then measured about 7 to 10 days after sampling when ^{224}Ra has been largely decayed. The long-lived ^{226}Ra and ^{228}Ra isotopes are traditionally measured by a gamma spectrometry (Moore, 1984) or in recent years by Thermal Ionisation Mass Spectrometry and Multicollector-Inductively Coupled Plasma Mass Spectrometry (Van Beek et al., 2010). The isotope of ^{228}Ra can be also detected with RaDeCC via ^{228}Th ingrowth (Moore, 2008; Kiro et al., 2012, 2013) and ^{226}Ra can be measured with a portable radon detector after it is in secular equilibrium with ^{222}Rn (Kim et al., 2001; Lee et al., 2012). Compared to a gamma spectrometer and other modern machines, RaDeCC and radon detectors are cost-effective and can be brought to the field. The procedures to analyse the entire radium quartet of a water sample are relatively complex and time consuming. It takes at least 2 months to obtain a whole set of Ra data. The total time to obtain a full data set for SGD analyses can become long and this must be considered in the project planning.

Various studies were carried out to study the uncertainties of radium measurements using the above approaches. The uncertainty for ^{224}Ra and ^{223}Ra is 7% and 12%, respectively (Moore and Arnold, 1996; Garcia-Solsona et al., 2008; Moore and Cai, 2013), the uncertainty for ^{226}Ra ranges from 3% to 15% (Lee et al., 2012) and that for ^{228}Ra is about 7% (Moore, 2008). Although these previous statistics can provide some ideas about the general uncertainties, one should always make an attempt to investigate the uncertainties for the specific conditions of a study.

For a well-designed SGD study, ideally piezometers should be installed in the intertidal zone to collect samples representative of groundwater immediately inland from the coastline. For shallow systems, a pushpoint sampler, which is a small-diameter hollow steep pipe with a sharp end, can be manually driven into the beach sediments and then water samples are pumped from the pipe using a peristaltic pump.

Radon sampling and measurement. As ^{222}Rn decays to ^{218}Po by alpha-decay, it can be measured using alpha spectroscopy, and a number of portable devices are commercially available. The required sample volume can be as small as 250 ml of water, depending on the ^{222}Rn activity. As radon in water can escape quickly into air, in the sample bottle must be filled and closed under water (Figure 7.13).

The measurement of the decaying alpha particles is not done in the water sample itself, but in air that has equilibrated with the sample. To this end, air is bubbled though the water sample in a closed loop system. As the solubility of Rn is temperature dependent, the temperature must be monitored. The time to reach equilibrium depends on the volume of

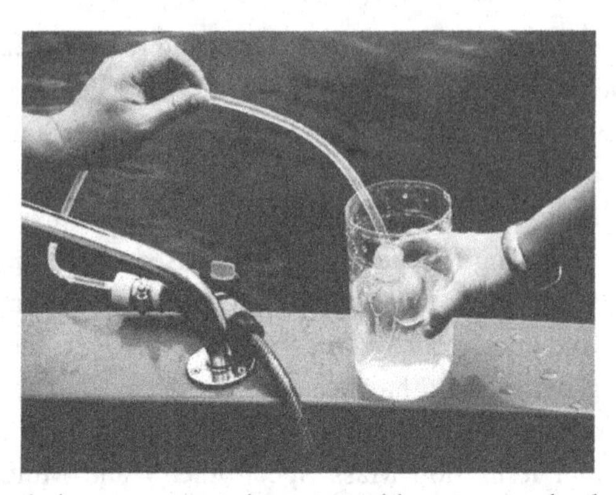

Figure 7.13 Sampling technique to prevent air contact with water sample: the sampling bottle is held inside the beaker and water is fed to the bottom of the bottle until overflow occurs. When the water rises well above the bottle, the bottle is flushed for some time and closed firmly underwater.

the sample relative to the gas inside the closed loop. Once the air has reached equilibrium a measurement can be started. When the Rn activity is high enough, a single measurement cycle can be 5 minutes long, and an average and standard deviation are calculated from 3 or 4 cycles. At low Rn activities, 5 minutes may not be enough and the number of disintegrations per minute determines the optimal cycle length. For very low activities, such as in open seawater, the measurement time may become unpractically long, or the accuracy unacceptable. In that case, specialised set-ups, using multiple instruments in series, are required in order to still use portable radon devices (Stieglitz et al., 2010).

7.5.3.3 Using Short-Lived Radium Isotopes to Estimate Water Mass Ages

Various transport time scales such as flushing time, water mass age and residence time are involved in describing the movement and retention of water or tracers within a water body (Monsen et al., 2002). The flushing time of a water body can be defined as the volume of water in a system divided by the volumetric flow rate through the system (Geyer et al., 2000). This estimate assumes that the flow is dominated by one-dimensional advection. If various other processes such as dispersion or diffusion are considered, the flushing time can be much longer.

Water mass age is the time that a water parcel has spent to travel from a boundary where it entered a water body to its location within that water body. Residence time is the time that has been spent for a water parcel to travel from an entering boundary to the outlet where it exits the water body (Zimmerman, 1988; Monsen et al., 2002). Water mass age or residence time apply to a water parcel, which is different from the flushing time that is for the entire body.

A detailed discussion about different approaches to estimate the flushing time, residence time and water mass age can be found in Monsen et al. (2002). Only the estimation of water mass age by ^{223}Ra and ^{224}Ra is presented here. Water mass age is crucial to constraining nutrient budgets in coastal zones as well as the sensitivity of coastal ecosystem to contaminant loadings (Brooks et al., 1999). The age can be calculated from the radioactive decay equation of constant of ^{223}Ra and ^{224}Ra, as derived by Moore (2006) and Houngham (2007):

$$^{223}\mathrm{Ra}_{sw} = {}^{223}\mathrm{Ra}_{gw} F_{gw} e^{-\lambda_{223} t} \tag{7.1}$$

$$^{224}\mathrm{Ra}_{sw} = {}^{224}\mathrm{Ra}_{gw} F_{gw} e^{-\lambda_{224} t} \tag{7.2}$$

where λ_{223} $[T^{-1}]$ and λ_{224} $[T^{-1}]$ are the decay constant of ^{223}Ra and ^{224}Ra and have the values of 0.0606 d^{-1} and 0.189 d^{-1}, respectively, t is the time (in days) and F_{gw} is the fraction of groundwater contained in the seawater samples collected. The subscripts of *sw* and *gw* represent seawater and groundwater, respectively. The above two equations can be combined by equating F_{gw} to obtain the following equation:

$$[^{224}\mathrm{Ra}/^{223}\mathrm{Ra}]_{sw} = [^{224}\mathrm{Ra}/^{223}\mathrm{Ra}]_{gw} [e^{-\lambda_{224} t}/e^{-\lambda_{223} t}] \tag{7.3}$$

where $[^{224}\mathrm{Ra}/^{223}\mathrm{Ra}]_{sw}$ and $[^{224}\mathrm{Ra}/^{223}\mathrm{Ra}]_{gw}$ represent radium activity ratio in the seawater and coastal groundwater measured in samples taken in the aquifer, respectively. Solving for water mass age *t*, Eqn (7.3) yields

$$t = \frac{\ln[^{224}\mathrm{Ra}/^{223}\mathrm{Ra}]_{sw} - \ln[^{224}\mathrm{Ra}/^{223}\mathrm{Ra}]_{gw}}{\lambda_{223} - \lambda_{224}} \tag{7.4}$$

It should be noted that the water mass age estimated from the above equation assumes that the activity of radium is only controlled by radioactive decay described by Eqns (7.1) and (7.2). However the radium concentration is also diluted by water mass mixing. Thus, the water mass age estimated from Eqn (7.4) is an apparent age (Moore, 2000). The above approach is based on the following assumptions. (i) In the study area, the radium activity in groundwater is at least one or two orders of magnitude greater than those in seawater and radium input from river discharge is assumed to be negligible. (ii) $[^{224}\mathrm{Ra}/^{223}\mathrm{Ra}]_{gw}$ is assumed to be both temporally and spatially stable. This means that the production rate of ^{223}Ra and ^{224}Ra must be essentially constant, which requires the aquifer to be geochemically and hydrochemically homogeneous. (iii) The open sea has negligible ^{223}Ra and ^{224}Ra. (iv) Once the radium enters the seawater from groundwater, there is no additional radium contribution from sediments or suspended particles. Before this approach is used, radium activities in river discharge, groundwater and the open sea should be measured and compared to test the validity of these assumptions. Hougham et al. (2007) indicated that the neglecting of the open sea contribution is valid for relatively slow flushing embayments with flushing times larger than 2 days.

The assumptions that there has no radium input from rivers and sediments are not always met (Moore and Krest, 2004). Therefore, it is better to measure the ^{224}Ra/^{223}Ra in the vertically mixed zone nearshore, instead of inside the aquifer, and express the water age as the time elapsed as the water moved from this zone to the stratified open ocean. This approach can be used where there is significant radium input from river flow and sediments (Moore and Krest, 2004; Peterson et al., 2016).

7.5.3.4 Using Long-Lived Radium Isotopes to Estimate the SGD Flux

Mass balance models of both long-lived isotopes of ^{228}Ra and ^{226}Ra in the nearshore seawater can be employed to estimate SGD. Only the mass balance model of ^{226}Ra is discussed here. It is assumed that the system is in steady state, that is: the total gain of ^{226}Ra in the system equals the total loss of ^{226}Ra from the system.

Radium can be added to near-coastal seawater by Ra desorption from seafloor sediment (G_{des}), Ra diffusion from the seafloor sediment (G_{diff}), erosion of terrestrial sourced sediment (G_{ero}), river supply (G_{riv}), and SGD (G_{SGD}). The loss is due to decay of the isotope (L_{dec}) and tidal mixing (L_{mix}) between the seawater in the modelled area and open seawater (Moore et al., 2006). The general mass balance equation then is

$$G_{des} + G_{diff} + G_{ero} + G_{riv} + G_{SGD} = L_{mix} + L_{dec} \tag{7.5}$$

The unit of these terms is dpm per time (dpm means disintegrations per minute) and a measure of the intensity of the source of radioactivity (Davis, 1955). The term G_{SGD} can be calculated from Eqn (7.5) if all other terms are known. After G_{SGD} is estimated, and if the groundwater Ra activity is known, the volumetric flow rate of the SGD can be calculated as

$$Q_{SGD} = G_{SGD}/Ra_{gw} \tag{7.6}$$

where Ra_{gw} (dpm l^{-1}) is the radium activity in groundwater [dpm l^{-1}] and Q_{SGD} (l d^{-1}) is the volumetric SGD rate.

It can be useful to employ both ^{226}Ra and ^{228}Ra to carry out the above balance analysis so that sources of radium can be better defined. In addition to radium sourced from SGD, ^{226}Ra may be more controlled by benthic diffusion, while ^{228}Ra is influenced by terrestrial fluxes derived from riverine loading and particulate desorption. The SGD estimated from these two models can be cross-checked (Moore, 1996, 2003).

7.5.3.5 Using Radon Isotopes to Study SGD Dynamics and Fluxes

Similar to the budget analysis of Ra, a Rn mass budget approach can be also used to estimate SGD. This approach was first detailed by Burnett and Dulaiova (2003) and has been widely used since then (Burnett et al., 2008; Tse and Jiao, 2008; Luo et al., 2016). The assumptions are that (i) there is no river runoff into the coastal area, (ii) radon decay and atmospheric evasion are the main loss of radon and (iii) the water body is well mixed and there is no water stratification in terms of radon distribution.

Various sources and sinks can influence the concentration of ^{222}Rn in the coastal area, such as ingrowth from ^{226}Ra decay, tidal pumping, loss to the atmosphere, diffusion from sediments, mixing losses to the open sea and input from SGD (Figure 7.14). The ^{222}Rn activity change over a time interval is the net balance between these sources and sinks

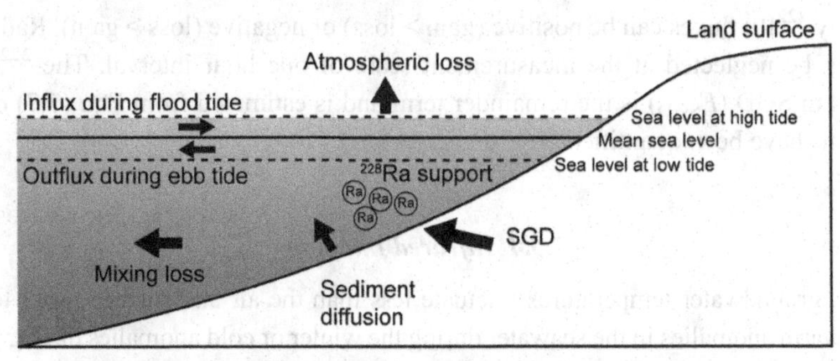

Figure 7.14 Sources and sinks of ^{222}Rn in coastal waters (Tse and Jiao, 2008).

during that period. When the ^{222}Rn activity in seawater is continuously measured (usually hourly) to calculate the ^{222}Rn difference between two successive measurements, a radon balance model can be written as

$$F_t = (F_{\text{SGD}} + F_{\text{sed}} + F_i) - (F_o + F_{\text{atm}} + F_m) \tag{7.7}$$

where F_t is the rate of ^{222}Rn change inferred from two successive samples, F_{SGD} is the ^{222}Rn flux attributed to SGD, F_{sed} is the diffusion flux from sediments, F_i is the flux entering with the incoming tide, F_o is the flux leaving with the outgoing tide, F_{atm} is the atmospheric loss and F_m is the mixing loss (flux out of the system) caused by mixing of nearshore high radon seawater with offshore low radon seawater (Burnett and Dulaiova, 2003). By solving the equation for F_{SGD} and dividing F_{SGD} by the groundwater ^{222}Rn concentration, the Q_{SGD} is obtained.

Using the above approach to estimate groundwater fluxes requires hourly Rn measurements in seawater over a period of one to two days to obtain the temporal fluctuation of ^{222}Rn activities such as in Figure 7.2. The main aim of the measurements is to calculate the ^{222}Rn inventory, which is the product of ^{222}Rn activities and water column height, assuming that the ^{222}Rn in the column is uniformly distributed. The ^{222}Rn inventory changes with tidal height because ^{222}Rn can be removed from the water column with the outgoing seawater on the ebb tide (F_o) and added to the column with the incoming seawater during the flooding tide (F_i).

Atmospheric loss (F_{atm}) to the air from the water surface can be estimated by an empirical equation (MacIntyre et al., 1995) after the wind speed, the concentrations of radon in water and in air above the sea surface have been measured. However, atmospheric losses due to turbulence, e.g. breaking waves, are more difficult to quantify. The diffusive flux (F_{sed}) from the seafloor sediments can be estimated by another empirical equation (Martens et al., 1980) after radon concentrations in pore water in the sediments and seawater above the sediments, porosity of the sediments, molecular diffusivity coefficient of radon have been estimated.

The mixing loss (F_m) is estimated indirectly by inspecting the temporal change in the estimated ^{222}Rn inventories after being corrected for atmospheric loss (Burnett et al., 2008).

The hourly ^{222}Rn fluxes can be positive (gain > loss) or negative (loss > gain). Radioactive decay can be neglected at the measurement scale of one hour interval. The ^{222}Rn flux attributed to SGD (F_{SGD}) is the remainder term and is estimated from Eqn (7.7) once the other terms have been quantified.

7.5.4 Infrared Imaging

Typically, groundwater temperatures fluctuate less than the air and surface water temperatures, so warm anomalies in the seawater during the winter or cold anomalies during summer can indicate groundwater outflow. Measurements should be taken when the contrast between groundwater temperature and seawater temperature is the greatest. Seawater temperatures can be measured directly, or be inferred from infrared images (Scientific Committee on Oceanic Research, 2004). Although infrared satellite images are widely available, their spatial resolution is usually too low to be useful for SGD studies. For identification of SGD, ideally infrared images with a resolution of about 1 m are needed.

Identifying SGD outflow zones using infrared images (Mulligan and Charette, 2006; Johnson et al., 2008) is efficient because it can map the SGD distribution at large scale quickly when done from the air. However infrared images require very careful analysis because surface water temperature is sensitive to sunlight, reflection and shadows of clouds and coastal vegetation. To avoid the influence of sunlight, infrared images are best taken early in the morning. A site should be mapped at least twice so that non-groundwater-related factors can be filtered out. The tidal stage must be considered as SGD is usually strongest during falling tide. So far thermal measurements have been used mainly to identify potential SGD locations and aid in planning of the fieldwork. It is difficult to quantify the amount of SGD based on thermal images.

7.6 Limitations, Uncertainties and Future Research in SGD Estimation

SGD estimation involves field sampling, laboratory measurements and mass budget calculations. All these involve errors and uncertainties. Problems may also arise from various other factors such as the heterogeneous nature of the coastal aquifer system, complex flow patterns of the coastal groundwater and seawater, and physical and chemical processes in the transition zone of fresh groundwater and saline water.

7.6.1 Mass Budget Analysis Based on Isotope Tracers

The SGD flux is typically estimated based on the budget equation of a tracer such as Eqn (7.5) or (7.7), which involves the quantification of many components. After the other sources and sinks of the tracer, such as from rivers, sediments, air, the mixing loss due to tidal exchange, and tracer decay are identified and estimated, the tracer input from SGD is left as the remainder term to keep the equation balanced. As a result, all the errors and uncertainties in other components propagate into the SGD estimate.

To convert the tracer flux to a SGD flux, the tracer concentration (e.g. Ra_{gw} in Eqn (7.6)) of the SGD entering the sea across the aquifer-seawater interface, or the groundwater end-member, is required. The estimation of the SGD flux is thus very sensitive to errors in the denominator Ra_{gw}. The concentration of Ra_{gw} should be measured right at the exit point of the groundwater. For small-scale SGD studies, groundwater can be sampled using seepage meters, pushpoint samplers or specially designed multi-level sampling systems immediately near the shoreline to obtain radium, radon and other solute concentrations of interest for the groundwater end-member. Since for larger-scale studies this information is typically not available, the tracer concentration in groundwater samples from existing drinking-water wells, which are often within freshwater zones and fairly away from the shoreline, are sometimes used to approximate the concentrations of SGD at the aquifer-seawater interface. Some SGD studies use the Ra and Rn in wells of tens of metres or even kilometres away from the shoreline to represent the groundwater end-member. The errors in the analysis of the tracer concentration will of course greatly influence the accuracy of the estimated SGD flux.

Moreover, the isotopic and chemical composition of the groundwater may change significantly over a short distance in the transition zone where the geochemical characteristics vary greatly. It has been noted that groundwater Ra can vary spatially (and temporally) by several orders of magnitude (Gonneea et al., 2013). Radium activities are sensitive to salinity and pH (Kraemer and Reid, 1984; Beck and Cochran, 2013), both of which change significantly along the flow path in the transition zone.

The isotopes of Ra and Rn behave differently in the transition zone. Figure 7.15a shows a coastal unconfined aquifer and the possible changes of Ra and Rn along a horizontal line AB and vertical line CD. When the subsurface production rate by radioactive decay is constant, Rn will become progressively more concentrated along the flow path (Figure 7.15b). In the immediate vicinity of the exit point to the sea, Rn concentration levels may drop due to evasion of gaseous Rn or mixing with Rn-depleted seawater. The change of Ra along the path, however, is more complicated. In freshwater, a portion of the radium is adsorbed on the aquifer sediments. Radium concentrations are elevated in the saltwater zone because the sediment-bound Ra will desorb. Just like with Rn, the concentration of Ra may decrease again immediately near the exit point. Wells of different horizontal locations and depths will thus yield different levels of Ra and Rn, as shown in Figures 7.15b and c.

Figure 7.15 is a simplified conceptual model and does not consider other factors, such as pH and redox conditions, that may also affect Ra activities. If the aquifer is heterogeneous, which is always the case for a real coast, the spatial distributions of Rn and Ra can be much more complicated. Hydraulic heterogeneity causes a more complex flow field, and geochemical heterogeneity further adds to the complexity because the production rates of Rn and Ra vary spatially. Figure 7.16 shows the spatial distribution of ^{224}Ra, ^{223}Ra and ^{228}Ra obtained from a multi-level monitoring system in the beach aquifer of Tolo Harbour, Hong Kong. The spatial change of ^{224}Ra is most significant. Within the domain of 3 m by 100 m, its concentration spans over two orders of magnitude. Even when such high-resolution data are available, there is no general guideline about how to use the data in mass balance calculations. A rule of thumb is to take the average or the median of all the measurements in the intertidal

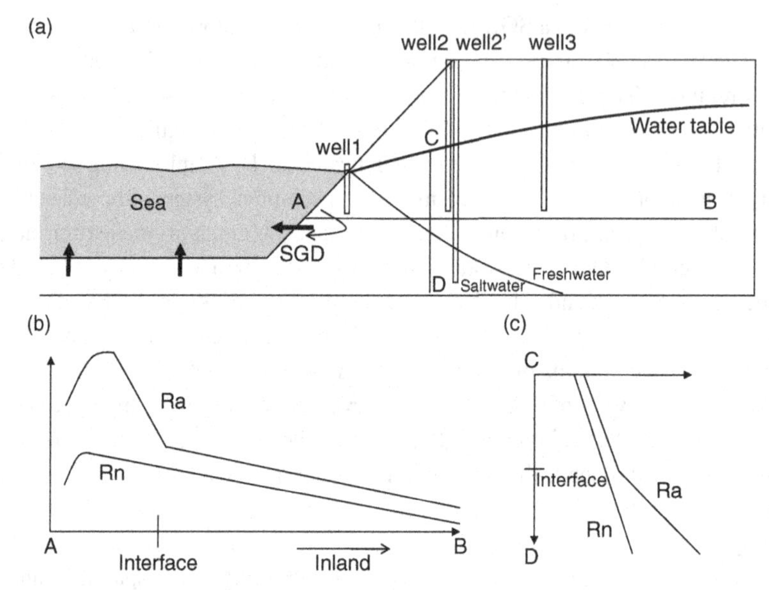

Figure 7.15 A conceptual model showing (a) an interface and piezometers installed at or near the interface in a coastal unconfined aquifer, (b) Ra and Rn distributions along the horizontal line B to A and (c) vertical line C to D.

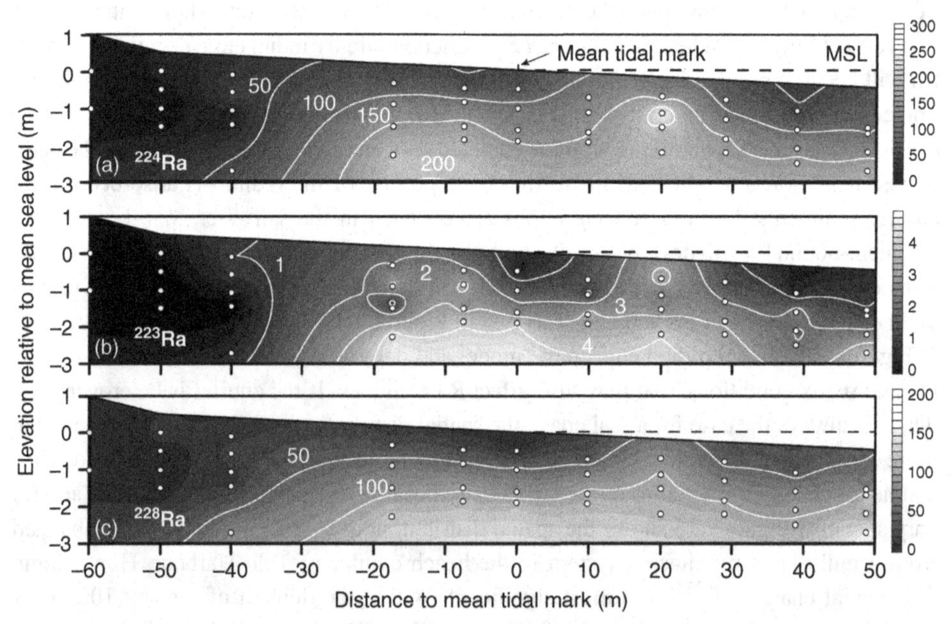

Figure 7.16 The spatial variation of ^{224}Ra, ^{223}Ra and ^{228}Ra (dpm l^{-1}) on 25 April 2016 measured in samples from a multi-level monitoring system in the beach aquifer of Tolo Harbour, Hong Kong.

zone as the groundwater end-member. It is suggested that the ratio of ^{228}Ra/^{226}Ra may be a better tracer than ^{228}Ra or ^{226}Ra because the spatial variation of the ratio is more stable than the activities of the two isotopes alone and this ratio in groundwater is usually much higher than seawater. Using both ^{228}Ra and ^{226}Ra can also help determine which groundwater sources are driving the Ra budget (Moore, 2003; Peterson et al., 2016).

For SGD studies at the continental shelf scale, however, it is impossible to conduct detailed groundwater sampling along the shoreline because the study area is too large. As a result, large uncertainties arising from the non-representativeness of samples for the groundwater end-member are unavoidable. Even when detailed sampling data are available at some parts of the shoreline, there is no general guideline about how to upscale the data to the entire shoreline. SGD-derived nutrient fluxes to the ocean are evaluated by multiplying the SGD rates by the average nutrient concentration in the groundwater wells (e.g. Slomp and Van Cappellen, 2004; Moore et al., 2006; Hwang et al., 2010; Moore, 1996; Santos et al., 2008). Consequently, the uncertainty of both the SGD flux as well as the nutrient concentration in groundwater will all contribute to that of the estimated nutrient fluxes.

7.6.2 Temporal Variability of SGD

Coastal aquifers are heterogeneous and groundwater flow patterns are dynamic, so SGD varies spatially and temporally. The variability of tracer and nutrient inputs to the ocean by SGD is caused by heterogeneity of the aquifer properties, physical and chemical reactions of the tracers along flow paths, and the dynamic character of the weather- and tide-driven freshwater–seawater interaction in the transition zone. To understand the effects of these controls and their interplay on tracer activities in the transition zone, a dedicated groundwater monitoring network of multi-level observation wells is required. Since such studies are much more time and resource demanding than seawater-based SGD sampling, they remain limited in number (Michael et al., 2011; Gonneea et al., 2013).

The different behaviours of Rn and Ra in coastal aquifers discussed before suggest that the SGD estimated from these two tracers may represent different contents of SGD. Fresh SGD does not contain as much Ra as saline SGD. Thus, fresh SGD may not be adequately quantified by means of the Ra approach, as the SGD estimate is likely to be dominated by circulated seawater. Radon, however, behaves independent of groundwater salinity, and is therefore representative of the total SGD (Charette, 2007).

It has been demonstrated that there is typically a time lag between SGD and the seasonal fluctuations of inland recharge that drive seaward groundwater flow (Michael et al., 2005). The delay can be significant in a coastal aquifer with low hydraulic head gradients and low-permeability materials. Because of this delay, the peak flux in groundwater-derived chemicals may lag behind the wet season.

Isotopic activities may fluctuate with water table, which changes the soil types in contact with the groundwater, and vary with groundwater flow velocity and thus residence time.

In dry seasons when groundwater level is low and the sea tide is high, seawater intrudes into the aquifer and the salinity increase induces radium, and possibly also ammonium, desorption from the sediment (Charette and Buesseler, 2004).

Radium isotopes of different half-lives have different temporal characteristics and the short-lived isotopes (^{224}Ra and ^{223}Ra) will be more sensitive to the transient changes of the flow system. At least one year monitoring of the dynamics of the flow, nutrients, and radon and radium activities in the coastal wells or piezometers are required to understand the seasonal changes.

Mass balance calculations often suggest that a large mass of natural isotope tracers from the aquifers is required to balance the equations (e.g. Eqns (7.5) and (7.7)). So far there is no direct study to prove if the aquifer can provide that much tracer. Isotopic activities in aquifers depend on the balance between the rate with which water circulates through the sediments and the renewal rate of the isotope from its parent isotope. If the regeneration rate of the tracer is low relative to the water circulation rate, further circulation of the water may not always be able to remove more tracers such as Rn and Ra from the aquifer and bring them to the sea. This calls for studies to understand the generation, movement and distribution of tracers in groundwater, especially in the transition zone. More specifically, these studies are needed to understand the release of radon and radium from the aquifer materials, the rates of which are dependent on the aquifer material properties and the subsurface chemical environment, and the pathway and time scales of the transport of these tracers inside the aquifer towards the discharge boundary at the sea. Such studies focusing on spatial and temporal variability of tracers inside freshwater-seawater transition zones, however, are very limited. So far there have been only a few field coastal sites with extensive temporal and spatial sampling to understand tracer behaviour and its influence on SGD estimation (Smith et al., 2008; Gonneea et al., 2013; Heiss and Michael, 2014; Liu et al., 2018).

7.6.3 Movement and Distribution of Tracers in Seawater

In seawater-based studies, SGD can be estimated by a diffusion model, assuming the advection in the seawater can be ignored, or a mass balance model, assuming that the offshore water system is steady and there is no significant stratification in the water body (Moore, 1996, 1999). These models have various limitations. Mass balance analysis may overlook some SGD sources in a multiple source situation if there are insufficient observation points (Moore, 2003). Seawater is very dynamic and currents, waves and tides all contribute to the transport and mixing of natural tracers. Budget analysis usually cannot provide information on the spatial and temporal variations of the SGD.

It will be a great breakthrough if a coupled groundwater–surface water model were developed that simulates the reactive transport process of Ra or Rn, including the release from the aquifer, adsorption and its migration and mixing in the ocean. With such models the generation and fate of SGD tracers can be better understood.

8

Coastal Palaeo-Hydrogeology

8.1 Introduction

It was discussed in Chapters 3 and 6 how a steady state seawater wedge can form in coastal aquifers. This will only be achieved, however, if the hydrological conditions and sea level remain stable over a time period longer than the time required for the groundwater system to adjust to the prevailing environmental conditions. The propagation of pressure changes in response to a perturbation of a hydrological forcing is a relatively fast process in aquifers, but adjustment of the solute distribution to the perturbation tends to be much slower and may take thousands to millions of years in systems with low-permeability layers. Consequently, the present-day salinity distribution in coastal aquifers is often found to be out of equilibrium with the present-day sea level and coastline position (Kooi and Groen, 2003; Post et al., 2013). This is borne out by old seawater residing in aquifers on land, and relatively freshwater present in aquifers below the seafloor. This chapter explores the geological and hydrogeological conditions which may cause the formation of such relic fresh or relatively fresh groundwater below seabed and residual saltwater in onshore aquifers. Examples in Europe, America and China are used to describe the occurrence and distribution of the palaeo-waters in different hydrogeological environments. With concerns over sea level rise mounting, a proper understanding of the control of the sea level on coastal groundwater systems becomes increasingly important. Conversely, the redistribution of water between the continents and the ocean basins influences sea level (Reager et al., 2016). The connection between coastal aquifers and the marine environment and the dynamics over time thereof needs to be understood.

In this chapter, the abbreviations ka and Ma refer to *one thousand* and *one million* years, respectively if they indicate a time in the past. If they indicate a duration they are written as ky or My. The abbreviation BP stands for *before present* and AD for *anno domini*, the number of years since the Christian year zero.

8.2 Sea Level Change on a Geological Timescale

From a human perspective it is tempting to think of the current sea level position and the land-ocean distribution as fixed. Sea level is, however, highly dynamic, and fluctuates at the

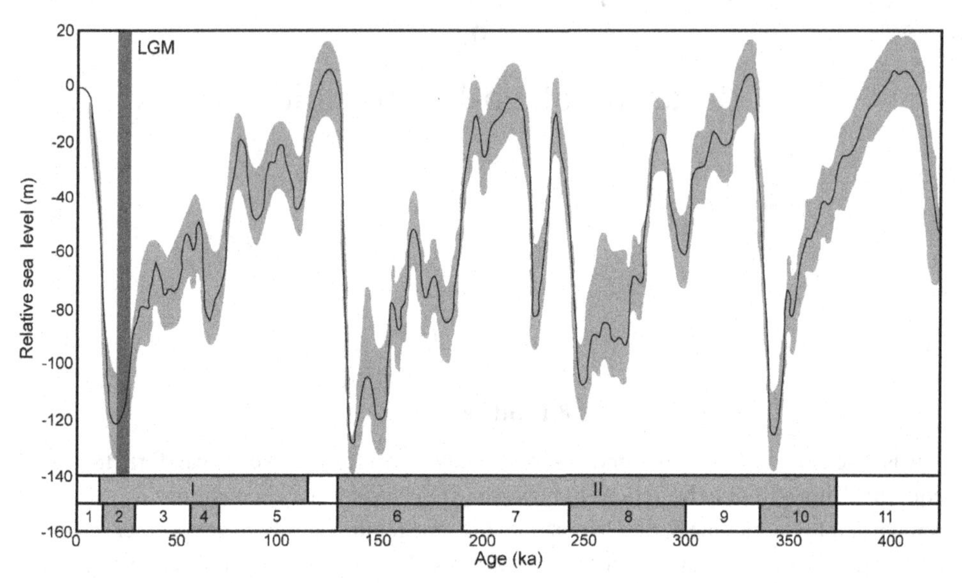

Figure 8.1 Sea level variations over the last 432 ka based on oxygen isotopic ratios in benthic foraminifera from the North Atlantic and Equatorial Pacific Ocean (Waelbroeck et al., 2002). The shaded area along the curve corresponds to the confidence bands reported by Waelbroeck et al. (2002). The vertical shaded bar indicates the Last Glacial Maximum (26.5–19 ka BP). Along the bottom of the graph, the horizontal bars with roman numerals indicate the alternation of glacial (white) and interglacial (grey) periods. The horizontal bars with Arabic numbering indicate the marine isotope stages (MIS). For naming, see Table 8.1.

timescale of hours, as discussed in Chapter 4 on tides, to thousands of years. On a geological timescale, sea levels are mainly driven by glacial cycles. During glacial periods, enormous volumes of the Earth's water are sequestered in glaciers and ice caps, and consequently, sea levels fall.

Since the onset of the Pleistocene Epoch 2.6 Ma BP, the Earth's climate has been dominated by repeated glacial cycles of which the strongest have been occurring since 800 ka ago. The duration of glacial periods that have occurred since then was around 100 ky, with the interglacial periods lasting between 10 and 30 ky (PAGES, 2016). The interglacial periods are relatively warm and have a reduced ice cover compared to glacial periods. The strongest interglacial periods occurred around 400 ka BP and between 130 to 115 ka BP (Figure 8.1; Table 8.1). The present interglacial period coincides with the Holocene epoch, which is the subdivision of the geological timescale that started 11.7 ka BP.

Different definitions of glacial and interglacial periods exist in the literature, and the nomenclature varies between regions (Table 8.1). The terms stadial and interstadial are used to indicate cooler and warmer phases within glacial periods. In palaeo-climate studies a subdivision of the Pleistocene into marine isotope stages (MIS) has been developed based on the oxygen isotope ratio in planktonic foraminifera from deep-sea sediments, which has

Table 8.1 *Terminology Used in the European and American Literature to Denote Glacial and Interglacial Periods during the Quaternary*

Epoch name	Approximate equivalent MIS	North America	Northern Europa	Alps	Great Britain
Holocene	1				Flandrian
Pleistocene	5d–2	Wisconsin	Weichselian	Würm	Devensian
	5e	Sangamonian	Eemian	Riss–Würm	Ipswichian
	10–6	Illinoian	Saalian	Riss	Wolstonian
	11	Yarmouthian	Holstein	Mindel–Riss	Hoxnian
	12	Kansan	Elsterian	Mindel	Anglian
	21–13	Aftonian	Cromerian	Günz–Mindel	Cromerian

Note. Shaded rows indicate glacial periods.

become adopted for ice core, speleothem and terrestrial palaeo-climate records as well. In this system, even numbers represent cooler and odd numbers represent warmer periods.

8.2.1 Eustatic Sea Level Change

When the volume of ocean water changes due to a change of the ice volume, it is referred to as glacial eustasy. Eustatic sea level change can also occur by the addition of juvenile water (i.e. water released during volcanic activity) and the storage of water in sediments (Woodroffe and Horton, 2005), but the associated changes tend to be small and slow. The seasonal redistribution of liquid water between the ocean and the continents on the other hand is much stronger, and is estimated to cause variations around the mean sea level of up to 15 mm with significant inter-annual variations (Reager et al., 2016).

Tectono-eustasy denotes changes to the volume of the holding capacity of the oceans caused by processes like sediment compaction and tectonics. Because this is a slow process, the associated changes in sea level are on the order of less than 0.1 mm yr^{-1} (Mörner, 1996), but over millions of years the associated changes in sea level are hundreds of metres (Lambeck and Chappell, 2001). Expansion or contraction of the ocean water also contributes to a change of its volume. If they are associated with a change of water temperature, they are referred to as thermosteric changes, whereas halinosteric changes are related to salinity (Church et al., 2013). Thermosteric changes are a major component of contemporary sea level rise, with halinosteric changes contributing to a lesser degree (Woodroffe and Horton, 2005; Church et al., 2013).

During the previous interglacial (130 to 115 ka BP), global sea level is estimated to have been at least 3 m above the present-day value (Cuffey and Marshall, 2000). During the

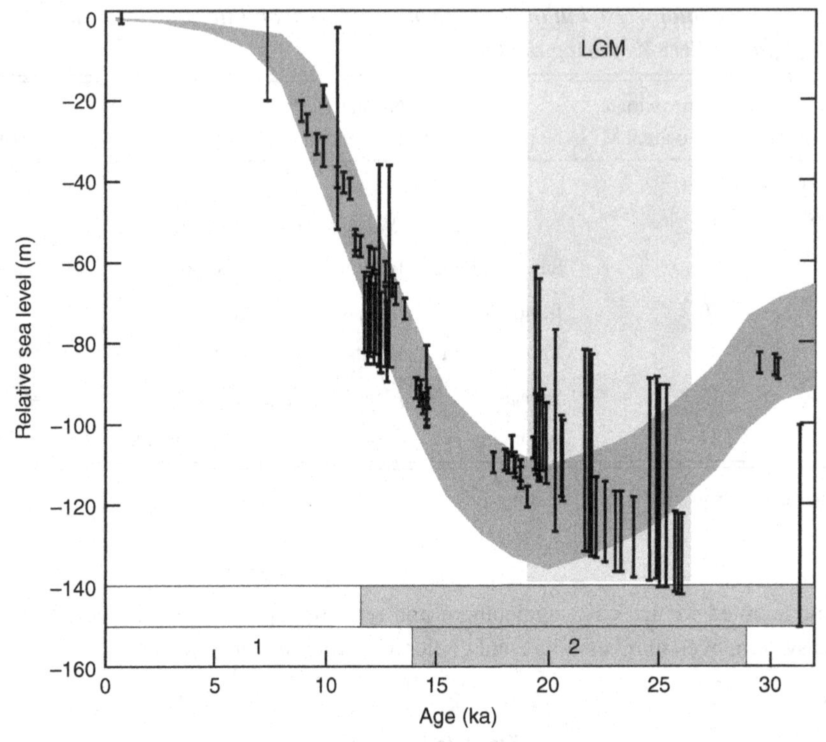

Figure 8.2 Same sea level versus time curve as in Figure 8.1, but zoomed in to the last 32 ka. Bars are the reconstructed sea level data from dated coral species in Barbados (Peltier and Fairbanks, 2006). The length of the bar indicates the uncertainty due to the range of water depths at which a particular coral species is found.

coldest parts of the glacial periods global sea level fell by more than 100 m (Figure 8.1). When the ice sheets reached their maximum extent during the Last Glacial Maximum (LGM), from 26.5 to about 19 ka BP years ago (Clark et al., 2009), sea levels were lower by as much as 130 m relative to the present-day sea level, and the ice volume was up to ~52 × 10^6 km^3 greater than today. At the onset of this period, sea levels fell approximately 40 m in a period of only 2000 years (Lambeck et al., 2014).

The melting of the ice caps caused sea level to rise since the LGM (Figure 8.2). During the main phase of deglaciation between ~16.5 and ~8.2 ka BP, the average global mean sea level rose at an average rate of 12 m ky^{-1} (Lambeck et al., 2014). The rise in sea level was punctuated by jumps caused by the catastrophic collapse of ice-dammed lakes in North America. Locally, such as in the Rhine-Meuse delta in north-western Europe, these melt-water pulses caused abrupt sea level increases of up to 2.1 m (Tornqvist and Hijma, 2012), and similar jumps have been identified in Asian river deltas (Hori and Saito, 2007). The high rates of global mean sea level rise that prevailed around the start of the

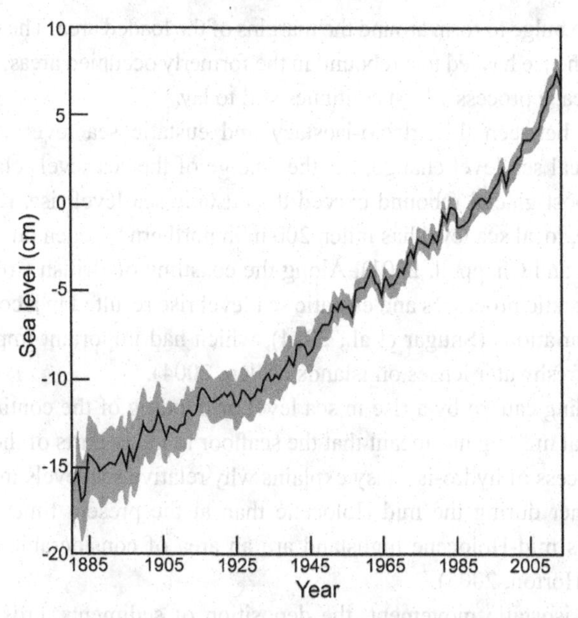

Figure 8.3 Change in eustatic sea level (relative to the year 1990) between 1880 and 2014 inferred from tide gauge measurements. Shaded area corresponds to the confidence bands. Data and a description of the methodology were originally published by Church and White (2011). Updated data set was obtained from Legresy (2016).

Holocene epoch decreased, and between 7 to 3 ka BP, mean sea level rose by 2 to 3 m to approximately the present-day level (Church et al., 2013).

Rates of change since 3 ka BP were on the order of tenths of mm yr^{-1}, but acceleration started to occur again to values on the order of mm yr^{-1} during the first half of the twentieth century, but possibly already during the second half of the nineteenth century. Studies of tidal gauges suggest that the rate was 1.5 to 1.9 mm yr^{-1} between 1901 and 2010, and that the rate has been increasing during the last two centuries (Figure 8.3). Variations occur on the timescale of decades, for example between 1920 and 1950 rates were comparable to the relatively high rates observed today. Based on satellite altimetry data, it is estimated that the rate of sea level rise was 2.8 to 3.6 mm yr^{-1} between 1993 to 2012 (Church et al., 2013).

8.2.2 Relative Sea Level Change

Superimposed on eustatic sea level changes are local changes, which can have a variety of causes. A principal factor that influences sea levels on the timescale of glacial cycles is isostatic adjustment (Lambeck and Chappell, 2001), i.e. the deformation of the Earth's surface due to the redistribution of ice and water masses. It is caused by the movement of material in the mantle away from areas that are being loaded. At high latitudes, the growth of the ice sheet pushed away mantle material during the last glacial period, and the displacement principal caused a bulge to form around the margins of the loaded area. The unloading caused by the retreat of the ice has led to a rebound in the formerly occupied areas, and a subsidence in the bulging area, a process which continues still today.

The interplay between this glacio-isostacy and eustatic sea level variations largely determine the local sea level change, i.e. the change of the sea level relative to the land. Where rates of post-glacial rebound exceed the eustatic sea level rise, relative sea levels fall. For example, local sea level has fallen 200 m in northern Sweden during the Holocene epoch (Lambeck and Chappell, 2001). Along the coastline of British Columbia the interplay between isostatic processes and eustatic sea level rise resulted in a complex history of local sea level variations (Shugar et al., 2014), which had important implications for the development of freshwater lenses on islands (Allen, 2004).

The mass loading caused by a rise in sea level or flooding of the continental shelves in response to glacial melting has meant that the seafloor in many parts of the world has been sinking. This process of hydro-isostasy explains why relative sea levels in the Indo-Pacific region were higher during the mid-Holocene than at the present time. The timing and magnitude of this mid-Holocene highstand are an area of considerable scientific debate (Woodroffe and Horton, 2005).

In addition to isostatic movement, the deposition of sediments, erosion and tectonic activity can also cause vertical motion of the land relative to the sea. Other factors include changing ocean currents, water temperature and inputs of freshwater from the continents (Church et al., 2013). Changes in the Earth's gravitational field by changes in ice mass also lead to a redistribution of water in the oceans because the ocean surface must remain at an equipotential (Mitrovica and Peltier, 1991). As a result, locally sea level can fall near where ice is melting, but rise further away from where melting occurs. Moreover, the elevation of the land surface may subside in areas where subsurface fluids like groundwater or fossil fuels are abstracted on a large scale (Galloway and Burbey, 2011) or where land is drained to control the position of the water table (Oude Essink et al., 2010; Erkens et al., 2016).

All these factors combined make that local sea level trends can deviate considerably from global trends. Along the same coastline, the response may also show wide variability (Shennan and Horton, 2002; Shugar et al., 2014). For example, along the shoreline of the south-eastern North Sea coastal plan, which spans across Belgium, the Netherlands and Germany, the Holocene sea level curves are offset by metres, even though the physiographic characteristics of the coastal zone are very similar. The reason for the large discrepancy is that the magnitude of the post-glacial isostatic adjustment decreases from the north to the south: Northern Germany has experienced 7.5 more isostatic adjustment than Belgium since 8 ka BP (Vink et al., 2007).

One of the most dramatic examples of the effect of the interplay between eustatic sea level change and local topography is the more than 1500 m drop in the level of the Mediterranean Sea that occurred during an event known as the Messinian salinity crisis. The formation of a sill due to tectonic activity and a drop in eustatic sea level made that the water mass of the Mediterranean became isolated from the Atlantic Ocean. Complete isolation became established around 5.59 My ago (Krijgsman et al., 1999). The inflow of freshwater to the closed basin was less than the evaporation which resulted in desiccation and the deposition of massive evaporite deposits across the Mediterranean. Tectonic subsidence, aided by erosion and global sea level rise, opened up the present Gibraltar Strait around 5.33 My ago. This caused the so-called Zanclean flood, which was associated with extraordinary discharges estimated to have been on the order of 10^8 m^3 s^{-1} and sea level rise of more than 10 metres per day (Garcia-Castellanos et al., 2009). The relevance of the Messinian salinity crisis to present-day coastal hydrogeology is that the base level drop had a strong influence on the development of karst networks in the limestone rocks around the Mediterranean, which will be further discussed in Section 8.6. Moreover, the presence of evaporite beds and brines from this episode in geological history constitute a source of saline groundwater around the Mediterranean (Re and Zuppi, 2011).

8.2.3 Shoreline Migration

Sea level change is also a primary factor in controlling shoreline migration. Over the course of the Quaternary, coastline positions have moved by hundreds to thousands of kilometres across continental shelves in response to eustatic sea level rise. A landward migration of the coastline is called a transgression, whereas a regression is a seaward migration of the coastline. For a given rate of sea level rise, the slope of the land surface is the principal determinant of the rate of lateral movement of the coastline. The vertical position of the continental shelf will vary depending on the height of the water column due to isostasy. This complicates the reconstruction of former coastline positions, as this effect has to be inferred from models (Lambeck and Chappell, 2001). Figure 8.4 shows an example of how the coastlines evolved during the Holocene in the North Sea region.

The rate of sea level rise is an important factor in the development of river deltas. The deceleration of the Holocene sea level rise was a major factor that caused the progradation of deltas worldwide between 8.5 and 6.5 ka BP (Stanley and Warne, 1994). For example, a major progradation of the Song Hong (Red River) delta occurred during this period, which was followed by a period between 6 and 4 ka BP when mangroves developed on the delta plain. When relative sea levels dropped by up to 3 m after 4 ka BP, the plain emerged above sea levels. Similar changes occurred in deltas elsewhere, although the timing may vary.

In addition to sea level, sediment supply is another key factor that controls the position of the coastline in depositional areas. Frihy and Khafagy (1991) found that the shorelines of the Nile delta were prograding during the nineteenth century when sediment supply was high, whereas retrograding shorelines, with regression rates as high as 53–58 m yr^{-1}, occurred when lower floods brought less sediment to the sea during the twentieth century. Relative sea level rise caused a regression of the shoreline of the Bohai Sea in China of up to 100 km after 7 ka BP (Deng et al., 2016). Here, the increase in sediment supply of the Yellow River also played a major role, which was partially due to deforestation during the Zhou dynasty (1046–256 BC).

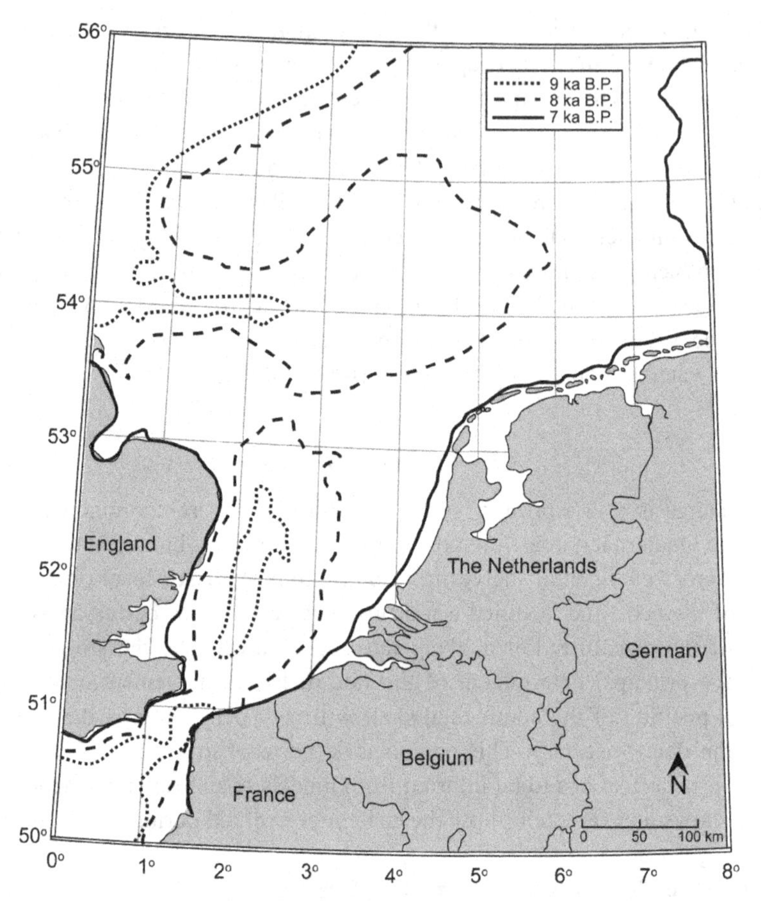

Figure 8.4 Map of the southern North Sea showing coastline positions at 9, 8 and 7 ka BP (Post, 2004).

Studies of the Pearl River Delta in China have shown that the evolutionary history of the delta in the past 9 ky was an interplay between sea level, river discharge, and human activities (Zong et al., 2009). Rapid sea level rise during the early Holocene created the available accommodation space for the delta to expand. When the sea level rise slowed down significantly around 7 ka BP, monsoon-induced river runoff delivered large volumes of sediment to the delta. Between 6.8 and 2 ka BP, the progradation of the delta slowed because the runoff was reduced. During the last 2 ky, accelerated coastline advance has been occurring as a result of human activities. Farmers reclaimed tidal flats by trapping sediments behind walls built along the low tide mark. Because coastline changes have a direct effect on inhabitation and settlements, archaeological sites can be indicators of palaeo-shorelines. Figure 8.5 shows the estimated position of

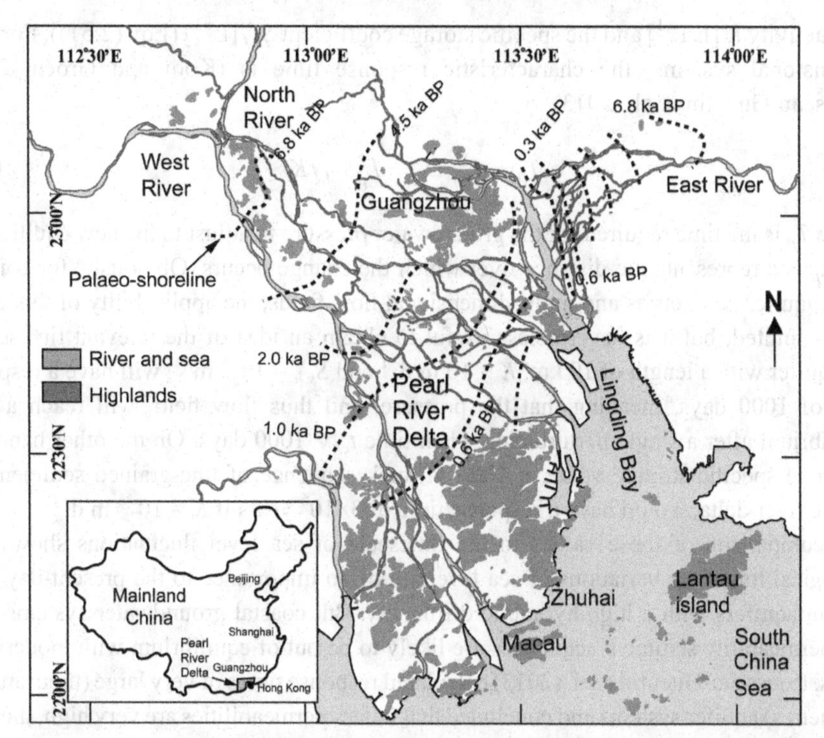

Figure 8.5 Palaeo-shorelines in the Pearl River Delta estimated based on archaeological evidence and historical records (Kuang et al., 2016).

the advancing shorelines in the Pearl River Delta based on historical records and archaeological finds such as shell mounds, village relics and tomb sites (Li and Qiao, 1982; Li and Chen, 1991).

8.3 Reorganisation of Coastal Groundwater Systems

The response of the groundwater system to the continually changing environmental conditions at the land surface will depend on its hydrological properties. While the pores of aquifers are flushed multiple times over geological timespans, the much slower response of low-permeability units means that their pore water chemistries record changes that have occurred over periods of millions of years. An understanding of the response times is thus required to study the impacts of sea level change on coastal groundwater systems.

8.3.1 Coastal Aquifer Response Times

The time required for fluid pressures in an aquifer or aquitard to adapt to a perturbation is controlled by the hydraulic diffusivity D_h [L^2 T^{-1}], i.e. the ratio of the hydraulic conductivity K [L T^{-1}] and the specific storage coefficient $S_{s,f}$ [L^{-1}] (Eqn (2.51)). For one-dimensional systems, the characteristic response time is (Kooi and Groen, 2003; Rousseau-Gueutin et al., 2013)

$$T_p = L_p^2/D_h = L_p^2 S_{s,f}/K \tag{8.1}$$

where T_p is the time required for the groundwater pressure to adjust to the new equilibrium and L_p is a representative distance over which the change occurs. Obviously, for complicated aquifer geometries and multi-dimensional flow fields, the applicability of this equation is limited, but it is nevertheless useful to obtain an idea of the relevant timescales. An aquifer with a length of 10 km, $K = 10$ m d^{-1} and $S_{s,h} = 10^{-4}$ m^{-1}, will have a response time of 1000 days, meaning that the pressure, and thus flow field, will reach a new equilibrium after a single perturbation when time $t \gg 1000$ days. On the other hand, for the same specific storage value, a 1000 m thick sequence of fine-grained sediments in a large river delta, would have a response time of 3×10^5 years if $K = 10^{-6}$ m d^{-1}.

A comparison of these values to the timescale of sea level fluctuations shows that geological timescale variations of sea level are of no importance to the present-day flow field in aquifers with a high hydraulic diffusivity. But coastal groundwater systems with low-permeability sediment sequences are likely to be out of equilibrium with modern sea levels. Rousseau-Gueutin et al. (2013) calculated response times of very large (thousands of kilometres) aquifer systems and concluded that unless permeabilities are very high, they are in a constant state of disequilibrium with respect to changes in recharge and sea level. This is further compounded by the loading caused by the accumulation of sediments, and anomalous fluid pressures are likely to develop where sedimentation rates have been high during the Holocene (Kooi and Groen, 2003). Situations can even exist where the sedimentation rates are so high, and fluid pressures cannot dissipate fast enough, that hydraulic gradients are created that are on the same order of magnitude as those that drive topographic flow systems.

8.3.1.1 Diffusion

The same equation as Eqn (8.1) can be used to establish a timescale for solute movement in low-permeability layers, like clay and shale in which flow is restricted and transport is dominated by diffusion, by replacing the hydraulic diffusivity D_h by the effective molecular diffusion coefficient of solutes D_e. The latter is on the order of 10^{-10} m^2 s^{-1} for a solute like chloride, and so response times for solutes can quickly become on the order of hundreds of years or more. For example, for a 10 m thick clay layer, the response time will be over 3×10^4 years. Due to the long response times associated with diffusive transport, low-permeability layers often still contain water related to former flow regimes. These thereby record the palaeo-hydrological history of the coastal groundwater system, and can form a source of salinity to the groundwater in adjoining aquifers (Dakin et al., 1983).

The salinity in a confining unit of infinite extent that is exposed to a step change in salinity at position $x = 0$ and time $t = 0$, will change according to (Groen et al., 2000)

$$C(x,t) = C_0 + (C_s - C_0)\,\mathrm{erfc}\left(\frac{x}{2\sqrt{D_e t}}\right) \tag{8.2}$$

where C [M L^{-3}] is the solute concentration, C_0 [M L^{-3}] is the initial solute concentration, C_s [M L^{-3}] is the new solute concentration at $x = 0$ and $t > 0$. This equation is useful for obtaining a first-order estimate of the spatio-temporal evolution of dissolved solute concentrations in a low-permeability unit that is flooded during a transgression. However, purely diffusive systems are rare and more complex conditions generally prevail due to the presence of advective flow, compaction or sediment deposition.

The effect of an advective component can be described with the following formula (Groen et al., 2000):

$$C(x,t) = C_0 + \frac{(C_s - C_0)}{2}\left[\mathrm{erfc}\left(\frac{x - vt}{2\sqrt{D_e t}}\right) + \exp\left(\frac{vx}{D_e}\right)\mathrm{erfc}\left(\frac{x + vt}{2\sqrt{D_e t}}\right)\right] \tag{8.3}$$

where v [L T^{-1}] is the mean groundwater velocity.

More complex formulations have been published as well. An expression including the effect of sedimentation was developed by Groen et al. (2000). They used it to simulate the history of transgression and regression in the coastal plain of Suriname by fitting the model results to measured chloride concentrations in Holocene clay layers. They inferred sedimentation rates and the timing of regression, and these were found to be consistent with the known geological history of the area (Groen, 2002). Volker and Van der Molen (1991) presented a solution which assumes that the initial concentrations vary with exponentially with depth, rather than being constant. They used this equation to fit measured concentration-depth profiles at several locations and treated the upward seepage flux as a fitting parameter. This allowed them to map the upward flux across the bottom of the IJsselmeer lake in the Netherlands.

Chlorine has two stable isotopes, ^{35}Cl and ^{37}Cl, and diffusion of dissolved chloride is accompanied by significant fractionation that can be used to infer complex solute transport histories (Eggenkamp and Coleman, 2009). Eggenkamp et al. (1994) applied Eqn (8.3) to measured chloride concentrations and δ^{37}Cl values in the sediments below Kau Bay in Indonesia, where the advective flow velocity was replaced by the sedimentation rate. By assuming that the chloride concentration in the Bay increased 10 ka BP, when saltwater from the Pacific Ocean entered the lake across a 40-m deep (with respect to current sea level) sill, they were able to fit the model to the measured data. Tokunaga et al. (2010) also considered the effect of diffusion and sedimentation to explain the observed patterns of chloride and δ^{37}Cl values in volcanic rock underlying a Holocene marine clay layer. Based on indications provided by groundwater ages from ^{14}C measurements it was inferred that the extent of fresh groundwater circulation extended much further beyond the present coastline before the clay layer was deposited, and that stagnant conditions now prevail in the offshore portion of the aquifer underlying the marine clay (Figure 8.6).

Beekman et al. (2011) used a numerical model to simulate the sequential stages of erosion, mixing and deposition in an environment that saw a complex palaeo-geographical

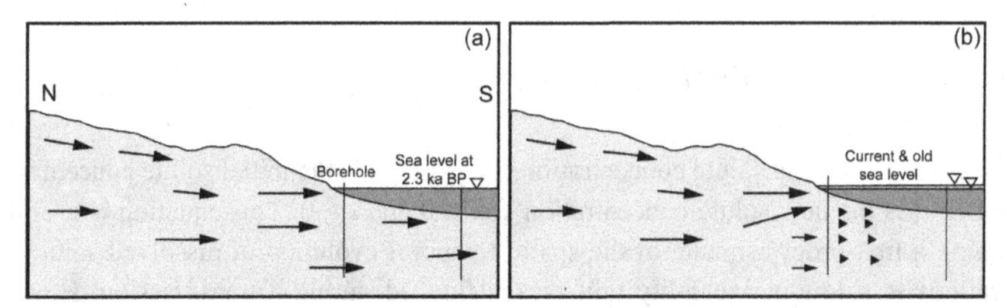

Figure 8.6 Conceptual model of the evolution of the flow pattern of freshwater (black arrows) towards the Yatsushro Sea, Japan (Tokunaga et al., 2010). The situation depicted in (a) represents the conditions before the deposition of marine clay, which commenced circa 2.3 ka BP; the situation in (b) represents the current conditions, with almost stagnant conditions below the seafloor.

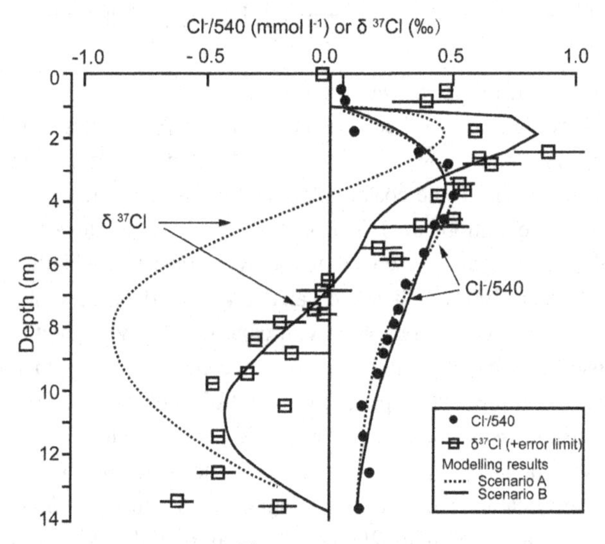

Figure 8.7 Chloride concentration and δ^{37}Cl values versus depth of pore water below Lake Markermeer, the Netherlands (Beekman et al., 2011). Model scenario B differs from scenario A in that it considers the effect of sedimentation and erosion. Both models fit the chloride concentrations well, but only model B matches the measured δ^{37}Cl values.

evolution. The studied sites are in what is now a freshwater lake, but the environment at these locations consisted of a peat bog with freshwater lakes between 3.4 and 2 ka BP. Due to erosion of the peat, the lakes became larger and more brackish between AD 0 and 1570. In the latter year, a severe storm created a permanent connection with the sea, resulting in permanently saline conditions until the construction of a closure dam in AD 1932. Despite the complexity of the geological history and the inherent uncertainties involved in models spanning several centuries, Beekman et al. (2011) could reproduce the measured chloride and δ^{37}Cl values as shown in Figure 8.7. The value of using δ^{37}Cl as a tracer was that it turned out to be much more sensitive to erosion and sedimentation than the Cl^- concentration alone, and therefore provided understanding of the variety of the processes that caused salinisation and freshening in the study area.

Kuang et al. (2015) and Kuang et al. (2016) studied the depth profiles of Cl^- concentrations as well as δ^2H and δ^{18}O values in the aquifer–aquitard system of the Pearl River Delta, China. They used a one-dimensional model in which the sedimentation process was simulated by considering a moving upper boundary. The sedimentation was continuous, but the sedimentation rate varied. The model could be fitted to the data with effective diffusion coefficients of the aquitard and the aquifer of 5.0×10^{-11} m^2 s^{-1} and 2.0×10^{-10} m^2 s^{-1}, respectively (Kuang et al., 2015). Inclusion of advective transport in the model tended to underestimate Cl^- concentrations in the aquitard and overestimate Cl^- concentrations in the basal aquifer. It was therefore concluded that diffusion was the dominant vertical transport mechanism, both for the single borehole studied by Kuang et al. (2015) as well as for the nine additional sites spread across the Pearl River Delta (Kuang et al., 2016).

8.3.1.2 Advection

For advection, the response time T_a is the ratio of the length of a flow path L_p, and the flow velocity v (Eqn (2.22)). Using the example of an aquifer with $L_p = 10$ km and $K = 10$ m d^{-1} as before, and assuming that the porosity is 0.25 and the hydraulic head gradient 1×10^{-4}, by Darcy's law, $v = 10/0.25 \times 10^{-4} = 4\times10^{-3}$ m d^{-1}, and hence $T_a = 10\,000/4 \times 10^{-3} = 2.5 \times 10^6$ days $= 7 \times 10^3$ years. This is on the same order as the duration of the Holocene epoch, and for many coastal aquifers, T_a is long relative to the time since the regression of the coastline, which is why relic seawater remains even though the sea retreated thousands of years ago.

Just like flushing of seawater by freshwater is governed by T_a, the intrusion of seawater into a freshwater aquifer is associated with the same timescale. The only difference is that the length L_p is the distance over which the seawater moved and, in the case of seawater intrusion under natural conditions, the driving force is due to density effects. During a transgression, the upper limit of the advective velocity of the interface between fresh- and saltwater beneath the continental shelf can be approximated by (Kooi et al., 2000)

$$v_{swi} = \frac{\kappa \rho_s g}{\mu n} \tan\beta = \frac{K_s}{n} \tan\beta \tag{8.4}$$

where v_{swi} [L T^{-1}] is the rate of horizontal interface displacement, κ [L^2] is the intrinsic permeability, n is the porosity, μ [M L^{-1} T^{-1}] the dynamic viscosity, ρ_s [M L^{-3}] the density of seawater and β is the slope of the land surface over which the transgression occurs. $K_s = \kappa\rho_s g/\mu$, i.e. the hydraulic conductivity of an aquifer filled with seawater. A representative value of tan β for continental shelves is tan β = 10^{-3} (Kooi et al., 2000) so for $K_s = 10$ m d^{-1}, $v_{swi} = 0.01$ m d^{-1} = 4 m yr^{-1}.

This value is very small in comparison to the rate of horizontal movement of the coastline that occurred during the periods of high sea level rise at the onset of the Holocene. For example, based on the former positions of the coastline of the North Sea as shown in Figure 8.4, the rate of transgression was on the order of 10^2 m yr^{-1}. This result in a situation where the sea transgresses over aquifers that are still filled with freshwater. When this occurs, the resulting density stratification is potentially unstable, and convective fingering may occur (Section 6.2.5.2). The rate of descent of the salt fingers was found to be (Wooding, 1969)

$$v_z = \frac{\kappa(\rho_s - \rho_f)g}{4\mu n} \tag{8.5}$$

where ρ_f [M L^{-3}] is the density of freshwater. Post and Kooi (2003) investigated this relationship in more detail using numerical models and confirmed its validity but also found that there is a deceleration of the plumes and that the depth at which this starts occurring is a function of the permeability. This deceleration occurs because plumes lose mass by diffusion, which decreases the driving force for flow. Finger coalescence, on the other hand, creates larger fingers which make that the driving force is maintained. Complex flow dynamics thereby result, and the rate of aquifer salinisation by convective fingering is still an area of active research. Nonetheless, it seems that convective fingering is a process by which permeable formations can become salinised to depths of hundreds of metres within timespans of decades (Post, 2004).

8.3.2 Definition of Palaeo-Water

Numerous occurrences of high-salinity groundwater inland or low-salinity groundwater below the seafloor exist, which are not explainable by the present patterns of groundwater flow and recharge (Post et al., 2013). This indicates the existence of different environmental conditions or a different configuration of flow systems in the past. Within the context of coastal groundwater, it was already acknowledged in the nineteenth century (Braithwaite, 1855) that the salinisation of wells may not be due to active seawater intrusion, but by drawing in remnants of seawater that remained in the pores of rocks formed in a former marine environment. Occurrences of water trapped since the time when rocks were formed are termed connate (from the Latin word *connatus*, which translates as *born at the same time*).

Fossil groundwater is water that infiltrated usually millennia ago, often under climatic conditions different to the present, and that has been stored underground since that time (Margat et al., 2006). In the literature the term fossil (from the Latin word *fossilis*, meaning *excavated*) groundwater is used interchangeably with palaeo-groundwater, or simply palaeo-water (palaeo derives from the Latinised form of the Greek word for ancient *palaios*). The terms have been used since the 1960s when studies of groundwater in arid and semi-arid areas showed that groundwater that were found to be thousands of years old (based on ^{14}C), had different stable water isotope compositions than modern waters (Edmunds et al., 2001), and therefore had to have formed under different climatic conditions than the present day (Gat, 1983).

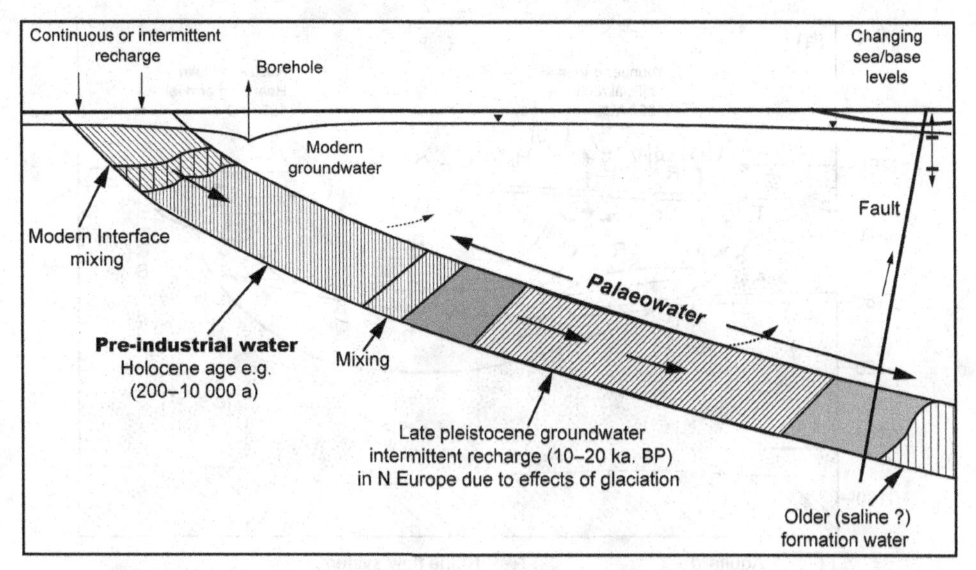

Figure 8.8 Schematic cross section of a confined aquifer showing the definitions of modern and palaeo-water (Edmunds et al., 2001).

Just like Margat et al. (2006), others often associate palaeo-water with a different climate and a particular timespan. For example (Figure 8.8), Edmunds et al. (2001) defined palaeo-water as 'all groundwaters that can be clearly identified in terms of radiocarbon age, or another isotopic or noble gas signature, as originating in colder climatic conditions of the Late Pleistocene'. The same authors identified modern water as being younger than 50 years. The latter is a practical definition that has been adopted by many others, because this is an age range within which tritium (^{3}H) and its daughter product ^{3}He can be used to constrain the time that elapsed since groundwater was recharged.

A less restrictive definition defines palaeo-waters as waters which are remnants of former hydrogeological regimes and are largely unaffected by natural circulation at the present day (Darling et al., 1997). It is broader in the sense that it does not necessarily involve a change of climate, but it excludes palaeo-waters that have been caught up in modern flow systems. During the initial formation of a new groundwater flow system, the contemporary recharge water must displace ambient groundwater, and this may take longer than the establishment of a new flow configuration. Stuyfzand (1999) therefore distinguished between hydrological maturity, which is reached when the flow becomes steady, and hydrochemical maturity, which is reached when the flow system is occupied only by water originating in the recharge area.

This is illustrated in Figure 8.9 which shows the configuration of groundwater flow systems and the distribution of waters of a particular origin (hydrosomes) in the freshwater lens beneath the coastal dune in the Netherlands. The deep flow system between the North Sea and the deep polder area Haarlemmermeer formed in 1853 when the former lake was

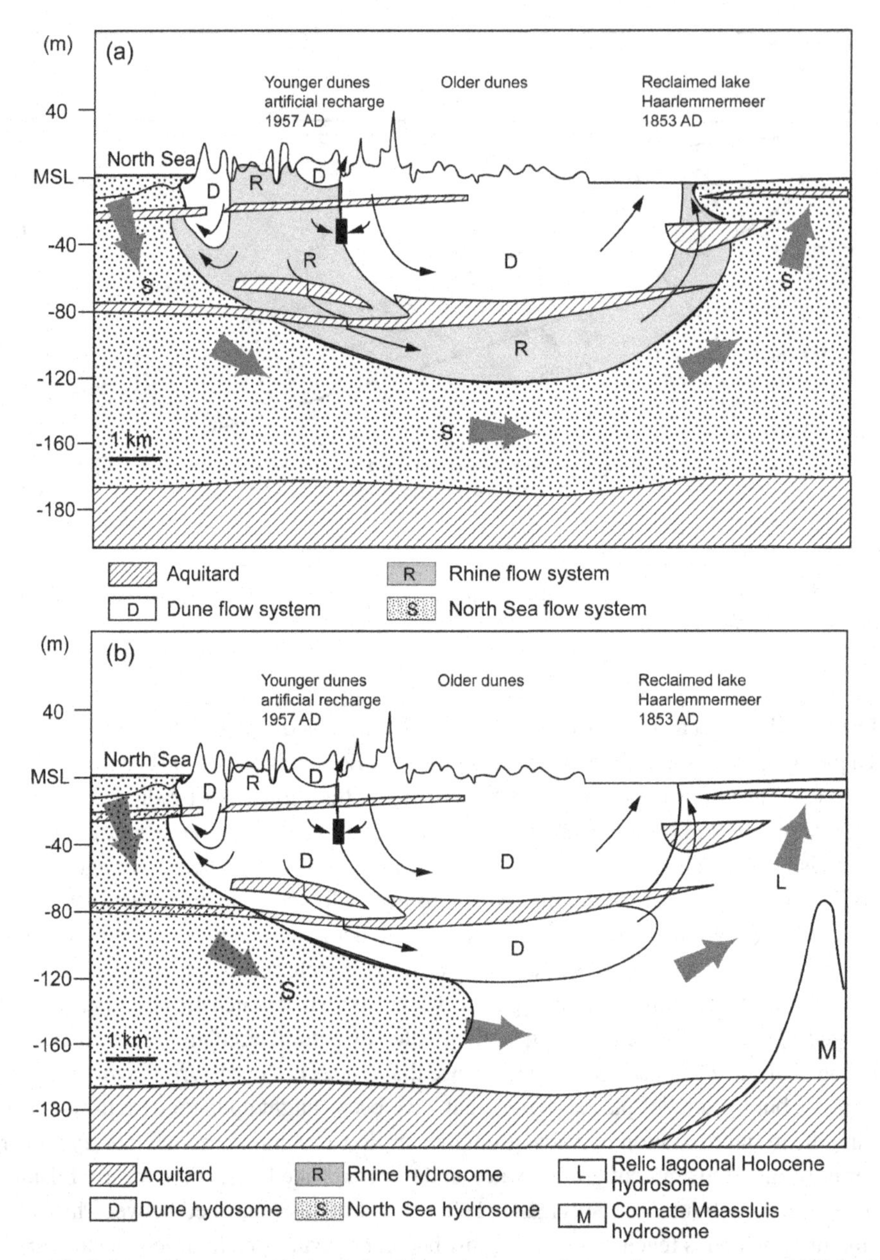

Figure 8.9 Schematic cross sections of the (a) current groundwater flow systems and (b) hydrochemical systems below the Dutch coastal dunes (Stuyfzand, 1999). While the flow systems have reached steady state, the hydrochemical systems are still readjusting to the new flow configuration.

drained. Based on time series of hydraulic heads it can be inferred that the flow system has reached a stationary state (Stuyfzand, 1999). From the recharge zone below the seabed to the discharge point in the polder in the east, the water along a flow line ranges from intruded North Sea water that is actively intruding, to North Sea water that had intruded prior to the flow system's existence to brackish groundwater that was recharged before the coastal dune landscape existed, when the area was still occupied by lagoons and tidal marshes.

The definition for palaeo-water by Groen (2002) is similar to that by Darling et al. (1997). He defined it as 'as meteoric groundwater formed by an old flow system that was controlled by conditions of sea level, topography or climate different from today, or any meteoric groundwater that cannot be explained by the present flow systems'. For coastal aquifers this definition excludes seawater intruded during past transgressions when the coastline was at a different position than the present day, like in the previous example (Figure 8.9). Therefore, in this book, palaeo-water refers to any groundwater formed under hydrological boundary conditions that no longer exist at the present time. This definition encompasses meteoric water as well as seawater, does not require that climate was different during the time of emplacement of the water, and does not impose any restrictions in terms of age.

The discussion above about the most appropriate definition may seem academic in nature, but a precise understanding of the origin of palaeo-water is essential to water management. The problem of depletion of non-renewable reserves of palaeo-water is well known (Margat et al., 2006), and in coastal areas an incorrect understanding of the salinisation history of a groundwater system may misinform the conceptual model that is used as a basis for future management decisions.

It must be borne in mind that the solutes contained in the water, as well as the isotopes of water, can migrate between flow systems by dispersion and diffusion (Verruijt, 1971). This is a well-known problem for age dating studies (Sanford, 1997), and it modifies the original composition of the palaeo-water. This complicates the interpretation of the origin of water types found in an area. This problem tends to become more severe with increasing age of the water because, the older the water, the less is known about its original composition as well as the hydrological and geological history since its time of formation. In groundwater discharge zones, waters of different ages and origins converge and the waters emerging in springs and streams, for example, are often mixtures of palaeo-water and groundwater from active flow systems.

Finally, it is important to distinguish between connate seawater and palaeo-seawater. Connate seawater is found in the pores of rocks of marine origin and has resided there since the time of formation of the rock. Palaeo-seawater is seawater that intruded when patterns of fresh and saline groundwater circulation differed from that of the present day, and it is younger than the rock matrix through which it migrated. Also, its presence is not restricted to formations of marine origin. The distinction is particularly important for the understanding of how groundwater evolved chemically, as well as for studies of diagenesis.

Some very old marine-derived groundwater has been identified in diffusion profiles. For example, the measured pore water chemistry of an almost 500 m deep profile in the Upper Cretaceous chalk aquifer in eastern England revealed the presence of connate seawater in the deepest part of the sequence (Bath and Edmunds, 1981). Sanford et al. (2013) encountered groundwater at depth of more than 900 m in a 45 My old meteorite impact crater below Chesapeake Bay (USA), which they interpreted to represent seawater from the Early Cretaceous North Atlantic ocean. Using ^{4}He as well as stratigraphic constraints as an indicator of age, the water was determined to be between 145 and 100 million years old. The measured pore water concentrations could only be fit to a model including diffusion and advection by assuming an initial Cl^- concentration of 38 g l^{-1}, i.e. about double that of modern seawater (Table 5.1). This would suggest that the earliest manifestation of the Atlantic Ocean had a much higher salinity than in the open ocean, which Sanford et al. (2013) attributed to evaporation.

8.4 Case Studies of Coastal Aquifer Evolution

8.4.1 Llobregat Delta, Spain

The delta of the Llobregat river, south of Barcelona, Spain, is a classical example of a groundwater system where freshening has occurred following the progradation of the delta (Figure 8.10). The lithological variation and the geological structure of the delta is complex, but in principle two aquifers can be distinguished, which are separated by an aquitard. The deepest aquifer is made up by permeable sediments of fluvial and marine origin which were deposited during the sea level lowstand of the last glacial period. At the apex of the delta it is connected to the shallow, narrower aquifer that consists of the coarse-grained alluvial deposits of the Llobregat river (Figure 8.11). The deep aquifer is between 5 to 15 m thick, and it thins and dips towards the coast. At the coastline it is found at a depth of 50 to 70 m. Offshore, it crops out in the seafloor at a distance of approximately 4–5 km from the coast where the water depth is 100 m (Iribar and Custodio, 1993; Custodio, 2012a).

A shallow, unconfined aquifer is present in the upper part of the delta. It is dominated by sandy facies, but strong lithological variations occur due to the spatial variations of the

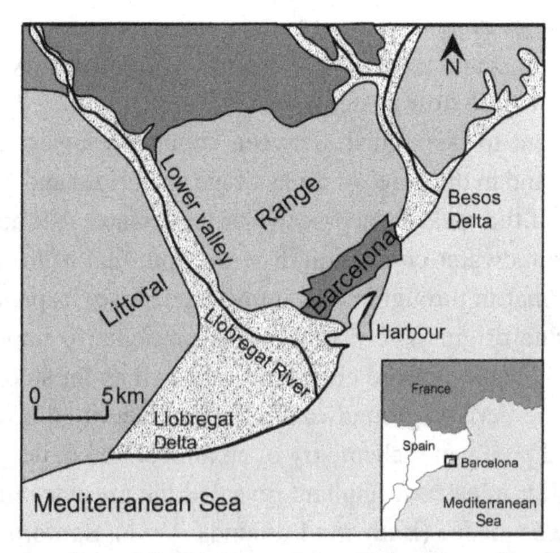

Figure 8.10 Location and outline of the Llobregat River Delta (Custodio, 2012a).

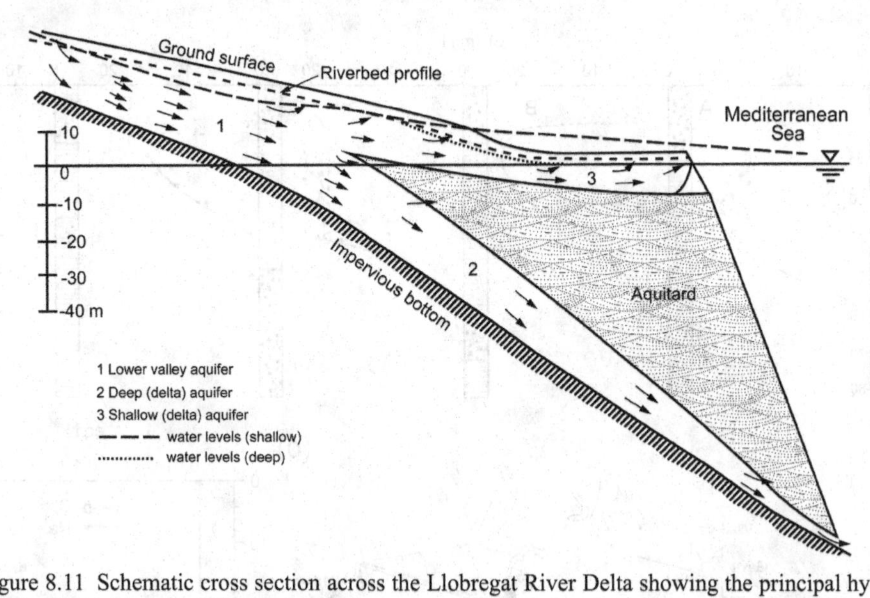

Figure 8.11 Schematic cross section across the Llobregat River Delta showing the principal hydrostratigraphic units and the flow patterns before groundwater abstraction (Custodio, 2012a).

former depositional environments (Custodio, 2012a). The shallow aquifer is up to 20 m thick and can be partially confined locally. The two aquifers are separated by an up to 40 m thick wedge of clay, silt and fine sand sediments. These were deposited in the pro-delta environment during the progradation of the delta since about 8 ka BP (Iribar and Custodio, 1993). It forms an aquitard that thins out towards the delta boundaries.

Due to over-abstraction, seawater intrusion occurred in the deep aquifer during the twentieth century, but before that, the seaward flow of freshwater had completely flushed out the seawater that had permeated the sediments during the Holocene transgression. The response time for advective flow can be estimated using data reported by Iribar and Custodio (1993) ($K = 40$ m d^{-1}) and Custodio (2008) ($dh = 8$ m, $n = 0.2$). A representative flow path length between the delta inlet and the offshore outcrop is 12 km (Custodio, 2008) and thus the advective flow velocity is on the order of $v = 8/12\,000 \times 40/0.2 \times 365 = 50$ m yr^{-1}. The timescale of advective transport is then $T_a = 12\,000/50 = 240$ years. Due to its high hydraulic conductivity and the relatively steep hydraulic gradient, the deep aquifer has a short response time relative to the time elapsed since the regression of the shoreline caused by the delta progradation. This explains why before active seawater intrusion in the twentieth century, the aquifer contained only freshwater, which was discharging offshore where the aquifer crops out in the seabed.

The aquitard on the other hand still contains relic seawater (Manzano et al., 1993), which is evident from the elevated chloride concentrations and the stable water isotopes (Figure 8.12). The shape of the pore water profiles can be explained by a combination of diffusion and advective transport driven by the upward flow of freshwater from the deep

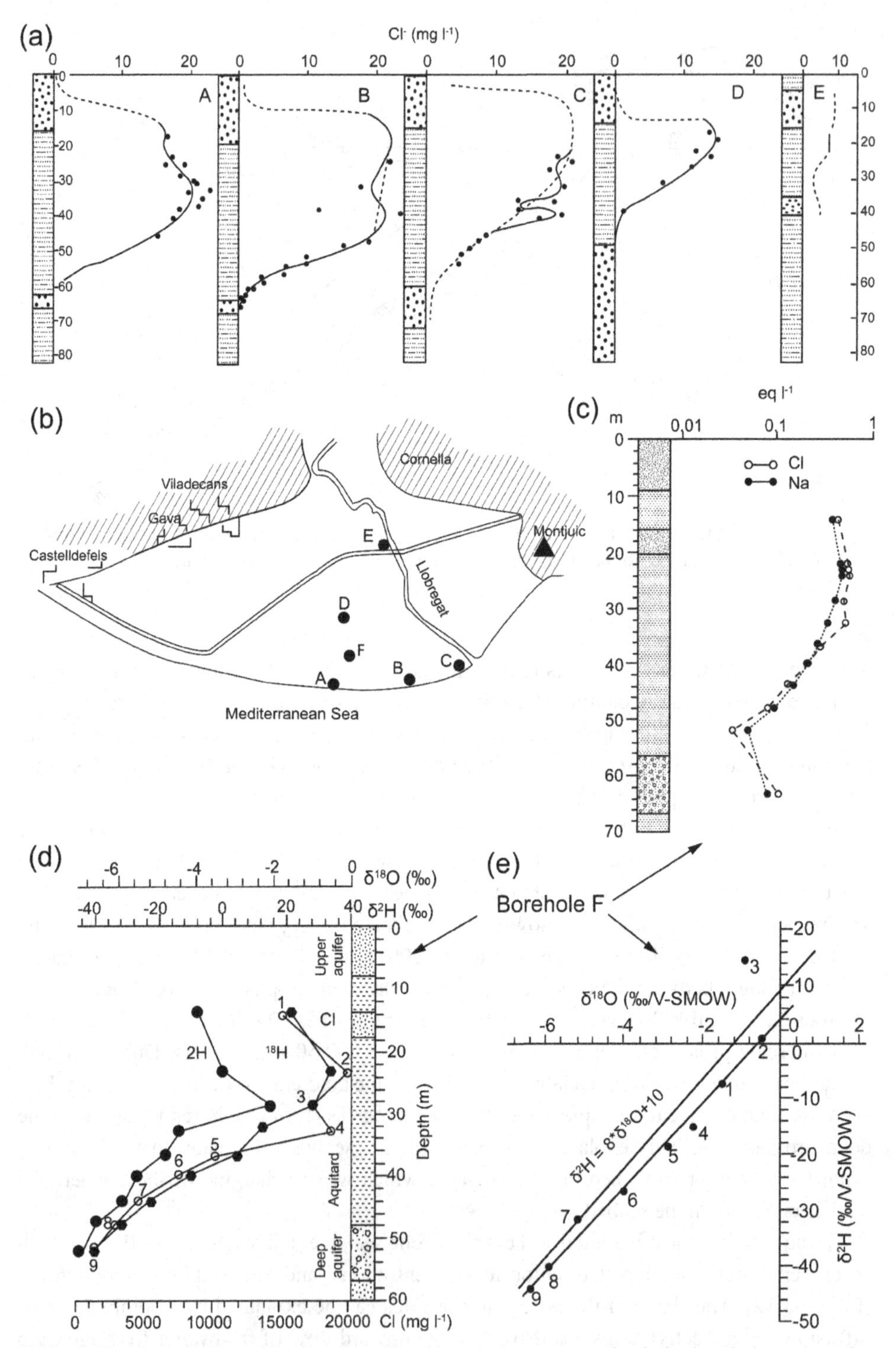

Figure 8.12 (a) Vertical profiles showing Cl⁻ concentrations versus depth for boreholes across the aquitard in the Llobregat River Delta (Custodio, 2012a). Locations are shown in (b), while (c), (d) and (e) show Cl⁻ and Na⁺ concentrations as well as stable water isotope data for borehole F.

aquifer due to the hydraulic head gradient prevailing before abstraction started. The most inland part of the aquitard has seen the highest removal of seawater, whereas the most seawater has been preserved in the aquitard near the coastline. This is because of the thickening of the aquitard towards the coast, and the decreasing upward flux of freshwater from the deep aquifer.

The complete flushing of the deep aquifer and the preservation of palaeo-seawater in the aquitard shows the contrasting response times of different units in this groundwater system. In the Llobregat Delta, palaeo-seawater has generally not been preserved in aquifers, except in isolated areas where the hydraulic head gradients are low, as indicated by radiocarbon ages (Custodio, 2012a).

8.4.2 The Netherlands

The coastal area of the Netherlands belongs to the North Sea coastal plain, which is a low-lying area that extends more or less continuously from northern Denmark to the north of France (Figure 8.4). The morphology is typically flat, and the elevation is near or below sea level. The natural landscape before human settlement consisted of salt marshes and extensive peat areas, dissected by estuaries, arms of the rivers Rhine and Meuse, as well as local rivers. Sea level fluctuations and changes in sediment supply from the large rivers and the North Sea, combined with changes in climate have made that the area saw numerous major coastline shifts and landscape changes over the last 2 My. During the Early Pliocene, almost the entire Netherlands was covered by the sea (Figure 8.13), but by 1.6 Ma BP the sea had retreated and fluvial deposition dominated (Zagwijn, 1989). The sediments of the major rivers built a delta and there is no evidence for seawater influence for a period of over 1 My. During the ice ages of the Pleistocene, the landscape was heavily modified, with the ice eroding deep valleys which are flanked by highs consisting of sediments pushed aside by the moving ice. The glacial valleys were the first areas that became inundated by the sea during subsequent interglacial periods, in particular the Eemian interglacial period (Figure 8.13). When eustatic sea level rose during the Holocene, marine influence became established again by 7.5 ka BP in the form of a system of tidal flats and lagoons (Figure 8.13, Calais transgression). But due to a deceleration of sea level rise, and the formation of barrier islands, freshwater peat bogs and lakes covered most of the area during the period 3 to 1 ka BP. Where these areas were eroded the sea could enter, sometimes temporary but also more permanently (Figure 8.13, Duinkerke transgression). Promoted by a lowering of the landscape by drainage and excavation of peat, the marine erosion intensified and only owing to the coastal defence structures that were erected during the course of the last 800 years has the coastline been fixed at its present position.

The repeated presence of the sea in the area means that there is the potential for seawater of different ages to still occur in the aquifers. The present-day salinity of the groundwater is highly variable in the area that experienced marine influences during the Holocene. This area extends up to 60 km inland locally. The influence of the groundwater flow system created by the artificially maintained surface water levels in the polders is manifested by

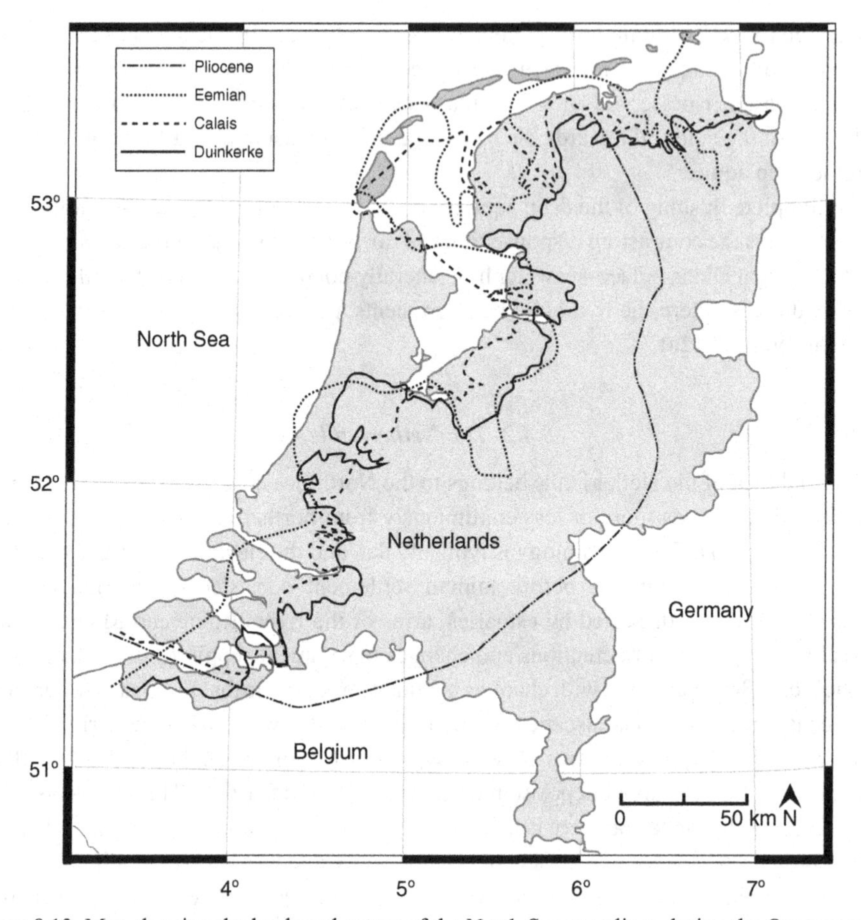

Figure 8.13 Map showing the landward extent of the North Sea coastlines during the Quaternary.

large-scale up-coning beneath the deepest polders (Figure 8.9). Freshwater lenses are present below the coastal dunes and locally in areas that had a high elevation when peat bogs existed in the landscape (Beekman, 1991).

Based on the high ^{14}C activities of the brackish and saline groundwater, and on calculated rates of vertical density-driven flow, Post and Kooi (2003) and Post et al. (2003) concluded that most of the saline groundwater intruded by vertical intrusion of seawater during the Holocene transgressions. Delsman et al. (2014a) later confirmed this finding based on a regional scale variable-density model. As downward migration of seawater by convective fingering can be on the order of metres per year, virtually complete salinisation of the sand aquifers up to 150 m deep occurred within decades. Nevertheless, due to the continual change of hydrological conditions at the surface, the groundwater system in the western part of the Netherlands did not once reached steady state during the last 8.5 ky.

Low-permeability strata at depth such as clay layers impeded the downward flow of seawater, but in the model by Delsman et al. (2014a), no layers thick or extensive enough were present to protect the underlying strata from salinisation. The low-permeability layers only worked to delay salinisation, rather than to prevent it. Where low-permeability strata are discontinuous, salt may first migrate downward through the aquitard window, resulting in a steep boundary between saline and freshwater. The subsequent rotation of the interface results in horizontal salinisation of the freshwater trapped underneath the low-permeability layer, which may progress laterally by several kilometres over the course of centuries (Post, 2004). It was shown in laboratory experiments (Post and Simmons, 2010) that freshwater trapped underneath a low-permeability lens can move up by buoyancy effects, counter-acting the diffusion and density-driven flow caused by the saltwater above it. It is even possible that the vertical upward movement of the freshwater through a low-permeability layer drags behind it saltwater, causing the salinisation of the low-permeability layer from below, even though the source of the saltwater sits above it. Complex patterns of fresh- and saltwater flow may thus develop.

The deepest parts of the aquifer system in the Netherlands consist of fine-grained marine sediments dating back to the Pliocene and Early Pleistocene. It is generally believed that these still contain connate seawater, although dilution by meteoric freshwater has reduced their salinity. However, another possible explanation for the presence of saline groundwater in these strata is that they were salinised during the Eemian interglacial period, after they had been partially or completely flushed by meteoric water during the Pleistocene when the area was under the influence of freshwater circulation for hundreds of thousands of years. There is every reason to believe that the vertical salinisation that occurred during the Holocene also operated during the Eemian transgression. Connate seawater from the Pliocene/Early Pleistocene cannot be distinguished from Eemian relic seawater based on hydrochemical composition or ^{14}C activity, and in the absence of other tracer data this remains a possibility that needs to be entertained. This shows the difficulty in reconstructing the palaeo-hydrological history of coastal aquifers.

8.4.3 Laizhou Bay, China

The formation of hyper-saline water is a common occurrence in coastal regions and in areas where the coastline formerly occupied a more inland position, remnant brines often occur. In the aquifer system bordering Laizhou Bay, which is part of the Bohai Sea, the volume of brine is so large that it is exploited for commercial purposes (Han et al., 2011). They occur within 10 km of the shoreline, and are encountered at different depth intervals down to approximately 75 metres (Figure 8.14). Their occurrence is controlled by the confining beds which inhibit vertical migration and mixing.

The Quaternary aquifer system in the Laizhou Bay area consists of unconsolidated fluvial and marine sediments. Unconfined conditions exist in the southern part where alluvial sand and gravels border the basement rocks, but conditions change to more confined towards the north as the grain size decreases and clay and silt layers become more prevalent. The brines

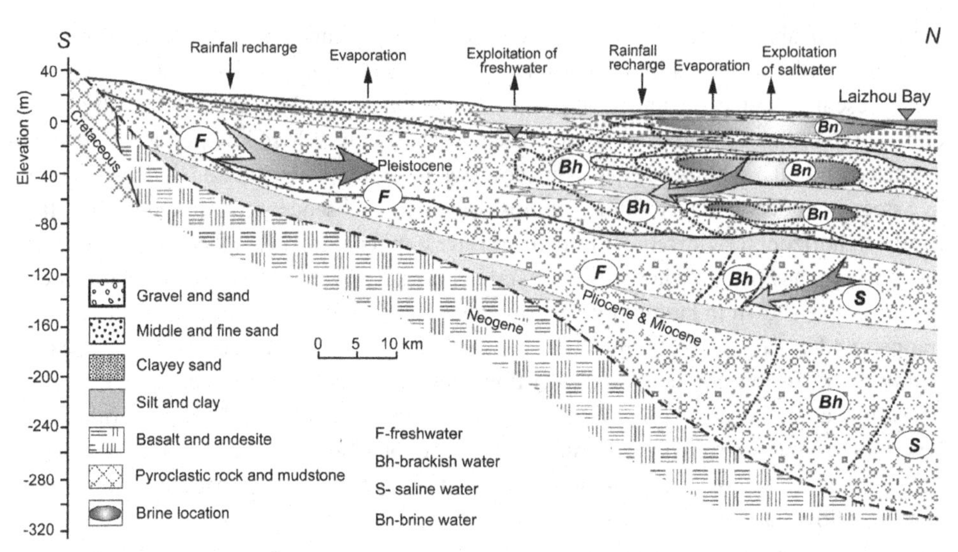

Figure 8.14 Schematic cross section of the Laizhou Bay area in China showing the occurrence of aquifers and aquitards, flow directions and groundwater salinity classes (Han et al., 2011).

have a chloride concentration of approximately 35 g l^{-1} or more, and their Na^+/Cl^- and Br^-/Cl^- ratios are close to that of seawater, indicating that the dissolved solids do not derive from dissolution of halite or other salt minerals. Instead, they are believed to have formed as a result of evaporation of seawater (Du et al., 2016). Based on an analysis of the $\delta^{37}Cl$ and $\delta^{81}Br$ values of the brine samples, Du et al. (2015) concluded that the brines formed in an intertidal region where complete desiccation periodically resulted in the precipitation of dissolved chloride as halite, which was dissolved again as seawater re-flooded the area.

The stable water isotope (δ^2H and $\delta^{18}O$) values indicate that the brines are depleted relative to seawater. As evaporation of seawater under open water conditions would result in an enriched concentrate, the brines most likely underwent mixing with meteoric water after their formation (Han et al., 2011). It is likely that this occurred in the subsurface, during convective downward migration of the brine. That mechanism was proposed to explain the occurrence of saline groundwater with similar stable water isotope values in Spain (Sola et al., 2014). Active systems of this kind are found along the southern coastline of Australia where numerous salt lakes are found near the shore. They are periodically flooded by seawater and receive inflow of freshwater as well, resulting in complex mixing relationships of the groundwater found in the vicinity of these lakes (Marimuthu et al., 2005).

The occurrence of relic brines is not restricted to arid- or semi-arid climate zones. Groundwater with chloride concentrations above that of seawater has been encountered, for example, in the Netherlands and Denmark. Post (2004) attributed these to the presence of salt marshes where strong evapo-concentration can occur during summer months. Due to density-driven flow in permeable sands or diffusion into confining layers, the concentrated

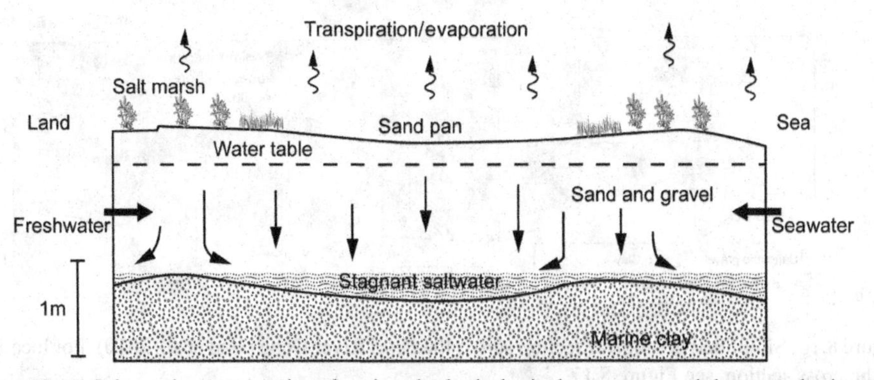

Figure 8.15 Schematic cross section showing the hydrological processes and downward migration and pooling of hyper-saline groundwater (Jørgensen, 2002).

seawater can become preserved as high-salinity groundwater. The feasibility of this proposed mechanism was confirmed by observations by Jørgensen (2002), who found evapo-concentrated seawater pooled in depressions at the bottom of a shallow aquifer beneath a salt marsh in Demark (Figure 8.15).

Evapo-concentrated solutions or groundwaters that obtained their salinity from rock salt dissolution are usually easily identified based on a combination of their chloride concentration and stable water isotope composition. When they have a salinity higher than seawater then they are easily identified based on their salinity alone, but when they are diluted to seawater concentrations or lower, their presence might not be revealed if salinity is the only parameter measured. This highlights the importance of multi-tracer approach for developing conceptual models of coastal groundwater systems.

8.4.4 Pearl River Delta, China

The Pearl River Delta, located in the coastal area of South China, contains a large Quaternary aquifer–aquitard system up to 50 thick (Jiao et al., 2010), which consists of four stratigraphic units from the bottom to the top: an old terrestrial unit (T2), an old marine unit (M2), a younger terrestrial unit (T1) and a younger marine unit (M1). At about 7000 years BP, the transgression reached its maximum and the coastline at this period was much further northwest (Figure 8.16). The younger marine unit was mainly deposited between 7000 and 2000 years BP, whereas the other three units were formed during the Pleistocene. The bottom terrestrial unit is dominated by sand and gravel and forms the basal aquifer. Due to the low permeability of the marine units and the flat topography, regional groundwater flow is sluggish (Jiao et al., 2010).

The TDS concentrations of the groundwater in the basal aquifer (Figure 8.17) range from 1 g l^{-1} in inland areas up to 75 km from the coastline, to 26.8 g l^{-1} near the south-eastern shore (Wang and Jiao, 2012). Using Cl^{-} as a conservative tracer, the seawater fraction in the

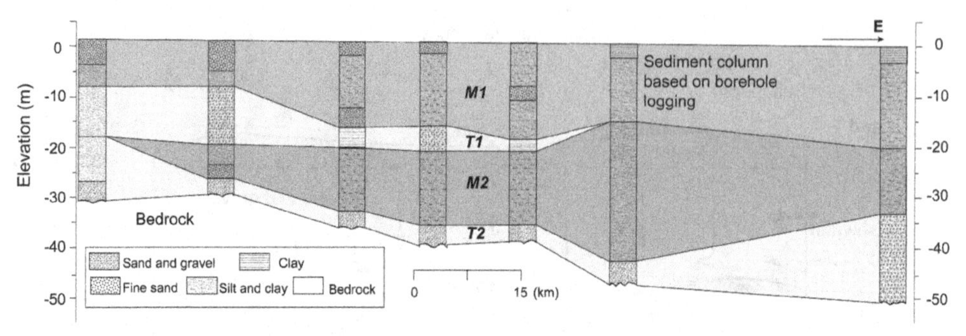

Figure 8.16 Simplified geological section across the Pearl River Delta (Jiao et al., 2010). For location of the cross section, see Figure 8.17.

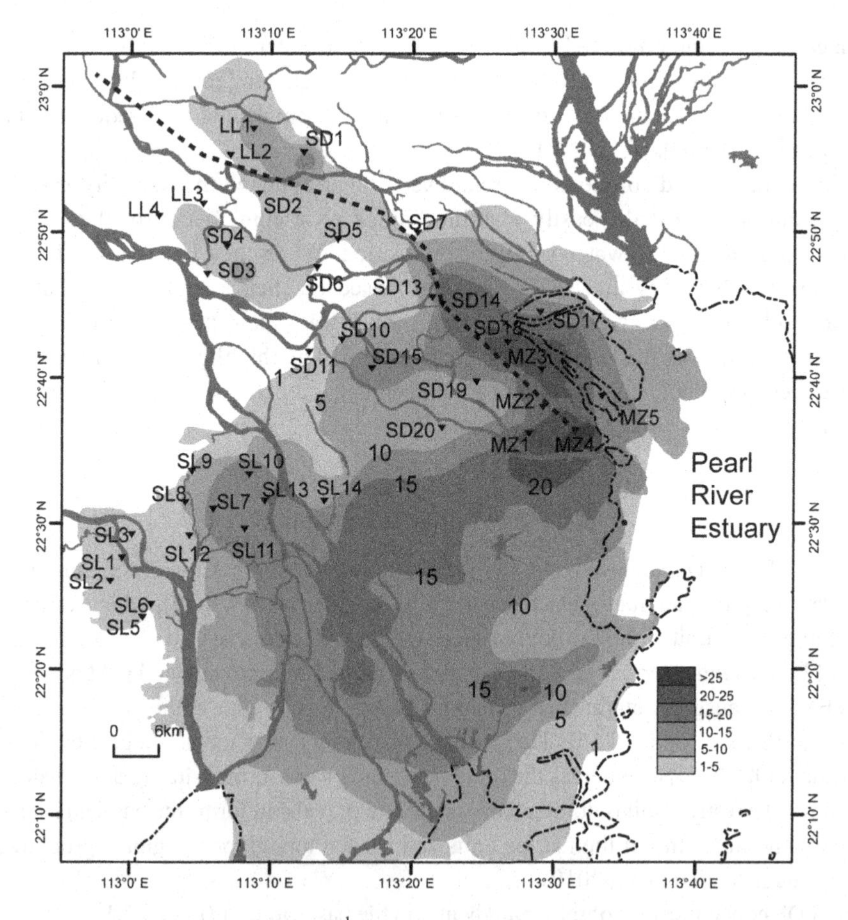

Figure 8.17 The distribution of TDS (g l^{-1}) in the basal aquifer in the Pearl River Delta, China (Wang and Jiao, 2012). Triangles show the locations of the sampling wells. The thick broken line represents the approximate location of the cross section of Figure 8.16.

groundwater samples was estimated to vary from less than 1% in the inland to 74% near the shoreline. Analyses of ^{14}C in groundwater samples from seven boreholes in the basal aquifer indicate an apparent age of 8.8 ± 1.3 to 2.2 ± 0.2 ka. Although ^{14}C activities are influenced by various geochemical processes (Aravena et al., 1995; Clark and Fritz, 1997) which make that the apparent age overestimates the true age, it is believed that the groundwater in the basal aquifer is largely influenced by the palaeo-seawater of a few thousand years old which was trapped in the aquifer and aquitard system during the Holocene transgression.

8.4.5 Armorican Massif, France

Because metamorphic and igneous rocks tend to form poorer aquifers than sedimentary rocks and limestone, unless they are heavily weathered or fractured, relatively few observational data exist that document the effect of sea level fluctuations and coastline migration on the salinity in such systems. Recent studies, however, have highlighted the relationship between groundwater salinity and former marine transgressions in the Armorican basement in western France (Aquilina et al., 2015; Armandine Les Landes et al., 2015). This basement consists of metamorphic (schist, sandstone, micaschist and gneiss) and igneous rocks (granites and basalts). During the Jurassic and Cretaceous periods, continental conditions prevailed, but subsequent to this, three main sea level highstands have been identified during the (i) transition between the Miocene and Pliocene (5.3 ± 0.8 Ma BP) (ii) Pliocene (2.7 ± 0.3 Ma BP) and (iii) the Pleistocene (1.8 ± 0.2 Ma BP) periods.

The approximate sea levels ranged between 90, 60 and 30 m above the present mean sea level for these three periods respectively. The chloride concentrations in samples taken from wells across the basement area show a marked increase below this depth interval, whereas the groundwater chloride concentrations above it are low (Figure 8.18). The origin of the chloride above the transgression interval can be explained by meteoric input (rainfall) and evapo-concentration, with some recent contribution from anthropogenic sources (Armandine Les Landes et al., 2015). The elevated chloride concentrations at greater depth are linked to the past transgressions, which is confirmed by hydrochemical data such as the Br^-/Cl^- and SO_4^{2-}/Cl^- ratios, as well as boron and sulphur isotope data.

In the conceptual model proposed by Aquilina et al. (2015) seawater first entered the subsurface by density-driven flow within the major faults (Figure 8.19). It subsequently entered the pores of the unweathered rocks by diffusion. After the retreat of the sea, the seawater that entered the basement was flushed out by the circulation of freshwater through the fault and fracture zones. The circulation of groundwater was deeper during the sea level lowstands of the glacial periods, and restricted to within the top 100 metres during the present sea level elevations.

Armandine Les Landes et al. (2015) tried to infer the timescales associated with the flushing out of the seawater that entered during transgressions. Because the oldest

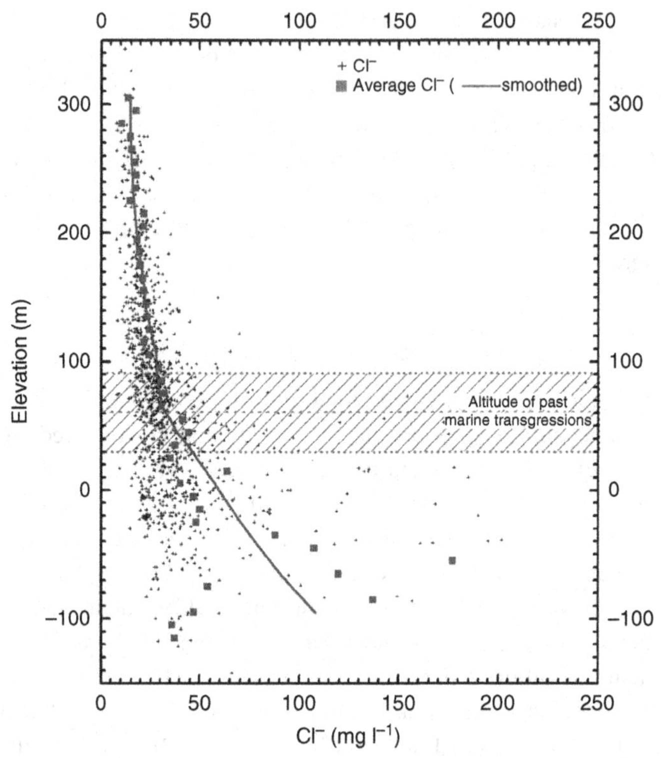

Figure 8.18 Chloride concentration of groundwater samples versus depth in the Armorican Massif (Armandine Les Landes et al., 2015).

transgression was the highest and most widespread, and each of the two subsequent inundations was at a lower sea level and covered a smaller area than the previous, it was possible to relate the average chloride concentration of each of the three inundated areas to the time elapsed since the transgression (Figure 8.20). The data points were fitted to the following exponential relationship

$$C = C_{in} \exp(t_e/t_c) + C_0 \tag{8.6}$$

where C is the current chloride concentration, C_{in} the concentration injected at time t_e, i.e. the time since the transgression event, t_c a characteristic timescale and C_0 the background concentration. By fitting Les Landes et al. (2015) found C_{in} = 100 mg l^{-1} and $t_c = 2.3 \times 10^6$ years. The high value of t_c indicates that the times required to flush seawater is on the order of millions of years. The value of C_{in} is very low relative to the chloride concentration in seawater. In the interpretation of Armandine Les Landes et al. (2015), this was attributed to the fact that the advective circulation of seawater enters is only through the most permeable zones, whereas in the bulk of the rock

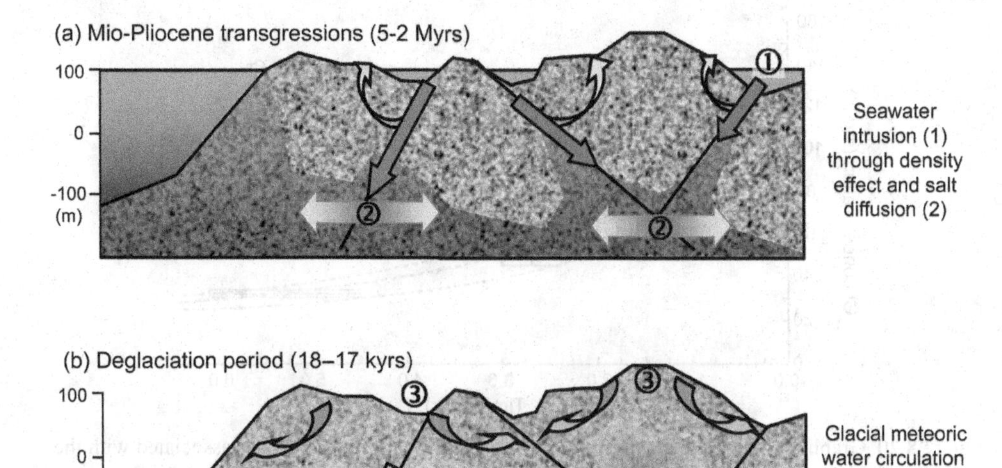

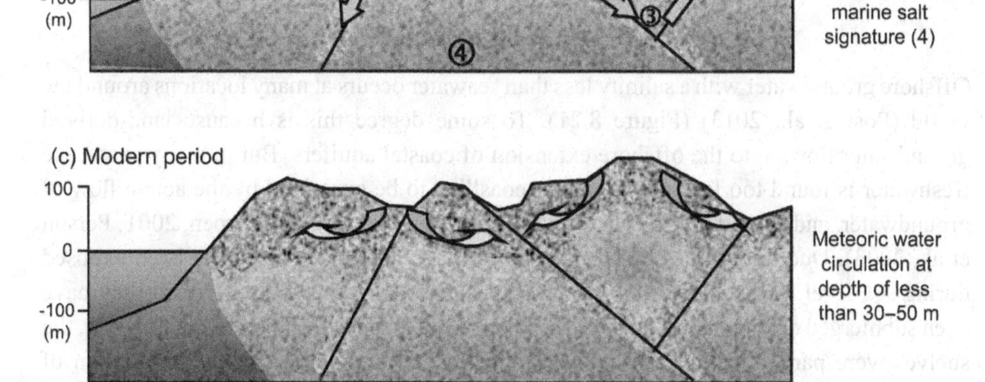

Figure 8.19 Schematic cross sections showing the conceptual model of salinisation and freshening stages in the Armorican Massif (Aquilina et al., 2015). During transgressions, seawater intrudes along faults and spreads in the deeper part of the system by density-driven flow. During glacial periods, deep circulation of meteoric water is promoted by low sea levels and results in dilution. Seawater from the former transgression resides in the immobile regions where diffusion dominates. During the present time, sea level is higher than during the glacial periods, but not as high as during past transgressions. Groundwater circulation is restricted to the shallower parts.

volume salinisation is by diffusion. This would have resulted in a significant dilution of the concentrations, because rather than displacing the ambient freshwater by seawater, this process leads to a slow mixing and evening out of the concentrations. Armandine Les Landes et al. (2015) further speculated that if deeper samples became available, the value of C_{in} would probably increase because of the systematic increase of chloride concentrations with depth.

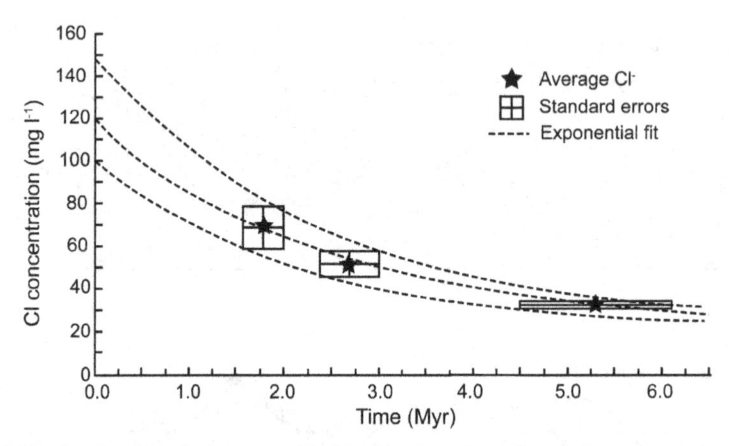

Figure 8.20 Graphs showing the average chloride concentration of groundwater associated with the three transgressions versus time (Armandine Les Landes et al., 2015).

8.5 Offshore Palaeo-Groundwater

Offshore groundwater with a salinity less than seawater occurs at many locations around the world (Post et al., 2013) (Figure 8.21). To some degree this is because land-derived groundwater flows into the offshore extension of coastal aquifers. But in many cases, the freshwater is found too far away from the coastline to be explained by the active flow of groundwater, and should thus be classified as palaeo-water (Kooi and Groen, 2001; Person et al., 2003). Due to eustatic sea level variations, continental shelves have been exposed during sea level lowstands during glacial stages/periods for much longer than they have been submerged during interglacials. This means that for the largest part of their history, the shelves were part of the terrestrial hydrological cycle and subject to the circulation of freshwater. From that perspective the occurrence of low-salinity groundwater in offshore aquifers is not at all surprising, but the scientific investigations of continental shelf aquifer systems are in their infancy and exactly how the circulation of fresh and seawater occurs in these environments remains a topic of future research.

Studies of the geochemistry and stable isotopes of submarine pore water have confirmed that the freshwater is of meteoric origin. The sea level lowstands that have occurred throughout the Plio- and Pleistocene are believed to have provided a greater topographic driving force to flush presently offshore aquifers with freshwater. Moreover, meteoric recharge occurred on the exposed continental shelves. The volume of continental shelf freshwater appears to be inversely correlated with mean continental shelf water depth because depth controls the width of shelf sub-aerial exposure of the shelf during seawater lowstands (Cohen et al., 2010).

Groundwater bodies with a salinity less than one-third of that of seawater that can be traced over a spatial extent of more than 10 km have been found beneath the continental shelves off the coasts of Canada, Greenland, Nigeria, South Africa, China, Indonesia and Australia (Post et al., 2003). The occurrences that are the best-

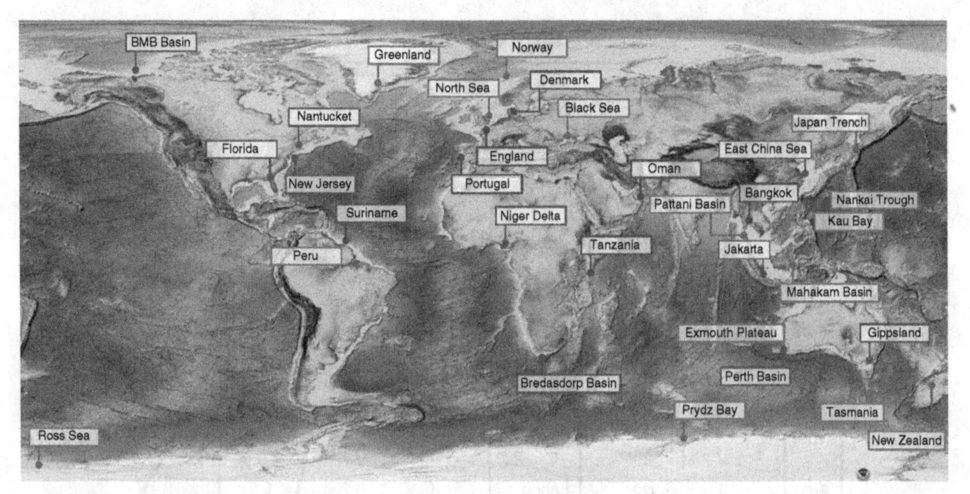

Figure 8.21 World map showing known occurrences of offshore fresh and brackish groundwater (after Post et al., 2003). Continental landmasses are displayed as grey areas. The shading in the oceans indicates the depth, with dark being shallowest and white being deepest.

documented are those located along the Atlantic continental shelf of North and South America (Hathaway et al., 1979; Groen et al., 2000; Cohen et al., 2010). The first discovery was that by Hathaway et al. (1976) along the Atlantic continental shelf east of New Jersey. Based on pore water samples obtained from squeezing sediment samples from boreholes, the study demonstrated the presence of water with a chloride concentration less than 5 g l^{-1} beneath much of the continental shelf (Figure 8.22). Later drilling confirmed that the offshore extent of the wedge is around 130 km from the New Jersey coast (Malone et al., 2002).

Groen et al. (2000) presented a case study of meteoric water wedged between shallow and deep saline waters in the onshore and offshore sediments of Suriname, in South America. In permeable Tertiary formations, they identified a stagnant body of fresh and moderately brackish groundwater of meteoric origin with chloride concentration of 250–1000 mg l^{-1}, which extends over 25 km offshore on the continental shelf (Figure 8.23). The Quaternary deposits overlying the Tertiary aquifers have a thickness ranging from 5 to 30 m and the clay in the deposits has a vertical hydraulic conductivity of about 10^{-4} m d^{-1}. Radiocarbon dating shows that groundwater in the coastal plain becomes progressively older northward along the flow direction, ranging from a few hundreds to 30 ka.

High-resolution sampling on the New Jersey shelf has revealed a distinctly layered structure of the salinity distribution, with freshwater occurring preferentially in fine-grained layers and saltwater in coarse-grained layers (Van Geldern et al., 2013). This finding implies that permeability variations exert an important influence on the circulation patterns. Modelling studies to date have not considered this, but the effect of immobile zones has

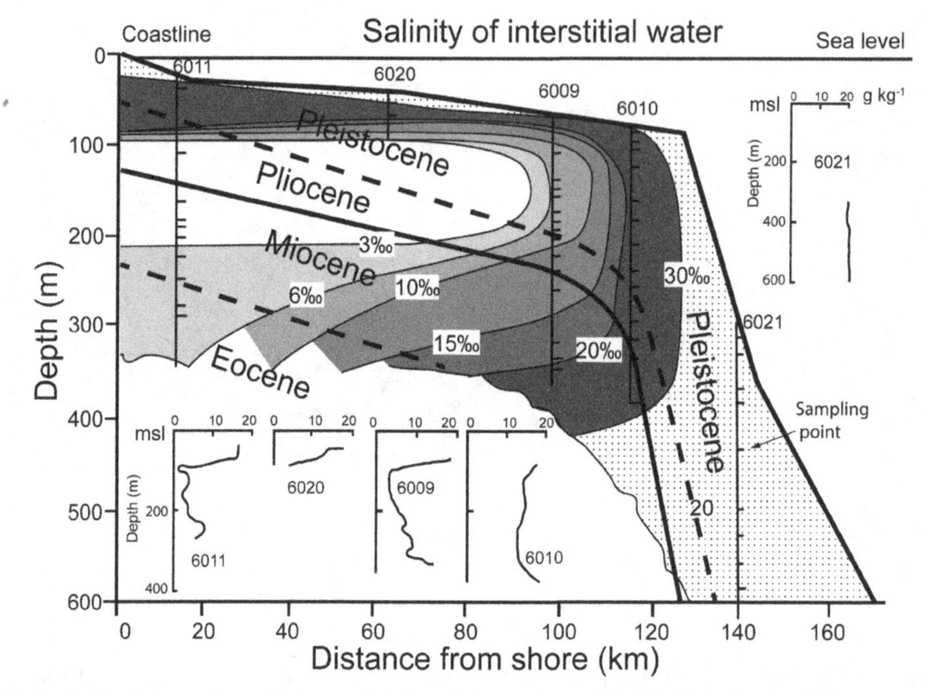

Figure 8.22 Cross section of the coastal plain and the continental shelf in New Jersey, USA, showing contour lines of the chloride concentration (g kg^{-1}) of the pore water (Hathaway et al., 1976).

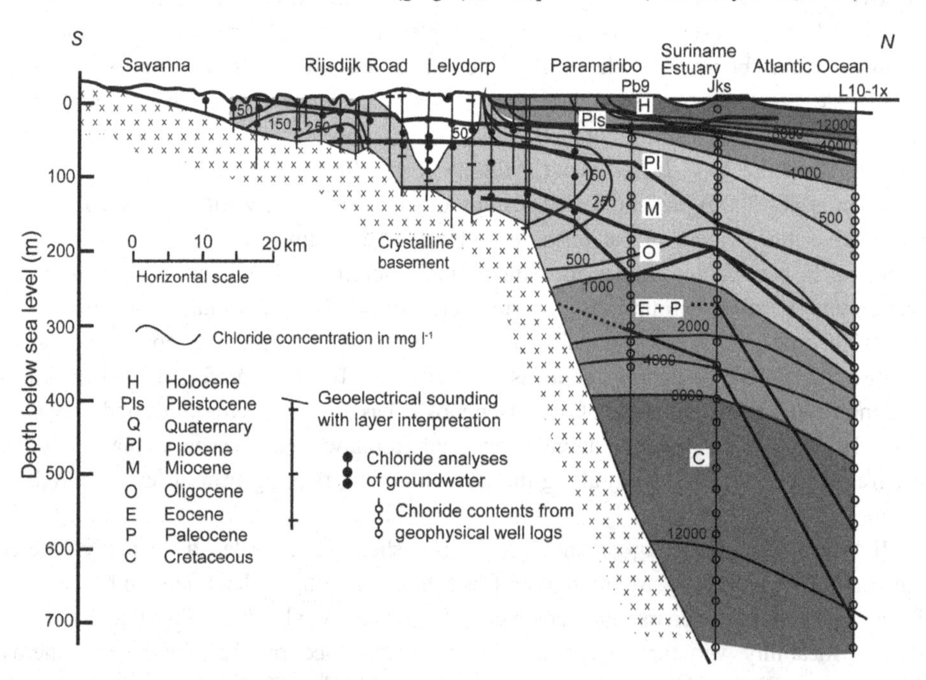

Figure 8.23 Cross section showing groundwater chloride concentrations in the coastal plain of Suriname (Groen, 2002). Chloride concentrations in the inland boreholes were obtained directly from groundwater samples, but those of the three offshore boreholes were inferred from geophysical well logs.

been shown to be important in the development of fresh-saltwater transition zones in onshore aquifers (Lu et al., 2009), as well as to have an effect on vertical salinisation by free convection (Lu et al., 2016).

8.5.1 Emplacement

In both the New Jersey and the Suriname cases, model calculations show that the low salinity of the groundwater cannot be explained by submarine flow systems originating onshore that exist under the present-day conditions (Kooi and Groen, 2003; Person et al., 2003). This is supported by radiocarbon data of groundwater samples near the coastline, which suggest that the waters date back to the Pleistocene. This provides a strong indication that the low-salinity water was recharged when sea levels were much lower, and the continental shelves were exposed.

On the basis of the cross section in Figure 8.22, Meisler et al. (1984) constructed a numerical model and demonstrated that the transition zone between seawater and the low-salinity water is not in equilibrium with present sea level. Later, more sophisticated models of the continental shelf off the coast of New England considered the effects of flow, heat and solute transport, ice sheet loading, and sea level fluctuations during the Pleistocene (Cohen et al., 2010). This model supports the hypothesis that freshwater was emplaced during Pleistocene sea level lowstands when the shelf was exposed to meteoric recharge. A key finding was that the hydraulic conductivity of the confining unit exerts a major control on the offshore salinity distribution. Freshwater remained limited to the area close to the shoreline in models where the confining unit acted as a strong barrier to flow. But due to the outcropping of shallow sands in submarine canyons, discharge points of low-salinity water could exist during sea level lowstands, which enabled the flushing of the aquifer by the water that fed the canyon springs.

Cohen et al. (2010) found that significantly more freshwater was recharged in model simulations that included the effect of glacial loading than in those that did not. Recharge beneath ice sheets is possible due to basal melting caused by the pressure exerted by the overlying ice mass (Person et al., 2003). The effect remains restricted to the proximal parts of the ice sheet, whereas further afield, the increased shore-perpendicular gradients due to the lower sea level are more important. The presence of proglacial lakes (Person et al., 2012) has further been suggested to play a role in the emplacement of freshwater below the continental shelf of New England.

Van Geldern et al. (2013) found that the stable water isotope composition of the freshwater resembled that of modern recharge onshore. This finding is contradictory to the inferred recharge of the freshwater during glacial periods, because in that case, a more depleted isotopic signature would be expected due to the prevailing cooler temperatures (Gat, 1983). Van Geldern et al. (2013) suggested that this contradiction might be solved by assuming that freshwater recharge was mainly restricted to warmer interstadial stages during sea level lowstands. Indeed, recharge has been found to have been intermittent during glacial periods elsewhere. For example, there is a distinct absence of palaeo-groundwaters

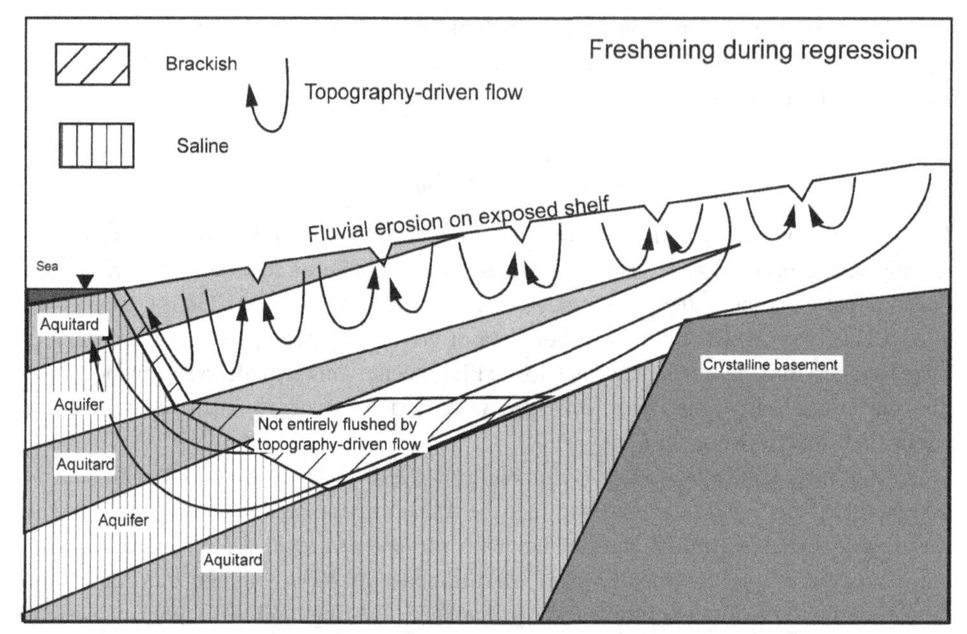

Figure 8.24 Schematic cross section along the continental shelf showing the development of local topography-driven flow systems and the existence of a regional flow system during sea level low-stands (Groen, 2002).

in northern Europe between 20 and 10 ka BP (Edmunds et al., 2001), which is attributed to the pervasiveness of permafrost conditions that impaired the infiltration of water into the soil. However, generic model simulations of continental shelf aquifers with heterogeneous permeability distributions have demonstrated that it is indeed possible that continental fresh groundwater flows penetrate beneath the seafloor across tens of kilometres (Michael et al., 2016) (Figure 6.6). It may therefore be that, under favourable conditions, a sea level lowstand is not a prerequisite for the emplacement of fresh groundwater offshore. But past sea level variations are a given, and the evolution of offshore groundwater must be a complex function of both changes in recharge and sea level.

The topography created on the continental shelf during the sea level lowstands is another key factor for the emplacement of meteoric water. The fall in sea level lowered the base level, which made that rivers and streams formed incisions in the exposed surface on the shelf (Figure 8.24). This led to the development of topography-driven flow systems that are of a smaller extent than the regional flow systems that would develop beneath a flat continental shelf. Due to the shorter flow paths, the residence times in these local systems is much lower than in the regional systems, which led to vigorous flushing of the subsurface by meteoric water. In the case of Suriname, the existence of local flow systems was a prerequisite to be able to explain the presence of meteoric groundwater in the most distal parts of the continental shelf (Groen, 2002).

8.5.2 Preservation

Fresh groundwater emplaced during sea level lowstands becomes subject to salinisation after the land surface becomes flooded during a transgression. As the rate of shoreline migration generally exceeds the rate of horizontal movement of the transition zone, the underground freshwater becomes inundated by seawater at the land surface. The potential for the freshwater to remain preserved is mainly a function of the permeability of the subsurface. Where the freshwater is beneath a laterally extensive layer of low permeability which limits groundwater flow, the mode of salinisation is dominated by molecular diffusion. As discussed in Section 8.3.1.1 this is a slow process and depending on the thickness of the low-permeability structure and the freshwater body below, it might take thousands of years or more to reach complete salinisation. In reality, boundary conditions change continually over such time spans, and equilibrium will never be attained.

Where the permeability of the subsurface is high, the presence of seawater over freshwater represents an unstable density stratification and salinisation by downward vertical density-driven flow will occur (Figure 8.25). This is a rapid process and can lead to the loss of freshwater reserves over the course of decades, as exemplified by the coastal aquifer system in the Netherlands (Section 8.4.2). This process will also occur in aquifers covered by low-permeability strata provided that enough time has elapsed for sufficient salt to diffuse through the confining layer. Salinisation of the aquifer will not be as severe as when it is in

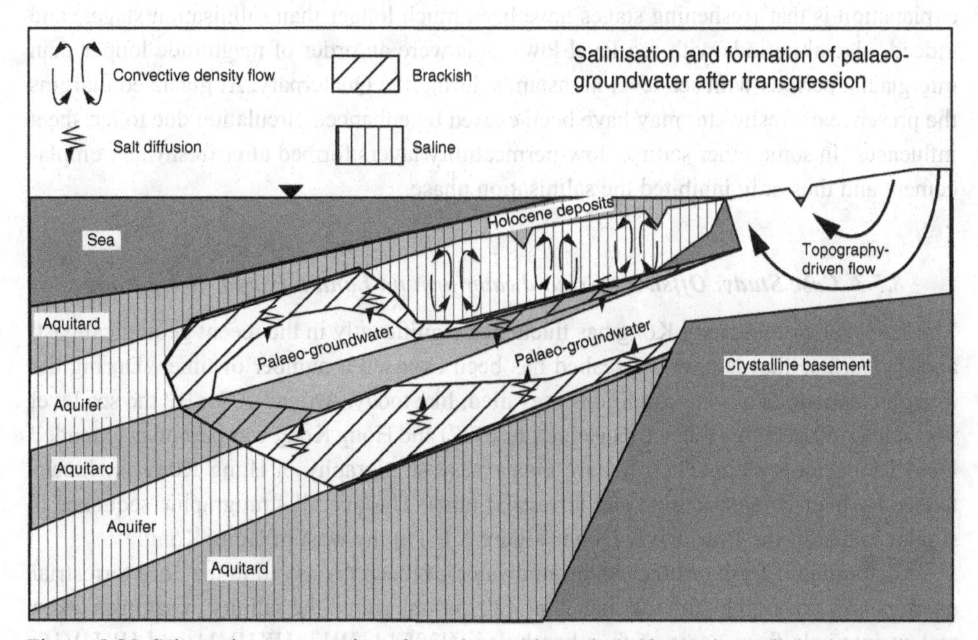

Figure 8.25 Schematic cross section along the continental shelf showing salinisation processes during sea level highstands (Groen, 2002).

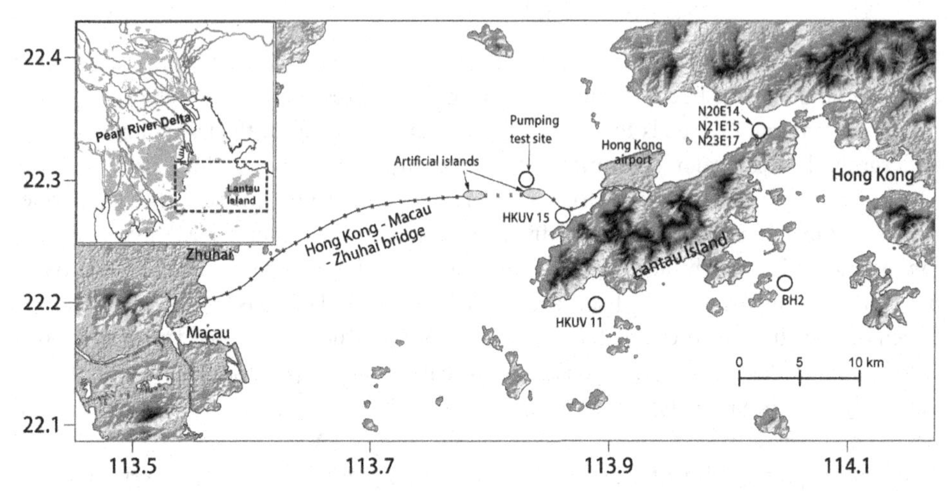

Figure 8.26 Locations of boreholes around Lantau Island, Hong Kong. The inset shows the location of Lantau Island relative to Pearl River Delta.

direct hydraulic contact with the sea, but enhanced rates of downward vertical flow of saline or brackish water will occur nonetheless.

The same low-permeability strata that inhibit salinisation also counteract the emplacement of freshwater during periods of low sea level, which forms a conundrum. One explanation is that freshening stages have been much longer than salinisation stages, and indeed, glacial periods with sea level lowstands were an order of magnitude longer than interglacial periods with sea level highstands during the Quaternary. At glaciated margins the presence of freshwater may have been caused by enhanced circulation due to ice sheet influences. In some other settings low-permeability layers formed after freshwater emplacement and thus only inhibited the salinisation phase.

8.5.3 Case Study: Offshore Groundwater around Lantau Island, Hong Kong

The sea level around Hong Kong has fluctuated significantly in the recent geological past, and the surface that is now the seabed has been exposed a number of times. During the interglacial periods marine conditions prevailed, like today. Around 10 ka BP the sea level was at least 60 m below today's (Shackleton, 1987) and Hong Kong was part of an extended Pearl River flood plain. The offshore Quaternary stratigraphy in Hong Kong is characterised by inter-layered marine and terrestrial units. The overall stratigraphic sequence is similar to that of the Pearl River Delta (Figure 8.16) to the west of Hong Kong.

To determine if fresh or brackish groundwater has been preserved in the once terrestrial aquifers now covered by the sea, Jiao et al. (2015a) measured the salinity of samples taken at 1 m intervals from cores of four boreholes (N20E14, BH2, HKUV11 and HKUV15; Figure 8.26) drilled over different time periods in the offshore area around Lantau Island.

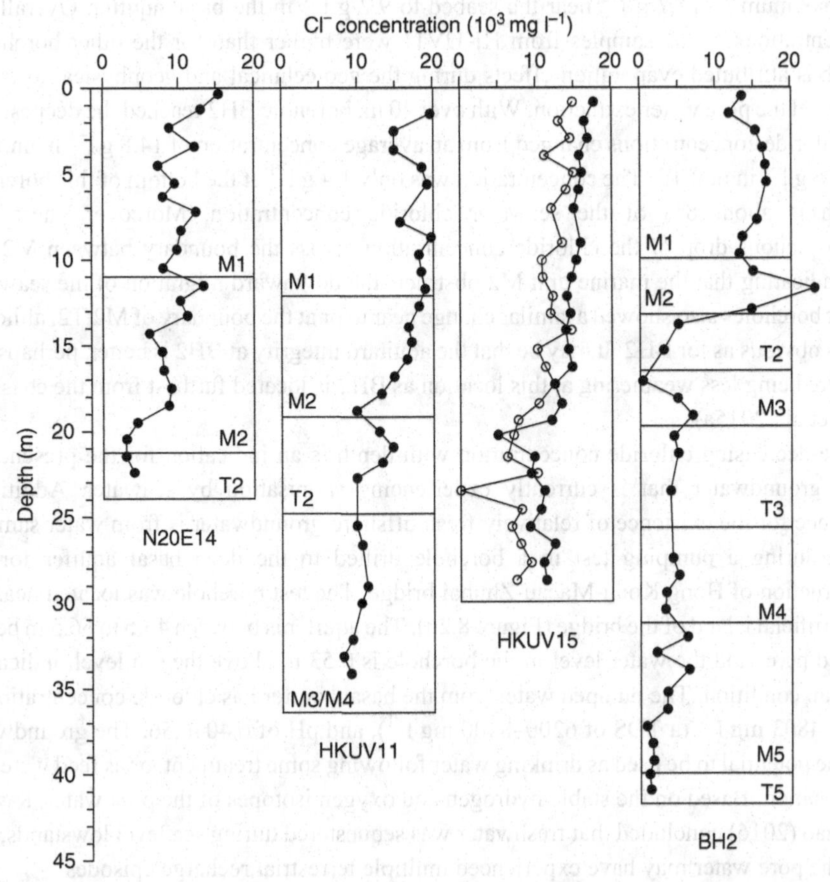

Figure 8.27 Chloride concentration profiles in N20E14, HKUV11, HKUV15 and BH2 (Jiao et al., 2015a). For HKUV15, the concentrations were obtained by centrifuging fresh core samples (open circles) and using a soil water extraction method (solid circles). The stratigraphic units are similar to those in Figure 8.16, but more marine and terrestrial units are distinguished for HUKV11 and BH2 (Yang et al., 2008). No stratigraphic information is available in HKUV15.

Since the cores were taken for the purpose of geotechnical investigations, there were some difficulties with the analysis of the chloride concentrations, which are described in detail by Jiao et al. (2015a) and Kwong and Jiao (2016). Despite this, the chloride concentration profiles for all four boreholes showed a clear and consistent decreasing trend with depth (Figure 8.27).

For core N20E14, the chloride concentration in the units M1, M2 and T2 showed an overall decreasing trend with depth and was much lower than the seawater chloride concentration of ~17.1 g l^{-1}. In borehole HKUV11, the chloride concentrations, which were measured in pore water obtained by centrifuging, also showed a decreasing trend from the maximum of 17.7 g l^{-1} near the seabed to 9.2 g l^{-1} in the basal aquifer. Overall, the concentrations of the samples from HKUV11 were higher than for the other boreholes, which is attributed evaporation effects during the geotechnical and geophysical tests that preceded the pore water extraction. With over 40 m, borehole BH2 reached the deepest, and the chloride concentrations changed from an average concentration of 14.8 g l^{-1} in unit M1 to ~3.6 g l^{-1} in unit T2. The concentration was only 1.4 g l^{-1} at the bottom of this borehole, which is about 8% of the seawater chloride concentration. Moreover, there was a conspicuous drop of the chloride concentration across the boundary between M2 and T2, indicating that the marine unit M2 obstructs the downward migration of the seawater. Other boreholes also showed a similar change near to or at the boundary of M2/T2, although not as obvious as for BH2. It may be that the aquitard integrity at BH2 is better, perhaps due to there being less weathering at this location as BH2 is located furthest from the coastline (Jiao et al., 2015a).

The decreasing chloride concentration with depth is an indication for the presence of fresh groundwater that is currently experiencing salinisation by seawater. Additional evidence for the existence of relatively fresh offshore groundwater is from water samples taken during a pumping test in a borehole drilled to the deep basal aquifer for the construction of Hong Kong-Macau-Zhuhai bridge. The test borehole was located near the east artificial island of the bridge (Figure 8.26). The aquifer is between 46.5 to 66.5 m below seabed here, and the water level in the borehole is 0.53 m above the sea level, indicating artesian condition. The pumped water from the basal aquifer has chloride concentration of 3722–4803 mg l^{-1} (a TDS of 6200–8400 mg l^{-1}), and pH of 6.40–6.56. The groundwater has the potential to be used as drinking water following some treatment, or as feed water for desalination. Based on the stable hydrogen and oxygen isotopes of the pore water, Kwong and Jiao (2016) concluded that freshwater was sequestered during sea level lowstands, and that the pore water may have experienced multiple terrestrial recharge episodes.

8.6 Sea Level Change and Aquifer Development

The history of local sea level rise and palaeo-climate in areas with limestone areas has an important influence on the hydraulic properties of coastal aquifers because of their control on chemical weathering and the development of secondary porosity. This is particularly the case where karstification leads to the formation of connected conduits (openings greater than 1 cm created by dissolution of the bedrock) (Field, 2002) and even cave systems. The position of the zone of active karstification is strongly related to the water table elevation (Mazor, 2004; Milanovic, 2004). As sea level changes lead to changes in the regional hydrological base level, different horizons of karstification linked to previous water table elevations can often be discerned in coastal karst aquifers (Fleury et al., 2007; Mylroie, 2013).

Figure 8.28 schematically shows the effect of the change in base level on the formation of karst networks, starting off with a situation of a high sea level. Unless a low-permeability layer dictates the outlet elevation of the karst drainage network, the sea prescribes the

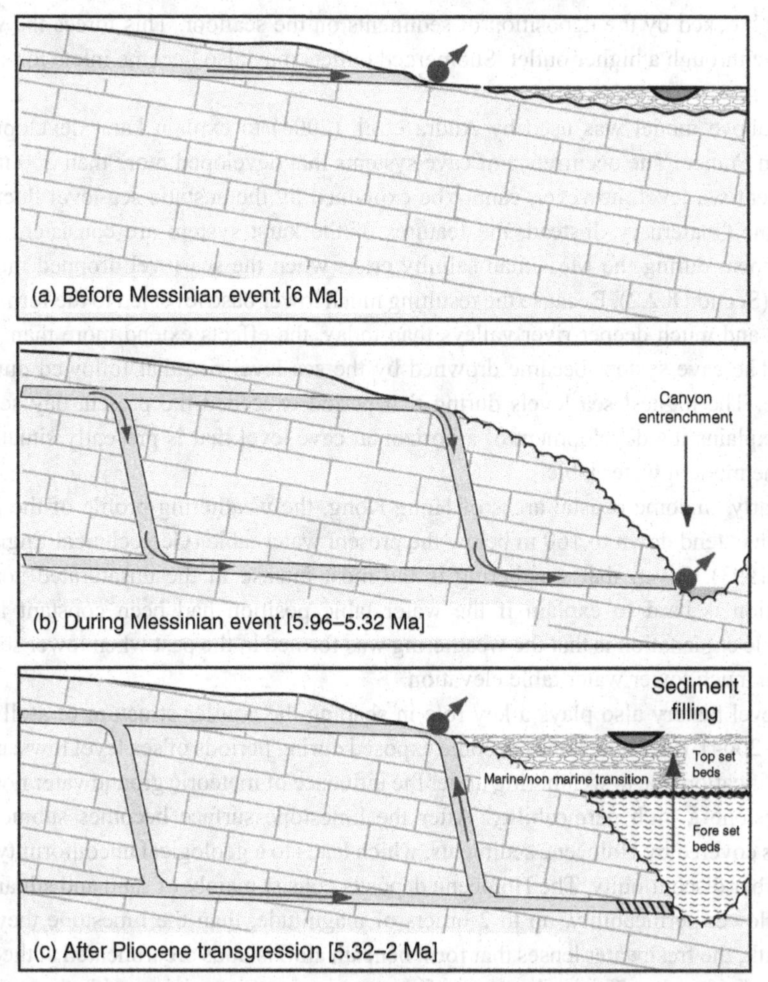

Figure 8.28 Schematic cross sections of a limestone aquifer showing the relation between karst drainage evolution and sea level change (Audra et al., 2004). (a) Karst drainage network and coastal spring when sea level is high. (b) A deeper drainage network develops when sea level falls. (c) The shallow network becomes reactivated when sea level rises. The deeper network may be a conduit for seawater intrusion or submarine groundwater discharge, depending on the flow direction, unless the outlet is blocked by impermeable marine sediments.

regional base level and the points where springs are located (Fleury et al., 2007). When sea level falls, the karst base level drops too and a new network develops in the deeper parts of the limestone. The original drainage network is abandoned, although it may be reactivated periodically during periods of high flow. When sea level rises again, the deeper conduit network becomes submerged and becomes part of the saturated zone. The discharge of groundwater through the outlet of the deep karst drainage may continue, or cease if the outlet is blocked by the deposition of sediments on the seafloor. This forces the water to discharge through a higher outlet. Submerged outlets may also become inlets for seawater intrusion.

This above model was used by Audra et al. (2004) to explain karst development in Southern France. The occurrence of cave systems that developed more than 200 m below the present sea level, however, cannot be explained by the eustatic sea level fluctuations during the Quaternary. Instead, the features of the karst system are consistent with its development during the Messinian salinity crisis when the sea level dropped more than 1500 m (Section 8.2.2). Because the resulting much lower base level led to the formation of canyons and much deeper river valleys than today, the effects extend more than 100 km inland. The cave system became drowned by the sea level rise that followed during the Pliocene. The highest sea levels during that period exceeded the present-day sea level, which explains the development of a horizontal cave level that is presently situated high above the modern water table.

Similarly, in some coastal areas of Hong Kong, the weathering profile of the igneous rocks can extend down to 160 m below the present water table (Geotechnical Engineering Office, 1993). Given that weathering is the most intense in the unsaturated zone, this observation is hard to explain if the water table position had been constant in time. A possible explanation is that the weathering was formed in the past when lower sea levels created a much lower water table elevation.

Sea level history also plays a key role in shaping the aquifer structure of atoll islands (Vacher, 2004). Reef limestones become exposed during periods of sea level lowstands and the karstification due to weathering under the influence of meteoric groundwater flow gives the limestone a high permeability. After the limestone surface becomes submerged, it becomes covered by Holocene sediments, which leads to a geological unconformity, called the Thurber discontinuity. The Holocene deposits consist mainly of sand and silt and have a much lower permeability, up to 2 orders of magnitude, than the limestone they cover. As a result, the freshwater lenses that form beneath atoll islands are truncated at the bottom at the unconformity. This is because of the more vigorous mixing with the permeable limestone, as well as due to the refraction of the flowlines caused by the permeability contrast.

9

Impact of Land Reclamation on Coastal Groundwater Systems

9.1 Introduction

Land reclamation is the formation of new land from areas that were once under water. Land has been created in many coastal areas around the world and this has been important in coastal urban development. Land reclamation represents a main human interference in the natural coastal aquifer systems and may considerably alter the regional groundwater flow system, causing changes to the water level, discharge zone locations, seepage towards the ground surface, regional groundwater divide and the transition zone between seawater and fresh groundwater.

This chapter first introduces the basic concepts and the major hydrogeological problems related to land reclamation. It then presents analytical solutions to study the alteration of water level and interface by land reclamation in unconfined aquifers. A numerical case study of the response of groundwater level and pollutant transport in Penny's Bay, Hong Kong, is also included.

The treatment in this chapter focuses on the so-called large-scale land reclamation, which is loosely defined as the land reclamation whose reclaimed area is comparable to the area of the original groundwater catchment. The theory and methods discussed here can be also used for land reclamation around major lakes and rivers.

9.2 Land Reclamation

9.2.1 Occurrence and Scale

Land reclamation along coasts has been practised since the early history of civilisations in various regions of the world, such as China, Korea, Japan and the Netherlands (Chan, 2001; Kim et al., 2006b; Hoeksema, 2007). The first land reclamation in Hong Kong can be traced back to the early Western Han Dynasty (206 BC–AD 24) (Chan, 2001). The early reclamation in Hong Kong was for activities such as agriculture, mariculture and salt production, but the reclamation after the 1950s has been mainly for urban expansion (Lumb, 1976; Seasholes, 2003; Bi et al., 2012). By 2002, over 30% of the urban area in Hong Kong had been gained from the sea. Hong Kong International Airport and a major theme park are all largely built on reclaimed land. In mainland of China, an area of about 22 000 km^2, or 50%

of the tidal wetland area has been reclaimed between 1949 and 2001 (Wang et al., 2014). As a result, over 200 islands in Zhejiang Province and 300 islands in Guangdong Province in China have disappeared because they have become part of the land (Jiang, 2016). In South Korea, 38% or 1550 km^2 of coastal wetlands had been reclaimed by 2006 and 45% of Korea's population lived on reclaimed land. In Singapore, about 20% of the original size, equal to 135 km^2, had been reclaimed by 2003. Some coastal cities such as New Orleans, Washington DC, Helsinki and Cape Town are largely built on reclaimed land. In many coastal areas, man has now become the most dominant factor controlling geological processes along the coast.

In no country in the world has land reclamation been more important in shaping the nation than in the Netherlands, both from a physiographic and a political governance perspective. The constant fight against the water from both seas and rivers required special forms of collaboration and organisation. The country has a long history of water management which started to influence the landscape on a large scale since approximately AD 800 with the drainage of peat areas to make them suitable for agriculture (Van de Ven et al., 2004). This resulted in compaction and oxidation of the subsoil, and an irreversible land subsidence process was started that continues to the present day (Erkens et al., 2016). Peat was also excavated for fuel, and lakes and inland seas were reclaimed to create agricultural land. Over the course of centuries, the mean elevation of the coastal plain was thereby lowered from a few metres above sea level to largely below sea level (Van de Ven et al., 2004). At present, 25% of the country lies below sea level, with surface elevations as low as more than 6 metres below mean sea level at some locations.

9.2.2 Land Reclamation Methods

There are two distinctly different approaches in carrying out land reclamation. Land can be claimed by dumping fills on the seaside along the coast, creating new land that is typically a few metres above the sea level. This is called landfilling approach. Land can also be reclaimed by enclosing land by a dam which prevents it from being flooded by the sea (or rivers). This is called the empoldering method.

9.2.2.1 Landfilling Method

Reclamation in most of the coastal regions in the world is presently carried out by the landfilling approach. The technique was summarised by Lumb (1976, 1980), and case studies for Hong Kong can be found in Plant et al. (1998). Since 1841, there has been a relentless creation of land by reclamation and levelling of hills in Hong Kong. Sea walls are constructed first, and the water enclosed by the original land and the sea wall is then filled. The fill materials have varied with time. Before the middle of the twentieth century, mainly public waste was used (Lumb, 1976). By 1955, due to a rapid population increase and a great demand for factory space, the required fill material started to be sourced from the nearby hills. The fill was usually the residual soil mantle of decomposed igneous

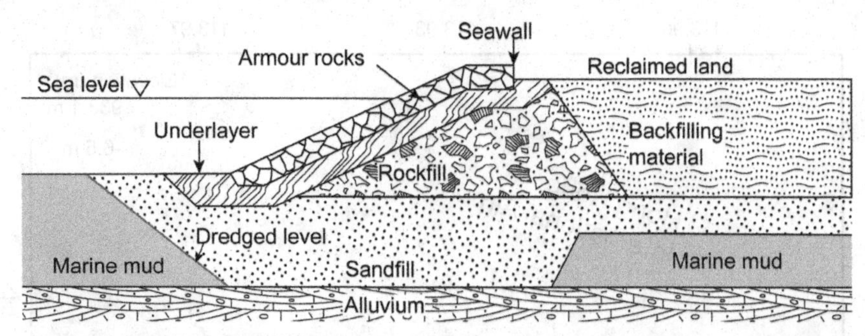

Figure 9.1 Cross section of a typical reclamation site in Hong Kong created by the landfilling method. The armour rocks, underlayer and rockfill consist of rock boulders and rock fragments of various sizes. The marine mud is partially removed to construct the seawall.

rocks, which is typically silty coarse sand. Since 1973, Hong Kong Government introduced the New Town Programme, aiming to provide land in its New Territories for public housing development and by the 1990s, many large bays and coves were reclaimed for creating these towns. In addition to weathered igneous rock, the fill materials used in these projects included sand extracted from the seabed as well as sand dredged from the Pearl River Delta in the mainland of China.

Figure 9.1 shows the internal structure of a typical deep-water reclamation in Hong Kong. The usual procedure is to construct seawalls along the perimeter of the area to be reclaimed, leaving a gap to be completed at the last stage of the work. The seawall construction includes the following stages: dredging out a trench in the soft seabed mud, constructing a foundation by filling the trench with sand or decomposed granite and rubble, and then placing large berm stones or precast concrete blocks over the foundation (Lump, 1976; Plant et al., 1998). A seawall can also be constructed without dredging the mud (Yeung, 2016). Filling in the site proceeds from the landward boundaries inwards, by end-tipping from trucks, dumping progressively up to final formation.

For some reclamation projects, mud on the seabed was removed before reclamation to avoid the problems of slow consolidation later, but this can be difficult and often a mud layer of 5 to 10 m thick remained (Lumb, 1980). At some reclamation sites, the mud is intentionally not displaced because, if the mud is removed, more fill materials are needed and a site is needed to dispose the mud. Both mud dredging and disposal create contamination problems. Thus at some reclaimed sites the mud was reworked and very little was driven off the site.

In the past, reclamation could take more than 10 years to complete and then the site would have been left without construction for years for the soil to be consolidated. Modern land reclamation is usually completed in 2 to 3 years. Various geotechnical techniques such as surcharge and vertical drains are used to speed up the soil consolidation (Plant et al., 1998). For example, the 12.5 km^2 Chek Lap Kok Hong Kong International Airport was created by excavation of two existing small islands and 9.4 km^2 of land was reclaimed from

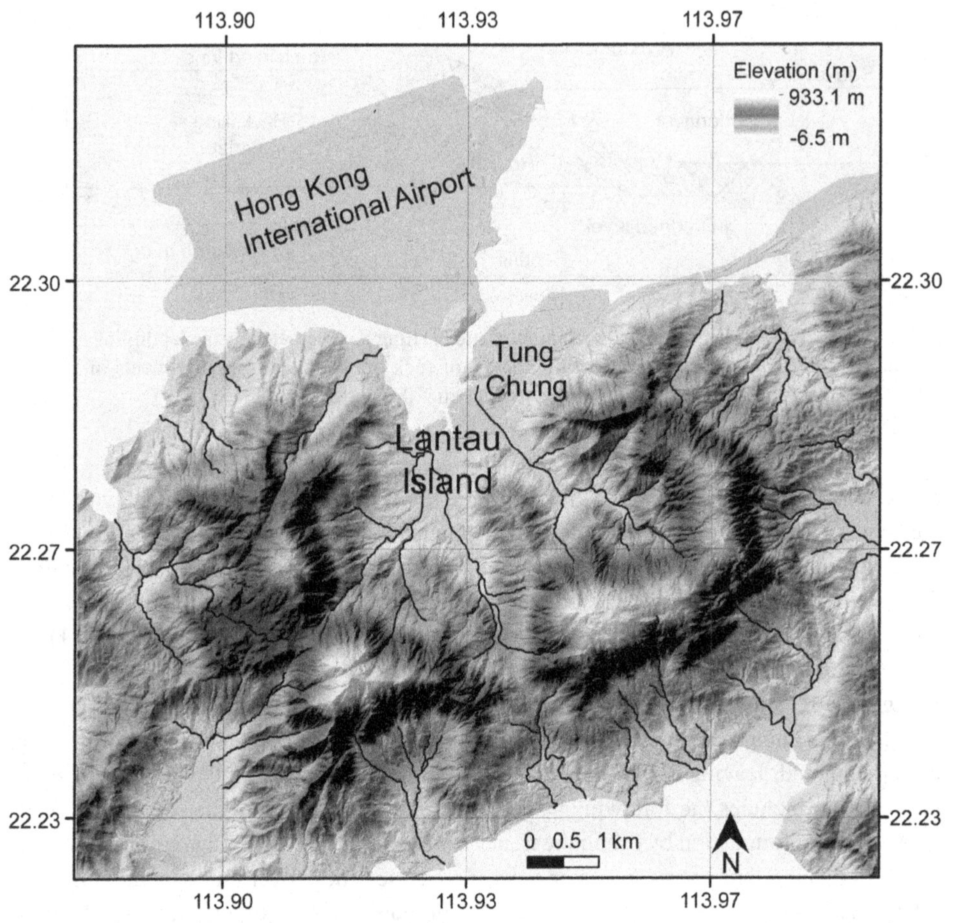

Figure 9.2 Hong Kong International Airport and the connected groundwater catchment. The airport is reclaimed around two islands. It sits over the original discharge zone of the groundwater flow system originating on Lantau Island. The dark lines indicate the stream network.

the sea (Figure 9.2) within 32 months. Land was created at the rate of 20 000 m^2 d^{-1} at the peak of activity (Pickles and Tosen, 1998; Plant et al., 1998). Usually the reclaimed land is linked to the mainland, but a shallow water zone was left deliberately between this airport and Lantau Island for management of the airport security. The groundwater below the airport, however, is linked to the subsurface flow system of Lantau Island, below which a water table exists that causes upward groundwater flow beneath the airport (Wang and Jiao, 2005).

In the early days, the cost of reclamation in Hong Kong was relatively low, primarily because ample sources of rock and soil fill were close at hand, and the depth of water being filled in was relatively small. But land reclamation in Hong Kong has been becoming

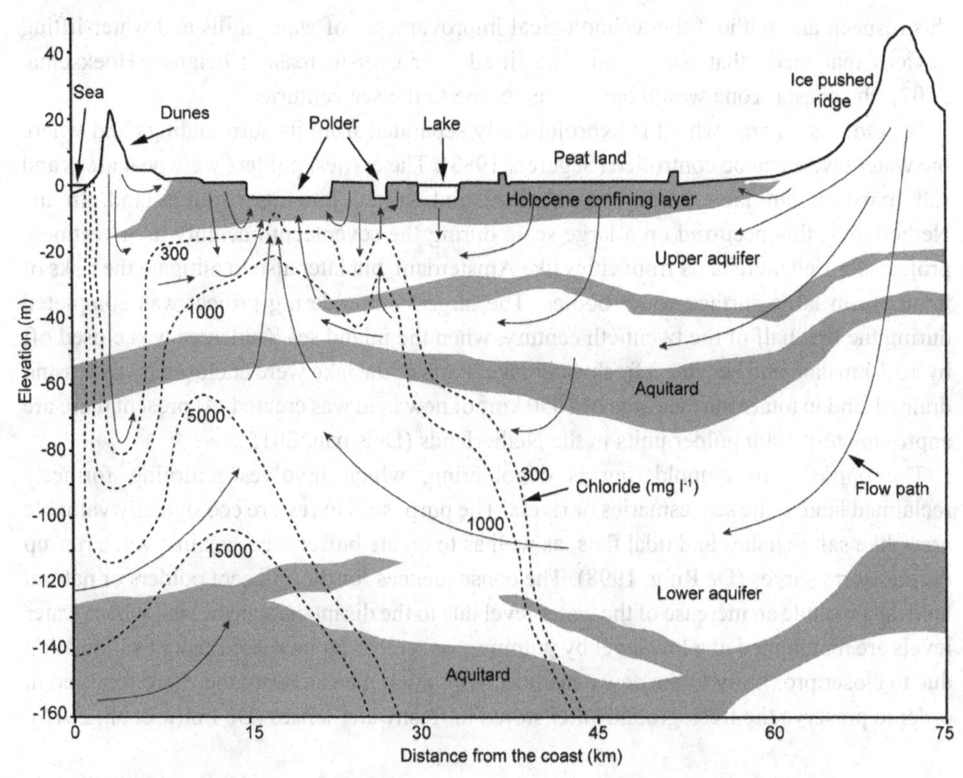

Figure 9.3 Schematic cross section of flow systems and chloride concentrations in the western part of the Netherlands (De Louw et al., 2010).

progressively expensive because fills are not easily available anymore and the sea to be reclaimed has ever greater depths. There is also an increasingly societal outcry against land reclamation as the public is better aware of its adverse environmental effects. Land reclamation remains very popular though in many other countries.

9.2.2.2 Empoldering Method

Land reclamation by empoldering has been practised extensively in the low-lying coastal zones along the North Sea since the thirteenth century (Van de Ven, 2004). By that time, the draining and excavation of peat areas had led to ever more frequent flooding by the sea and coastal defence works became a necessity. The first dykes that were built in the Netherlands were primitive structures that were in constant need of repair. Initially, the surface elevation of the areas enclosed by dykes was just above mean sea level and drainage occurred via streams and small rivers, most of which were natural. Gradually land subsidence made that drainage became more difficult, and in areas that sank below mean sea level, drainage became possible only during low tide. When this became impossible too, lifting of the excess water became necessary. The windmill proved to be indispensable in this respect, and without the technological improvements of water mills and water-lifting devices that made that water could be lifted over ever-increasing heights (Hoeksema, 2007), the coastal zone would have fallen prone to the sea centuries ago.

A polder is an area which is hydrologically separated from its surroundings and where the water levels can be controlled (Segeren, 1983). The earliest polders were peat areas and salt marshes, but later lakes were drained and turned into agricultural land. In the Netherlands, this occurred on a large scale during the seventeenth century as investment projects for rich merchants from cities like Amsterdam, but later also to mitigate the risks of floods from large surface water bodies. The largest empoldering project was completed during the first half of the twentieth century, when the inland sea Zuiderzee was closed off by a 32-km dam and became a freshwater lake. Parts of the lake were enclosed by dykes and drained, and in total a surface area of 1650 km^2 of new land was created. At present there are approximately 3000 polder units in the Netherlands (Delsman, 2015).

The opposite of empoldering is depoldering, which involves returning formerly reclaimed land to the sea, estuaries or rivers. The purpose is to restore ecologically valuable areas like salt marshes and tidal flats, as well as to create buffer areas against wave run-up during storm surges (De Ruig, 1998). The consequences for the adjacent polders or natural land area include an increase of the water level due to the disappearance of land where water levels are maintained at a low level by pumping, as well as an increased risk of salinisation due to closer proximity to the new coastline. Mitigation measures are therefore required in order to preserve the fresh groundwater stored in freshwater lenses (De Louw et al., 2016).

9.2.3 Hydrogeological Issues Related to Land Reclamation

While land reclaimed from the sea can mitigate the pressure of land use, land reclamation also creates various environmental, ecological and engineering problems such as changes of sediment deposition and erosion patterns, habitat loss, alteration of the marine environment due to dredging of sand or mud, and coastal flooding after destroying wetlands, which act as a buffer between the ocean and the land and absorb much of wave energy (Barnes, 1991; Noske, 1995; Ni et al., 2002; Terawaki et al., 2003). Reclamation also affects groundwater systems, to an extent that is dependent on the scale of land reclamation. Coastal zones form the terminus of regional aquifer systems and the interaction between groundwater and seawater will be altered by land reclamation. However, this phenomenon trends to be ignored because the reaction of a subsurface flow system to reclamation is typically sluggish and not so noticeable.

Broadly speaking, it is expected that the landfilling method leads to a build-up of the water level in the aquifer connected to the reclaimed area, while the empoldering method leads to a reduction in water level. The impact of land reclamation on coastal groundwater flow system, therefore, depends on the ways that the land is reclaimed.

In the case of empoldering (Figure 9.3), to protect the land from flooding, an extensive network of dams, dykes, drainage canals and pumping stations is put in place (Van Dam, 1999). The maintenance and running costs in the Netherlands are in the billions of Euros annually and are expected to increase in the future, but without this system in place, about 65% of the country would be flooded by the sea or rivers during periods of high water level. Ongoing land subsidence and sea level rise form a significant threat (Oude Essink, 1999). Another problem related to the creation of the polder landscape is the upward seepage of saline groundwater (Van Dam, 1999). The high salt load is a problem for agriculture, and the water's high nutrient load has an adverse effect on the quality of surface water (Griffioen, 1994).

Because large volumes of seawater are still present in the sediments dating back to the time when the area was under the influence of the sea (Post et al., 2003), the upward seepage of saltwater will be a very long-lived phenomenon, on the order of centuries and more (Oude Essink, 2001a). It cannot be stopped and the drainage system needs to be designed to separate the brackish and freshwater flows as best as possible (Van Dam, 1999). The elevation difference between individual polders, which are on the order of decimetres to metres, drive local groundwater flow systems and results in strong spatial differences of the groundwater salinity (Figure 9.3). The deepest polders act in the same way as abstraction wells and this is where up-coning of saline groundwater occurs (De Louw et al., 2013) (Section 6.2.3).

The discussion hereafter focuses mainly on the landfilling approach. Usually the most favourable area for land reclamation is near a river mouth because of the shallow bathymetry due of sedimentation of the river, such as Hong Kong International Airport (Figure 9.2). In Hong Kong, huge concrete box culverts are usually installed to direct the river water to the sea, which affects the hydraulic gradient of the upstream part of the river. While engineering schemes tend to consider the modification to the river in detail, the effects on the groundwater flow system do not always receive proper consideration.

A few conceptual models are presented here to demonstrate these effects (Jiao, 2000). Figure 9.4 shows the impact of reclamation by landfilling on an unconfined aquifer system. The seaward groundwater discharge will be reduced because the fill increases the subsurface flow path length, causing an increase of the water level elevation in the upstream part of the aquifer. For an unconfined aquifer, the increase may not be very significant if groundwater can discharge through springs that form when water table reaches the land surface. These springs typically form along the former coastline. For a confined aquifer, if the outlet of the aquifer at the seabed is obstructed by reclamation as shown in Figure 9.5, water level can be increased significantly over a large area.

Figures 9.4 and 9.5 assume that the divide of the groundwater system is so far from the coast that it remains unaffected by reclamation. If the extent of reclaimed land approaches or exceeds that of the pristine subsurface water catchment, as often the case in Hong Kong, the modification of the groundwater system can be more significant (Jiao, 2000). As shown in Figure 9.6, since the discharge to the sea is hampered by the fill after reclamation, more water will move to the opposite side of the hill. As a result, the reclamation causes the displacement of the groundwater divide. Depending on the hydraulic properties of the fill and the scale of reclamation, the effects of the modifications may only manifest themselves after years.

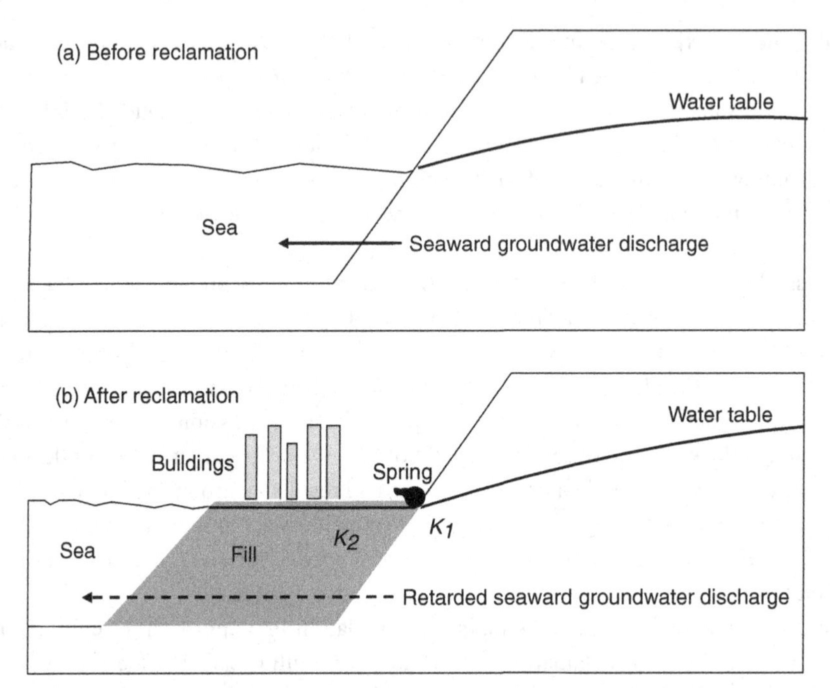

Figure 9.4 Modification of a coastal unconfined aquifer in a coastal extensive landmass (a) before and (b) after land reclamation (Jiao, 2000). The grey area indicates reclaimed land.

The hydraulic conductivity of the fill material is a key factor controlling the alteration of groundwater flow system. The hydraulic conductivity varies with the nature of the fill materials and the method of placement and can be very heterogeneous (Jiao, 2002). A fill of marine sand and completely weathered igneous rock, typical for Hong Kong, has a hydraulic conductivity on the order of 10^{-5} or 10^{-4} m s^{-1}. If the fill materials are compacted, the hydraulic conductivity ranges between 10^{-8} and 10^{-6} m s^{-1} (Geotechnical Engineering Office, 1993). Below the reclamation sites, there usually exists a layer of soft marine mud, of which the hydraulic conductivity is typically between of 10^{-9} and 10^{-7} m s^{-1} (Kwong, 1997). This layer of mud gradually becomes less thick and less permeable as consolidation continues.

The subsurface heterogeneity of land reclaimed from the sea may be greater than any naturally formed land. For older reclamation sites, the fill may include domestic and construction waste, as well as soil excavated from nearby hills. Even in a well-designed modern reclamation site, the hydraulic conductivity of the actual fill materials may be still very different from, and very often lower than, designed due to inevitable admixture of marine mud. With ongoing reclamation, seawalls may become locked in by land areas created during the next expansion phase. Thus, the material in reclaimed coastal zones may range from clay particles to objects of metres, such as boulders, berm stones and concrete blocks. This makes the hydraulic properties of the site in general unpredictable and site investigation very difficult.

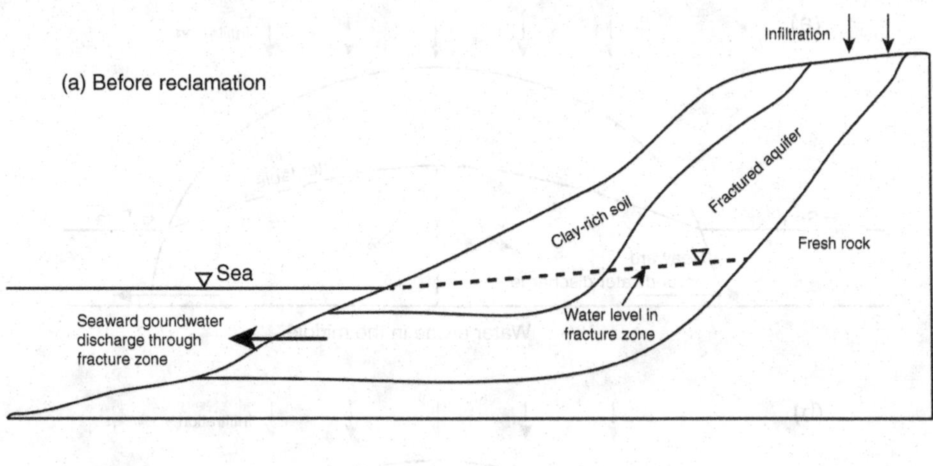

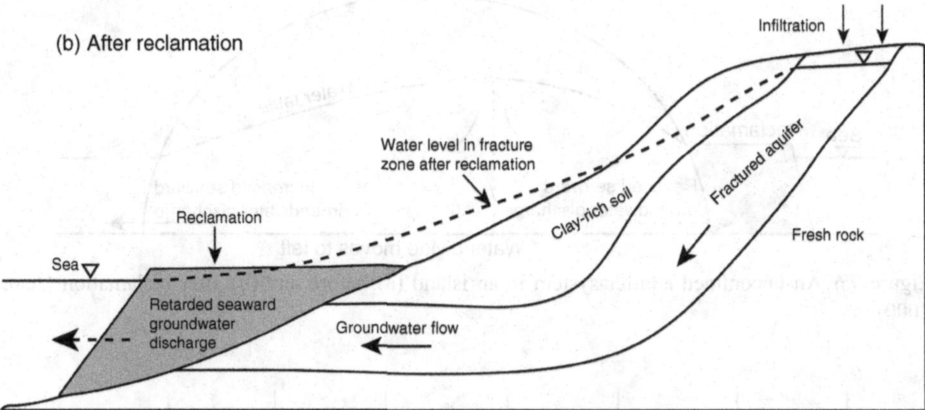

Figure 9.5 Impact of land formation on a confined groundwater system (Jiao, 2000). The confined aquifer consists of fractured rock and is confined at its base by unweathered rock and at the top by clay-rich soil.

9.3 Analytical Solutions for Land Reclamation Effects

To illustrate the hydrogeological effects of reclamation by landfilling, two coastal aquifer configurations are considered: continental coastal aquifers (i.e. an extensive landmass bordered on one side by the coast) and strip island aquifers (Section 3.4). First, the long-term effects of reclamation on the flow system will be demonstrated by comparing the steady state situation before and after reclamation. Second, timescales will be assessed based on transient solutions.

In all cases, it will be assumed that the groundwater originates from precipitation recharge and that the aquifer is unconfined. The Dupuit assumption applies, and both the fill materials and the original aquifer are assumed to be uniform and isotropic (Guo and Jiao, 2007). Where the solution involves an interface, the Ghijben–Herzberg principle applies.

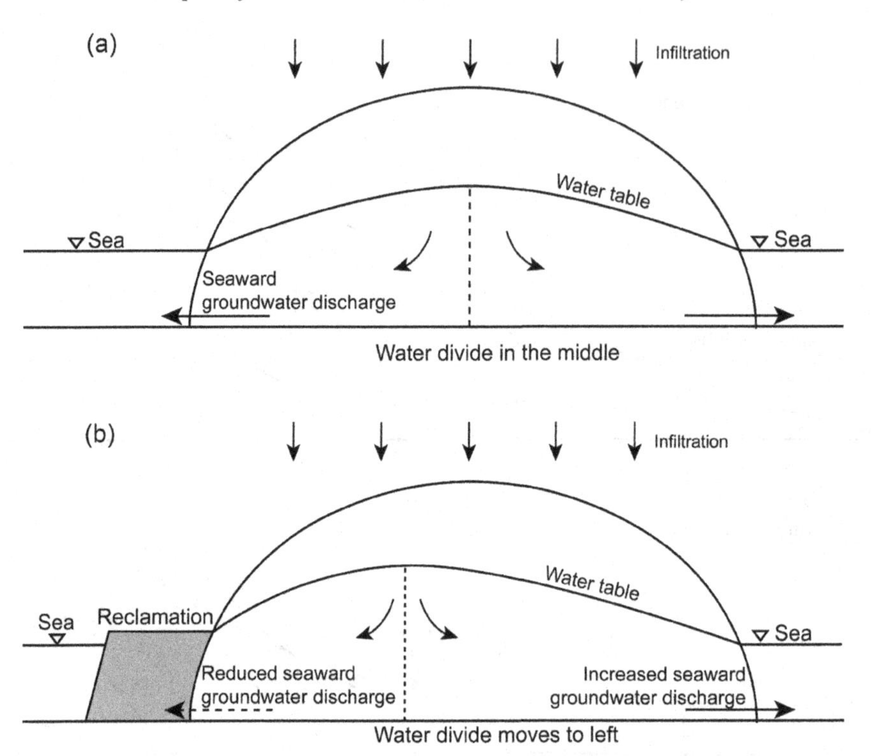

Figure 9.6 An unconfined aquifer system in an island (a) before and (b) after reclamation (Jiao, 2000).

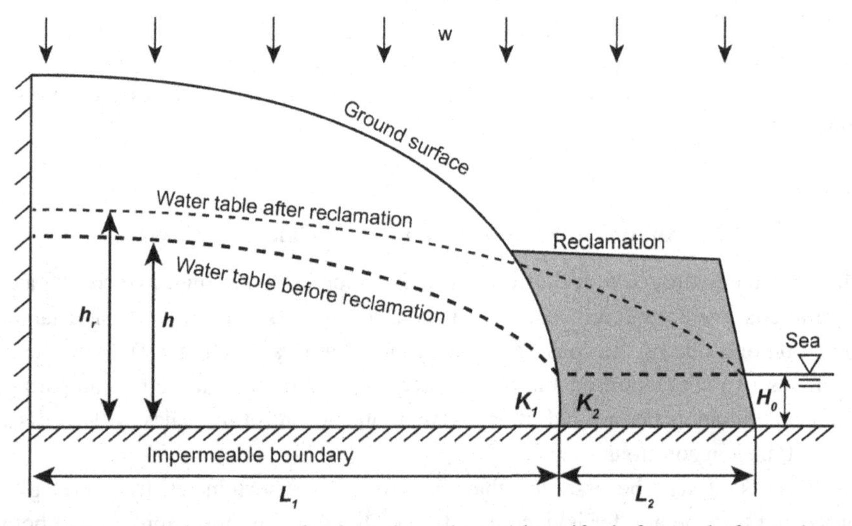

Figure 9.7 Water table in an unconfined continental coastal aquifer before and after reclamation. The shaded zone indicates the reclaimed land. The origin ($x = 0$) of the system is located in the left bottom corner (Hu et al., 2008).

9.3.1 Unconfined Continental Coastal Aquifer without an Interface

This example of the effect of land reclamation on subsurface flow for the assumptions stated above was published by Jiao et al. (2001). The density difference between freshwater and seawater will be ignored.

An outline of the situation considered is shown in Figure 9.7. The original aquifer has a hydraulic conductivity of K_1 [L T^{-1}]. There is a water divide on the left-hand side. The distance from the sea to the divide is L_1 [L]. The meteoric recharge flux is w [L T^{-1}]. The impermeable bottom of the aquifer is chosen to be the datum of the system. The sea level is H_0 [L]. Due to reclamation, the coastline is extended to the sea by L_2 (which is noted as reclamation length hereafter) using fill materials that have a hydraulic conductivity K_2.

9.3.1.1 Situation before Reclamation

The steady state groundwater flow problem can be expressed by the following governing equation and boundary conditions:

$$\frac{d}{dx}\left(K_1 h\frac{dh}{dx}\right) + w = 0, (0<x<L_1) \tag{9.1}$$

$$\frac{dh}{dx} = 0, (x = 0) \tag{9.2}$$

$$h = H_0, (x = L_1) \tag{9.3}$$

The solution to Eqns (9.1)–(9.3) is

$$h^2 = H_0{}^2 + \frac{w}{K_1}(L_1{}^2 - x^2) \tag{9.4}$$

Obviously, water level has a maximum at the water divide at the left boundary where $x = 0$:

$$h^2 = H_0{}^2 + \frac{w}{K_1}L_1{}^2 \tag{9.5}$$

9.3.1.2 Situation after Reclamation

After reclamation, the flow can be expressed by the following governing equation and boundary conditions:

$$\frac{d}{dx}\left(Kh_r\frac{dh_r}{dx}\right) + w = 0, (0<x<L_1 + L_2, x \neq L_1) \tag{9.6}$$

$$\frac{dh_r}{dx} = 0, (x = 0) \tag{9.7}$$

$$h_r = H_0, (x = L_1 + L_2) \tag{9.8}$$

subject to the continuity constraints

$$K_1 \frac{dh_r}{dx}\bigg|_{x=L_1-0} = K_2 \frac{dh_r}{dx}\bigg|_{x=L_1+0}, (x = L_1) \tag{9.9}$$

$$h_r|_{x=L_1-0} = h_r|_{x=L_1+0}, (x = L_1) \tag{9.10}$$

where

$$K = \begin{cases} K_1, (0 < x < L_1) \\ K_2, (L_1 < x < L_1 + L_2) \end{cases}$$

In the above equations, h_r denotes the groundwater level after reclamation. Equation (9.6) assumes that the recharge rates in the original and in the reclaimed land are the same. Integrating Eqn (9.6) with respect to x from 0 to x and using boundary condition Eqn (9.7) and the continuity conditions Eqns (9.9) and (9.10) leads to

$$\frac{1}{2}\frac{dh_r^2}{dx} = -w\frac{x}{K}, (0<x<L_1 + L_2, x \neq L_1) \tag{9.11}$$

Integrating Eqn (9.11) once more with respect to x from x to $L_1 + L_2$ and using boundary condition Eqn (9.8) yields

$$H_0^2 - h_r^2(x) = -2w\int_x^{L_1+L_2} \frac{x}{K}dx \tag{9.12}$$

If $0 \leq x \leq L_1$, then $\int_x^{L_1+L_2} = \int_x^{L_1} + \int_{L_1}^{L_1+L_2}$, which leads to

$$h_r^2(x) = H_0^2 + w\left[\frac{1}{K_1}(L_1^2 - x^2) + \frac{1}{K_2}(2L_1 + L_2)L_2\right] \tag{9.13}$$

If $L_1 \leq x \leq L_1 + L_2$, one has

$$h_r^2(x) = H_0^2 + w\frac{1}{K_2}[(L_1 + L_2)^2 - x^2] \tag{9.14}$$

The change of water table between 0 and L_1 due to reclamation can be calculated by subtracting Eqn (9.4) from Eqn (9.13):

$$h_r(x) - h(x) = \sqrt{H_0^2 + \frac{w(L_1^2 - x^2)}{K_1} + \frac{(2L_1 + L_2)L_2 w}{K_2}} - \sqrt{H_0^2 + \frac{w(L_1^2 - x^2)}{K_1}} \tag{9.15}$$

This equation shows that there is always an increase in water table after reclamation since the right-hand side of Eqn (9.15) is always greater than 0. The maximum and minimum changes will occur at the old coast ($x = L_1$) and at the water divide ($x = 0$), respectively:

$$h_r(L_1) - h(L_1) = \sqrt{H_0^2 + \frac{w}{K_2}(2L_1 + L_2)L_2} - H_0 \tag{9.16}$$

$$h_r(0) - h(0) = \sqrt{H_0^2 + \frac{wL_1^2}{K_1} + \frac{(2L_1 + L_2)L_2 w}{K_2}} - \sqrt{H_0^2 + \frac{wL_1^2}{K_1}} \tag{9.17}$$

Equation (9.16) shows that the maximum water level change at the old coast is independent of K_1, but depends on H_0, L_1, L_2 and the ratio of the recharge rate w and K_2. If the recharge is large and the hydraulic conductivity is low, the reclamation has a significant damming effect and the hydraulic head will build up. Equation (9.16) further shows that the water table will increase as both L_1 and L_2 increase, but the water table will increase much more significantly as L_2 increases. Since for a given coastal aquifer, the parameters w and L_1 are both fixed, Eqn (9.15) or (9.17) can be used to estimate the water table change in response to fill materials of different hydraulic conductivity and scale of reclamation (Jiao et al., 2001).

9.3.1.1 Example

The impact of reclamation on water level in the aquifer system at a coast is demonstrated by a hypothetical example. Assume that the distance from the groundwater divide to the coast is $L_1 = 1000$ m, the aquifer hydraulic conductivity is 0.1 m d^{-1}, the recharge rate is 0.0006 m d^{-1} and H_0 is 10 m. Figure 9.8 illustrates how the water level change at the old coastline as a function of the hydraulic conductivity K_2 for different reclamation lengths (L_2) calculated using Eqn (9.16). Water level build-up at the old coastline declines with decreasing L_2,

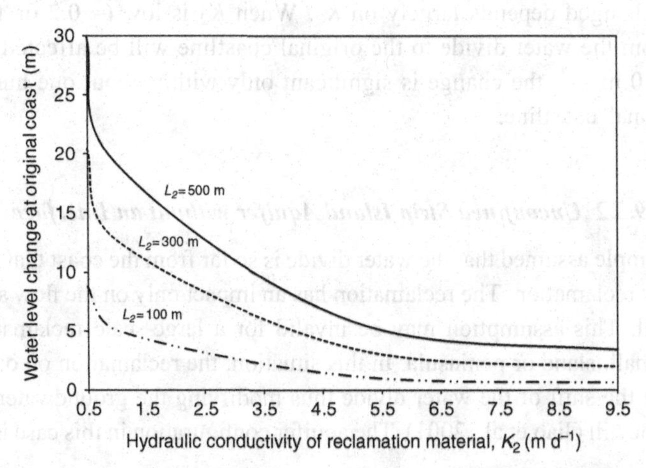

Figure 9.8 Relation between groundwater level increase at the former coastline position and the fill hydraulic conductivity for different reclamation lengths (modified from Jiao et al., 2001).

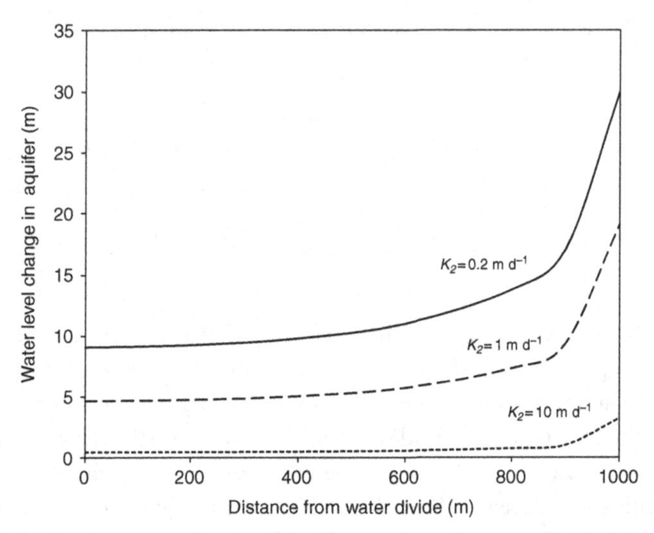

Figure 9.9 Relation between water level and the distance from the water divide for various values of the fill hydraulic conductivity (K_2) when $L_1 = 1000$ m, $L_2 = 500$ m and $K_1 = 0.1$ m d^{-1} (modified from Jiao et al., 2001).

while it increases as K_2 decreases. For low K_2, the water level build-up can be large and is very sensitive to the change of K_2. If K_2 increases, the water level increase becomes less sensitive to K_2.

Figure 9.9 shows the water level increase with distance from the divide to the old coast for different K_2 at a given L_2 of 500 m calculated using Eqn (9.15). The water level change is the greatest at the original shoreline and decreases towards the water divide. The inland distance from the original coast over which the water level will be significantly changed depends largely on K_2. When K_2 is low (= 0.2 or 1 m d^{-1}), the entire area from the water divide to the original coastline will be affected. When K_2 is increased to 10 m d^{-1}, the change is significant only within about one hundred metres from the original coastline.

9.3.2 Unconfined Strip Island Aquifer without an Interface

The above example assumed that the water divide is so far from the coast that its position is not affected by reclamation. The reclamation has an impact only on the flow system of one side of the hill. This assumption may be invalid for a large-scale reclamation near the hillside of a small island or peninsula. In this situation, the reclamation on one side of the hill may cause the shift of the water divide thus modifying the groundwater situation on both sides of the hill (Jiao et al., 2001). The aquifer configuration in this case is as shown in Figure 9.10.

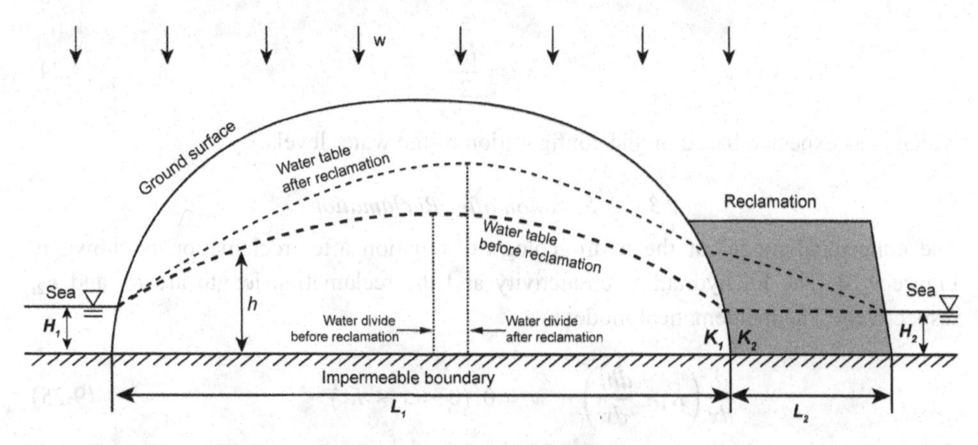

Figure 9.10 Water table in a strip island unconfined aquifer before and after reclamation. The origin ($x = 0$) of the system is located in the left bottom corner (Hu et al., 2008).

9.3.2.1 Situation before Reclamation

It is assumed first that the water levels in the two surface water bodies are different, with the water level in the left H_1 being higher than H_2 on the right. Flow in the aquifer can be written as

$$\frac{d}{dx}\left(K_1 h\frac{dh}{dx}\right) + w = 0, (0<x<L_1) \tag{9.18}$$

$$h = H_1, (x = 0) \tag{9.19}$$

$$h = H_2, (x = L_1) \tag{9.20}$$

The solution of Eqn (9.18) subject to Eqns (9.19) and (9.20) is (Bear, 1979)

$$h^2 = H_1{}^2 + \frac{x}{L_1}(H_2{}^2 - H_1{}^2) + \frac{wx}{K_1}(L_1 - x) \tag{9.21}$$

Differentiating Eqn (9.21) and using Darcy's law, the discharge per unit width q_w ($L^2 T^{-1}$) is

$$q_w = -K_1 h\frac{dh}{dx} = K_1\frac{H_1{}^2 - H_2{}^2}{2L_1} + wx - \frac{wL_1}{2} \tag{9.22}$$

If the water levels H_1 and H_2 in the two water bodies are the same, Eqn (9.22) becomes

$$q_w = wx - \frac{wL_1}{2} \tag{9.23}$$

At the water divide $q_w = 0$, so the divide location (x_d) is

$$x_d = \frac{L_1}{2} \tag{9.24}$$

which is as expected based on the configuration of the water levels.

9.3.2.2 Situation after Reclamation

The conceptual model of the hydrogeological situation after reclamation is shown in Figure 9.10. The fill hydraulic conductivity and the reclamation length are K_2 and L_2, respectively. The mathematical model is

$$\frac{d}{dx}\left(K_1 h_r \frac{dh_r}{dx}\right) + w = 0, (0 < x < L_1) \tag{9.25}$$

$$\frac{d}{dx}\left(K_2 h_r \frac{dh_r}{dx}\right) + w = 0, (L_1 < x < L_1 + L_2) \tag{9.26}$$

with boundary conditions

$$h_r(0) = H_1 \tag{9.27}$$

$$h_r(L_1 + L_2) = H_2 \tag{9.28}$$

and the continuity conditions of water level and flux at the interface ($x = L_1$) are

$$h_r|_{x=L_1-0} = h_r|_{x=L_1+0}, (x = L_1) \tag{9.29}$$

$$K_1 h_r \frac{dh_r}{dx}\bigg|_{x=L_1-0} = K_2 h_r \frac{dh_r}{dx}\bigg|_{x=L_1+0}, (x = L_1) \tag{9.30}$$

where as before h_r [L] indicates the hydraulic head after reclamation.

The general solutions for Eqns (9.25) and (9.26) are presented by Jiao et al. (2001). For the situation where $H_2 = H_1$ the solutions to the above equations are

$$h_r^2 = -\frac{w}{K_1}x^2 + \frac{wK_2}{(L_1K_2 + L_2K_1)}\left[\frac{L_1^2}{K_1} + \frac{L^2 - L_1^2}{K_2}\right]x + H_1^2, (0 < x < L_1) \tag{9.31}$$

$$h_r^2 = \frac{w}{K_2}(L^2 - x^2) + \frac{wK_1}{(L_1K_2 + L_2K_1)}\left[\frac{L_1^2}{K_1} + \frac{L^2 - L_1^2}{K_2}\right](x - L) + H_1^2, (L_1 < x < L_1 + L_2) \tag{9.32}$$

where $L = L_1 + L_2$. On the basis of Darcy's law,

$$q_{1r} = -K_1 h_r \frac{dh_r}{dx} = wx - \frac{wK_1K_2}{2(L_1K_2 + L_2K_1)}\left[\frac{L_1^2}{K_1} + \frac{L^2 - L_1^2}{K_2}\right] \tag{9.33}$$

At water divide $q_{1r} = 0$, the location of the water divide after reclamation (x_{dr}) can be found by letting Eqn (9.33) equal 0, which leads to

$$x_{dr} = \frac{K_1K_2}{2(L_1K_2 + L_2K_1)}\left[\frac{L_1^2}{K_1} + \frac{L^2 - L_1^2}{K_2}\right] \quad (9.34)$$

The water level change Δh [L] at the original coastline can be calculated as the difference between the head from Eqn (9.31) at $x = L_1$ and H_1:

$$\Delta h = h_r - H_1 = \sqrt{\frac{wL_1L_2(L_1 + L_2)}{L_1K_2 + L_2K_1} + H_1^2} - H_1 \quad (9.35)$$

As with Eqn (9.16), Eqn (9.35) shows that at the original coast the water level will always increase due to reclamation.

Due to reclamation the water divide will change its position. Equations (9.24) and (9.34) give the locations of water divide before and after reclamation, so by subtracting Eqn (9.24) from Eqn (9.34), the horizontal shift of the water divide is expressed as

$$\Delta x_d = \frac{K_1L_2(L_1 + L_2)}{2(L_1K_2 + L_2K_1)} = \frac{(1 + L_2/L_1)L_2}{2(K_2/K_1 + L_2/L_1)} \quad (9.36)$$

The above equation indicates that after reclamation, the water divide will move to the right (Figure 9.10). This is because flow to the sea on the right-hand side is impeded after reclamation. As a result a greater proportion of the total recharge is forced to flow to the sea on the left. Equation (9.36) shows that the displacement of the water divide is controlled by the geometry of the system (L_1 and L_2) and the ratio of the aquifer hydraulic conductivities (K_1 and K_2). The displacement is especially sensitive to the reclamation length L_2. When the geometry of the system is fixed, the displacement depends on the fill hydraulic conductivity. The displacement will be small if the fill hydraulic conductivity (K_2) is much higher than that of the original aquifer (K_1).

The discharge to the sea on the left, before and after reclamation, can be calculated from Eqns (9.23) and (9.33), respectively. The change of discharge Δq [L^2 T^{-1}] at $x = 0$ is therefore

$$\Delta q = q_{1r}(0) - q_0(0) = \frac{wK_1L_2(L_1 + L_2)}{2(L_1K_2 + L_2K_1)} = w\Delta x_d \quad (9.37)$$

As expected, Eqn (9.37) shows that the increase in the seaward discharge to the left shoreline after reclamation is equal to the recharge amount between the new and old water divide.

9.3.3 Unconfined Continental Coastal Aquifer with an Interface

The equations in previous sections did not consider the interface between fresh- and saltwater. Guo and Jiao (2007) derived solutions to study the changes in both water level

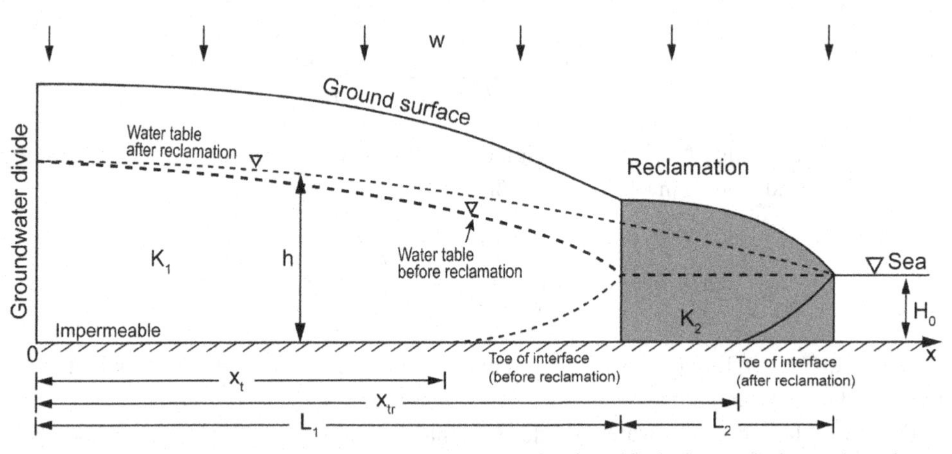

Figure 9.11 Water level and interface in a coastal unconfined aquifer before and after reclamation.

and the interface configuration in response to the land reclamation by considering the density difference between seawater and fresh groundwater. A schematic cross section representing a coastal unconfined aquifer with an interface before and after reclamation is presented in Figure 9.11.

The position of the water level and toe location before land reclamation were discussed in Section 3.4.1 (Figure 3.7). After reclamation (Figure 9.11), the new toe location of the interface becomes x_{tr}. Equations (3.17) and (3.15) are still valid for $x \geq L_1$ but with K_2 in place of K_1 and $L(= L_1 + L_2)$ in place of L_1, which leads to

$$h_r = \sqrt{\frac{w}{K_2}[(L_1 + L_2)^2 - x^2] + \frac{\rho_s}{\rho_f} H_0^2}, \qquad (L_1 \leq x \leq x_{tr}) \tag{9.38}$$

$$h_r = \sqrt{\frac{w(\rho_s - \rho_f)}{K_2 \rho_s}[(L_1 + L_2)^2 - x^2]} + H_0, \qquad (x_{tr} \leq x \leq L_1 + L_2) \tag{9.39}$$

Equation (9.38) defines h_r at $x = L_1$. Integrating Eqn (3.14) from $x = 0$ to $x = L_1$ subject to the boundary conditions

$$-K_1 h_r \frac{dh_r}{dx} = 0$$

at $x = 0$ and h_r as defined by Eqn (9.38) at $x = L_1$ leads to

$$h_r = \sqrt{w\left[\frac{1}{K_1}(L_1^2 - x^2) + \frac{1}{K_2}(2L_1 + L_2)L_2\right] + \frac{\rho_s}{\rho_f} H_0^2}, \qquad (0 \leq x \leq L_1) \tag{9.40}$$

Setting h_r equal to $\rho_s H_0/\rho_f$ (Eqn (3.13)) in either Eqn (9.38) or (9.39) at $x = x_{tr}$ yields

$$x_{tr} = \sqrt{(L_1 + L_2)^2 - \frac{K_2({\rho_s}^2 - \rho_s\rho_f)}{w{\rho_f}^2}{H_0}^2} \tag{9.41}$$

By comparing the water level after reclamation (Eqn (9.40)) to that before reclamation (Eqns (3.15) and (3.17)), the change in water level in the original aquifer ($0 \le x \le L_1$) equals

$$\Delta h = \sqrt{w\left[\frac{1}{K_1}({L_1}^2 - x^2) + \frac{1}{K_2}(2L_1 + L_2)L_2\right] + \frac{\rho_s}{\rho_f}H_0^2} - \sqrt{\frac{w}{K_1}({L_1}^2 - x^2) + \frac{\rho_s}{\rho_f}H_0^2},$$
$$(0 \le x \le x_t) \tag{9.42}$$

$$\Delta h = \sqrt{w\left[\frac{1}{K_1}({L_1}^2 - x^2) + \frac{1}{K_2}(2L_1 + L_2)L_2\right] + \frac{\rho_s}{\rho_f}H_0^2} - \sqrt{\frac{w(\rho_s - \rho_f)}{K_1\rho_s}({L_1}^2 - x^2)} - H_0,$$
$$(x_t \le x \le L_1) \tag{9.43}$$

where x_t is the interface toe before reclamation defined by Eqn (3.16). The largest water level change occurs at the original coast, and the equation to estimate this change can be obtained by setting $x = L_1$ in Eqn (9.43):

$$\Delta h = \sqrt{\frac{w}{K_2}(2L_1 + L_2)L_2] + \frac{\rho_s}{\rho_f}{H_0}^2} - H_0 \tag{9.44}$$

This equation demonstrates that the maximum water level rise increases with L_1, L_2 and the ratio of w and K_2, but is independent of K_1.

As a result of reclamation, the toe position of the interface moves seaward by a distance of $x_{tr} - x_t$, which is obtained by subtracting Eqn (3.16) from Eqn (9.41):

$$x_{tr} - x_t = \sqrt{(L_1 + L_2)^2 - \frac{K_2({\rho_s}^2 - \rho_s\rho_f)}{w{\rho_f}^2}{H_0}^2} - \sqrt{{L_1}^2 - \frac{K_1({\rho_s}^2 - \rho_s\rho_f)}{w{\rho_f}^2}{H_0}^2} \tag{9.45}$$

The above equation shows that the value $x_{tr} - x_t$ is always positive for practical problems, i.e. the toe is pushed seaward due to reclamation.

Using a hypothetical example, Guo and Jiao (2007) discussed the effect of the reclamation on the interface and the water level. They concluded that the water level change calculated using either the equations with (e.g. Eqn (9.44)) and without (e.g. Eqn (9.16)) the interface is small, suggesting that the presence of a seawater wedge has negligible effect on the water level change. The displacement of the interface toe increases with L_2 and decreases with K_2 and its value is close to L_2 (Eqn 9.45). The interface length increases roughly linearly with K_2. Reclamation increases both recharge and freshwater storage by extending the aquifer seaward. If K_2 decreases, the freshwater storage increases significantly because lowering K_2 leads to higher water levels and a shorter interface penetration.

9.3.4 Unconfined Strip Island Aquifer with an Interface

Guo and Jiao (2008a) also studied how the groundwater level and the interface are affected by reclamation in a strip island, considering a freshwater lens in a semi-infinite domain (no impermeable layer) or a lens bounded by an impermeable layer. Only the case without an impermeable layer is presented here (Figure 9.12). For the case of a lens bounded at the bottom, the readers are referred to Guo and Jiao (2008a) or Guo (2008).

It is assumed that the strip island has a freshwater lens between the water table and the interface at the bottom (Figure 9.12). As before, the original width of the island is L_1, the aquifer has a uniform hydraulic conductivity K_1 and w is the rainfall recharge per unit area. The flow below the interface is assumed to be negligible. The origin of the horizontal axis ($x = 0$) is set at the left coastline. Sea level is set as the datum of the water level h_f in the freshwater lens. The distance from the sea level to the depth of the interface is denoted as z.

Analytical expressions for the water level and interface position in a strip island have been presented in Section 3.4.4. Here it is assumed that land reclamation occurs on the right-hand side of the island (Figure 9.12). Assuming that the water divide shifts to the right, and that its new location becomes x_{dr}, the mathematical model can be described as

$$w(x - x_{dr}) = -K_1(1 + \delta)h_{1f}\frac{dh_{1f}}{dx}, (0 \leq x \leq L_1) \tag{9.46}$$

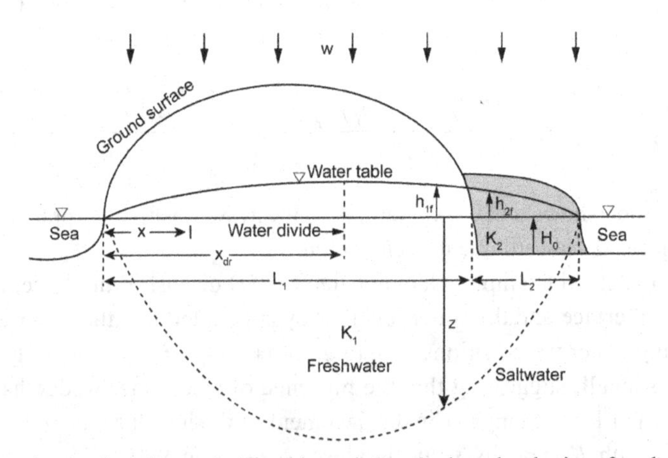

Figure 9.12 Freshwater lens in a strip island bounded at the base by the interface before and after reclamation.

$$w(x - x_{dr}) = -K_2(h_{2f} + H_0)\frac{dh_{2f}}{dx} - K_1(\delta h_{2f} - H_0)\frac{dh_{2f}}{dx}, (L_1 \le x \le L_1 + L_2) \quad (9.47)$$

subject to boundary conditions

$$h_{1f}(0) = 0 \quad (9.48)$$

$$h_{2f}(L_1 + L_2) = 0 \quad (9.49)$$

and the continuity conditions for water head and flux at $x = L_1$:

$$h_{1f}(L_1) = h_{2f}(L_1) \quad (9.50)$$

$$K_1(1+\delta)h_{1f}\frac{dh_{1f}}{dx} = K_2(h_{2f} + H_0)\frac{dh_{2f}}{dx} + K_1(\delta h_{2f} - H_0)\frac{dh_{2f}}{dx} \quad (9.51)$$

where h_{1f} and h_{2f} are the water level in the freshwater lens in the original aquifer and the reclaimed area, respectively. The solution of Eqn (9.46) subject to Eqn (9.48) is

$$h_{1f} = \sqrt{\frac{w}{K_1(1+\delta)}(2x_{dr}x - x^2)} \quad (9.52)$$

and the solution of Eqn (9.47) subject to Eqn (9.49) is

$$\begin{aligned} &\frac{K_2 + \delta K_1}{2}{h_{2f}}^2 + H_0(K_2 - K_1)h_{2f} \\ &= wx_{dr}x - \frac{w}{2}x^2 + \frac{w}{2}(L_1 + L_2)^2 - wx_{dr}(L_1 + L_2) \qquad (L_1 < x \le L_1 + L_2) \end{aligned} \quad (9.53)$$

By combining Eqns (9.50), (9.51), (9.52) and (9.53), the following equations are obtained:

$$\kappa\eta^2 + \nu\eta - (L_1L_2 + {L_1}^2) = 0 \quad (9.54)$$

where

$$\eta = \sqrt{\frac{w}{K_1(1+\delta)}(2x_{dr}L_1 - {L_1}^2)} \quad (9.55)$$

$$\kappa = \frac{L_1K_2 + L_2K_1 + K_1(L_1 + L_2)\delta}{wL_2} \quad (9.56)$$

$$\nu = 2\frac{H_0L_1(K_2 - K_1)}{wL_2} \quad (9.57)$$

The solution to Eqn (9.54) is

$$x_{dr} = \frac{L_1}{2} + \frac{K_1(1+\delta)\eta^2}{2wL_1} \tag{9.58}$$

$$\eta = \frac{\sqrt{\nu^2 + 4\kappa(L_1L_2 + L_1{}^2)} - \nu}{2\kappa} \tag{9.59}$$

The solutions for water level and the interface after reclamation are obtained by combining Eqns (9.52), (9.53), (9.58) and (9.59). As can be seen, the seawater depth H_0 does not appear in the final equations, suggesting that the sea level has no influence on the shape and the subsequent the volume of freshwater. However, as discussed by Guo (2008), in the case of an island sitting on an impermeable bottom, the volume of freshwater lens can be significantly influenced by the sea level.

From Eqn (9.58), the shift of water divide to the right is

$$\Delta x_d = x_{dr} - \frac{L_1}{2} = \frac{K_1(1+\delta)\eta^2}{2wL_1} \tag{9.60}$$

The increased groundwater discharge to the coast on the right is

$$w\Delta x_d = \frac{K_1(1+\delta)\eta^2}{2L_1} \tag{9.61}$$

9.3.5 Transient Flow Induced by Land Reclamation

The adjustment of the water level in a coastal aquifer in response to reclamation takes on the order of years to decades. To forecast the gradual change of the water level after reclamation and estimate the time needed for the aquifer to achieve a new equilibrium, transient analytical solutions are required. Hu et al. (2008) derived a transient solution to describe the change of the groundwater system in both continental and strip island aquifers. In modern reclamation projects, the time to complete the works is short compared to the time during which the resultant unsteady change of the flow system occurs (Plant et al., 1998). Therefore Hu et al. (2008) assumed that the land reclamation is completed instantaneously. Transient solutions are difficult to obtain because the governing equation is non-linear and the hydraulic transmissivity is a function of time and space. To linearise the equation, Hu et al. (2008) used an arithmetically averaged transmissivity.

Figure 9.13 shows the spatial distribution of the water level for different times since reclamation for an unconfined continental coastal aquifer (Figure 9.7) with a reclamation length and aquifer parameters typical for Hong Kong. The distance between the water divide and the coastline was set to 1000 m and the reclamation length is 500 m. The hydraulic conductivity of the original aquifer $K_1 = 0.1$ m d^{-1} and the storativity of both the fill materials and the background aquifer are 0.1. The average sea level is 10 m and the recharge is 0.0006 m d^{-1}.

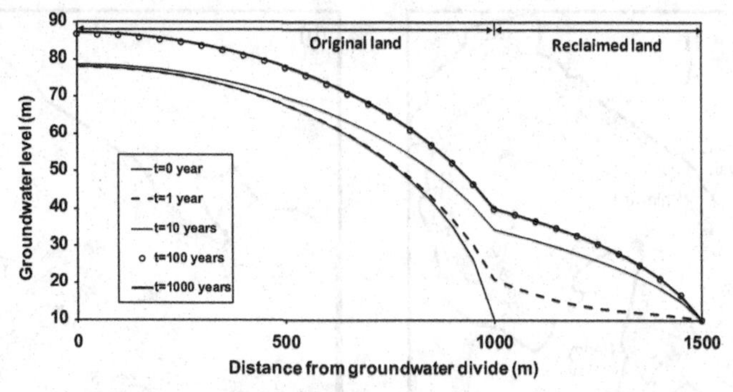

Figure 9.13 Change of water level with distance at different times for $L_2 = 500$ m and $K_2 = 0.5$ m d^{-1} (modified from Hu et al., 2008).

As Figure 9.13 shows the water level changes quickly within the reclaimed area but relatively little in the original aquifer shortly after the reclamation. With time though the water level increase slowly propagates into the original aquifer. The change in water level within the reclamation area becomes insignificant after about 10 years but the water level rise in the aquifer, particularly towards the water divide, remains noticeable until much later. After 100 years though, no significant changes in water level occur anymore.

As demonstrated by this example, water level changes may take tens of years. An implication is that once the impacts of the water level rise start to become apparent, people will not relate them to the long-forgotten reclamation (Jiao, 2000).

9.3.6 Limitations of Analytical Solutions

Of course, the analytical studies here are based on various assumptions. For example, the fill hydraulic conductivity is considered to be uniform, but in reality it varies both in time and space. This is because the fill materials are inevitably mixed with the mud at the original sea bottom, and both the fill and the mud experience gradual consolidation. Consequently the hydraulic conductivity in the reclaimed site decreases slowly with time. Another assumption is that the model does not take into account any limitations by the topography. New seepage zones or springs may develop after reclamation when the water table increases to intersect the land surface. In Hong Kong, this is widely noticed at the interface between the reclaimed site and the natural aquifer. The analytical solutions thus overestimate the build-up of the water level because they cannot include the effect of seepage to the ground surface.

9.4 Numerical Case Study of Groundwater Flow and Contaminant Transport after Land Reclamation

It is clear that analytical solutions can provide insight into the impact of land creation on groundwater flow and a first-order assessment of the water level rise, but their strict

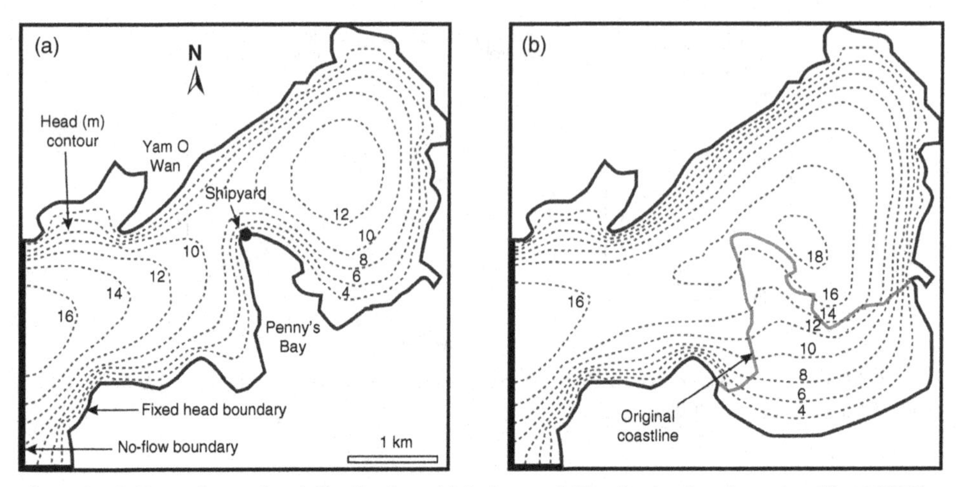

Figure 9.14 Groundwater head distributions (a) before and (b) after land reclamation (Jiao, 2002).

assumptions limit their use for real coastal aquifer problems. In these cases the problem is best studied by numerical models (Section 2.5). Usually a model is constructed for the original aquifer before reclamation and then again for the original aquifer plus the reclaimed area so that the effect of the reclamation on the groundwater system in the original aquifer can be examined. Groundwater case studies related to land reclamation are very limited (Jiao, 2002; Hu and Jiao, 2010). A case study on the change of flow system in Penny's Bay at Lantau Island, Hong Kong (Jiao, 2002), is presented in this section.

A land area of about 2.8 km^2 was reclaimed in Penny's Bay at Lantau Island for building a large theme park. The volume of fill material placed in the bay was over 70 million m^3. The fill materials consisted of fluvial sand imported from a river delta, and marine sand dredged from deeper sea, soil originated from igneous rocks, public landfill and construction waste (Jiao, 2002). The final elevation of the reclaimed land reached about 10 m above mean sea level. The natural Quaternary deposits underlying the fill consist of marine mud, firm to stiff silty clay, and sand and gravel. The bedrock around Penny's Bay is dominated by volcanic rock.

A shipyard was situated on the north-western shore of the bay (Figure 9.14a) and had been operated here since the 1960s. The soil below the yard was contaminated by organic solvents, heavy metals and oils. The site was excavated to remove the contaminated soil, but the spreading of legacy chemicals into the theme park remained a concern to the public and the government. A numerical modelling study using the code FEMWATER (Lin et al., 1997) was conducted to investigate the changes of the flow system caused by reclamation in Penny's Bay, and to estimate the possible migration pathway of the contaminants (Jiao, 2002).

Figure 9.14 presents the water table distributions before and after reclamation. Land reclamation led to an increase in water table below the entire land tongue, to the extent that it could become a problem for slope stability. The changes in the flow directions due to

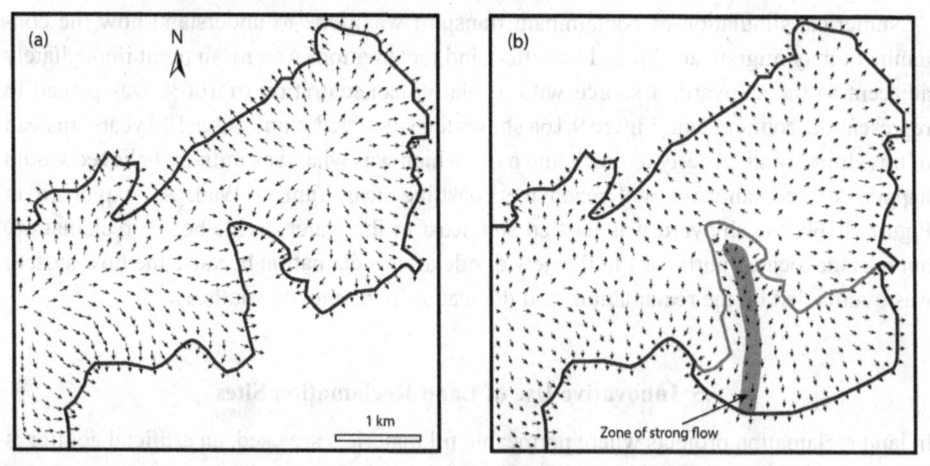

Figure 9.15 Groundwater velocity distributions in the land area around Penny's Bay (a) before and (b) after land reclamation. The zone of strong convergent flow is indicated by the shaded area (Jiao, 2002).

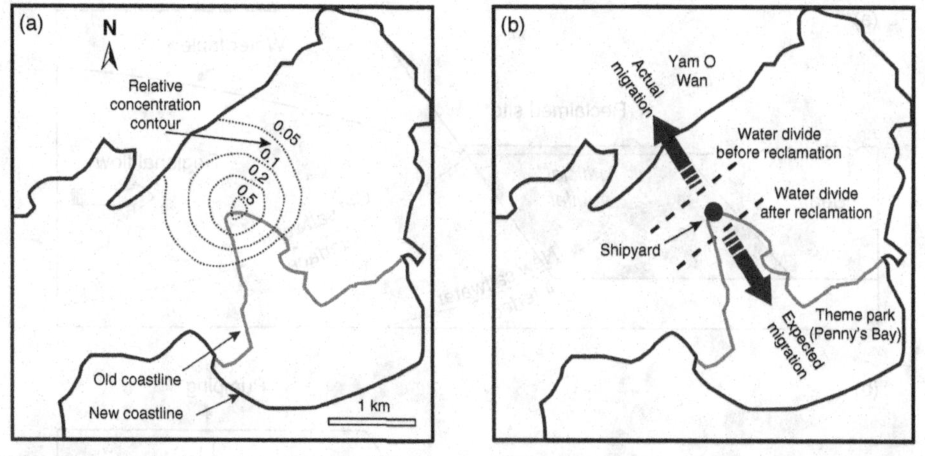

Figure 9.16 (a) Simulated contours of relative contaminant concentration 100 years after reclamation and (b) water divide shift in response to land reclamation and its effect on contaminant migration (Jiao, 2002).

reclamation are shown in Figure 9.15. The flow patterns are affected both near and beyond the bay. Near the centreline of the bay, there is a zone of strong convergent flow. In this example, the model results indicate that high-permeability fill materials in the strong convergent flow zone, which provides fast groundwater drainage to the sea, are most effective in reducing the water table rise. This shows how numerical models may be useful for designing measures that can be taken to minimise the impact of reclamation on the flow system.

Numerical simulation of contaminant transport was done to understand how the contaminant may migrate and spread out after land reclamation. At a mesh point immediately adjacent to the shipyard, a source with a relative concentration of 100% was placed to represent the contaminant. Figure 9.16a shows the simulated plume after 100 years. Instead of travelling south-easterly to the theme park, which was what was initially believed would happen, the contaminant will reach the coastline near Yam O Wan. As explained in Figure 9.16b, the shipyard was located southeast to the water divide before reclamation, but became located northwest to the new divide after reclamation because the flow system was overhauled by the reclamation, and the water divide moved southeast.

9.5 Innovative Use of Land Reclamation Sites

In land reclamation projects where permeable fill materials are used, an artificial aquifer is created (Figure 9.17a). The reclaimed land may thus benefit groundwater resource management and aquifer exploitation if the materials of the fills are chosen carefully and the reclamation project is designed properly. The water in the reclaimed land is initially saline

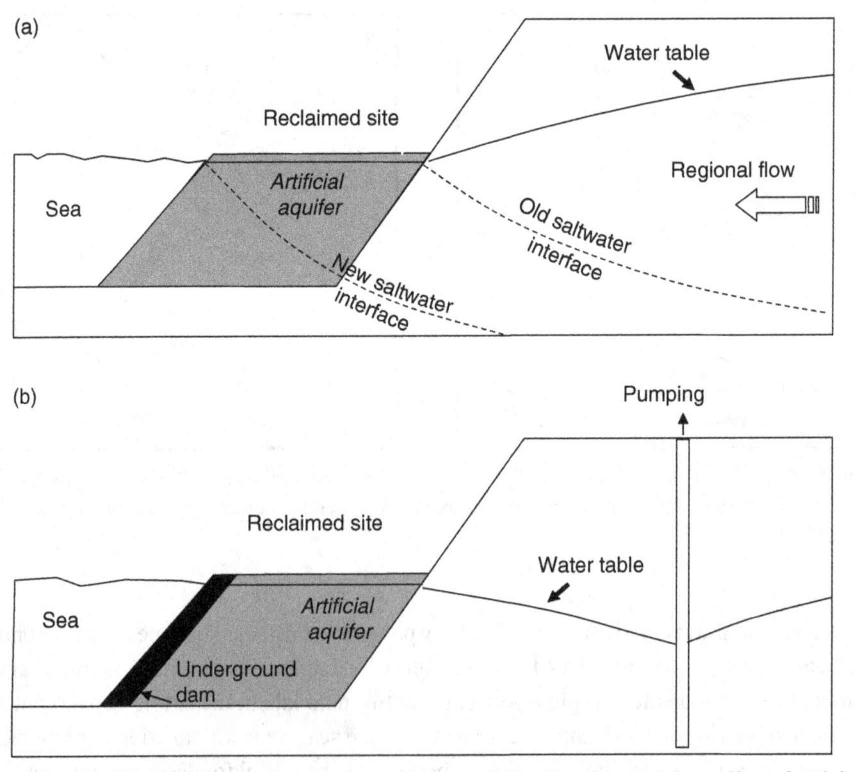

Figure 9.17 Possible (a) additional groundwater in the reclaimed land and (b) dam effect of reclaimed site.

but will gradually become fresh because the interface will move seawards when the aquifer receives freshwater recharge from above or from the adjacent natural aquifer.

In some coastal areas, underground dams are constructed to form a underground reservoir, or prevent fresh groundwater flowing to the sea or seawater from intruding into the aquifer (Shi and Jiao, 2014) (Section 12.4.4). If the materials are chosen properly, a landfilled reclamation site can provide additional land and, at the same time, may function as a dam to cut off the hydraulic connection between groundwater and seawater. Landfill materials with low hydraulic conductivity near the new shoreline may thus reduce the possibility of seawater intrusion (Figure 9.17b).

9.6 Guidelines of Hydrogeological Assessment of Land Reclamation

Environmental assessments of land reclamation projects do not always give due consideration to groundwater. The discussion in this chapter shows, however, that a hydrogeological study is crucial so that the project can be designed carefully. By choosing the fill materials, locations to dump the fill with different permeability, and the scale of reclamation, modifications to the flow in the aquifers can be minimised, and the subsequent engineering and environmental impacts can be better mitigated. These impacts include the reduced capacity of rain infiltration, flooding, stability of foundations and slopes, and impacts on the coastal marine environment (Jiao, 2000).

Before land reclamation is started, a hydrogeological survey should be carried out in the original groundwater catchment adjacent to the potential reclaimed land to provide baseline information. This includes hydrogeological mapping to understand the hydraulic conductivity distribution, water level, groundwater chemistry, spring locations, as well as the distribution of confined and unconfined aquifers. Monitoring wells should be installed and the water level should be measured over a period of at least one hydrological year to establish the dynamic response of the system to recharge, pumping and sea level variations. The baseline information is critical for assessing the change of the system in response to the reclamation.

In addition to surveys on land, the hydrogeology in the marine area to be reclaimed should be studied. For an unconfined coastal aquifer, groundwater discharge may primarily occur in a small zone along the coastline. If there is a confined aquifer, the offshore outlet of the aquifer or any springs should be mapped. If there is a major submarine spring in the sea, provisions should be made within the reclaimed land area so the groundwater can continue to discharge through the spring outlet, so that the spring water can potentially be used and the unnecessary water level increase by blocking the spring can be avoided.

With the hydrogeological information collected from the land and offshore areas, a numerical model should be constructed to predict the possible modification of the groundwater flow due to the land reclamation, including the water level rise, changes in the groundwater discharge to the reclaimed areas and other areas. The model area should include the entire groundwater catchment area behind the reclaimed land. The model should consider different land reclamation scenarios, considering factors such as the scale of reclamation, fill materials of different hydraulic conductivities, the distribution of the materials and whether the marine mud is being dredged or not. As previously discussed, the reclamation scale and the hydraulic conductivity of fill materials are two primary factors that control the modification of the flow system. The seepage velocity should be analysed too. The fast flow or convergent flow areas should be identified. Very permeable materials should be placed in the areas with strong groundwater flow so that groundwater can be discharged to the sea and thus help alleviate the effect of water level build-up.

The slope stability in the vicinity around the reclaimed area should be assessed because the stability is sensitive to water level increase. In low-lying regions, groundwater may be forced to flow to the land surface and enhance the flooding possibility after rainstorms. Monitoring wells should be installed in the reclaimed site. Both the wells in the reclaimed land and those installed in the original land before reclamation must be monitored for water level and water chemistry to identify any unanticipated water level and chemistry variations.

Seawater initially contained in the soil of the reclaimed area is gradually replaced by fresh groundwater; usually the fill materials and underlying the marine mud have distinctly different chemical characteristics, but this varies from site to site. Consequently there will be physiochemical reactions between the different waters and materials. The groundwater chemistry and the chemistry of the fills and the marine mud should be studied to assess potential chemical reaction among the groundwater, seawater, fill and marine mud (Section 5.7.8). Reclamation will affect groundwater discharge water characteristics to the sea and this may be a problem for some organisms. If acid sulphate soils may be formed as a result of reclamation, their impact on coastal ecology should be assessed.

10
Sea Level Change and Coastal Aquifers

10.1 Introduction

Sea level rise can affect coastal aquifer systems in a number of ways (Oude Essink, 1996; Barlow, 2003). The interface between fresh- and saltwater will be pushed further inland, leading to exacerbated seawater intrusion. There will be further upstream migration of seawater along river channels in coastal estuaries, which then impacts on the salinity of the adjacent groundwater. Aquifers in low-lying areas become more likely to be inundated by seawater, leading to seawater intrusion both laterally and vertically. Sea level forms the hydrological base level of a coastal aquifer, so a sea level rise will elevate the groundwater levels. On the other hand, if the rise of sea level leads to land loss, it will reduce the size of the area receiving freshwater recharge (Barlow, 2003); which could abate the trend of rising groundwater levels.

These differing effects and their interactions mean that predicting the impact of sea level rise on coastal groundwater systems is not straightforward. Moreover, sea level rise caused by global warming is likely to be accompanied by other climatological changes such as changes in rainfall and evaporation, which will simultaneously exert an impact on coastal aquifers. It should also be emphasised that the effect of sea level change on nearshore groundwater resources should be assessed together with human activities such as groundwater abstraction and land reclamation. Human activities usually dominate the seawater intrusion process in coastal groundwater systems. Sea level change is gradual and occurs at a scale of at least decades, but overexploitation of groundwater can cause seawater intrusion within a few months to years. In coastal areas with extensive groundwater pumping, the influence of sea level rise may be of secondary importance.

The discussion in this chapter primarily deals with the change to coastal aquifer systems caused by sea level rise on a decadal timescale, such as those expected to accompany future climate change. The impact of a sudden sea level rise or catastrophic flooding induced by tsunami or storm surges is discussed in Section 6.2.5.2, while Chapter 8 focuses on the effects of past, geological, timescale sea level change.

10.2 Climate Change and Sea Level Rise

Climate change is driven by various factors such as oceanic circulation, solar radiation, volcanic eruptions and plate tectonics, and, since the industrial revolution, human activities (Murtaza et al., 2015). Human activities such as deforestation and fossil fuel burning cause a rise in atmospheric greenhouse gas levels, mainly carbon dioxide (CO_2) but also methane (CH_4) and dinitrogen monoxide (N_2O), which enhance the heat retention by the atmosphere and lead to global warming. Climatological factors such as precipitation, temperature, and evapotranspiration affect groundwater recharge and discharge, which in turn impact water level, storage and the quality of groundwater. In coastal areas, climate change also impacts aquifer systems by changing the sea level. There are many publications on the influence of climate variations on inland groundwater systems (Eckhardt and Ulbrich, 2003; Bloomfield et al., 2006; Green et al., 2011), but there has been less investigation of the impact of these variations on coastal groundwater (Werner and Simmons, 2009; Nicholls and Cazenave, 2010). Ketabchi et al. (2016) provided a review of papers on impacts of sea level rise on seawater intrusion in coastal aquifers, which were mainly published after 2009.

The processes that controlled past sea level variations were discussed in Section 8.2, and those that are responsible for sea level change over the next decades are graphically summarised in Figure 10.1. Figure 10.2 shows the relative contributions of various processes to projected global sea level rise in the twenty-first century. The largest contributors will be thermal expansion of ocean water and the melting of land ice (Church et al., 2013). Due to the poorly understood response of the Greenland and Antarctic ice sheets to global warming, the contribution of land ice melting to global sea level rise is associated with the largest uncertainty. Uncertainty further stems from the future development of atmospheric CO_2 concentrations and the radiative forcing. The range in predicted global mean sea level

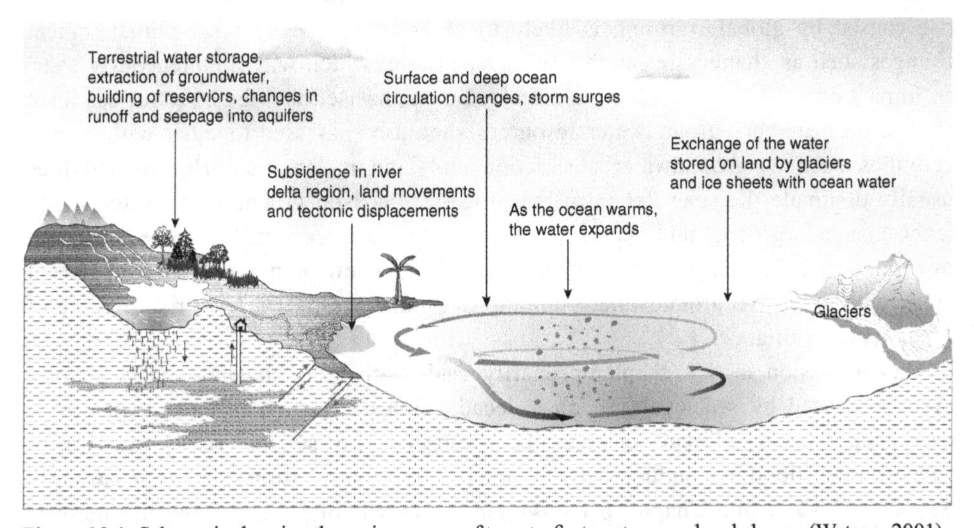

Figure 10.1 Schematic showing the major causes of twenty-first century sea level change (Watson, 2001).

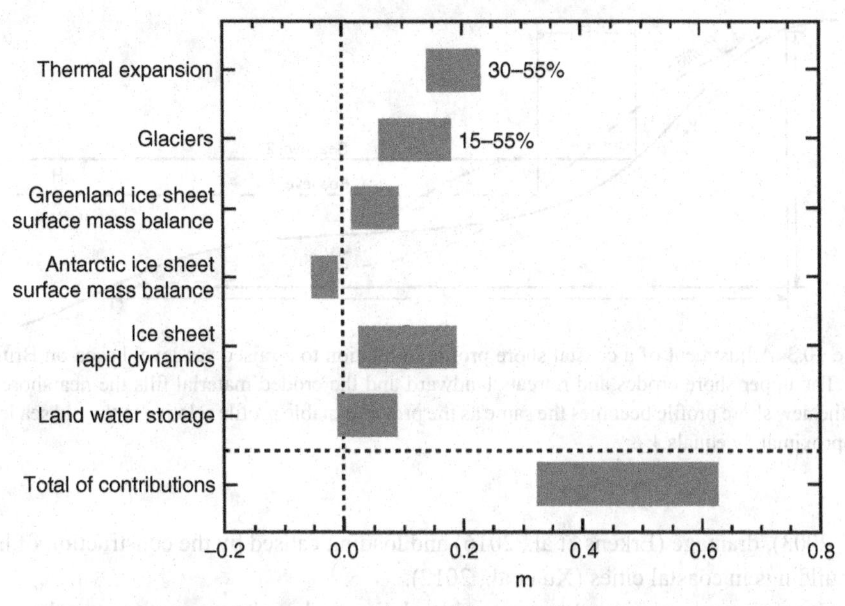

Figure 10.2 Contribution of various processes to projected global sea level rise in the twenty-first century (Church et al., 2013).

for the year 2100 is 0.28 to 0.98 m (Church et al., 2013), but this could be more by as much as a metre if the Antarctic ice cap enters a state of runaway meltdown (Feldmann and Levermann, 2015; DeConto and Pollard, 2016). A relatively new insight is that groundwater storage depletion by abstraction may also contribute to sea level rise to a certain degree (Konikow, 2011b; Wada et al., 2012), as will be discussed in Section 10.6.

It is expected that sea level will rise in 95% of the ocean area, the exception being areas near large ice masses where sea levels will fall when their melting causes them to exert less gravitational pull. About 70% of the world's coastlines will experience a change in sea level that will be within 20% of the global average, but locally important deviations will occur depending on other processes that control relative sea level such as isostatic rebound or tectonic movement (Church et al., 2013).

10.3 Impact of Sea Level Rise on Coastal Zones

10.3.1 Shoreline Retreat and Flooding

Accelerated sea level rise during the twentieth century (Kopp et al., 2016) has led to a higher incidence of minor flooding events along the coast of the USA (Sweet et al., 2014), and rising temperatures lead to a greater risk of cyclone-related storm surges (Grinsted et al., 2013). It is expected that millions more people will experience more floods as a result of the rise in sea level (IPCC, 2007). Land subsidence exacerbates damage from coastal flooding. Land subsidence can be caused by over pumping of coastal aquifers (Chen

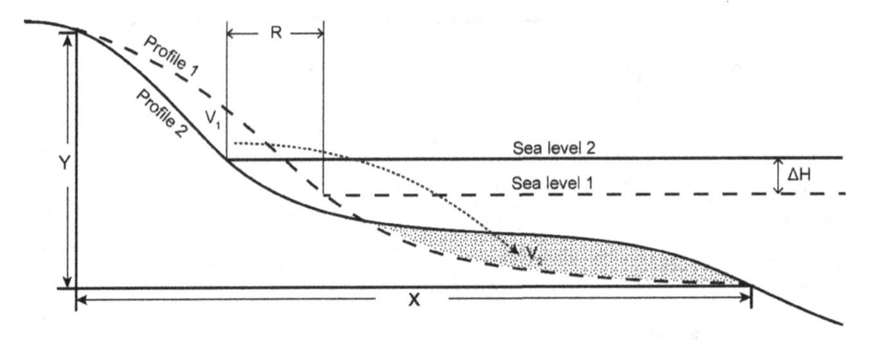

Figure 10.3 Adjustment of a coastal shore profile in relation to a raised sea level based on Bruun's Rule. The upper shore erodes and retreats landward and the eroded material fills the nearshore sea until the new shore profile becomes the same as the previous stable profile relative to the old sea level. V_1 approximately equals V_2.

et al., 2003), drainage (Erkens et al., 2016) and loading caused by the construction of high rise buildings in coastal cities (Xu et al., 2012).

Sea level rise not only inundates coastal land directly but also promotes coastal erosion and landward shoreline retreat (Wong et al., 2014). Natural systems are likely to be resilient to some degree and respond in a dynamic fashion to sea level rise (Lentz et al., 2016). Evidence is also emerging that low-lying coral atolls, which are considered to be among the most vulnerable environments to sea level rise, have a natural ability to adapt to sea level rise (Webb and Kench, 2010; Kench et al., 2014). However, where human activities interfere with natural processes, or where infrastructure is in place, the ability of the natural system to adapt may be compromised.

Coastal erosion driven by processes such as wave action usually lead to a retreat of the coastline and land loss several times greater than the area that was initially inundated (Oude Essink, 1996). Adjustments to the coastal topography profile do not just occur at the shoreface above the sea surface but also to a water depth of over 10 m (Cooper and Pilkey, 2004). Unless it is due to catastrophic events like earthquakes (Sherrod et al., 2000), sea level change and the adjustment of coastal processes to it, tend to be gradual.

10.3.1.1 Bruun's Rule

Insight into the effects of sea level rise on the coastal profile can be gained by analysing subsequent steady states. For example, Bruun's rule (Bruun, 1962) predicts how a sandy beach profile will adjust to a new sea level (Figure 10.3). Following sea level rise, the sea initially inundates part of the seashore. The coastal profile, however, has a tendency to retain its shape relative to the sea level. If there is no external supply of sediment, wave erosion of the beach provides the materials to raise the foreshore sea bottom. The volume of the eroded material V_1 [L^3] equals the volume V_2 of the material deposited over the seabed (Ramsay and Bell, 2008). It is for this reason that this usually leads to a landward shifting of the backshore after the sea level rise.

Bruun's rule states that the landward shift of the coast equals (Bruun, 1962)

$$R = \frac{\Delta H \times X}{Y} \tag{10.1}$$

where R [L] is the landward shift of the coast, ΔH [L]is the sea level increase and X and Y [L] are the horizontal and vertical dimensions of the active profile, respectively (Saye and Pye, 2007). For most sandy coasts X is much greater than Y, and Bruun's rule predicts that the retreat of the shoreline ranges from 50 to 100 times the sea level increase (Addo et al., 2011). Bruun's rule is often used because it is easy to apply, but it is subject to strict assumptions. For example, longshore sand transport and deposition are ignored, and the rule only applies to the case where wave direction is dominantly transverse to the coast (Selivanov, 1993; Oude Essink, 1996). The actual shoreface changes can be much more complex and the shore profile may not keep the same configuration as before the sea level rise. Instead the shoreface responds to sea level rise in many different ways, which depend on grain size of the beach material, sediment supply, coastal geology, wave conditions, wind- and tide-induced currents, and the interactions between bottom currents and waves (Cooper and Pilkey, 2004). Nevertheless, Eqn (10.1) shows that the shoreline retreat can add to the landward migration of the coastline, and thus of the interface.

Most investigations on the influence of sea level changes on nearshore aquifers assume that the coastline remains unchanged (Carneiro et al., 2010; Guha and Panday, 2012). There are only a few studies that considered the change of shorelines (Oude Essink, 1996; Melloul and Collin, 2006; Yechieli et al., 2010). The models that ignore these coastal topographic processes may underestimate the severity of the seawater intrusion induced by sea level increase. For a comprehensive study on the interaction among various processes such as currents, waves, morphology, sediment transport, water quality and ecology, more complicated numerical codes exist. Such models can provide key insights into the behaviour of the coastal morphology. For example, Melloul and Collin (2006) showed that, for a 1 m increase in sea level, a steep shoreline would retreat approximately 100 m if the subsurface consists of sand, but only about 60 m in the case of more resistive material. Reliable long-term prediction of the impact of sea level changes on coastal aquifer systems requires interdisciplinary collaboration between coastal hydrogeologists and geomorphologists, combining investigations of sea level change, topography and bathymetry adjustment, as well as seawater intrusion.

10.3.2 Upstream Migration of a Saltwater Wedge in Rivers Connected to the Sea

Just like in a coastal aquifer (Figure 1.3), there exists a dynamic saltwater wedge in rivers and estuaries due to the density difference between seawater and freshwater (Figure 10.4). The inland intrusion of this saltwater wedge is controlled by the river discharge and gradient. A high discharge pushes the saltwater and freshwater boundary towards the river mouth. As the discharge of rivers varies seasonally, so will the extent of the wedge. The term 'salty

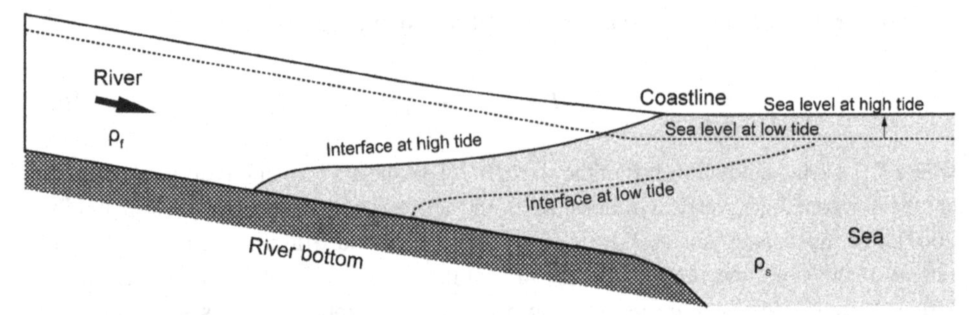

Figure 10.4 Interface in estuaries and rivers before and after sea level rise (modified from Oude Essink, 1996).

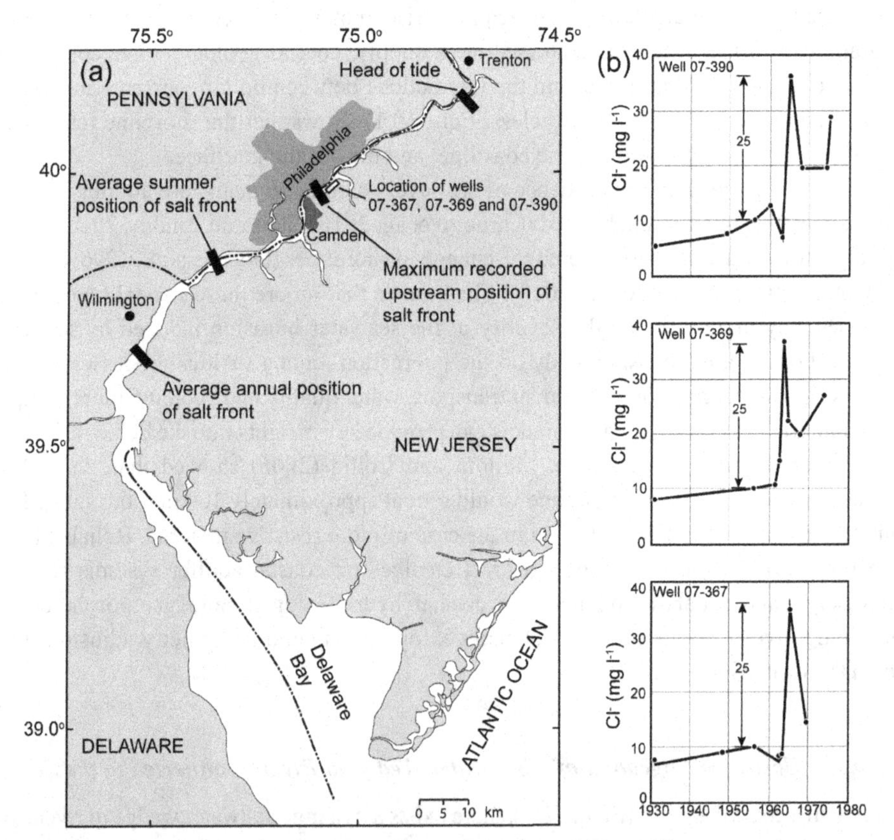

Figure 10.5 (a) Salt front location in the Delaware River, New Jersey. Shown are average annual, average summer and maximum recorded upstream positions. (b) Short-term chloride concentration increase in wells influenced by a salty tide (Barlow, 2003).

tide' refers to the phenomenon that seawater migrates upstream along the river channel when the sea level is high, especially during astronomical high tides, and the river discharge is low. In the Pearl River Delta, China, the seawater can migrate inland along the channel up to 50 km (Mok et al., 2011). Along the Gambia river in Western Africa, the seawater is known to migrate up to over 230 km upstream from the sea (Oude Essink, 1996). Salty tides can cause shortages of fresh surface water supply as the surface water quality becomes unusable (Barlow, 2003; Lu et al., 2007), and may furthermore cause saltwater intrusion along rivers, especially where groundwater is abstracted to substitute for the river water in the dry season.

As an example, Figure 10.5a shows the locations of the salt front and head of tide (i.e. the most upstream location where a river is influenced by tidal fluctuations) in the Delaware River and Estuary in New Jersey, USA. Indicated are the average annual, average summer and maximum recorded upstream positions (Barlow, 2003). The change of the chloride concentration with time in the wells close to the river shown in Figure 10.5b indicates that the wells may have withdrawn some saltwater from the river reaches that were temporarily saline due to seawater intrusion, as indicated by the higher than normal chloride concentrations in the well water during the drought period (Barlow, 2003).

If sea level rises, salty tide in rivers and estuaries will become more frequent and migrate further inland. Salty tides can further be enhanced by human activities such as dam construction or water use for irrigation in the upper river reaches, or sand dredging in the delta regions which lead to sediment depletion and river channel incision.

10.4 Impact of Sea Level Rise on Coastal Groundwater Systems

Barlow (2003) contended that sea level rise during the past decades may have caused an increase in coastal groundwater levels at some locations. As an example he cited a study by McCobb and Weiskel (2003) who reported a mean groundwater level rise of 2.1 mm yr^{-1} between 1950 and 2000 in a well installed in Cape Cod, Massachusetts, which is very similar to the relative sea level increase of 2.5 mm yr^{-1} observed at the Boston tide gauge in the period between 1921 and 2000. In general though, the magnitude of sea level change is small compared to water level changes brought about by other factors such as recharge variability, pumping and land drainage, and it is very difficult to separate the contribution of individual processes, especially since multi-decadal time series are not available for many coastal areas.

The impact of sea level rise on coastal aquifers depends on their characteristics such as the degree of confinement, hydraulic properties and recharge conditions. The slope of the shoreface also plays an important role (Figure 10.6). If the shoreface is vertical and does not retreat due to coastal processes, there is no shoreline movement or direct inundation. Sea level rise can only impact groundwater by raising the head at the vertical boundary. If the shoreface has a slope, and assuming this slope will remain fixed, there will be shoreline retreat and seawater will inundate the land surface. Seawater can intrude into

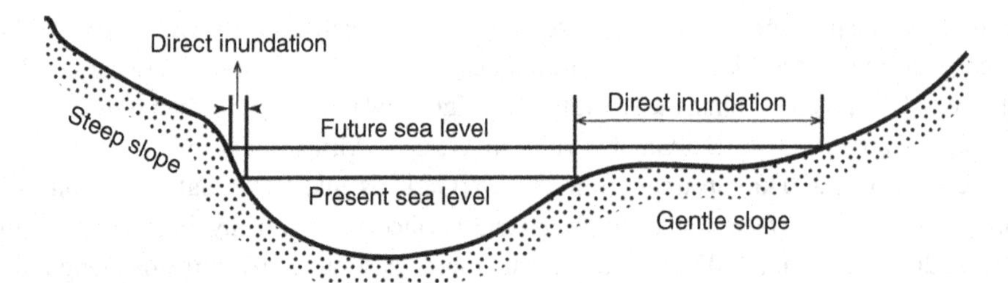

Figure 10.6 Schematic cross section of land inundation by sea level rise. The shoreline movement greatly depends on the land slope (modified from Kana et al., 1984).

the aquifer laterally and also from the land surface. As shown in Figure 10.6, the gentler the slope, the greater the shoreline retreat.

To illustrate the impact of sea level rise on the steady state interface position, a total of four analytical solutions are discussed in this chapter. The first is for the unconfined continental coastal aquifer configuration with a vertical shoreface where no inundation occurs. The second is for the same configuration but with a sloping land surface which becomes inundated due to sea level rise. The third is for an unconfined island aquifer without inundation. The fourth is a confined continental coastal aquifer configuration without inundation.

It is expected that an unconfined aquifer with a gently sloping surface is most sensitive to sea level rise because there will be transgression over the sloping land surface (e.g. Ataie-Ashtiani et al., 2013; Chesnaux, 2015; Morgan and Werner, 2016). The basic assumptions for all cases are that both the Ghijben–Herzberg principle and the Dupuit assumption apply (Section 3.3.1), and there is no change in the shape of the shoreface. Under these assumptions, the equations for the groundwater level and toe location of the interface after sea level rise can be easily obtained by modifying the corresponding analytical equations presented in Chapter 3.

10.4.1 Unconfined Continental Coastal Aquifer without Inundation

The conceptual model of an unconfined continental coastal aquifer is shown in Figure 10.7. It is assumed that the shoreface is vertical, so there is no inundation after a sea level rise of magnitude ΔH [L]. The location of the new toe of the interface is x_{tn} [L].

The equations for water level and saltwater toe location for this system were derived in Section 3.4.1 (Eqns (3.15)–(3.17)). Equivalent equations for the new sea level can be obtained simply by replacing H_0 by $H_0+\Delta H$ in Eqns (3.15), (3.17) and (3.16):

$$h = \sqrt{\frac{w(\rho_s - \rho_f)}{K_1\rho_s}(L_1{}^2 - x^2) + (H_0 + \Delta H)}, (x_{tn} \leq x \leq L_1) \tag{10.2}$$

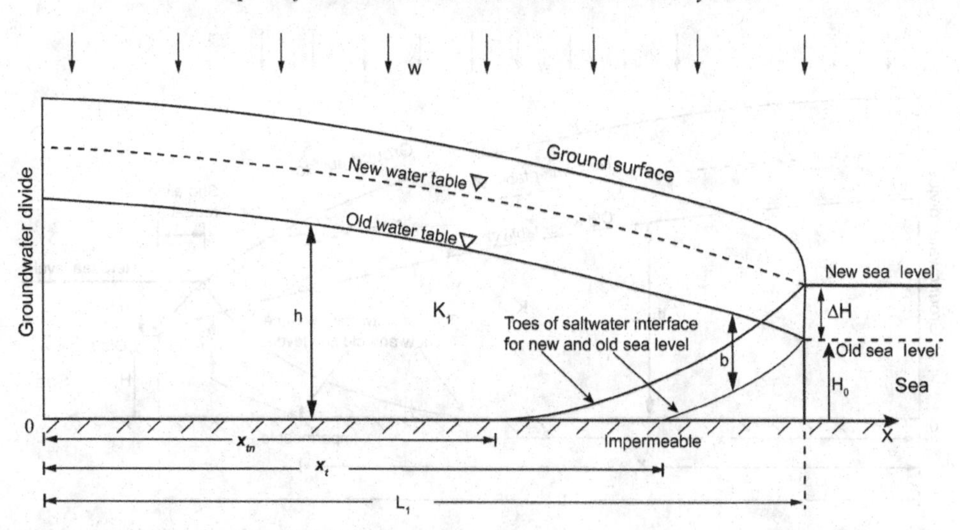

Figure 10.7 Water level and interface in an unconfined continental coastal aquifer with a vertical shoreface after sea level rise without inundation.

$$h = \sqrt{\frac{w}{K_1}(L_1{}^2 - x^2) + \frac{\rho_s}{\rho_f}(H_0 + \Delta H)^2}, (0 \le x \le x_{tn}) \tag{10.3}$$

$$x_{tn} = \sqrt{L_1{}^2 - \frac{K_1(\rho_s{}^2 - \rho_s\rho_f)}{w\rho_f{}^2}(H_0 + \Delta H)^2} \tag{10.4}$$

Eqn (10.2) shows that the water level between x_{tn} and L_1 is simply increased by the same amount as the sea level increases, which is similar to the case of a confined aquifer (Section 10.4.4). There is also an increase in water level between 0 and x_{tn} but the increase is not linear (Eqn (10.3)). Compared to Eqn (3.16), Eqn (10.4) shows that x_{tn} becomes smaller after a rise in sea level, which means that the toe moves inland, as expected.

10.4.2 Unconfined Continental Coastal Aquifer with Land Inundation

When sea level rises and the land surface slopes, the shoreline retreats. This consequently reduces the area over which natural groundwater recharge occurs. Seawater intrudes into the aquifer not only laterally but also vertically from the seawater inundated area. The conceptual model for this problem is shown in Figure 10.8.

Assuming the coastal line moves inland over a distance R, Eqns (10.2)–(10.4) can be expressed for the new situation as

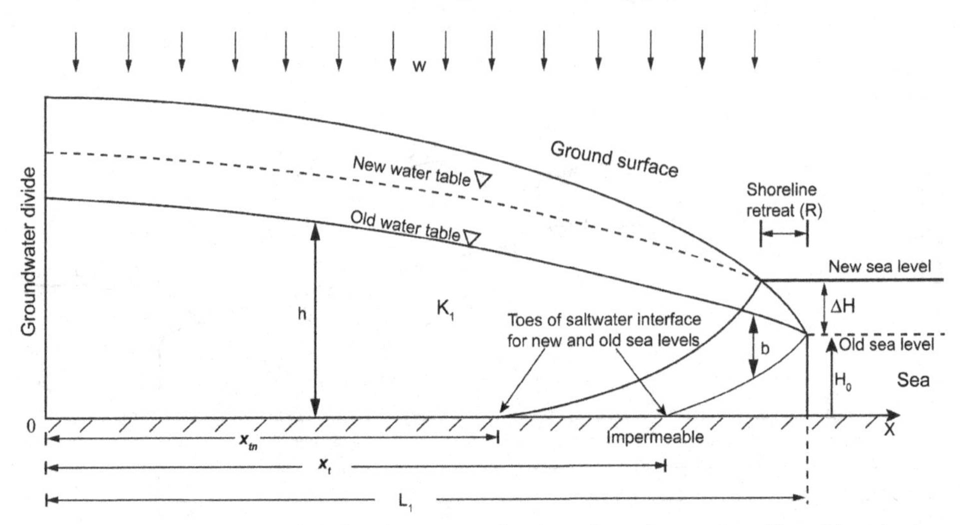

Figure 10.8 Water level and interface in an unconfined continental coastal aquifer with a sloping shoreface after sea level rise with inundation.

$$h = \sqrt{\frac{w(\rho_s - \rho_f)}{K_1\rho_s}[(L_1 - R)^2 - x^2)] + (H_0 + \Delta H)}, (x_{tn} \le x \le L_1) \tag{10.5}$$

$$h = \sqrt{\frac{w}{K_1}[(L_1 - R)^2 - x^2] + \frac{\rho_s}{\rho_f}(H_0 + \Delta H)^2}, (0 \le x \le x_{tn}) \tag{10.6}$$

$$x_{tn} = \sqrt{(L_1 - R)^2 - \frac{K_1(\rho_s{}^2 - \rho_s\rho_f)}{w\rho_f{}^2}(H_0 + \Delta H)^2} \tag{10.7}$$

Equation (10.7) shows that the toe moves further inland due to both sea level rise and inundation, and that, as expected, the inland migration is larger than in the case of a vertical shoreface because the shoreline has shifted and the recharge area has shrunk. The water level changes from Eqns (10.5) and (10.6) are not straightforward because the term with ΔH increases the water level but the term with $(L_1 - R)$, which expresses the reduction of the recharge area, leads to a decrease in water level. For most coastal aquifers, there will be a net increase in water level as long as R is much smaller than L_1.

10.4.3 Unconfined Island Aquifer in a Strip or Circular Island

For an unconfined aquifer in an island with a configuration as shown in Figure 10.9, the interface depth below sea level is computed using Eqn (3.26) for circular island or Eqn (3.32) for strip islands. These equations remain valid to calculate the steady state interface

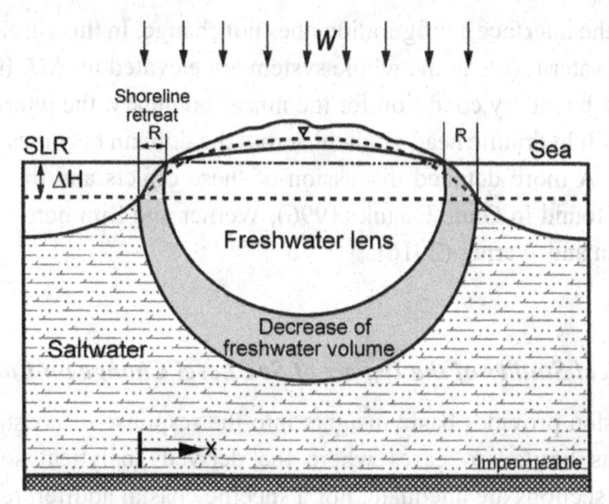

Figure 10.9 Water level and interface in an unconfined strip island aquifer after sea level rise with inundation (modified from Oude Essink, 1996).

position after a discrete amount of sea level rise, provided that the water table rise can be accommodated in the unsaturated zone. As soon as the water table rises to the land surface, the top of the saturated zone becomes fixed and the equations lose their validity. In that case, with the elevation of the water table fixed at a level defined by the land surface, every centimetre of sea level rise should result in a loss of 40 cm of freshwater at the base of the lens according to the Ghijben–Herzberg principle (Eqn (3.4)).

On small islands with shorefaces sloping at a low angle, the land loss due to inundation can be significant relative to the size of the area receiving precipitation recharge. The thickness of the freshwater body will then be reduced. The new maximum thickness of the freshwater body can be easily calculated using the same Eqn (3.26) or (3.32) but with a smaller width of the island. The water level and saltwater toe equations can also be obtained this way.

Ketabchi et al. (2014) studied the impact of sea level rise on freshwater lenses analytically and numerically for an island consisting two layers of different hydraulic properties. They concluded that, while coastal inundation significantly impacts the freshwater lens, the volume of the freshwater lens is also very sensitive to factors such as recharge, aquifer thickness and hydraulic conductivity.

10.4.4 Confined Continental Coastal Aquifer

The water level and position of the interface in a confined continental coastal aquifer have been discussed in Section 3.4.5 (Figure 3.12) and the equations for the interface depth below the top of the aquifer (H) and the toe location (x_t) are expressed by Eqns (3.35) and (3.36). As can be seen from these equations, when the freshwater flow from inland q_0 is constant, both H and x_t are independent of the sea level above the top of the aquifer (A), which means that the interface configuration does not change. In this case, when sea level is increased by ΔH, water levels in the whole system are elevated by ΔH. If a specified head were chosen as the boundary condition for the inland boundary, the interface position will change as the overall hydraulic head gradient across the domain becomes smaller when sea level is increased. A more detailed discussion of these effects and the role of boundary conditions can be found in Oude Essink (1996), Werner and Simmons (2009), Chesnaux (2015) and Morgan and Werner (2016).

10.4.5 Numerical Studies of the Impact of Sea Level Change on Coastal Aquifers

The above discussion provides basic insights into the response of coastal aquifers to sea level rise but it is limited to cases where the derived analytical solutions for two-dimensional cross sections are adequate. For a specific coastal aquifer, regional numerical models are needed to evaluate the salinisation process and inland propagation of the sea level rise. For example, using a density-variable numerical flow model, Carneiro et al. (2010) studied the impact of climate change in a shallow aquifer in the Mediterranean coastal area in Morocco. They considered both sea level rise and recharge variation. Due to decreasing inflow from a neighbouring aquifer and the recharge decline, the salinity was foreseen to increase up to 30 g l^{-1} in the immediate vicinity of the coastline.

Oude Essink et al. (2010) constructed a three-dimensional numerical flow and solute transport model of variable density to estimate the impacts of sea level rise, land subsidence and recharge changes on the flow system in the low-lying Dutch polder area. In this system the groundwater levels are highly controlled by the surface water levels that are kept fixed across much of the area to drain the land. Their results show that there is an increase in hydraulic heads due to sea level rise but this effect is only limited to a zone within 10 km of the shoreline. In the major part of their model area though, the heads will decrease due to land subsidence caused by peat oxidation and clay shrinkage, which necessitates a concomitant lowering of the controlled surface water levels.

Guha and Panday (2012) developed an integrated surface water and groundwater model to study the changes in salinity and groundwater level in an aquifer along the coast of South Florida, USA due to sea level increasing by 0.6 to 2.1 m by 2100. Their results showed that locally the increase of the groundwater levels can be 4% to 15% and that in some wells the increase in the chloride concentrations can reach 100% to 600%.

Yechieli et al. (2010) simulated the processes in the aquifers along the coast near the Dead Sea and Mediterranean Sea, taking into consideration the effect of coastal topography. After the sea level rises by 1 m, a coastal land surface slope of 2.5‰ led to a modelled inland shift of 400 m of the interface. There was no shift in the interface in the case of a vertical coastline, consistent with the outcomes based on the analytical solutions for unconfined systems in Section 10.4.1. The landward shift in the interface was significantly greater in the models that considered groundwater overexploitation and reduced recharge due to the change in climate. These studies highlight the fact that sea level rise does not occur in isolation. Predicting the future state of coastal aquifers must therefore be based on a careful and comprehensive analysis of the natural and anthropogenic factors and processes that change at the same time.

10.5 Management Approaches to Compensate for the Impact of Sea Level Rise

To evaluate or offer expert opinions on impacts of sea level rise on groundwater salinisation, hydrogeologists must have knowledge of the current and future state of the system, including the magnitude and rate of local sea level rise, the extent to which it will propagate into the aquifer, the adaptation of the coastal topography, and changes to the local climate as well as human activities. They also need practical methods to assess the problems. Based on this, policies or guidelines to mitigate the problems generated by sea level increase can be formulated.

The first step is to identify the areas that are most susceptible to sea level rise. The topography and bathymetry should be mapped with high-resolution to identify the critical regions which may be inundated by seawater and the pathways along which seawater can migrate inland (Melloul and Collin, 2006). With airborne light detection and ranging (LIDAR) technology, this can nowadays be achieved in a cost-effective manner for large areas (Wozencraft and Millar, 2005) and with geographic information systems the areas most affected by sea level rise can be mapped (Ramasamy et al., 2010). Critical regions do not only include low-lying areas with gentle slopes, such as coastal wetland and deltaic environments, but also steep coasts that are susceptible to erosion. Geological knowledge is therefore also prerequisite.

There are several countermeasures that can be taken to compensate the loss of fresh groundwater caused by sea level rise (Oude Essink, 1999). Land can be reclaimed along the coastline, enhancing the zone of freshwater influence by moving the coastline seaward. In areas where land reclamation is already planned for other purposes the effects on the groundwater system should be considered with the potential benefits as a buffer against sea level rise effects in mind (Glaser et al., 1991; Hu and Jiao, 2010) (Section 9.5). Along sandy coasts, thickness of freshwater lenses can be enhanced by increasing the width of the coastal dune belt. As sand dunes also serve as coastal protection structures, beach nourishment to prevent erosion of the dunes is a practice that is already in place in many locations to stabilise the coastline. Water management and coastal protection then go hand in hand, highlighting the need for integrated approaches. Not all countermeasures are cost-effective (Bear, 1999), and for some areas it is possible that the effects of sea level rise are so severe that there is no economically effective way to cope with salinisation of the coastal hydrogeological system. One example hereof is probably the option to inundate low-lying areas with freshwater to raise the groundwater level and push the interface to a greater depth (Oude Essink, 1999). This solution is usually not feasible because of the large surface area requirement, and the associated costs of acquiring land.

10.6 Groundwater Usage and Sea Level Rise

While it is well known that sea level changes modify the groundwater regimes, it is relatively less known that usage of groundwater will in turn also change the sea level. Groundwater is pumped for irrigation, drinking and industrial purposes. While a small portion of the pumped groundwater returns to the groundwater system, a much larger part enters the atmosphere by evaporation, falls back to the land as precipitation, and finally finds various pathways to the sea, which ultimately contributes to the increase in sea level.

Konikow (2011b) studied the possible relation between sea level rise and global groundwater depletion. In the period from 1900 to 2008 the total cumulative groundwater depletion in the world was estimated to be about 4500 km^3, which is equivalent to a global sea level increase of 12.6 mm. This contributes 6% to 7% of the sea level rise observed since 1900 (Figure 8.3). The depletion in groundwater is accelerating with time.

Wada et al. (2012) contended that the increase in groundwater depletion from 1900 to 2000 was mainly caused by the progressively growing demand for water, but that the projected rise in depletion from 2001 to 2050 will be largely driven by climate change, which reduces groundwater recharge and surface water availability as a result of greater evaporation and increased temperature. They also found that the contribution of groundwater extraction to global sea level rise was 0.035 mm yr^{-1} in the year 1900, but increased to 0.57 mm yr^{-1} in the year 2000. They further predicted that the contribution of groundwater depletion would increase to 0.82 mm yr^{-1} by 2050. The main source of uncertainty in these projections is that there are very limited reliable large-scale and long-term data about groundwater depletion.

Using an integrated model, Pokhrel et al. (2012) quantified the sea level fluctuations induced by terrestrial water storage changes related to human activities and found that terrestrial water storage can contribute significantly to sea level increase and that the unsustainable groundwater pumping accounts for most of the terrestrial storage depletion. Recent studies demonstrated that climate variability has led to a significant increase of water stored on the continents, especially groundwater (Fasullo et al., 2013; Jiao et al., 2015b). This significantly slowed down the sea level rise in the period from 2002 to 2014 (Reager et al., 2016).

Groundwater usage may impact sea level rise indirectly by releasing carbon dioxide contained in the underground water. Most carbon dioxide in groundwater originally comes from rainfall and decomposition of organic matter in soil. Zhang et al. (1997) estimated that the CO_2 released from groundwater abstraction in the North China Plain is about 10^6 tonnes yr^{-1}. A case study based on the Qijia aquifer system in China was carried out by Jia et al. (2002). They estimated that the average partial CO_2 pressure of groundwater is 1.28×10^4 Pa, which is much higher than atmospheric partial CO_2 pressure of 32.03 Pa. As a result, about 11 mmol of CO_2 will be emitted from every 1 kg of groundwater to the atmosphere when the groundwater from the aquifer reaches the ground surface. For the total annual groundwater yield of 3.65×10^7 m^3 in the well field in the Qijia aquifer system, 1.84×10^4 tonnes yr^{-1} of CO_2 are released to the atmosphere.

CO_2 added to the atmosphere as a result of outgassing of water in aquifers is eight times more than the average yearly CO_2 generated from volcano eruption, but this volume of groundwater-generated CO_2 is still very small compared to the emissions generated by burning fossil fuels (Macpherson, 2009). For example, Wood and Hyndman (2017) found that groundwater depletion in the USA is responsible for a CO_2 emission of 1.7×10^6 tonnes yr^{-1}, which is about 3000 times less than the 5×10^9 tonnes yr^{-1} estimated to be released by fossil fuel combustion.

11

Tide-Induced Airflow in Unsaturated Zones

11.1 Introduction

In inland areas, airflow in the vadose zone can be induced by changes in the atmospheric pressure (Nilson et al., 1991; Elberling et al., 1998; Ellerd et al., 1999). In coastal areas, water table fluctuations induced by tides are much more important in producing subsurface airflow than atmospheric effects as the amplitude and frequency of the sea tides are much greater than those of the atmospheric pressure oscillations (Parker, 2003; Jiao and Li, 2004). Water table fluctuations induced by tides have been investigated since the late nineteenth century (Veatch, 1906; Jacob, 1950; Jiao and Tang, 1999), as discussed in Chapter 4, but the fact that air pressure in the vadose zones can also oscillate due to the tide is not generally recognised. When the water table rises, there is an increase in the air pressure within the vadose zone and the air is driven out of the ground. This process is reversed and air is sucked into the vadose zone when the water table falls (Jiao and Li, 2004). This ventilation process is usually not noticeable by humans. The range of pressure fluctuation induced by tide is typically less than a few kPa, i.e. a few percent of the standard atmospheric pressure of 101.3 kPa and occurs usually less than 100 m from the shoreline. The impact can be significant enough though to have engineering, environmental and ecological effects (Elberling et al., 1998; Ellerd et al., 1999; Jiao and Li, 2004; Kuang et al., 2013).

If the ground surface is capped by materials of low (air) permeability, or the pores of the superficial soil are fully saturated after rainfall, the air exchange between the soil and atmosphere is impeded or even stopped. As a result, when the tide falls or rises, very low or high soil air pressures respectively can be generated. A low-permeability layer at the ground surface can be formed naturally or artificially. For example, the stratigraphy of a river delta typically consists of inter-layered high- and low-permeability strata so it is common to have a superficial less-permeable layer siting over a deeper more permeable layer. In many coastal zones, infrastructure such as roads, sea- and airports, container terminals and coastal defence works have been built right up until the coastline. These sites have areally extensive asphalt or concrete with a permeability much lower than the underlying soil materials. To support heavy vehicles such as big trucks and airplanes, gravel or coarse rock fragments with lots of voids are usually used as fill materials. This artificial soil stratigraphy forms a typical configuration of an unsaturated zone with substantial air storage capacity capped by a low-permeability ground surface.

In most investigations of the tide-induced interaction between seawater and groundwater in coastal unconfined aquifers, the airflow in vadose zones is ignored, i.e. the air phase is regarded to have no influence and be at constant pressure (Guo and Jiao, 2010). However, there are situations in which the airflow effects cannot be ignored, as demonstrated in the case study in Section 11.6. Quantitative studies specifically dealing with tide-induced airflow are very limited. More general studies on air–water flow in vadose zones have been carried out in fields such as soil remediation by soil vapour extraction (Pedersen and Curtis, 1991; Massmann and Madden, 1994) and rainfall infiltration (Touma et al., 1984; Barry et al., 1995; Weeks, 2002; Guo et al., 2008). Airflow has also been widely studied in connection to the release of CO_2 from soils by processes such as microbial respiration, decomposition of organic material, and root respiration (Luo and Zhou, 2006). A comprehensive review of airflow in vadose zones generated by various natural forcings such as rainwater infiltration, topographic effect and atmospheric pressure fluctuations and its applications can be found in Kuang et al. (2013).

This chapter starts with an introduction of some basic concepts such as air-confined and air-unconfined aquifers and basic equations for single-phase airflow and two-phase air and water flow. Spatial and temporal characteristics of air pressure induced by tidal fluctuation in air-confined systems are examined. A brief discussion of the possible engineering, environmental and ecological significance of airflow in coastal unsaturated zones is included. Finally a case study of the tide-induced abnormally high pressure of air in the reclaimed land below the Hong Kong International Airport is presented.

11.2 Water-Confined, Air-Confined and Air-Unconfined Aquifers

When an aquifer is overlain by a confining unit and the water level is higher than the contact surface between the two layers, this aquifer is called a confined aquifer, or water-confined aquifer. Water within the confined aquifer cannot escape through the confining unit, but in practice, only semi-confining units (and thus semi-water-confined aquifers) occur because there is no geological formation that is completely impermeable.

An aquifer is referred to as air-confined if the air in the unsaturated part cannot exchange with the atmosphere. This occurs when there is an air-impermeable layer at the top of the unsaturated zone (Figure 11.1). As with water, no geological formation is absolutely air impermeable so in practice the term confined means semi-confined. If the air in the unsaturated zone cannot freely exchange with the atmosphere, a pressure higher or lower than the atmospheric pressure, which is also referred as positive or negative pressure, will be generated when the water level rises or falls. Most unconfined aquifers are air-unconfined, although air-confined conditions may temporally develop in most of the shallow aquifers when intensive rainfall saturates the pores of the superficial soil, which thereby become impermeable to air (Weeks, 2002).

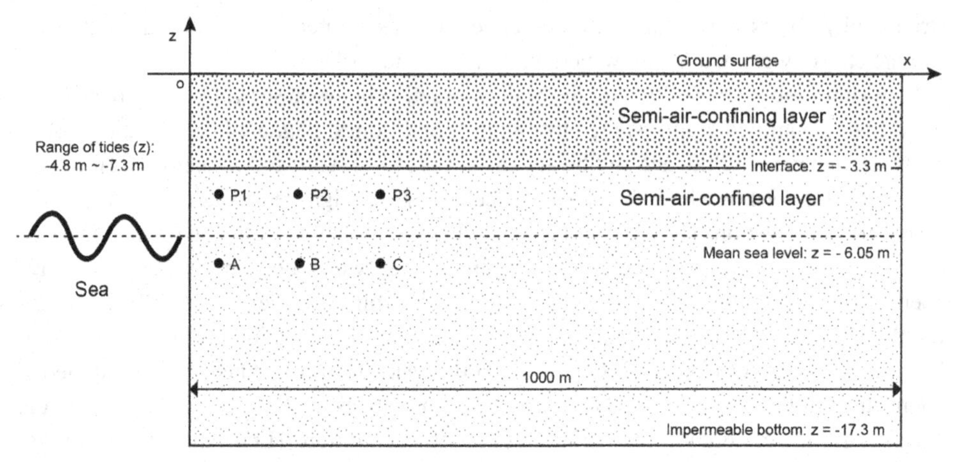

Figure 11.1 A two-layer air-confined system bounded by the sea ($z = 0$ m at the land surface; $x = 0$ m at the shoreline) (modified from Guo and Jiao, 2008b).

Air pressures in air-confined aquifers depend on the air volume, the fluctuation amplitude and rate of the water table, and the air permeabilities of the capping layer relative to that of the aquifer. The air volume is determined by the thickness of the unsaturated zone and the soil moisture content of the unsaturated zone. The rate of water level fluctuation controls the magnitude of pressure generated and the air permeability contrast between the unsaturated zone and the capping layer controls the pressure that can be dissipated through the upper layer. Studies by Jiao and Li (2004) and Jiao and Guo (2009), have demonstrated that for significant air pressure to be generated, the air permeability of the air-confining layer should be at least two orders of magnitude lower than the aquifer.

11.3 Equations for Airflow in Unsaturated Zones

If the focus of a study is the air pressure and airflow in the vadose zone, a single-phase (air) model can be used. Under some assumptions, analytical solutions may be developed to study the airflow in the vadose zone (Li and Jiao, 2005; Li et al., 2011a). If both airflow and water flow are to be considered, the governing equations describing two-phase immiscible fluid flow must be used (Vachaud et al., 1973), and numerical methods are needed. These two types of models were reviewed by Kuang et al. (2013).

11.3.1 Single-Phase Model

The governing equation for single-phase airflow in porous materials can be obtained based on the mass conservation principle and Darcy's law (Section 2.2.1). The assumptions made include (Kidder, 1957; Lu and Likos, 2004): the material is isotopic and homogeneous, the air behaves an ideal gas and has a constant composition and viscosity, the flow of the air obeys Darcy's law, and thermal and buoyancy effects can be ignored.

The conservation equation can then be expressed as (Kidder, 1957; Massmann, 1989)

$$\frac{\partial p_a}{\partial t} = \frac{\kappa_a}{n_a \mu_a} \frac{\partial}{\partial z}\left(p_a \frac{\partial p_a}{\partial z}\right) \tag{11.1}$$

where p_a [M L T^{-2}] is the air pressure, κ_a [L^2] is the air permeability, n_a is the air-filled porosity, t [T] is the time, μ_a [M L^{-1} T^{-1}] is the viscosity of air and z [L] is the vertical coordinate.

Multiplying both sides of Eqn (11.1) by p_a leads to

$$p_a \frac{\partial p_a}{\partial t} = \frac{\kappa_a p_a}{n_a \mu_a} \frac{\partial}{\partial z}\left(p_a \frac{\partial p_a}{\partial z}\right) \tag{11.2}$$

Introducing air diffusivity (D_a) [L^2 T^{-1}]as

$$D_a = \frac{\kappa_a p_a}{n_a \mu_a} \tag{11.3}$$

then

$$p_a \frac{\partial p_a}{\partial t} = D_a \frac{\partial}{\partial z}\left(p_a \frac{\partial p_a}{\partial z}\right) \tag{11.4}$$

This equation is equivalent to

$$\frac{1}{2}\frac{\partial p_a^2}{\partial t} = \frac{1}{2} D_a \frac{\partial}{\partial z}\left(\frac{\partial p_a^2}{\partial z}\right) \tag{11.5}$$

or

$$\frac{\partial p_a^2}{\partial t} = D_a \frac{\partial^2 p_a^2}{\partial z^2} \tag{11.6}$$

The dependent variable in Eqn (11.6) is p_a^2. Because the air diffusivity D_a is a function of the air pressure p_a, Eqn (11.6) is a nonlinear partial differential equation. To linearise Eqn (11.6) p_a in the air diffusivity term can be replaced by the mean atmospheric pressure p_{a0} (Fukuda, 1955; Weeks, 1978; Massmann, 1989; Shan, 1995; Neeper, 2002). Since the magnitude of the pressure oscillations usually does not exceed 10% of p_{a0} the linearisation is believed not to generate much error (Massmann, 1989; Shan, 1995). Moreover, if the deviation of the air pressure from the mean pressure is small, Eqn (11.6) becomes (Fukuda, 1955; Weeks, 1978; Massmann, 1989; Kuang et al., 2013)

$$\frac{\partial p_a}{\partial t} = D_a \frac{\partial^2 p_a}{\partial z^2} \tag{11.7}$$

This is a linear differential equation with p_a as the dependent variable.

More general equations subject to less stringent assumptions are available. For example, the equation for three-dimensional airflow in isotropic and homogeneous porous materials can be found in Muskat (1934) and Kirkham (1947); the equation for three-dimensional airflow in anisotropic and heterogeneous porous materials was given by Bear (1972).

Equations (11.6) and (11.7) are diffusion equations that can be solved analytically subject to a set of initial and boundary conditions analogous to classical heat flow problems (Carslaw and Jaeger, 1959). Based on Eqn (11.7), Li and Jiao (2005) derived the first analytical solution to examine the vertical airflow induced by water table fluctuations in the lower layer of a tidally controlled two-layer system. The flow was simplified as being one-dimensional in the vertical direction. The ground surface was chosen as the upper boundary with constant atmospheric pressure. The lower boundary was controlled by the water table fluctuation, which was assumed to be synchronous to the sea tide. This solution was expanded by Li et al. (2011b) to also consider the fluctuations of atmospheric pressure, which demonstrated that the influence of atmospheric pressure oscillations on subsoil air pressure is mainly controlled by the air-filled porosity difference between the two layers and the air permeability of the low-permeability upper layer. Their results also suggested that atmospheric pressure oscillations have negligible influence on the water table in a tidally controlled aquifer system.

Analytical solutions based on Eqn (11.7) have been used to estimate the air permeability of asphalt in the laboratory and unsaturated zones in the field. Li et al. (2004) developed an equation to determine the air permeability of asphalt using a falling-pressure laboratory test. Li and Jiao (2005) calculated the air permeability of the marine fill materials below Hong Kong International Airport by fitting the fluctuations of tidal level and subsurface air pressure calculated from their analytical solution to observations.

11.3.2 Air–Water Two-Phase Model

Equations that describe the coupled air–water two-phase flow problem are needed to simulate the interaction between air and water flow in the vadose zones. In addition to the assumptions for the development of the governing equation for single-phase airflow, it is assumed that both air and water are immiscible and that the air is compressible but water is not (McWhorter, 1971; Vachaud et al., 1973). With these assumptions, the water and air mass conservation equations (see also Section 2.2.3) for one-dimensional flow can be written as (Kuang et al., 2013)

$$\frac{\partial}{\partial t}(\rho_w \theta_w) = -\frac{\partial}{\partial z}(\rho_w q_w) \tag{11.8}$$

$$\frac{\partial}{\partial t}(\rho_a \theta_a) = -\frac{\partial}{\partial z}(\rho_a q_a) \tag{11.9}$$

where θ_a and θ_w are the volumetric air and water contents which satisfy the relation of $\theta_a + \theta_w = n$, ρ_a and ρ_w [M L^{-3}] are the air and water densities and q_a and q_w [L T^{-1}] are the air and water fluxes, respectively. The above two equations form a coupled set of nonlinear partial differential equations.

As before, Darcy's law is assumed to be applicable to describe the flow of both the air and the water. For flow in the vertical direction only and with the coordinate z being oriented positive downward, it becomes (Kuang et al., 2013)

$$q_w = -\frac{\kappa k_{rw}}{\mu_w}\left(\frac{\partial p_w}{\partial z} - \rho_w g\right) \tag{11.10}$$

$$q_a = -\frac{\kappa k_{ra}}{\mu_a}\left(\frac{\partial p_a}{\partial z} - \rho_a g\right) \tag{11.11}$$

where κ [L^2] is the intrinsic permeability of the aquifer materials, k_{ra} and k_{rw} are the relative permeability of the porous medium to air and water, μ_w[M L^{-1} T^{-1}] is the water viscosity and p_w [M L T^{-2}] is the water pressure.

According to Bear (1972) and McWhorter (1971), the capillary pressure (p_c) can be expressed as the difference between the air and water pressures:

$$p_c = p_a - p_w \tag{11.12}$$

The above equation provides a theoretical relationship to couple airflow and water flow. Given Eqns (11.8)–(11.12), the following relationships must be provided (Kuang et al., 2013):

$$k_{rw} = k_{rw}(\theta_w) \tag{11.13}$$

$$k_{ra} = k_{ra}(\theta_w) \tag{11.14}$$

$$p_c = p_c(\theta_w) \tag{11.15}$$

$$\rho_a = \rho_a(p_a) \tag{11.16}$$

Equations (11.8)–(11.16) form a set of equations to be solved subject to specific boundary and initial conditions (McWhorter, 1971). The governing equations are strongly nonlinear and have to be solved numerically. The following discussions of two-dimensional flow systems are all based on numerical modelling.

11.4 Spatial and Temporal Tide-Induced Pressure Fluctuations

The following subsections will illustrate the effect of tide-induced airflow and air pressure variations based on numerical simulations by Guo and Jiao (2008b). Their conceptual

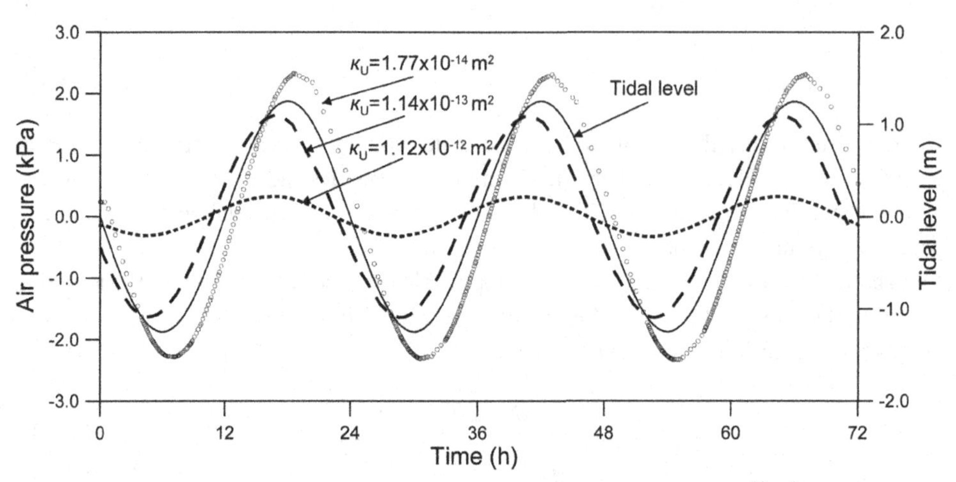

Figure 11.2 Temporal air pressure change at point P1 in the aquifer with $\kappa_L = 10^{-11}$ m^2 when the upper layer permeability (κ_U) is 1.12×10^{-12} m^2 (dotted curves), 1.14×10^{-13} m^2 (dashed curves) and 1.77×10^{-14} m^2 (circles) (Guo and Jiao, 2008b).

model consisted of an upper layer of low permeability (κ_U) and a lower layer of high permeability (κ_L) (Figure 11.1). The left boundary is the seawater-land boundary, and ground surface forms the upper boundary. The inland boundary at 1000 m from the coastline is treated as no-flow boundary, which implies that no airflow caused by sea level fluctuation can be noticed at that distance. The bottom of the domain is also a no-flow boundary. A constant atmospheric pressure of 101.3 kPa was specified at the ground surface and the seaside boundary above the mean sea level. A single-constituent sinusoidal diurnal tide (Eqn (4.40) in Chapter 4) was used as the tidal boundary. Variable-density effects on the groundwater flow due to differences in salinity were ignored. The permeability of the aquifer (κ_L) was set to a constant value of 10^{-11} m^2.

11.4.1 Air Pressure in Air-Confined and -Unconfined Systems

The simulated air pressures were analysed at observation points P1, P2 and P3, located at a depth of 4.2 m and a distance from the coast of 10, 50 and 100 m, respectively. The change of air pressure with time at P1 in the bottom layer is presented in Figure 11.2. When the upper layer permeability (κ_U) was set to 1.77×10^{-14}, 1.14×10^{-13} or 1.12×10^{-12} m^2, the amplitude of the pressure oscillation became 2.33, 1.64 and 0.33 kPa, respectively (or, 2.3%, 1.6% and 0.3% of the standard atmospheric pressure).

This result shows that, as expected, the air permeability of the top layer relative to that of the bottom layer controls the amplitude of the air pressure fluctuation. When κ_U is low, air cannot flow out or into the ground easily as the water table rises or falls. Consequently, the air pressure

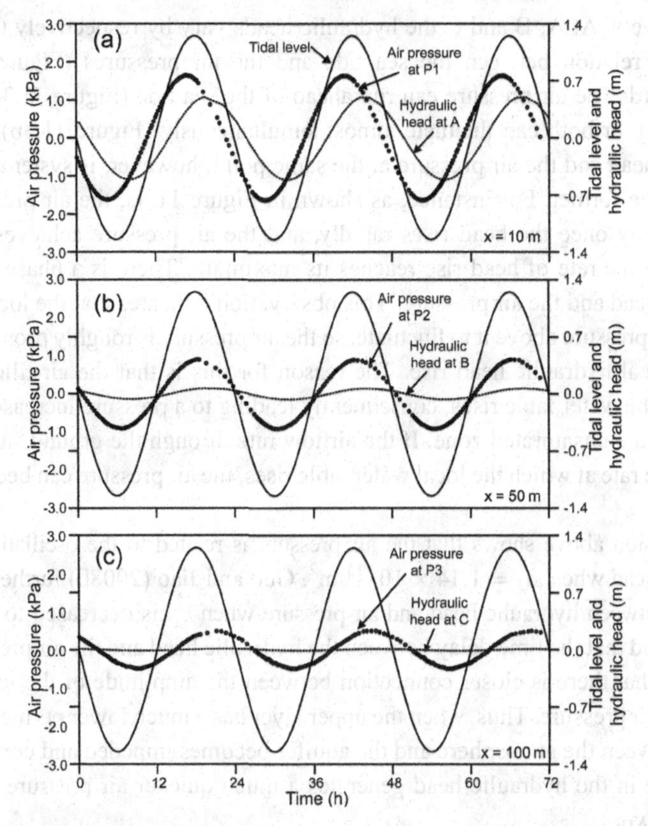

Figure 11.3 Temporal changes of air pressure in the unsaturated zone and hydraulic head in the saturated zone (Guo and Jiao, 2008b). Note that the head fluctuations in (c) are imperceptibly small at the chosen vertical scale.

in the vadose zone increased noticeably as the tide rises. An interesting observation from Figure 11.2 is that the air pressure fluctuation can either be phase-delayed (e.g. when $\kappa_U = 1.77 \times 10^{-14}$ m^2) or phase-advanced (e.g. when $\kappa_U = 1.14 \times 10^{-13}$ m^2) compared to sea level fluctuation. The responsible mechanism will be explained in Section 11.4.2. Such behaviour is very different from tide-induced water level fluctuation, which usually lags behind sea level (Chapter 4). Tidally induced air pressure changes are measurably dampened when κ_U is 1.12×10^{-12} m^2, which is only one order of magnitude smaller than κ_L.

11.4.2 Relation among Sea Level, Hydraulic Head and Air Pressure

The modelled hydraulic heads are analysed at observation points A, B and C, which are located in the saturated zone at the same distances from the coast as P1, P2 and P3 at a depth of 7.7 m. Figure 11.3 presents the simulated air pressures together with the heads at the observation points for the model with $\kappa_L = 10^{-11}$ m^2 and $\kappa_U = 1.14 \times 10^{-13}$ m^2. At P1, P2 and P3 the amplitude of the air pressure variation is 1.64, 0.61 and 0.1

kPa, respectively. At A, B and C the hydraulic heads vary by respectively 0.51, 0.07 and 0.02 m. The relation between the sea tide and the air pressures (Figure 11.3) is not straightforward: The air pressure can run ahead of the sea tide (Figure 11.3a), lag behind (Figure 11.3c), or both can fluctuate almost simultaneously (Figure 11.3b). The relation between the head and the air pressure at the same point, however, is systematic: the latter lags behind the former. For instance, as shown in Figure 11.3a, the air pressure starts to increase quickly once the head rises rapidly, and the air pressure achieves a maximum roughly when the rate of head rise reaches its maximum. There is a phase shift of 4.1 h between the head and the air pressure. This observation indicates that the local water table causes the air pressure above it to fluctuate, so the air pressure is roughly proportional to the rate of the local hydraulic head rise. The reason for this is that the air-filled pore space decreases as the water table rises, consequently leading to a pressure increase and pushing the air out of the unsaturated zone. If the airflow rate through the ground surface is much lower than the rate at which the local water table rises, the air pressure can become elevated considerably.

The discussion above shows that the air pressure is related to the oscillating rate of the groundwater head when $\kappa_U = 1.14 \times 10^{-13}$ m^2. Guo and Jiao (2008b) further investigated the relation between hydraulic head and air pressure when κ_U is decreased to 1.77×10^{-14} m^2. They found that the time delay between the hydraulic head and the air pressure is much reduced and that there is closer connection between the amplitude of the local hydraulic head and the air pressure. Thus, when the upper layer has a much lower permeability, the air exchange between the atmosphere and the aquifer becomes impeded and consequently the same increase in the hydraulic head generates a much quicker air pressure increase than with a higher κ_U.

On the basis of the temporal relationships among the air pressure, hydraulic head and sea tide, it can be summarised that the air pressure in the vadose zone is controlled by both the fluctuation rate and the amplitude of the aquifer hydraulic head. The local head, not the sea tide at the shoreline, is the direct driver for the oscillation of the air pressure, although it is the sea level variation that drives the head fluctuation in the aquifer.

11.4.3 Fluctuations of Air Pressure with Inland Distance

As demonstrated in the previous section, the air pressure oscillation attenuates inland because the fluctuation of the local hydraulic head in the aquifer, which is the direct driver of the air pressure above, attenuates with inland distance. However, the amplitude of the air pressure does not attenuate linearly with inland distance. Figure 11.4 shows the simulated the changes of the amplitude of the air pressure with depth and inland distance in the air-confining unit (Figure 11.1). The amplitude is zero at the shoreline, reaches a maximum at some distance away from the shoreline, then decreases gradually with inland distance. Near the left boundary, which is open to the air, the air inside the aquifer and that outside can exchange easily. At the point several metres away from the shoreline where

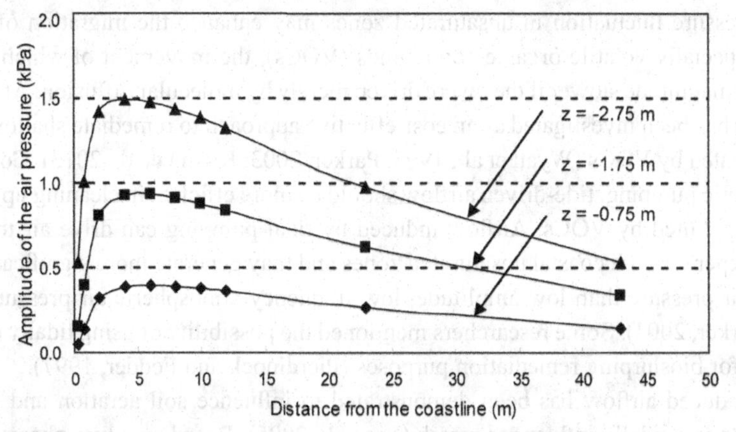

Figure 11.4 Change of amplitudes of the air pressure oscillation with inland distance at different depths in the unsaturated zone with $\kappa_U = 1.14 \times 10^{-13}$ m^2 and $\kappa_L = 10^{-11}$ m^2 (Guo and Jiao, 2008b).

the air exchange becomes weak but water table fluctuation driven by the tide is still fairly large, the pressure achieves a maximum.

The impact of the tide-driven water level fluctuations on the air pressure decreases when the depth of the observation point decreases. Near the ground, the atmospheric pressure that is assumed to be fixed at the land surface dominates the air pressure change. The air diffusivity of the vadose zone determines the attenuation of the fluctuation of the air pressure, which is driven by the water table changes from below. Consequently, the air pressure amplitudes decrease progressively upward.

11.5 Environmental, Engineering and Ecological Effects of Tide-Induced Airflow

The above discussion demonstrated that tidal fluctuations can induce considerable air pressure oscillation in the unsaturated zones in an air-confined aquifer near the sea. Understanding the underlying processes is of both practical and theoretical significance for certain engineering and environmental problems.

The air pressure generated by fluctuations of the water table may in turn influence the fluctuations of the head in the aquifer As discussed in Chapter 4, Jacob's (1950) equation (Eqn (4.26)) or variants thereof are commonly employed to estimate the diffusivity of coastal aquifers. Guo and Jiao (2010) demonstrated that for air-confined aquifers, ignoring airflow effects may lead to significant over-estimation of the hydraulic diffusivity, and the degree of over-estimation increases as the degree of the air-confinement increases. The reason is that in air-confined aquifers, the amplitude of the hydraulic head oscillations as well as the tidally influenced distance are increased. Hydraulic head fluctuations in an air-confined aquifer are amplified with increasing degree of confinement, as for a water-confined aquifer.

Air pressure fluctuation in unsaturated zones may enhance the migration of contaminants, especially volatile organic compounds (VOCs), the movement of which would be orders of magnitude slower if they were driven merely by molecular diffusion. Atmospheric pumping has been investigated as an cost-effective approach to remediate shallow aquifers contaminated by VOCs (Wyatt et al., 1995; Parker, 2003; Kuang et al., 2013). Compared to atmospheric pumping, tide-driven airflow should be more efficient in cleaning up nearshore aquifers polluted by VOCs. Airflow induced by tidal pumping can drive air to circulate much deeper through coastal unsaturated zones and may generate more significant fluctuation of air pressure than low-amplitude, low-frequency atmospheric air pressure fluctuations (Parker, 2003). Some researchers mentioned the possibility of using tidally controlled systems for bioslurping remediation purposes (Pierdinock and Fedder, 1997).

Tide-induced airflow has been demonstrated to influence soil aeration and plant root respiration in a tidally influenced marsh (Li et al., 2005). Based on a two-phase numerical model they also found that the entrapped air can have significant impact on water migration in the unsaturated zones as it reduces the permeability. Moreover, the better growth of some salt marsh plants close to tidal streams compared to the inner marsh regions is attributable to the better aeration near the creek where the tidal water table fluctuations are the highest. Continuous tide-induced 'breathing' is also believed to potentially impact organisms in intertidal zones as it speeds up nutrient transport and oxygen ventilation (Jiao and Li, 2004). However, this assertion remains to be confirmed.

The ground surface of reclaimed land is usually covered by concrete or asphalt, which is much less permeable than the underlying fill materials. Highways are usually designed only to support downward acting loads and may thus suffer from tide-induced heaving effects. Moreover, although the tide-induced loading is low, the material strength under cyclic loading is lower than that under monotonic loading. Tide-related processes have not yet been studied extensively by civil engineers.

While the heave damage by positive pressures can be obvious and easily appreciated (Jiao and Li, 2004; Leung et al., 2007), it is less obvious that a negative pressure can be also generated due to a rapid fall of sea level. Fine materials such as silt and clay particles can be dislocated by the suction. Underground cavities can be produced if this process is sustained for a long period of time. It has even been suggested that some subsidence or ground collapse commonly observed in the urbanised coastal reclamation areas in Hong Kong is due to this process, but detailed studies are yet to confirm this.

11.6 Case Study of Hong Kong International Airport

A striking example about the importance of tide-driven air pressure is the heave damage of the north runway at the Hong Kong International Airport (Figure 11.5) (Leung et al., 2007). It was demonstrated that the heave damage is initiated by the unusually high air pressure under the runway due to the combined effect of a high tidal amplitude and heavy rainfall (Jiao and Li, 2004; Leung et al., 2007). The problem was resolved by installing small-diameter pressure relief holes through the asphalt pavement (Leung et al., 2007). It was this

Figure 11.5 Dome-shaped heave damage at a runway of Hong Kong International Airport (Leung et al., 2007).

engineering problem of heave damage at this airport that triggered a series of numerical and analytical studies on airflow in coastal unsaturated aquifers (Jiao and Li, 2004; Li et al., 2004, 2011; Li and Jiao, 2005; Leung et al., 2007; Guo and Jiao, 2008b, 2010; Kuang et al., 2013).

11.6.1 Background

Hong Kong International Airport with a total area of 12.48 km^2 was created by levelling two small islands with an area of 3.1 km^2 and reclaiming 9.38 km^2 of land from the shallow sea (Plant et al., 1998) (Figure 9.2). A simplified cross section of the subsurface passing through the north runway and two taxiways is presented in Figure 11.6. It shows the asphalt pavements below the runway and taxiways, the fills of marine sand and rock fills, and a thin layer of geotextile between these. The fill materials, especially those consisting of coarse rock, are very porous and permeable. As shown by water level data from a borehole installed in the coarse fill at 285 m away from the shoreline (Plant et al., 1998), the fluctuation of the water table in the piezometer is practically identical to that of the sea tide.

Dome-shaped heave damage (Figure 11.5) was detected in the asphalt pavement of the runway and taxiways following heavy rainfall between 1999 and 2001 (Leung et al., 2007). As indicated by the grey bars in Figure 11.7, heave damage was observed four times from July to August in the summer of 2000. The figure also shows the rate at which the tide fluctuated, and the rainfall. Heaves formed when high rainfall and tidal rates coincided.

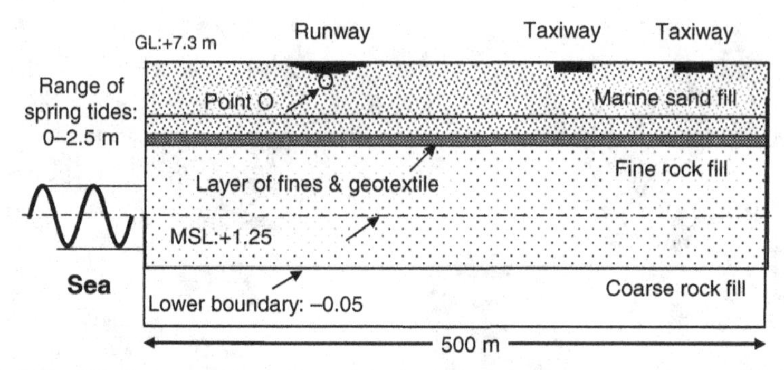

Figure 11.6 Simplified cross section of the subsurface of Hong Kong International Airport (Jiao and Li, 2004).

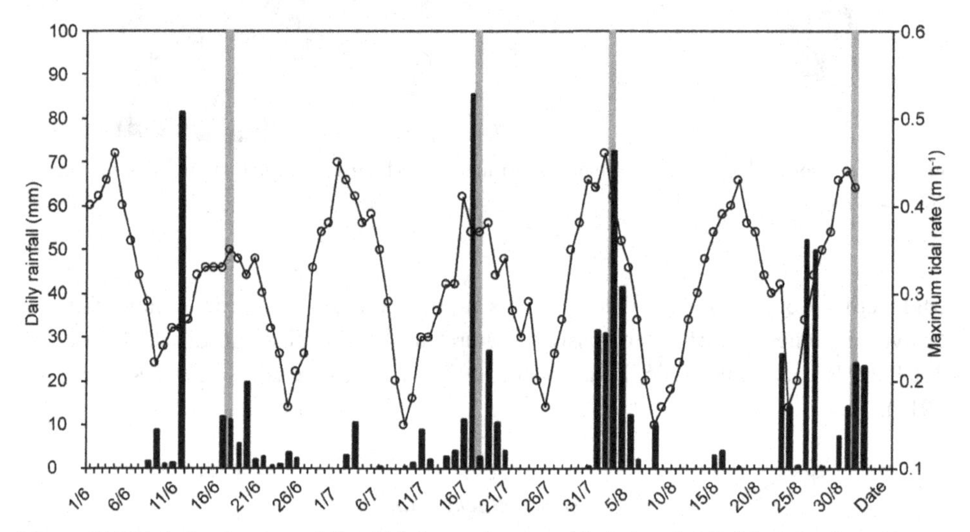

Figure 11.7 Relation between daily tidal fluctuation rate (circles), rainfall (black bars) and heave damage (grey bars) at Hong Kong International Airport in the summer of 2000 (Kuang et al., 2013).

Among the four days with heave damage, 18 July is an exception: the rainfall was only 2.4 mm. However, the 85.5 mm of rainfall on 17 July was the highest during that summer, so the soil was still very wet on 18 July. When rainfall was heavy but the tidal fluctuation rate was small, such as on 12 June, or when the tidal fluctuation rate was high but there was no or little rainfall, such as on 1 July or 18 August, there were no reports of heave damage.

11.6.2 Observed Air Pressures

The observed tidal level, atmospheric pressure and air pressure at observation point O below the runway from 7 to 9 February 2001 are shown in Figure 11.8. The amplitude

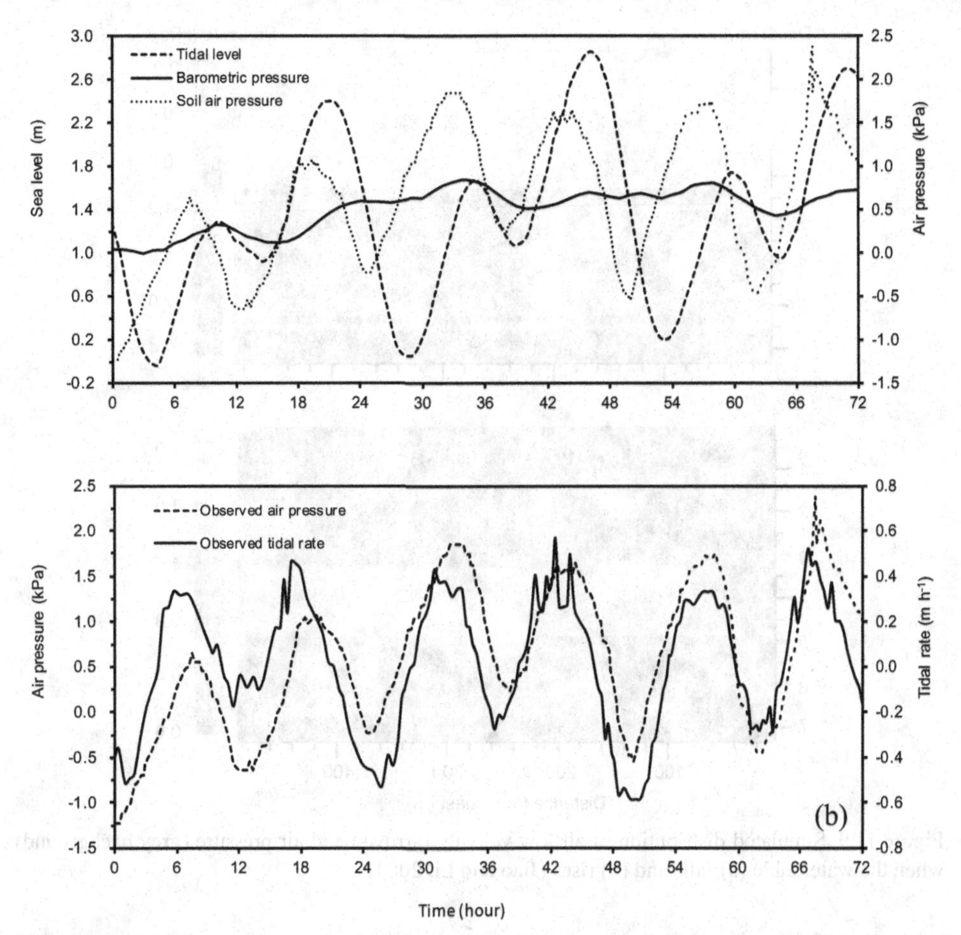

Figure 11.8 (a) Observed atmospheric pressure, soil air pressure and tide level over a 72-hour period from 7 to 9 February 2001. (b) Observed air pressure and tidal fluctuation rate in the same period (Jiao and Li, 2004).

and fluctuations of the atmospheric pressure are significantly smaller than those of the observed soil pressure, so the former could not be the main driver of the latter. Within each 24 hour cycle, there are two peaks in the tidal level, with one being markedly lower than the other. On the contrary, the two air pressure peaks during the same timespan have approximately the same magnitude, suggesting that there is no straightforward relation between air pressure and tidal height (Figure 11.8a).

The rate of the water level change, which is calculated as the water level change between two consecutive observation times divided by the time difference, however, does show a clear relation with the subsoil air pressure (Figure 11.8b). An important observation from Figure 11.8b is that the subsurface air pressure below the pavement is nearly proportional to the tidal fluctuation rate. This indicates that the extremes of the subsoil air pressure are controlled by the extremes of the tidal fluctuation rate, rather than the amplitude, of the tide. In other words,

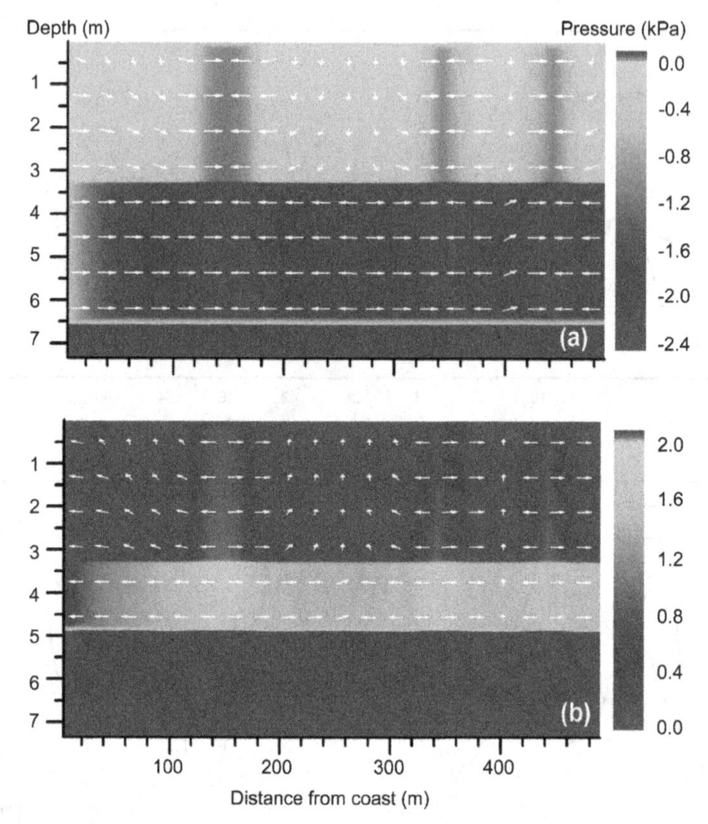

Figure 11.9 Simulated distribution of airflow velocity (arrows) and air pressure (grey background) when the water table (a) falls and (b) rises (Jiao and Li, 2004).

a high tidal amplitude does not give rise to a high subsoil air pressure if the tide rises slowly. Yet a sudden water level rise can generate a high air pressure even if the water level amplitude is not great. This observation helps to explain that subsoil air pressures of similar magnitude can be generated by both small and large peaks of the semi-diurnal tide (Figure 11.8a).

11.6.3 Numerical Modelling

Jiao and Li (2004) used a layered one-dimensional model (Figure 11.6) to better understand the mechanisms of tide-induced airflow. Each layer is assumed to have a uniform permeability. The air permeability of the asphalt pavement is based on laboratory measurements (Li et al., 2004), but those of other layers were found by calibrating the model to the observed air pressure below the runway. The details of the boundary and initial conditions are elaborated in the original paper by Jiao and Li (2004). The model results demonstrated that significant positive and negative subsurface air pressures are due to the combination of heavy rain and a quickly oscillating tide, similar to what the field data showed. Under no rain conditions, the modelled minimum and maximum soil air pressure below the north runway are respectively 1.64 and −0.56 kPa. However, when the rain is heavy, the maximum and minimum subsoil air pressure can be respectively 2–3 times higher or 5–9 times lower than the corresponding pressures without rain.

Figure 11.9 shows the modelled distributions of the air pressure (expressed in the grey background) and airflow velocity (expressed in arrows) at the maximum falling rate of the tide (Figure 11.9a) and maximum rising rate (Figure 11.9b) under no rain conditions. The subsoil air pressure approaches a minimum when the rate of the falling tide reaches a maximum. Extra pore space in the unsaturated zone is generated and air is sucked across the soil-atmosphere interface, as demonstrated by the air velocity field. Zones of low air pressure are generated underneath the three areas (the runway and taxiways) capped by the asphalt. The simulated air pressure in the model domain changes from 0 at the ground to −2.4 kPa close to water table.

When the rate of water table rise reaches a maximum, the air pressure approaches its peak. As is clearly shown by the velocity distribution of the airflow in Figure 11.9b, air above the water table is forced upward to the ground when it is compressed by the rising groundwater. Columns of high air pressure are generated below the areas capped by the asphalt pavement. The simulated air pressure varies from 0 close to the ground to 2.0 kPa at the water table.

Jiao and Li (2004) further used their model to investigate the air pressure under typical rainfall conditions and demonstrated that abnormally high or low air pressures can be formed under a favourable combination of the tidal fluctuations, rainfall duration and intensity and geological structure. Such abnormally high air pressures are sufficient to generate the heave damage.

12

Coastal Aquifer Management and Seawater Intrusion Control

12.1 Introduction

In the broadest sense, water management encompasses the policy framework and technical activities that regulate and enable the development, use and protection of water resources. Good management practices acknowledge both natural and socio-economic complexity, are adaptive and strive towards sustainability. Ideally, groundwater is managed within an integrated framework that acknowledges the inter-dependencies of the different parts of the natural system and crosses sectorial and institutional borders.

Coastal aquifers are linked into the larger hydrological cycle and are affected by the natural changes of and human interventions in that cycle. Since coastal zones are at the downstream end of river basins, the drivers for change that affect a coastal aquifer can be located far away from the coast, and are not seldom occurring outside of the jurisdiction of the local groundwater authorities. Identifying cause-and-effect relationships requires a proper understanding of the natural system, and the changes made to it by humans. This understanding is predicated upon measurements of the relevant hydrological processes, and thus data collection is the first step and the key prerequisite for any adequate management approach. Moreover, effective water management can only be achieved when the legislation is adequate, and when it is enforced by the responsible authorities.

This chapter discusses a broad array of groundwater management aspects in coastal zones. The ensemble of management activities always serves one or multiple objectives. For coastal aquifers, management practices tend to be aimed at minimising the risk of seawater intrusion, excessive depletion of the freshwater reserves, and adverse impacts on groundwater-dependent ecosystems (Werner et al., 2011). Since the topic is broad and multi-facetted, it is impossible to provide a comprehensive overview. The scope is therefore necessarily focused on those aspects most relevant to hydrogeologists, and the technical aspects related to management. Case examples are provided to illustrate seawater intrusion management strategies that have been studied or trialled in various parts of the world. Desalination is also discussed because this technology often accounts for a large proportion of the water budget in coastal zones, and impacts coastal aquifers in several ways.

12.2 Measurement and Monitoring

Knowledge of the hydrological functioning of an area is a prerequisite for, and integral part of, any successful coastal aquifer management strategy. The collection of such information requires a lot of resources and is a continuous and time-consuming task. The topography, drainage network and the geology and hydrogeological properties of the subsurface are among the most basic types of information required (Custodio and Bruggeman, 1987; Van Dam, 1999). These are relatively static, which is not to say that these don't change over management time scales (e.g. in areas with strong land subsidence), but usually noticeable change only occurs over a timespan of decades to centuries. Water levels, the chemical composition of the various water bodies, and water fluxes like precipitation, recharge, river discharge and pumping are more dynamic and can change at timescales of hours to days. A monitoring network specifically designed to capture the transient behaviour of the hydrological system is therefore needed.

This section will focus on the monitoring of water levels and groundwater salinity, because especially the latter is specifically unique to coastal groundwater investigations. The characterisation of the hydrogeological properties of the subsurface and the studies of river flows and groundwater recharge are in principle similar to other areas. These are discussed in works like Nielsen (1991), Kruseman and De Ridder (1994) and Brassington (2007), and this information is therefore not repeated here.

12.2.1 Head Measurements

Groundwater flow patterns are inferred from hydraulic head measurements. As the salinity of groundwater varies, so does the density and this complicates the interpretation of the hydraulic head data (Lusczynski, 1961; Post et al., 2007). When comparing the heads measured in observation wells containing either saline or fresh groundwater but located in the same aquifer, the values have to be normalised to a common density (Section 1.5). Obtaining reliable head data is further complicated by the fact that the density of the water column inside a piezometer can be variable. When an observation well is leaky, which may be the case for an old or improperly installed well, it will fill with water that is not representative of the groundwater at the location of the well screen. When the screen at the bottom of the well is in saltwater, but fresher water is leaking in at shallower intervals, a stratified water column will be present inside the piezometer, and the water level becomes an unreliable indicator of the groundwater pressure at the screen. This is illustrated in Figure 12.1, which presents an example of a piezometer in which the water level suddenly changed after the well was purged for chemical sampling. Before the purging, the standpipe was filled with water of varying salinity, which was lower than that of the native groundwater at the screen. After purging, the entire well became filled with the saline groundwater from the depth of the well screen. Consequently, the density of the water column increased, and because a shorter column of water was therefore needed to indicate the groundwater pressure at the depth of the screen, the water level fell by several decimetres. This drop is

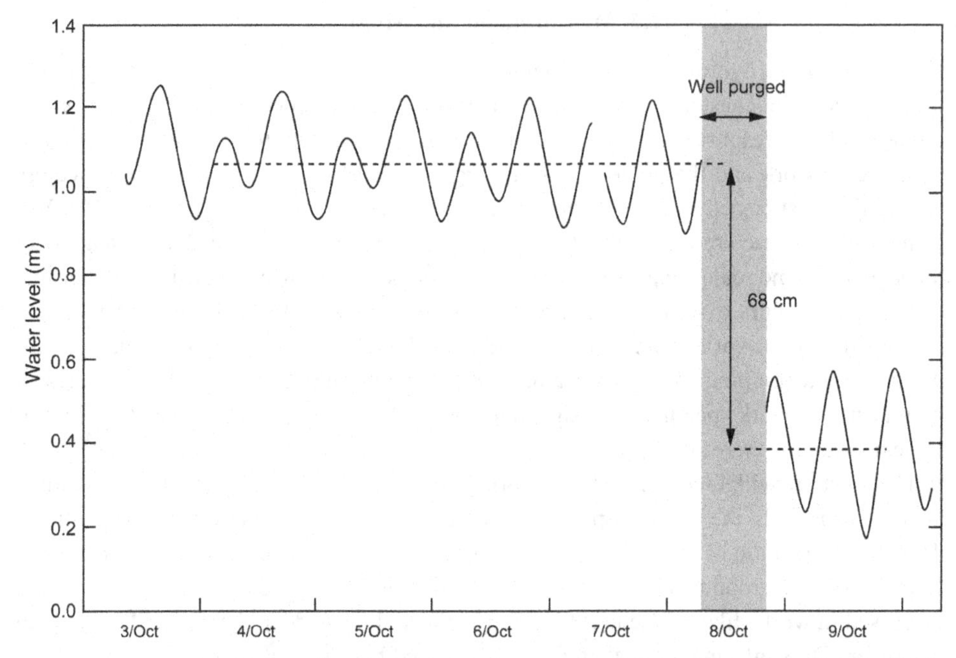

Figure 12.1 Water level versus time for a coastal observation well in northern Australia. The fall in the average head of 68 cm is caused by the purging of the stagnant water in the well, which had a lower average density before pumping than after pumping. This density difference was caused by a salinity stratification that existed before the well was purged. Due to the pumping, all the water inside the well became saline.

solely due to a change of the water properties inside the piezometer and has nothing to do with a change of the hydraulic head within the aquifer.

The example illustrates that regular revision and maintenance of the observation wells of a monitoring network is required. A salinity stratification can not only develop in leaky wells, but also in properly constructed wells when the salinity of the groundwater at the screen is variable in time, or when water seeps in along the top of the tube (e.g. rain, or water that floods an area to above the top of the piezometer). The problem was already recognised in the classical studies of seawater intrusion in Florida (Kohout, 1961), but the methodological aspects of measuring water levels in coastal aquifers received less attention than the other articles on the flow of fresh- and saline groundwater published around that time (Cooper, 1959; Cooper et al., 1964; Kohout, 1965).

12.2.2 Salinity Measurements

Salinity is a key measurement variable in coastal aquifer management. It is needed to delineate the extent of seawater intrusion, but as the previous example showed, head measurements in coastal areas cannot be interpreted without considering the density, which depends primarily on salinity. Salinity measurements can be made directly, by taking groundwater samples from wells, or indirectly, by deriving it from the electrical properties of the subsurface, which can be measured in boreholes or from the surface using geophysical techniques.

The quickest way to determine the salinity of a groundwater sample is by measuring the electrical conductivity of groundwater. This can be done by using handheld devices in the field and is therefore quick and cost-effective method to determine changes of salinity with distance and time. The electrical conductivity can be used to estimate the total dissolved solids concentration, as well as the density (Post, 2012). Since electrical conductivity varies with the water temperature, all measurements should be normalised with respect to a standard temperature, which is usually 25°C. This temperature-normalised value of the electrical conductivity is called the specific conductance (Section 5.2.1.1).

Instruments also exist to measure the electrical conductivity of water inside an observation well, but the results of these must be treated with great care as the salinity of the water inside the observation well may not represent that of the groundwater outside (Tellam et al., 1986). The salinity of coastal aquifers typically changes with depth and a single well to characterise the salinity distribution is therefore insufficient. Instead, multiple wells with short screens at appropriate depths are required to provide this information, which adds to the cost of the investigation. Wells with long screens (several metres or more) are sometimes used to investigate the characteristics of the transition zone. Their use should be avoided though, because flow of water through the borehole in the presence of vertical hydraulic gradients modifies the salinity distribution, and may even result in local salinisation of the freshwater lens (Shalev et al., 2009; Rotzoll, 2010). Rushton (1980) found that the transition from fresh- to saltwater inside open boreholes could be as much as 160 m above the transition zone in the aquifer.

Another disadvantage of long well screens is that a water sample obtained from a well represents a mixture of the vertically stratified groundwater in the aquifer. The proportions in which the waters of different salinity mix in the sample, however, are indeterminate. For an aquifer with a homogeneous permeability, the sample will be biased towards the fresher water near the top of the well screen, as illustrated in Figure 12.2. Kohout and Hoy (1963) attributed this to the fact that more energy is required to move the dense, saline water upward than to move freshwater horizontally into the well. They found that the lower the pumping rate, the fresher the sample compared to the saline water in the deepest part of the borehole. If however the permeability of the aquifer varies along the screened interval of the observation well, the inflow will be greatest from the intervals with the highest permeability (Kohout and Hoy, 1963). The most appropriate screen length is thus dependent on the local conditions (i.e. the degree of aquifer heterogeneity and the salinity stratification), but, generally speaking, screens longer than two metres are likely to result in samples affected by mixing.

Figure 12.3c shows an example of an optimised monitoring well design by the Amsterdam Water Supply Company (Waternet) in the Netherlands (Kamps et al., 2016). In this

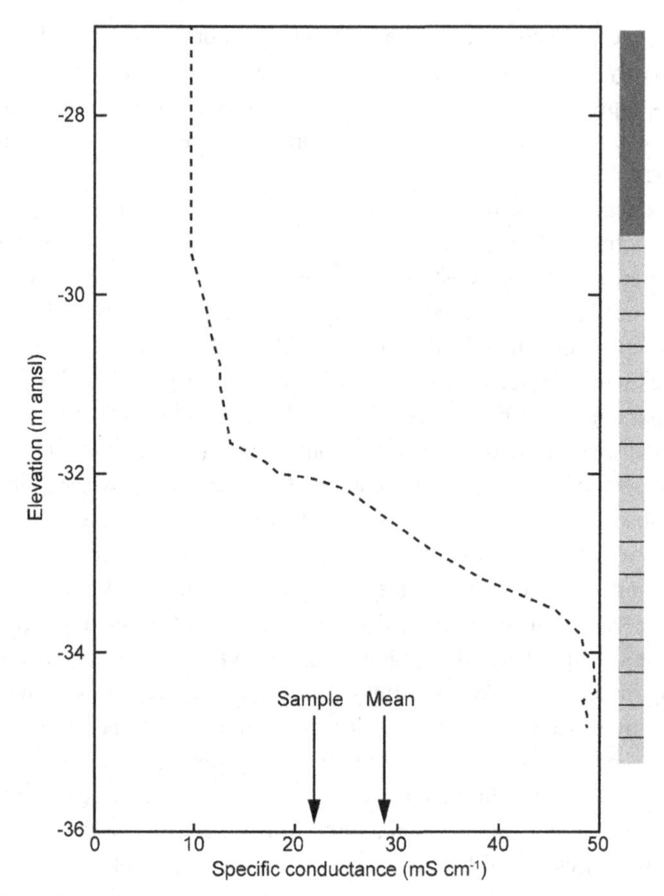

Figure 12.2 Example of biased sampling from a monitoring well with a well screen penetrating the transition zone (Post et al., 2018a). The screened section of the observation well is indicated to the right of the graph. The graph shows the specific conductance as a function of depth as measured using a downhole probe. The mean value of the specific conductance along the well screen (indicated by the arrow with the word 'Mean') based on the probe measurements is higher than the specific conductance of the water sample (indicated by the arrow with the word 'Sample') taken after purging the well at least three times using a submersible pump. This indicates that the value measured in the sample is biased towards the fresher water in the uppermost part of the well screen.

example, the borehole is equipped with three well screens and a series of 13 closely spaced mini-screens and electrodes, which serve to determine the position and width of the transition zone accurately. The observation wells used before this type was introduced, had seven screens in a single borehole. There is also a wide-diameter piezometer to measure the conductivity of the formation using a downhole induction instrument.

Figure 12.3 also shows a single well with a short screen, which is positioned above the transition zone and is therefore inadequate to detect seawater intrusion. The observation

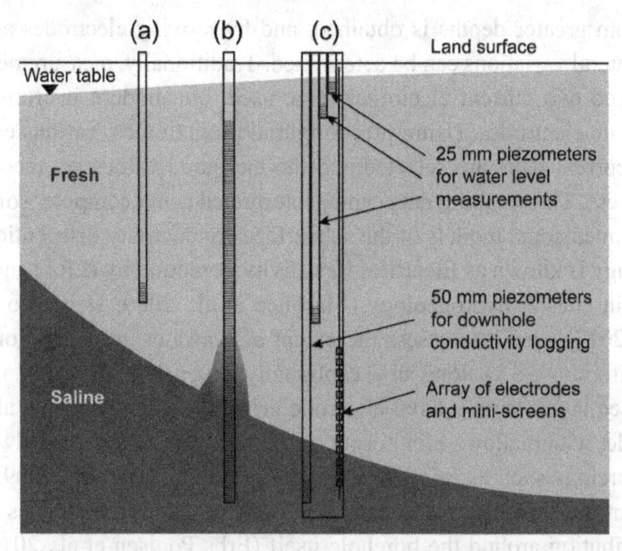

Figure 12.3 Three different monitoring well designs in a coastal aquifer: (a) single piezometer with a short well screen positioned above the wedge of intruded seawater; (b) observation well with a long well screen across both the freshwater and saltwater part of the aquifer, along which saltwater may be drawn upwards; (c) design of observation wells used by the Amsterdam Water Supply Company (Kamps et al., 2016). Inside a single borehole, three 25 mm piezometers are installed to measure water levels, alongside with an array of electrodes and mini-screens across the transition zone and a deep large-diameter observation well for downhole induction measurements.

well with the long well screen does intersect the seawater wedge, but in the presence of tides or a nearby abstraction well, it may act as a conduit and the transition zone can move upward, at least in the vicinity of the well.

Various geophysical techniques exist that provide measurements of the bulk electrical conductivity of the subsurface (Telford et al., 1990; Rubin and Hubbard, 2006). From these, the salinity of groundwater can be estimated because the bulk conductivity is at least partially a function of the electrical conductivity of groundwater. It is, however, also determined by the electrical properties of the rocks, as well as temperature, which introduces a degree of uncertainty and ambiguity in the conversion of bulk conductivity to salinity. Nevertheless, geophysical methods have been used in coastal aquifers for several decades (Stewart, 1999), and new technologies continue to extend the field of application (Loke et al., 2013; Micallef et al., 2018).

Geophysical measurements can be made using instruments at or above the Earth's surface, or in boreholes. For measuring the Earth's resistivity, two categories of measurements exist: geo-electrical and electro-magnetic. With the geo-electrical method, a current is introduced into the ground by means of electrodes. The resultant potential difference between two measurement electrodes is measured, and by taking several measurements at different locations and with varying electrode spacings, the subsurface conductivity distribution can be inferred. By increasing the spacing between the potential electrodes, information from greater depths is obtained, and by moving electrodes around between locations, the lateral variations can be determined. Traditionally, measurement set-ups with two potential and two current electrodes were used, but modern instruments consist of arrays of multiple electrodes. Using preconfigured programmes, various combinations of electrode and current pairs are selected and the measured values are recorded in a fully automated process. The resulting data can be interpreted using computer software to create two- or three-dimensional models of the subsurface conductivity distribution.

The technology is known as Electrical Resistivity Tomography (ERT) and its use is now commonplace in coastal hydrogeology (Martínez et al., 2009; Henderson et al., 2010). Goebel et al. (2017) used it to assess the extent of seawater intrusion along the coast of Monterey (California, USA) along an exceptionally long profile of 40 km, up to a depth of 280 m below sea level. With a fixed electrode array, measurements can also be taken in time-lapse mode, which allows monitoring of the response of the groundwater salinity to hydrological forcings such as rainfall or flooding (De Franco et al., 2009; Ogilvy et al., 2009). Arrays of multiple electrodes have also been installed in boreholes to measure the resistivity distribution around the borehole itself (Erbs Poulsen et al., 2010), or between different boreholes (Song et al., 2006). Such installations allow the tracking of the interface over long periods of time, making them particularly useful as early warning tools (Grinat et al., 2018).

With electro-magnetic (EM) methods, the response of the magnetic field induced by an electrical current is measured, which can be converted to a resistivity model of the subsurface at the measurement location. Several different measurement set-ups and configurations exist, but the set-up always consists of a transmitter and a receiver device. The great advantage of the method is that no direct contact between the ground and the instruments is required. This has led to development of airborne methods (Figure 12.4) that are becoming increasingly accurate and are suitable for the collection of large data sets over extensive areas (Siemon et al., 2009). In densely populated areas or in areas with abundant electrical installations, the method cannot be applied though.

Geophysical measurements must be converted to layered one- or multi-dimensional models of the subsurface, which are non-unique. That is, several models can provide equally good fits of the measurement data, which means that there is always an inherent degree of uncertainty. Sometimes the model outcomes can be validated by conducting independent measurements of the subsurface resistivity, using direct push technologies for example (Pauw et al., 2017). However, there is uncertainty in the relationship between resistivity and groundwater salinity too, because the former is not just a function of the latter but is also influenced by the rock type, pore connectivity, degree of cementation and temperature (Glover, 2016). This means that geophysical methods can only be effectively applied if also direct measurements of groundwater salinity are available. Nevertheless, geophysical data sets form invaluable complements to the point measurement data obtained from water samples from boreholes.

Brassington and Taylor (2012) made a comparison between different types of salinity measurements made in a borehole that penetrated the transition zone between freshwater

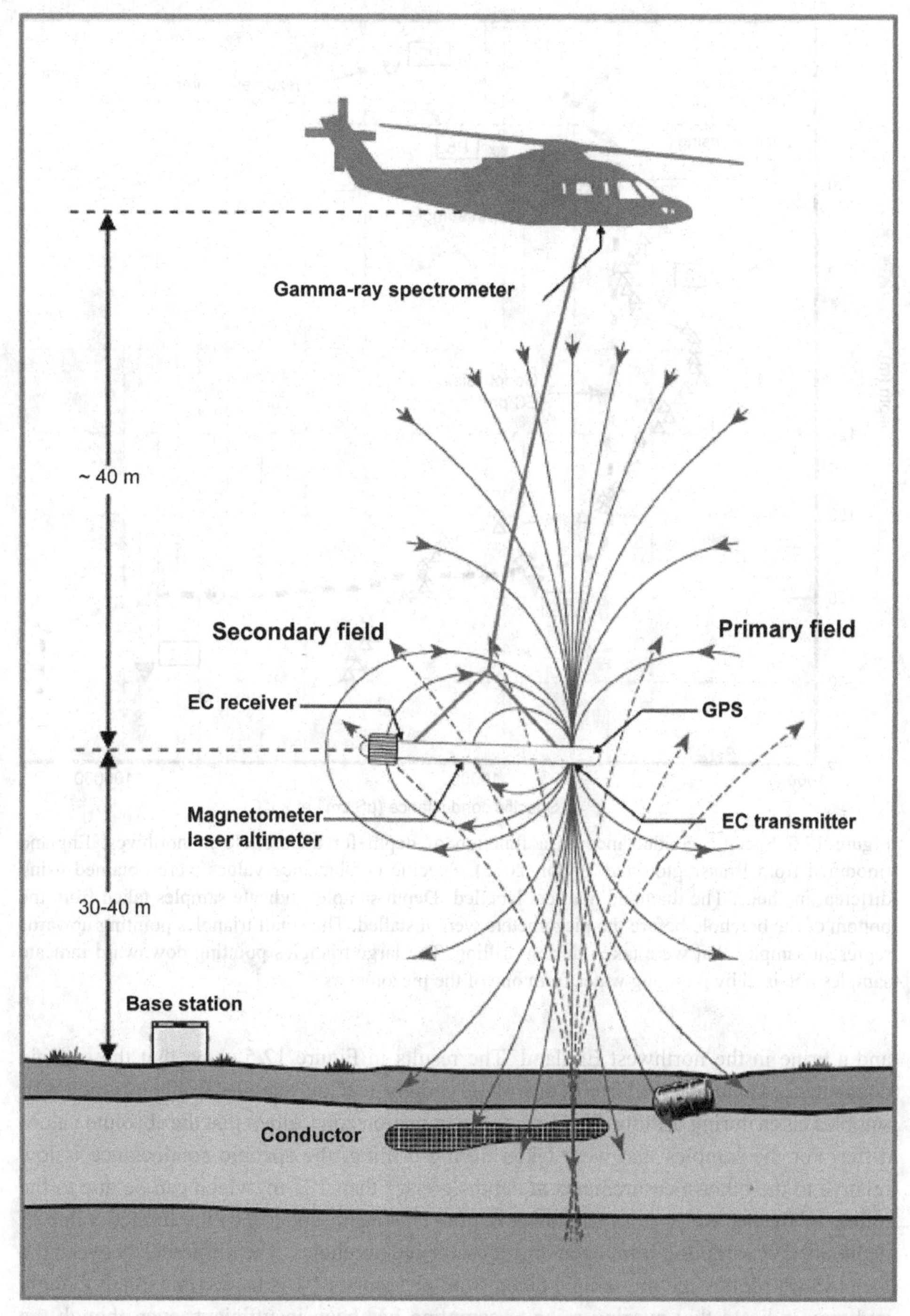

Figure 12.4 Principle of an airborne electro-magnetic system to measure subsurface conductivity (modified from Steuer et al., 2009).

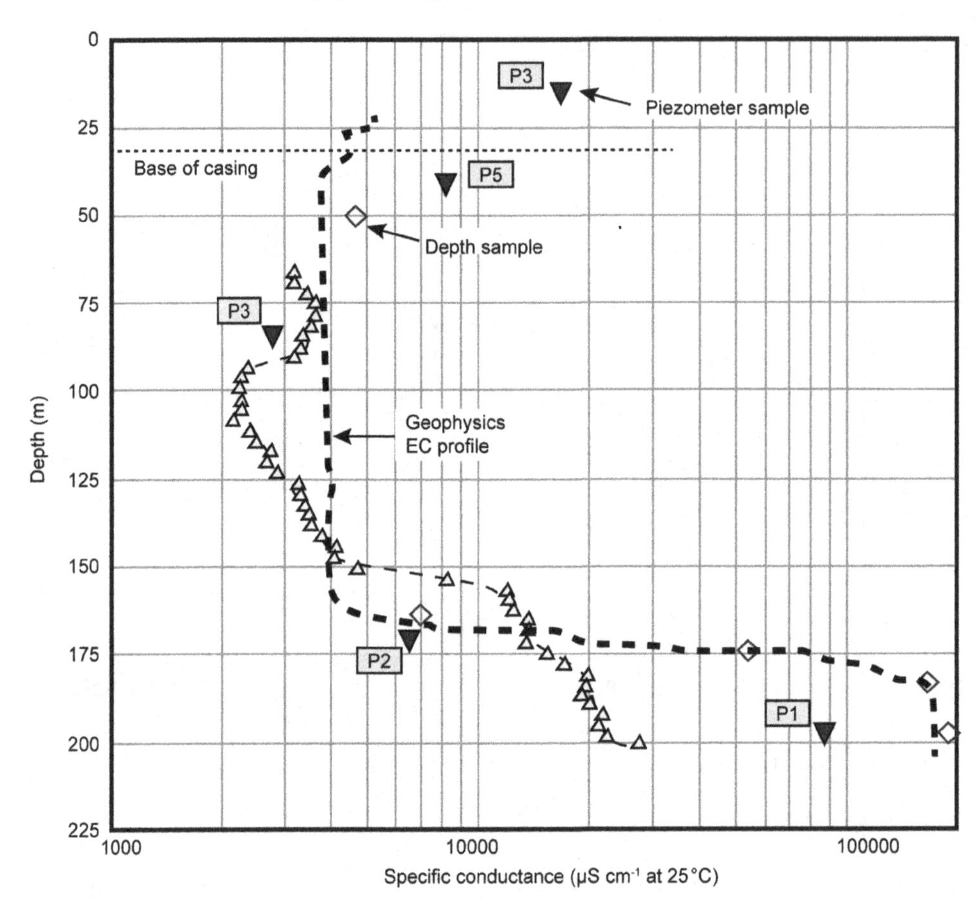

Figure 12.5 Specific conductance as a function of depth for a borehole in northwest England (modified from Brassington and Taylor, 2012). Specific conductance values were obtained using different methods. The diamond markers labelled 'Depth sample' indicate samples taken from the bottom of the borehole before the piezometers were installed. The small triangles pointing upwards represent samples that were taken during drilling. The large triangles pointing downward indicate samples obtained by pumping water from one of the piezometers.

and a brine in the northwest England. The results in Figure 12.5 show that the specific conductance values derived from a downhole geophysical log, samples from piezometers or samples taken during drilling all delineate the transition zone, albeit that the absolute values differ. For the samples that were taken during drilling, the specific conductance is low relative to the other measurements at depths greater than 175 m, which can be due to the inflow of fresher water from shallower depth. This again underlines the limited value of depth-specific sampling from open or fully screened boreholes. The difference between the geophysical profile and the sample taken from piezometer P1 is harder to explain though, and may indicate that purging prior to sampling has been insufficient, even though the prescribed minimum of three well volumes was removed.

12.3 Assessment of Coastal Aquifer Vulnerability

The detailed characterisation of coastal aquifers and an assessment of the extent of seawater intrusion based on field studies is resource-intensive, takes at least a number of years to complete and requires a high degree of expertise. For the case of large management regions, this may mean that only a few select areas can be studied in detail. Moreover, numerical modelling of large areas is often intractable because of limited data availability, and regional models are known to struggle with making predictions of salinisation processes at a local scale (Sanford and Pope, 2009). To provide an alternative, several studies have attempted to develop vulnerability indicators in which areas that are most at risk of seawater intrusion can be identified, based on a combination of data sets available for the management region of interest.

One group of vulnerability mapping methods involves calculating an index score based on various types of input data each weighted by some factor. These methods have their origin in mapping the vulnerability of aquifers to pollution, of which the DRASTIC method (Aller et al., 1987) is the most well known (Gogu and Dassargues, 2000). For coastal areas, the GALDIT method (Chachadi and Lobo Ferreira, 2001) has seen some application. The method takes into account the aquifer type, thickness, hydraulic conductivity, as well as the distance to the shore, the groundwater level and the current status of seawater intrusion. For each indicator, and for each aquifer in a multi-aquifer situation, a digital map is produced, and by summing the weighted scores for each indicator using geographical information system software, a final map of the vulnerability index is obtained.

The main strength of index methods is that they provide a means to systematically bring together and organise various types of data that exist for a coastal area. Moreover, the process of assembling the vulnerability indicator maps provides a platform for managers, policy makers, scientists and other stakeholders to exchange knowledge and information. The great weakness of the methodology is that some key determinants of seawater intrusion like aquifer heterogeneity, recharge and pumping are not explicitly considered. Moreover, the two-dimensional nature of the maps results in over-simplified representations of the system when multiple aquifers are present, or when aquifers are stratified.

Klassen and Allen (2017) developed a method that, in addition to vulnerability, also considered the risk to seawater intrusion. Risk is different from vulnerability in that it also takes into consideration the likelihood and the negative (both financial and ecological) consequences caused by an event. The vulnerability depended on coastal hazards and the susceptibility of the aquifer to salinisation. The latter depended on the distance to the coast as well as the slope of the land surface. The former was determined by natural hazards (sea level rise and storm surges, which depend on the topography) and anthropogenic hazards determined by pumping. As with other vulnerability mapping methods, the transient nature of seawater intrusion is not captured, nor is the localised nature of well salinisation, e.g. as would be the case for a fractured or karst aquifer.

Another weakness of the index-based method is that there is great subjectivity in assigning ranking scores and their weightings. Therefore, Werner et al. (2012) proposed to develop vulnerability indicators based on mathematical descriptions of seawater intrusion. They derived a set of indicators by determining partial derivatives of the analytical solutions developed by Strack (1976). These describe the extent to which a change in recharge, freshwater discharge to the sea or a change in inland head affect the volume of seawater in the aquifer or the position of the interface toe. While indicators based on a physical description of the natural system are arguably less subjective than index-based methods, the disadvantage of the method is that it relies on a highly simplified representation of the coastal groundwater system. Notwithstanding the assumptions and simplifications contained in the application of the Strack (1976) analytical framework, several studies have used it to assess seawater intrusion vulnerability at a nation-wide scale. This includes Spain (Wriedt and Bouraoui, 2009), Australia (Morgan et al., 2013) and the USA (Ferguson and Gleeson, 2012). Beebe et al. (2016) investigated the performance of analytical solutions as a screening tool for seawater intrusion for coastal aquifers in the USA and Australia. They found that a correct prediction of seawater intrusion causing salinisation of pumping wells could be achieved in approximately 60% of cases. This success rate was attributed to the common mismatch between the simplifying assumptions and more complex field settings.

Despite their limitations, vulnerability mapping techniques, provided the method applied is sufficiently descriptive, can be used to support land use planning, direct monitoring efforts and raise community awareness. However, they are prone to over-simplification, especially when applied at very large scales, and water management decisions should not be based on vulnerability mapping alone. Strategies to prevent or reverse seawater intrusion and to protect fresh groundwater resources can be successful only if they acknowledge the natural complexity of the coastal aquifer system, as well as the anthropogenic stresses acting on it. This requires process understanding, which can only be obtained by targeted field measurements and models that incorporate an appropriate level of complexity.

12.4 Seawater Intrusion Management

Measures to prevent or restore the consequences of seawater intrusion include a reduction of abstraction, increase of freshwater recharge, optimisation of pumping, engineered barriers, as well as land and surface water management (Van Dam, 1999; Oude Essink, 2001c; Abarca et al., 2006). Which set of measures is the most appropriate for a given area depends on the prevailing physical and socio-economic conditions. In coastal areas where seawater intrusion has been prevented or reversed, a combination of various measures has usually proven to be effective. The sections that follow discuss each of these broad categories in more detail, and provide examples of case studies in which experience with solutions has been acquired in practical applications.

12.4.1 Abstraction Reduction

Since the over-abstraction of groundwater is the leading cause of seawater intrusion in most aquifers (Ferguson and Gleeson, 2012), a reduction of pumping appears to be the most effective strategy to stop or reverse salinisation. Indeed, the closing of well fields or the severe reduction of the volume of pumped groundwater, has proven to be an effective instrument to restore salinity levels in some coastal aquifers (Han et al., 2015). For the implementation of measures to be effective though, alternative water sources or water-saving technologies must be available at the same time. Moreover, for policies to be put in place, strong governance structures and enforcement of the applicable legislation are essential prerequisites.

The first step in controlling water demand consists of identifying the most important water usages and user groups. Typical water uses in coastal regions include domestic, industrial, agricultural (irrigation) and tourism. A reduction of the demand for groundwater can be achieved in different ways, including raising public awareness, imposing water restrictions, adopting water-saving technologies, preventing distribution losses, pricing policies and finding alternative water sources (Van Dam, 1999; Marlow et al., 2013). The type and number of users and stakeholders determine how easy it is to successfully implement measures. When only a few private or public organisations withdraw groundwater, placing restrictions on well numbers and abstraction rates is easier than when a high number of private well owners is involved (Custodio, 2012b).

One critical step in determining the maximum admissible pumping rates is the quantification of a safe or sustainable yield. This concept has received ample attention in hydrogeology (e.g. Maimone, 2004). In the broadest sense, safe yield is the rate at which groundwater can be withdrawn sustainably without causing adverse effects. Determining which side effects are acceptable is a subjective choice. It relies on an understanding of the physical system and under ideal conditions involves some form of community consultation to try to establish a consensus. Examples of undesirably effects include damage to groundwater-dependent ecosystems, economic losses and land subsidence, but in coastal aquifer the salinisation of wells is of course the main threat to be avoided.

Determining the safe yield is prone to major uncertainty, primarily due to the difficulty of accurately quantifying the various component of the water balance. Invariably the safe yield is linked to the recharge and since some flow is required to prevent the saltwater to move too close to the point of abstraction, only a fraction of the recharge can be withdrawn from the aquifer. Which fraction is considered tolerable, depends on the management objectives. If the only objective is to maximise the withdrawal, a value can be sought that is just enough to prevent salinisation. But where other objectives are to be met as well, such as for example the protection of natural habitats and safeguarding minimum stream or spring flow rates, a lower value needs to be adopted (Maimone, 2004). Particularly important is also to ensure that a sufficiently large buffer of freshwater remains in the aquifer that is being exploited because recharge varies with time, which means that abstraction may exceed replenishment at times (Van Dam, 1999). This may occur sporadically, such as for example during a period of drought, or more regularly in areas with a strong seasonality of the climate.

Since the understanding of groundwater systems is always incomplete and conditions can change with time, e.g. recharge may change as a result of land use or climate, monitoring heads and groundwater quality is essential. The data should be reviewed regularly and new levels of water allocations must be established, making the concept of safe yield a dynamic one. A management plan should also foresee in establishing trigger-levels, which are certain pre-defined criteria based on head, subsidence or salinity measurements, which when exceeded, signal the need for actions, such as the reduction of abstraction (Werner et al., 2011).

12.4.2 Increase of Freshwater Recharge

In areas where sufficient water of suitable quality is available, the volume of fresh groundwater can be augmented by Managed Aquifer Recharge (MAR). MAR is the deliberate additional recharge of water to aquifers for subsequent recovery or environmental benefit (Dillon et al., 2009). The injection of water for storage and recovery from the same well is called aquifer storage and recovery (ASR). Aquifer storage, transfer and recovery (ASTR) refers to systems in which one well is used to inject water and another well is used to recover it. This guarantees a certain transit time of the water in the aquifer, which may have a beneficial effect in terms of the microbiological characteristics of the pumped water. Instead of pumping wells, infiltration basins or canals are often used for infiltration or abstraction of water.

An area where MAR has been successfully applied since 1957 is the drinking-water production area for the city of Amsterdam, located in the coastal dunes of the western part of the Netherlands. Groundwater abstraction started here in 1853, initially through a network of canals (Geelen et al., 2016). By 1880, a volume of 10×10^6 m^3 of water was produced annually, which was unsustainable as it equated to the precipitation excess (rainfall minus evapotranspiration), and as early as 1900, plans were proposed to use river water to compensate for the freshwater being abstracted. The discovery of a deeper aquifer though, led to the decision to withdraw groundwater using deep wells. The plan was fiercely opposed by the director of the water supply company, Pennink, who understood the risk of up-coning and was the first to demonstrate this process in the field (Houben and Post, 2017), but proponents wrongly believed the aquifer was being recharged laterally. Pennink was proven right when decades of unsustainable abstraction led to an upward movement of the bottom of the freshwater lens by tens of metres (Figure 5.14) over much of the dune area in the 1950s (Geelen et al., 2016).

The artificial recharge scheme that was started in 1957 aimed to restore the volume of freshwater in the deeper aquifer. Since the 1990s, the ecological restoration of the dune landscape became an additional objective, as many dune valleys, which contained water before groundwater abstraction began in 1853, had fallen dry. The scheme works by first pre-treating the river water, which is then infiltrated through several infiltration basins. The subsurface passage ensures that microbiological contaminants are removed. The river water has higher chloride and nutrient concentrations than locally recharged meteoric groundwater (Stuyfzand, 1999; Karlsen et al., 2012). The quality of the river water has a strong influence on the vegetation types that develop, which initially hampered ecological restoration efforts (Geelen et al., 2016).

Elsewhere, the introduction of allochthonous water has also led to issues with water quality. In Florida, for example, the injection of oxic potable water into an aquifer containing anoxic brackish water led to the mobilisation of arsenic through the oxidation of pyrite (Wallis et al., 2011). Other effects may include the enrichment of sodium due to cation exchange, or the release of iron and manganese (Zuurbier et al., 2016). Unlike arsenic, these elements are not hazardous, but an excess of sodium may render water unsuitable for irrigation purposes (Section 5.7.1), and iron and manganese may form precipitates that can clog pipes and wells. Water quality issues are particularly of concern when treated wastewater is used to augment freshwater supplies. Recharge of the Korba aquifer in Tunisia by treated wastewater since 2008 led to an improvement in terms of groundwater salinity, but the concentrations of nitrate and bacteria remained invariably high (Cherif et al., 2013).

The recovery efficiency is the volume of freshwater that can be recovered from an aquifer as a fraction of the volume injected. It is affected by several factors, such as the ambient groundwater flow and mixing between native groundwater and injected water at the fringes of the stored plume. In coastal aquifers containing brackish or saline groundwater the recovery is negatively impacted by buoyancy effects and up-coning (Oude Essink, 2001c; Ward et al., 2009; Zuurbier et al., 2017). These undesirable effects can be counteracted by optimising the position of the well screens used for injection and recovery. For example, the injection can take place in the deeper part of the aquifer, and the recovery in the shallower part, which minimises the risk of up-coning and upward losses due to buoyancy. Another option is to use horizontal wells. One concept that has been trialled in the Netherlands is to use a shallow horizontal well for injection and recovery of freshwater, in combination with a deeper one that scavenges the up-coning brackish groundwater (Zuurbier et al., 2017).

12.4.3 Optimisation of Pumping

12.4.3.1 Well Design

The placement and abstraction regimes of pumping wells have a strong effect on the migration saline groundwater. The risk can be lowered by positioning the well screens as far away from the transition zone, and as close to the water table as possible. Minimising the drawdown is key to preventing salinisation, but caution is required because while better well design can slow down the migration of saline groundwater, salinisation remains inevitable in the long run when pumping rates exceed the rates of freshwater recharge.

The drawdown caused by vertical wells placed above the saline groundwater will induce saltwater up-coning (Schmorak and Mercado, 1969). Various well configurations have been applied to minimise up-coning effects (Sufi et al., 1998), which have been summarised in Figure 12.6. Instead of using a single, high-discharge vertical wells it is better to distribute pumping across several wells with a lower pump rate. For example, on the German island of Langeoog, 20 small wells are run intermittently at pumping rates of only 10 $m^3 h^{-1}$ (Houben et al., 2014). A review of various field and numerical studies showed that multiple wells can yield more water of better quality than single-borehole wells for the same aquifer

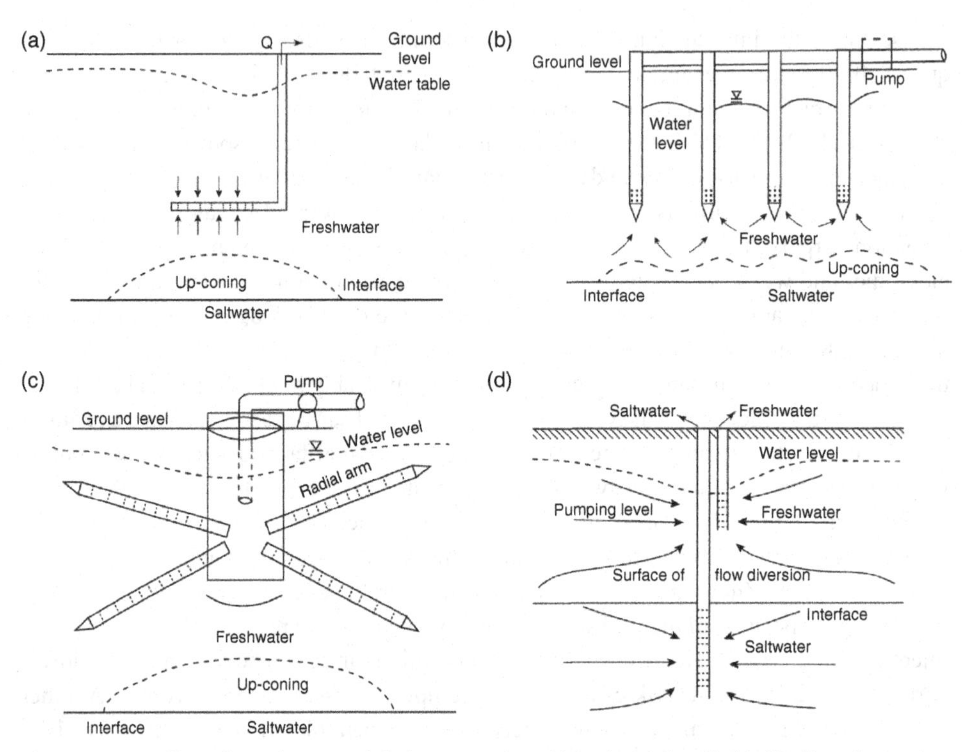

Figure 12.6 Well configurations to minimise up-coning: (a) horizontal well; (b) distributed pumping from multiple wells; (c) radial collector well; (d) scavenger well (modified from Sufi et al., 1998).

conditions (Aslam et al., 2016). These studies have also shown that limiting the duration of pumping and allowing for recovery periods are important to ensure that the salinity of the pumped water remains low. Moreover, traditional shallow, large-diameter dug wells were found to be less susceptible to salinisation than drilled vertical wells in a study in East Africa (Comte et al., 2016). This is because the large-diameter wells cause less drawdown than vertical wells, and the vertical wells have a greater depth and therefore their screens are positioned closer to the transition zone.

Horizontal wells have been applied extensively to abstract groundwater from thin freshwater lenses on atoll islands (White and Falkland, 2009). They are installed just below the water table and skim the water from the top part of the lens. In effect, they cause a reduction of the meteoric recharge, which leads to a thinning of the lens. This means that care must be taken that a large-enough buffer of fresh groundwater remains below the wells. The bottom of the lens will move up and down as recharge rates vary with time, and during prolonged droughts the brackish water transition zone may reach up into the wells and cause salinisation.

On Roi-Namur Island, part of the Kwajalein Atoll in the Republic of the Marshall Islands, a horizontal scavenger well is combined with a rainwater harvesting system to meet the water demand. The rainwater is collected from the 1370 m long runway by means of two concrete-lined basins, and provides enough water to meet the demand during the wet season between May and November. Groundwater is required during the dry season from December to April and is withdrawn using a 1000 m long horizontal well that runs parallel to the runway. Such wells are usually not drilled but are basically covered drainage ditches. The freshwater lens is artificially recharged through a swale above the horizontal well when there is more rainfall than what can be stored in the tank storage system. The system thus successfully combines different technologies to manage the freshwater lens, but due to the low elevation of the island, it is vulnerable to flooding by seawater (Storlazzi et al., 2018). A flooding event occurred in 2008 and caused salinity levels above potable limits for a period of 22 months after the event. During this time, water had to be brought in by barge, and extra treatment was required using a temporary reverse osmosis desalination system (Gingerich et al., 2017).

A radial collector well consist of a central large-diameter shaft of up to 30 m depth and several horizontal collectors screens that feed into it. The size, length, number and pattern of the collectors can be modified to achieve the optimum yield. Numerical and analytical studies have been carried out to study the flow to radial collector wells (Chen et al., 2003b; Bakker et al., 2005; Patel et al., 2010). The advantage over vertical wells is that extraction is distributed over a larger area and maximum drawdown is thus smaller, thus being less susceptible to up-coning. In recent years, directionally drilled horizontal wells have become available.

Scavenger wells are used to pump out saltwater to prevent it from reaching a production well. In one of the earliest published studies, Long et al. (1965) reported on an early successful field trial and demonstrated that this system can be effective for recovering freshwater floating over saline water. An overview of the development of scavenger well systems was provided by Aliewi et al. (1993), who further presented a numerical modelling study to identify the important parameters affecting the movement of saline water, such as recharge and pumping rates, the location of the well screen and the vertical hydraulic conductivity. Scavenger wells have been widely used in Pakistan (Sufi et al., 1998), where freshwater is being pumped from an aquifer with a stratified groundwater salinity distribution so that care must be taken not to draw the more saline water from greater depths into the wells screen.

The main problem with such compound well systems is the disposal of the pumped saltwater. One solution might be to combine a scavenger well with a reverse osmosis system to desalinate the abstracted saline water for additional water supply (Zuurbier et al., 2017). The remnant brine can be injected into a deeper aquifer with saline groundwater to minimise the environmental impact. A trial of this concept has been successfully conducted at a production well site in the Netherlands that had to be abandoned in 1993 because of salinisation. Fresh and brackish groundwater were both pumped at 50 m^3 h^{-1} so that the position of the transition zone remained stable. Chloride concentrations decreased across the aquifer after 8 months of pumping (Figure 12.7).

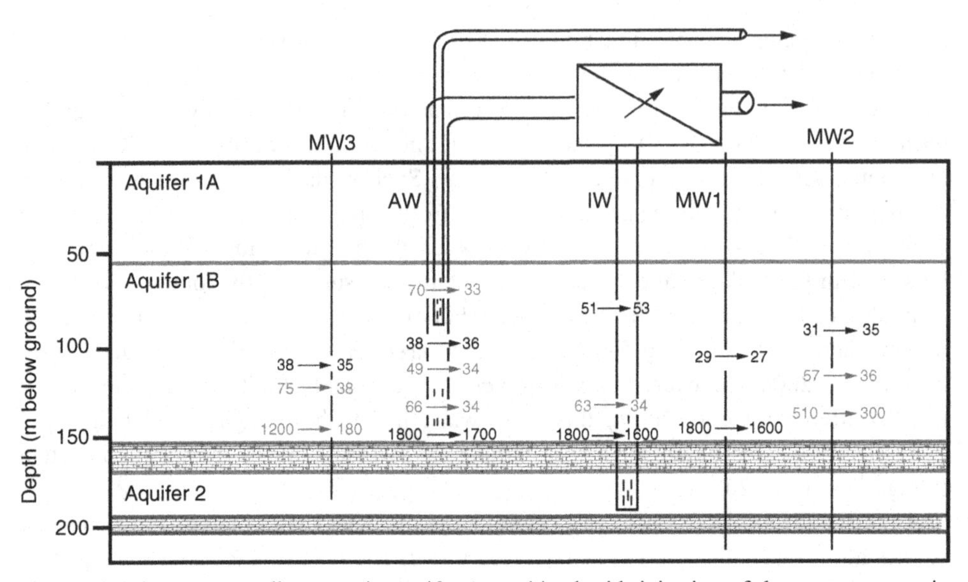

Figure 12.7 Scavenger well system in aquifer 1 combined with injection of the reverse osmosis concentrate into the deeper aquifer 2. Numbers show the chloride concentrations in mg l^{-1} at the start and after 8 months into the trial (Zuurbier et al., 2017). The layers with the dashed hatching denote aquitards. The transition zone is in aquifer 1B, and aquifer 2 is a confined, saline aquifer in which the remnant brine is injected. The aquifer below the deepest aquitard was not targeted. AW = abstraction well; IW = injection well; MW = monitoring well.

12.4.3.2 Pumping Regimes

The public water supply system in the South Downs region in southern England forms a well-known example of a coastal aquifer that has been managed sustainably over decades (Robins et al., 1999). Key to the success is the alternating pumping regime. The basic idea behind it is that by pumping groundwater from wells near the coast during winter, the fresh groundwater outflow to the sea is intercepted. Wells located more to the inland are only pumped during summer, which allows the storage volume of fresh groundwater to be replenished during the winter months. The strategy has allowed for an increase of the abstraction volumes without causing seawater intrusion. Integral to the water management of the region is a sophisticated monitoring system. The fissured nature of the aquifer makes that the salinity distribution is irregular and temporally variable, and the susceptibility of pumping wells to seawater intrusion is not a simple function of its distance to the coast. The operation of individual pumping wells thus needs to be fine-tuned, based on the knowledge about the salinity response to pumping, recharge and tidal influence. Moreover, the conceptual understanding of the hydraulics of the aquifer system becomes more refined as more data become available. Robins and Dance (2003) concluded that more groundwater was discharged laterally into rivers and streams, and less directly to the sea, than was previously thought. This demonstrates that coastal aquifer management is not a rigid framework, but a continuous, adaptive process that allows for further optimisation based on developing insights. It also needs to allow for changing environmental conditions, economic factors and changes in demand.

Optimisation of the temporal pattern of pumping may also reduce seawater intrusion. For example, in Camano Island in the USA, a computer-controlled system disperses the pumping over as long a time period as possible, while avoiding periods of high tide (Purdum and Engel, 2003). The system was installed after seawater intrusion was detected in 1994, based on summertime chloride concentrations of over 250 mg l^{-1}. The computer program optimises pumping based on input data about the demand, the tides and a set of pumping well operational rules, such as number of pumps, pumping rates, sequencing and timing. By increasing the above-ground storage capacity, the demand during peak periods such as weekends could be spread out by pumping over several days and storing the pumped water in large tanks. Taking into account the tides, and pumping during low tide only, was a key factor in lowering the chloride concentration of the pumped water. Chloride concentrations continued to fall even though consumption increased after an extra well was added.

Adaptive pumping has become more feasible with frequency-regulated pumps becoming more widespread. They allow easy adaptation of the pumping rate, usually by remote control, while earlier fixed-frequency pumps could only operate at a fixed pre-set rate.

The freshwater lens on the island of Bonriki, part of the Tarawa atoll in the Pacific Republic of Kiribati is the main water supply source for over 50 000 residents. It is being exploited by means of a system of horizontal wells and is monitored extensively to assess the state of the resource. The data have highlighted the strong dependency of the salinity of the pumped water on the preceding rainfall, which means that the water supply system is vulnerable to drought. Galvis-Rodriguez et al. (2017) conducted an extensive field and numerical investigation to determine optimal pumping strategies during drought periods. The study revealed that a redistribution of pumping between the various horizontal wells can be an effective strategy to protect pumping stations from salinisation and to safeguard the supply of freshwater during extended droughts. The graph in Figure 12.8 compares two scenarios: one in which pumping proceeds at the unmodified rates during droughts, and one in which pumping is reduced by switching off wells once they pump groundwater with a specific conductance of more than 1000 μS cm^{-1}. Clearly, the dynamic pumping regime can be successful in maintaining the salinity of the pumped water, albeit at the expense of a reduced water supply. The model has been used to recommend management responses and to integrate them into the government's drought response plan.

Pumping optimisation models form a separate category of management models and a summary review was presented by Singh (2014). The aim of these models is to find the maximum pumping rate subject to a set of constraints. These could be the minimisation of drawdown, salinity of the pumped water, volume of intruded seawater, inland toe distance or pumping costs. The difficulty faced with these models in coastal aquifers is that the dependence of salinity on the pumping regime results in a nonlinear optimisation problem (Abarca et al., 2006). Another major problem is the long run times of numerical models, which means that the execution of many model runs to solve the optimisation

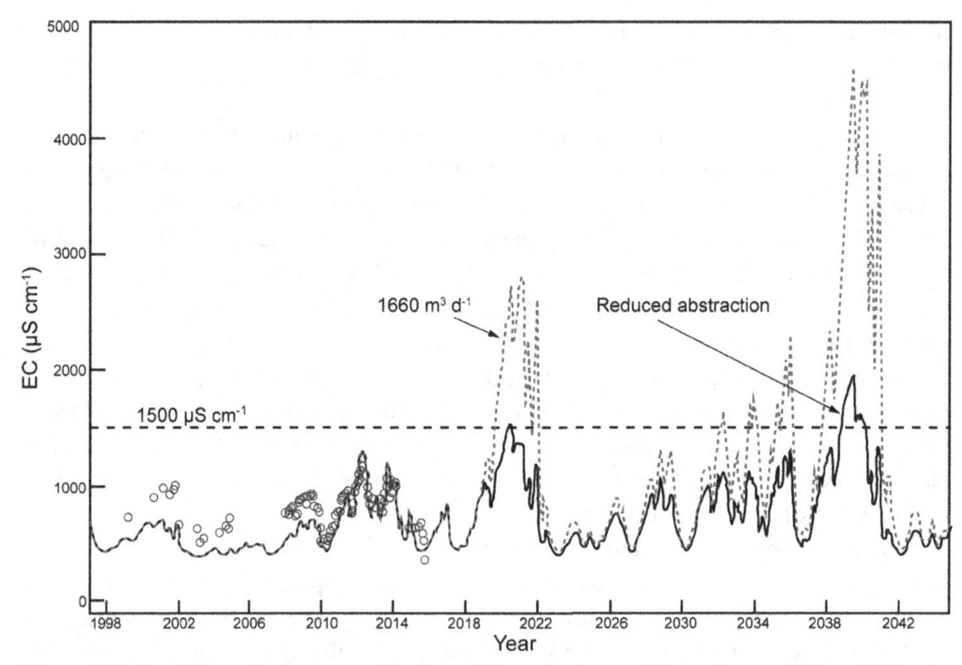

Figure 12.8 Specific conductance as a function of time based on a calibrated numerical model for the Bonriki Island freshwater lens (Galvis-Rodriguez et al., 2017). The dashed line represents the scenario in which pumping rates are constant, whereas the solid line is for a model in which pumping rates are adjusted dynamically based on salinity exceedance criteria.

problem becomes unpractical. To circumvent this problem, some optimisation models use analytical solutions (Section 3.5) or flow-only models, but these may not represent the true complexity of the aquifer system.

The introduction of water-saving measures is a very effective way to decrease the pressure on coastal groundwater resources. On the German island of Langeoog, water consumption from the local freshwater lens in 1983 was 452 000 m^3 yr^{-1}, while in 2011 the pumped volume was down to 333 000 m^3 yr^{-1}, despite a stable number of tourists visiting the island (Houben et al., 2014). This was achieved by the introduction of water-saving toilets and household-appliances.

12.4.4 Engineered Barriers

The basic concept behind a subsurface barrier is to minimise both the freshwater loss to the sea and the intrusion of seawater into the aquifer. Two general categories of barriers exist: hydraulic barriers and physical barriers (Figure 12.9). Hydraulic barriers work by pumping saline groundwater, whereas physical barriers are impermeable zones that cut off the hydraulic connection between the seawater and fresh groundwater.

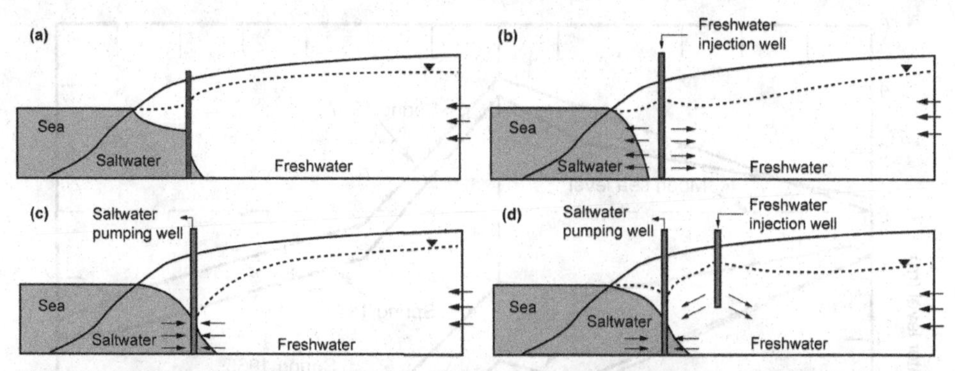

Figure 12.9 Types of barriers to control seawater intrusion (Pool and Carrera, 2010). Only unconfined aquifer situations are shown, but most of the concepts also apply for confined aquifers: (a) low-permeability subsurface barrier; (b) hydraulic ridge or positive hydraulic barrier by injection wells or infiltration ponds; (c) hydraulic trough or negative hydraulic barrier by pumping saltwater; (d) mixed barrier (injection-pumping system).

12.4.4.1 Injection Barriers

Injection barriers are created by injecting freshwater to form a high head zone between an inland pumping zone and the sea. The injected water can be river water, treated wastewater and rainwater. The quality of the injected water should not have any negative environmental impacts and quality requirements are bound to local legislation, which means that some form of pre-treatment is usually necessary. If the injected water ends up in nearby pumping wells, the distance between the injection and pumping locations must be large enough so that the injected water spends enough time underground for microbiological impurities to become neutralised.

One of the longest-operating and largest scale barrier system is that in the Central and West Coast Basins in Los Angeles county, USA (Reichard et al., 2003), where seawater intrusion had resulted in the abandonment of many wells during the first half of the twentieth century. The construction of the barrier started in 1951 with a pilot study of nine injection wells spanning about 1.6 km of coastline. It was followed by the extension of the system to 153 injection wells stretching over a 14.5 km transect parallel to the coast at a distance of at least 1.6 km from the shore. Since there are three aquifers in the area, the depths of the well screens are variable, ranging from 85 to 213 m. The distance between the wells varies from about 46 m to 259 m. The performance of the system is monitored by measuring water level and chloride concentrations, as well as other parameters, in a network of 276 observation wells (Johnson and Whitaker, 2004). The water level at midpoints between injection wells is monitored to identify any possible breaches. Figure 12.10 shows how the water levels in the aquifer changed after the installation of the injection barrier during the 1960s.

New injection wells had to be added where seawater intrusion was detected, and because the barrier divided the existing intruded seawater plume, a number of

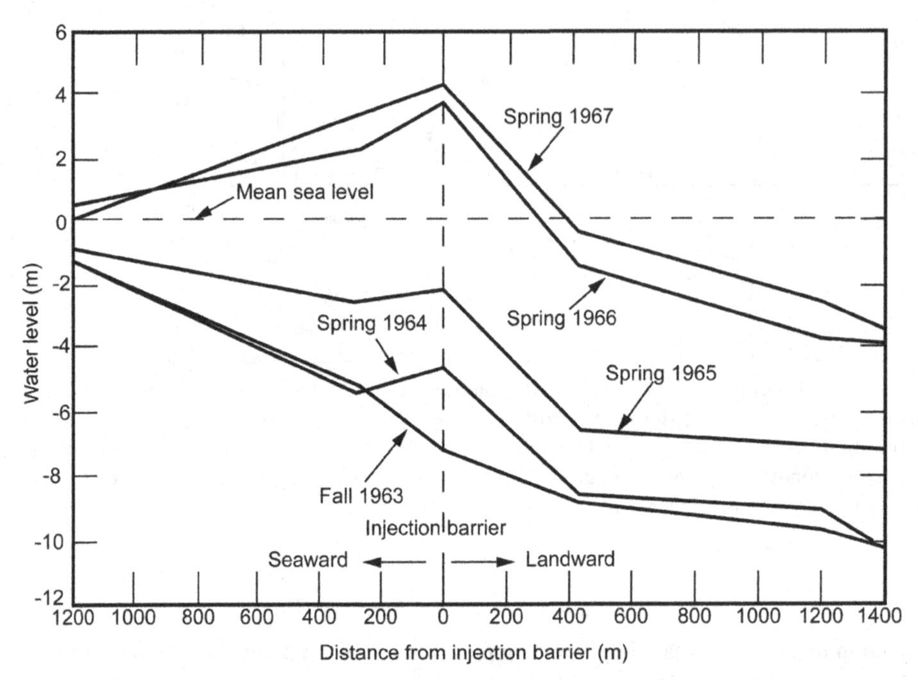

Figure 12.10 Water levels in the Silverado aquifer (California, USA) along a transect perpendicular to the seawater intrusion barrier at various time intervals demonstrating the establishment of the pressure ridge after initiation of the barrier at the end of 1963 (Todd, 1980).

extraction wells had to be installed on the landward side to pump out the saline groundwater. Before 1995, the injection water was imported from northern California and the Colorado River. Since 1995 treated wastewater has been used as well and has formed the sole source of injected water since 2009 (Johnson and Whitaker, 2004). There are strict requirements for the wastewater used for injection. The minimum residence time in the subsurface is 1 year, and the minimum travel distance between injection and the nearest drinking-water well must be 610 m. Also, the contribution of recycled water to an extraction well is not allowed to exceed 50% (Foreman, 2003).

Another example of an injection barrier is that in the Llobregat Delta aquifer in Spain, where treated wastewater is injected using a system of 15 wells, which are distributed along a 6 km line about 1 km inland from the coast (Ortuño et al., 2012). The wells reach down to 70 m and the water is injected into the confined part of the delta. Using an electronically controlled, remotely monitored system, each well injects water at a specific rate, which depends on the local conditions, to maintain the water level between 1 and 3 m above sea level. The water that is injected is tertiary-treated wastewater, which is further treated using ultrafiltration, reverse osmosis and UV disinfection at a water treatment plant with a production capacity of 15 000 m^3 d^{-1}.

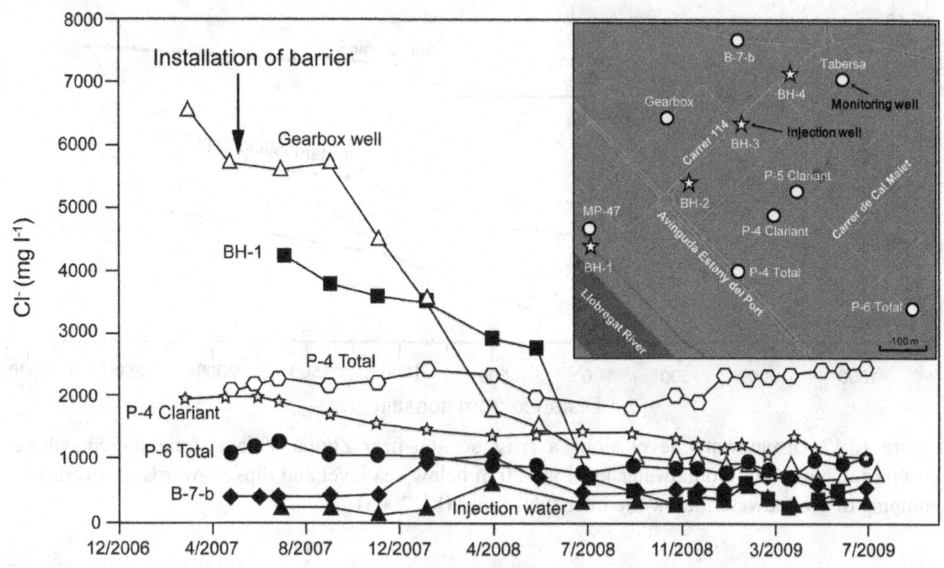

Figure 12.11 Chloride concentrations as a function of time for the monitoring wells during phase I of the hydraulic barrier pilot in the Llobregat Delta Aquifer (modified from Ortuño et al., 2012). The inset shows the locations of the injection wells (stars) and monitoring wells (circles).

The aquifer is monitored using a total of 37 monitoring wells, from which samples are taken for the measurement of major ions, nitrogen species, metals and chlorinated hydrocarbons. A pilot operation was started in 2007 and by mid-2010, about 4 million m^3 of water had been injected. In eight wells within 1 km from the four injection wells used during the pilot, a steady decrease of salinity has been observed with time (Figure 12.11), and the injection affects an irregularly shaped area extending 1 to 2 km from the injection wells. No clogging effects were encountered, and the costs of the scheme turned out to be more economic than seawater desalination (Ortuño et al., 2012).

12.4.4.2 Extraction Barriers

For aquifers in which a rise in water level by injecting water would cause flooding or damage to underground infrastructure, an extraction barrier may form an alternative (Pool and Carrera, 2010). This set-up consists of a line of wells installed near the shoreline to abstract the saline water and prevent it from reaching inland pumping wells. In a sense, the principle is similar to the idea of a scavenger well, except that multiple wells are combined to form a barrier parallel to the coastline. Theoretical investigations have shown that the pumping rates must be chosen carefully. If they are too high, too much freshwater is drawn towards the barrier wells. If too low, seawater can flow around and underneath the barrier wells and contaminate the inland pumping wells (Pool and Carrera, 2010).

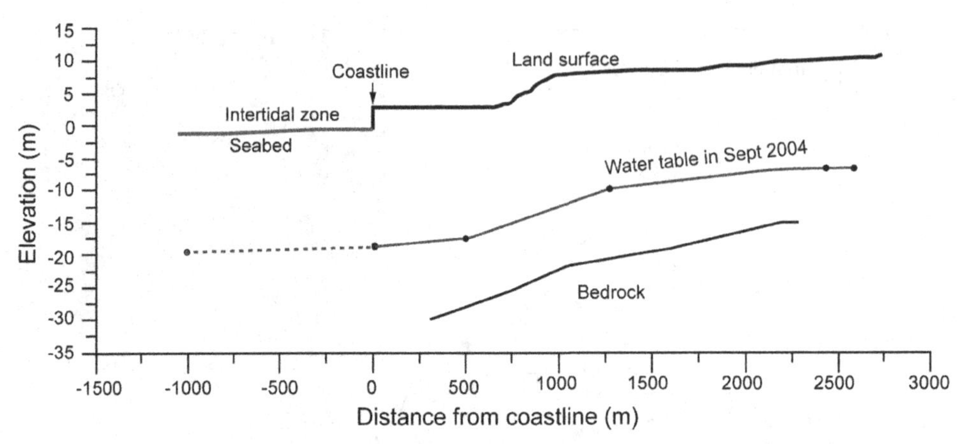

Figure 12.12 Groundwater level along a cross section near Zhujia Village, Laizhou, Shandong Province, China. The groundwater level is ~20 m below sea level and dips seawards as a result of pumping of groundwater below the inter-tidal zones (Li, 2005).

The water pumped in the saltwater zone is brackish and must be discharged somewhere where it does not contaminate freshwater. No known applications exist of negative barriers to prevent seawater intrusion, but the system appears to be applied unintentionally in aquaculture. In the Shandong Province in China fish farms have been pumping brackish water in the inter-tidal zones to feed turbot, which is native in marine or brackish waters, since the 1990s (Li et al., 2011c). Each farm consists of a greenhouse with wells that pump groundwater from 10 to 20 m below the seabed. Groundwater is preferred because it has a more stable quality and temperature than seawater (Lei et al., 2002). The wells have resulted in a lowering of the groundwater level to approximately 20 m below sea level in the intertidal zone (Figure 12.12). Abandoned large-diameter wells can cause seawater intrusion, and the farms have a detrimental effect on the above-ground environment as well. Some local governments have therefore banned the uncontrolled turbot aquaculture activities.

In the Ansedonia promontory in Tuscany, Italy, the production of various species of sea fish is a multi-million Euro industry, which relies on pumping of relatively warm (21°–24°C), saline water from a carbonate aquifer. The pumping creates an elongated cone of depression, where heads are a few metres below sea level, which is deepest after the summer months (Figure 12.13). The wells of the fish farms also draw in significant amounts of freshwater. Even though only the saline water is needed for their operations, Nocchi and Salleolini (2013) found that the fish farms were the largest user of freshwater in the area, much greater than irrigation and domestic use. The solution is to drill deeper wells into the part of the aquifer where the groundwater is completely saline. If done with care, more freshwater could be saved, and the wells could form an extraction barrier that protects the inland irrigation and domestic wells against seawater intrusion.

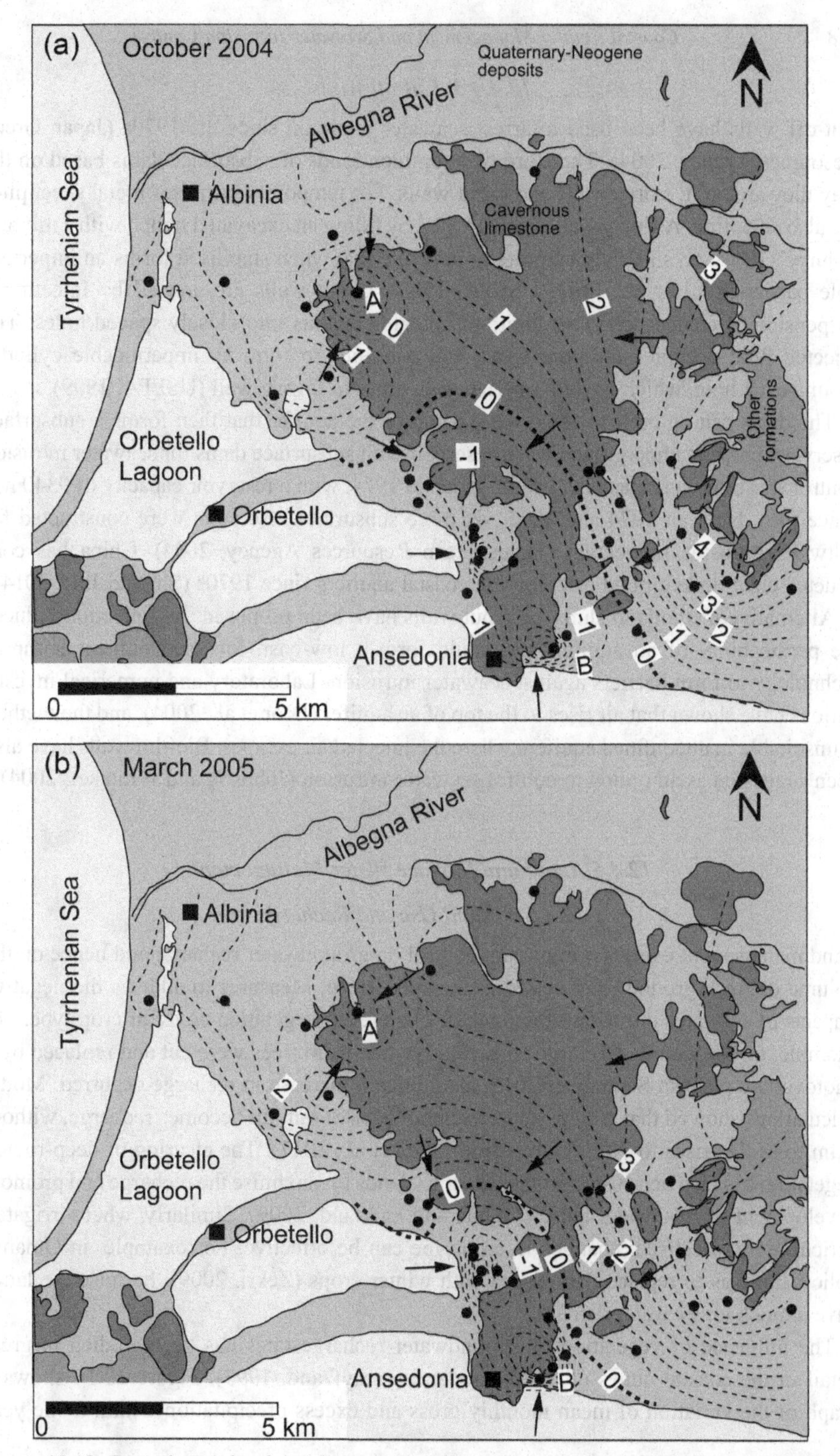

Figure 12.13 Groundwater level (m above sea level) contour map for the carbonate aquifer in the Ansedonia promontory, Italy, for (a) October 2004 and (b) March 2005 (modified from Nocchi and Salleolini, 2013). The arrows indicate flow direction.

12.4.4.3 Cut-off Walls

Cut-off walls have been used to arrest seawater intrusion since the 1970s (Japan Green Resources Agency, 2004). There are two common kinds of subsurface dams based on the way they are built: slurry walls and grout walls. For temporary purposes metal sheet piles are also an option. A slurry wall is constructed by filling an excavated trench with a mixture ('slurry') of water, soil and bentonite or concrete that when stabilised forms an impermeable barrier (Paul et al., 1992; USEPA, 1999). Grout walls are created by injecting a suspension of cement, clay, bentonite or silicate reagents into closely spaced holes. The injected fluid occupies the pore space and solidifies to form an impermeable cylinder around each hole, which have to overlap each other to form a wall (USEPA, 1999).

The dams can be built to enclose a portion of the aquifer that then forms a subsurface reservoir. The first underground water reservoir with subsurface dams for seawater intrusion control was built in Kabashima Island, Japan in 1974, with a reservoir capacity of 9340 m^3. Since then, between 1974 and 2003, six more subsurface reservoirs were constructed for saltwater intrusion prevention (Japan Green Resources Agency, 2004). China has constructed eight underground reservoirs in coastal aquifers since 1970s (Shi and Jiao, 2014).

Alternative methods to slurry or grout walls have been proposed. Air injection reduces the permeability of an aquifer, and could form a low-cost, low-environmental impact technology to form barriers against seawater intrusion. Laboratory and numerical investigations have shown that air rises to the top of an aquifer (Dror et al., 2004), and the method is unsuitable in unconfined aquifers where the injected air escapes. Biofilm walls have also been suggested as an option to control seawater intrusion (Johnson and Whitaker, 2004).

12.4.5 Land and Surface Water Management

12.4.5.1 Land Use and Recharge

Land management exerts an important control on groundwater recharge and hence on the volume of fresh groundwater stored in the subsurface. Measures to address the negative impacts of seawater intrusion often include changes in vegetation cover or crop type. For example, on the island of Bonriki in Kiribati, when palm trees were cut and replaced by a photovoltaic plant in September 2015, a significant increase in recharge occurred. Model calculations showed that with palm trees, about 54% of rainfall becomes recharge, without palm trees this rises to 76% (Galvis-Rodriguez et al., 2017). The clearing of deep-rooted vegetation has also been used in other Pacific islands to maximise the recharge and promote development of freshwater lenses (White and Falkland, 2009). Similarly, where irrigated agriculture is practised, changes in crop type can be effective. For example, in Oman a policy aim was to replace date palms with winter crops (Zekri, 2009), because the latter have a much lower water demand.

The influence of vegetation on groundwater recharge rates has been studied in great detail in the coastal dunes of the Netherlands (Stuyfzand, 1993). Figure 12.14 shows a graph of the variation of mean monthly gross and excess precipitation within a one year

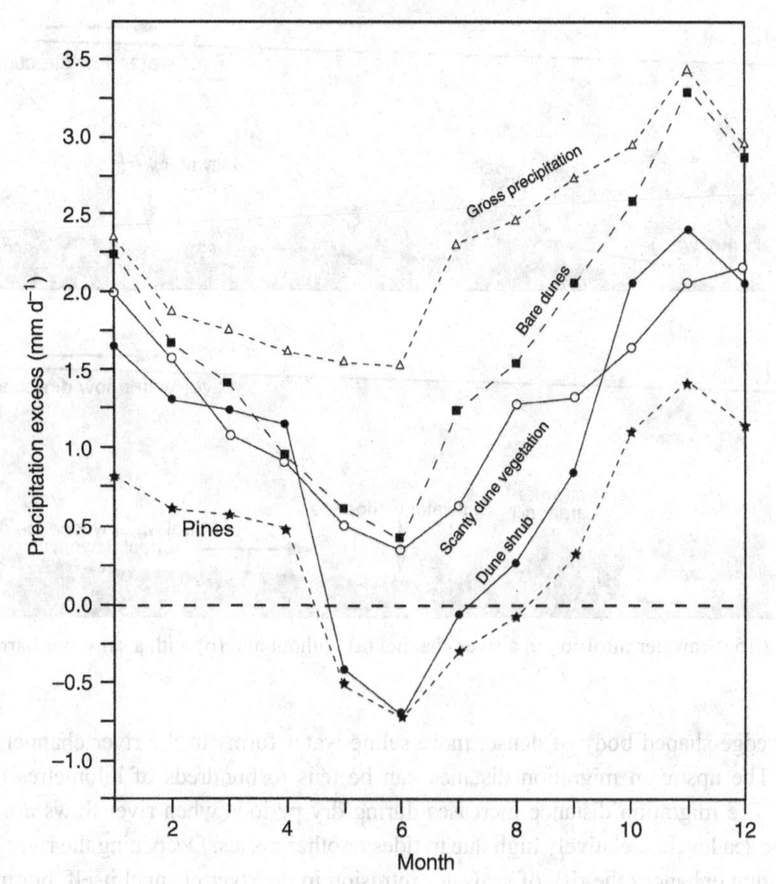

Figure 12.14 Mean monthly gross and excess precipitation for various vegetation types in the coastal dunes of the Netherlands (Stuyfzand, 1993).

period for various vegetation types based on lysimeter studies. The precipitation excess is the gross precipitation minus all (actual) evaporative losses. The differences are significant and mainly reflect changes in rainfall interception, which is the part of the rainfall that is evaporated from the canopy. The maturity of the vegetation plays an important role, and recharge decreases as the vegetation becomes more mature. While bare and sparsely vegetated sites receive recharge throughout the whole year, areas with shrub and pine trees do not receive much recharge in the summer. The negative values in the graph of Figure 12.14 indicate that soil moisture is depleted during this time.

12.4.5.2 Surface Water Salinity

In low-lying coastal areas, seawater can migrate upstream in river channels. If there is a good hydraulic connection with the groundwater, seawater may intrude into the aquifer via the riverbed. The mechanism of seawater intrusion in surface water is similar to that in aquifers in

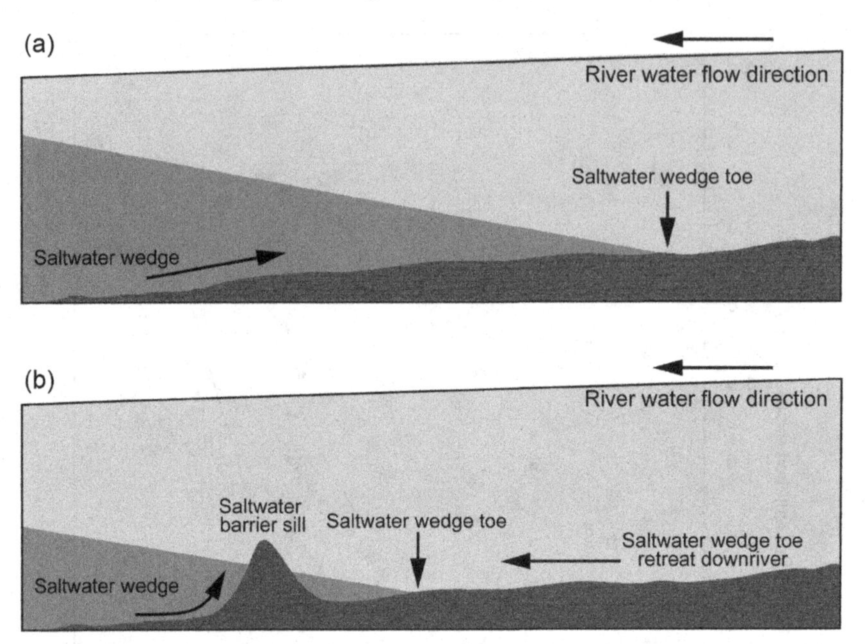

Figure 12.15 Seawater intrusion in a river channel (a) without and (b) with a saltwater barrier sill.

that a wedge-shaped body of dense, more saline water forms in the river channel (Figure 12.15). The upstream migration distance can be tens to hundreds of kilometres (Section 10.3.2). The migration distance increases during dry periods when river flows are low, or when the sea level is relatively high due to tides or other cycles. Deepening the river channel by dredging enhances the risk of seawater intrusion in the river channel itself, but may also increase the connectivity between the river and the subsurface, possibly leading to more groundwater salinisation. Erosional and depositional processes may also cause changes in the river bed hydraulic resistance, which varies with time as a result (Hatch et al., 2010).

A low hydraulic conductivity sill constructed at the river bed to the proper height above the river bottom can reduce saltwater flow and thus arrest the wedge, so that the freshwater intakes upstream of the sill can be protected (Figure 12.15). Saltwater sills are used in Mississippi River to limit upstream salinity intrusion (McAnally and Pritchard, 1997). Sluices can be constructed in rivers to prevent inland migration of seawater and to control river stages and groundwater levels, as practised in the canals in coastal aquifers in Florida, USA (Barlow, 2003). In the wet season, the gates can be opened to let river water flow to the sea to prevent flooding, whereas in the dry season, the gates can be closed to prevent seawater intrusion into the inland channel (Figure 12.16). In Singapore, a large-scale gated structure, or a barrage, separates the freshwater in the Singapore River from the seawater of the South China Sea (Khoo, 2009). The barrage consists of nine large steel gates that can release water to the sea when the river water level is high, or prevent seawater intrusion when sea level is high, by adjusting their angle.

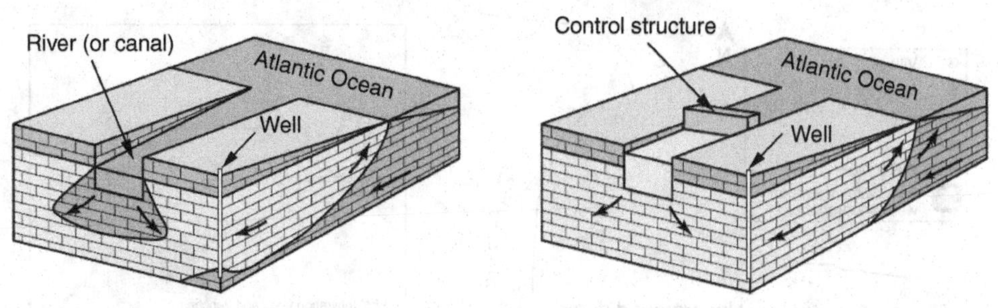

Figure 12.16 Coastal aquifer in contact with a river with and without a control structure: (a) periodic seawater intrusion in an uncontrolled river influenced by tidal seawater; (b) control structures (sluices) in a river prevent inland migration of seawater and provide flood protection as well as artificial recharge to the aquifer (Barlow, 2003).

Building structures to prevent seawater intrusion into rivers and estuaries has enormous ecological consequences as the natural salinity of the water decreases and tidal influences are diminished. When building the large flood defence structures in the south-western part of the Netherlands after the catastrophic 1953 flood, a project that took several decades to design and complete, the original emphasis was on maximum protection. Concerns over fisheries and changing views about the ecological significance of the largest estuary to be closed off, led to a change in the design of the dam. The barrier that was built ended up having 62 movable gates, which are only closed during extremely high seawater levels. The estuary thus maintained its connection to the sea.

12.4.5.3 Land Reclamation

The reclamation of land has been suggested as a way to increase the storage of freshwater in the subsurface and displace the saline groundwater seaward (Van Dam, 1999; Oude Essink, 2001c). The effects of land reclamation on freshwater lenses have been discussed in Chapter 9. It is an expensive option, which means that plans of creating land area for the sake of groundwater management alone are most likely not economically viable.

The development of a freshwater lens beneath land reclamation sites has been studied in Singapore where over 2000 ha were reclaimed for airport expansion and industrial development. As the development of the reclamation works progressed in various stages, differences in freshwater lens thickness could be identified between sites that were older and those that were younger. Sand-filled areas that were completed in 1997 had freshened down to a depth of 6 m, whereas those completed in 2004 had freshwater up until 4 m below the ground surface (Chua et al., 2007). The relatively high annual rainfall of 2.4 m aids in building the lens. Topography and the compaction and heterogeneity of the fill material play a further role in the development of the lens. Nakada et al. (2012) found that the high permeability of the material used to reclaim swamps on the Fongafale Islet in Tuvalu increased the connectivity of the aquifer with the ocean, and prevented the development of a freshwater lens. This illustrates the need for an integrated management approach, which

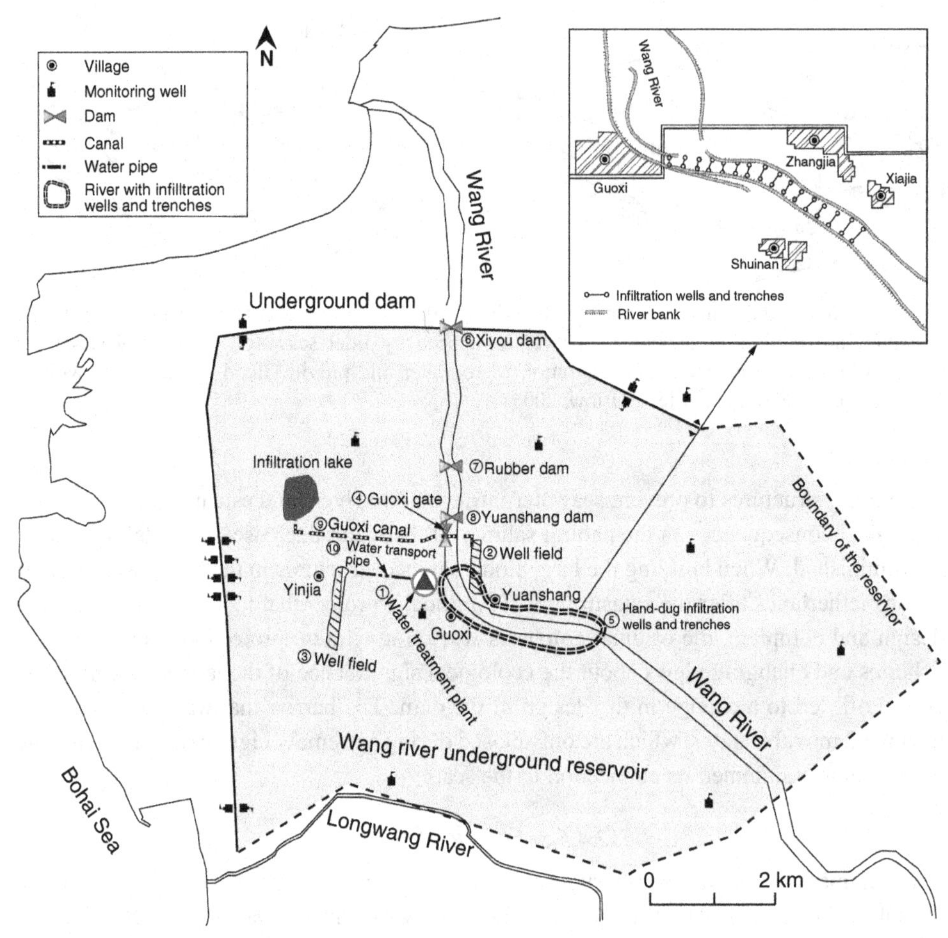

Figure 12.17 Locations of the underground reservoir, dam and other structures to control seawater intrusion and enhance river water infiltration. The inset shows the river section with installed infiltration wells and trenches. 1: water treatment plant; 2: no. 1 well field; 3: no. 2 well field; 4: Guoxi gate; 5: infiltration wells and trenches; 6: Xiyou Dam; 7: rubber dam; 8: Yuanshang dam; 9: Guoxi infiltration canal; 10: water transport pipe (based on Zhang, 2004; Xiong, 2011).

gives due consideration to groundwater when engineering works and other land management activities are proposed.

12.4.6 Case Study: Wang River Delta, China

The Wang river is located near Laizhou Bay in the Shandong Province of China (Figure 12.17). The delta of this river was affected severely by seawater intrusion, and represents a good example of an area where several engineering measures have been implemented to solve the problems. The case study is compiled based on the information from various sources (Zhang, 2004; Ma, 2005; Li, 2006; Xiong, 2011).

Due to the construction of a reservoir in the upstream part of the Wang River, there is limited discharge into the coastal section of the river except during heavy rainfall. The yearly precipitation is 604 mm on average, with a range of 356 to 1172 mm. Quaternary marine and river sediments overlie weathered, metamorphic bedrock with intrusions of granite and granodiorite. Depending on the location, three to four aquifers can be discerned, which are typically 10 to 20 m thick and have a hydraulic conductivity between 50 and 150 m d^{-1}. A fine-grained layer comprising clay and silt, with thickness ranging from 2 to 10 m, forms a superficial aquitard.

Groundwater forms the primary water supply for industrial, agricultural and domestic use. In the 1980s, the annual groundwater abstraction was 21 million m^3, while the renewable amount was estimated to be only 18 million m^3 yr^{-1}. This led to the development of a zone of heads below sea level and seawater intrusion into the Quaternary weathered bedrock aquifers, either laterally or by up-coning of palaeo-seawater. Seawater transported inland for commercial activities such as aquaculture and salt production also caused contamination of groundwater.

To address the seawater intrusion problems, an array of engineering measures were installed between the late 1970s and the year 2000. These include (i) river dams to stop seawater intrusion along the river channel and raise the river water level to enhance infiltration, (ii) an infiltration lake to recharge the shallow aquifer, (iii) infiltration wells and trenches to enhance infiltration of river water and (iv) underground dams to stop seawater intrusion as well as to create an underground reservoir.

The river dams are meant to store water and to elevate the river water level to different stages at different parts of the river. The Xiyou dam was aligned with the north underground dam (Figure 12.17) to not only impound river water but also to prevent seawater from migrating upstream through the river channel during high tide. The Yuangshang dam creates a reservoir with a storage capacity of 316 000 m^3. All the water in the reservoir, except for the loss due to evaporation, infiltrates into the ground. A topographic depression near Yinjia Village, originally formed by sand excavation, was turned into an infiltration lake that can hold up to 1 million m^3 of water. During the summer, excess storm water is directed from the Wang river to the lake via the Guoxi channel.

Infiltration wells and trenches were installed in the bed of the river and canal so that river water can bypass the superficial aquitard. Where the aquitard was thin enough, wells were dug by hand, but most wells were drilled through the aquifer and into the bedrock. All together, 244 dug wells and 1068 bored wells were constructed (Figure 12.18a). Moreover 187 trenches were constructed over the river bed (inset of Figure 12.17). The wells and trenches were back-filled with coarse material to provide rapid infiltration pathways for the river water.

The boundaries of the underground reservoir are shown in Figure 12.17. The surface elevation in the reservoir area ranges from 3.0 to 30 m. The total area of the reservoir is 68.5 km^2. To create the reservoir, underground dams were constructed along the

Figure 12.18 (a) A drilled infiltration well of about 20 m deep at the river bed. The well is capped with a slotted concrete cover and sits in a catch pit which has not yet been filled with rock fragments and coarse sand. Note the aquitard material near the surface. (b) A section of the underground dam, which was excavated at this site to demonstrate the quality of the dam.

north and west boundaries. In the east, the low hydraulic conductivity bedrock forms a natural impermeable boundary, while in the south, the regional groundwater flow recharges the reservoir through the Quaternary deposits. The length of the underground dams measures 6.5 km in the north and 8.1 km in the west. The elevation of the top of the dam is 1 to 2 m above sea level and penetrates into the bedrock 0.5 m, making it up to almost 37 m tall locally. To build the dam, boreholes were drilled 1.8 m apart and cement was injected into the holes. The average thickness of the dam is 0.18 m. Figure 12.18b shows a small section of the dam excavated for inspection of the performance of the grouting and quality of the dam.

As a result of the engineered structures a water table mound along the river bed has formed. Water levels rose up to 8 m, especially near the infiltration lake and the river, where locally the water table reached levels as high as 5 m above sea level (Zhang, 2004). Annually about 32 million m^3 of water is artificially recharged, of which 13 million m^3 is pumped out for irrigation. Chloride concentrations fell significantly, and the groundwater reached potable quality.

12.5 Desalination

Desalination technologies have advanced tremendously during the last few decades and play an increasing role in meeting the water demand of coastal areas (Elimelech and Phillip, 2011; Stein et al., 2016). Early plants were based on thermal desalination, which require huge amounts of energy to heat seawater. Multi-stage flash, for example, is a production process by which heated seawater is depressurised so that it evaporates, and the condensing water vapour forms the produced freshwater (Miller et al., 2015). Modern reverse osmosis systems, in which the seawater is pushed through semi-permeable membranes under high pressure (Section 5.7.7.1), are much more energy-efficient. The large-scale application of seawater desalination is important to coastal groundwater resources as it forms a source of water beyond what is naturally available, and can thus be important to relieve stress on groundwater resources.

With technology improvements, the price of desalinated water has come down as well, to the level that it is competitive with other technologies to produce potable water. The use of desalinated water for irrigation remains limited because of the high cost, but in Spain, Israel and the USA, it is being used to grow high-value crops in some areas (Stein et al., 2016). The low salinity of the water can result in higher yields because plants experience less salt stress, and the risk of soil salinisation is reduced. Besides these beneficial effects, the relatively high concentration of the phytotoxic element boron, and the lack of Ca^{2+}, Mg^{2+} and SO_4^{2-} may cause plant stress and result in additional cost because of the required nutrient augmentation. Finally the dominance of Na^+ is a concern from the perspective of soil sodicity (Section 5.7.1), due to which swelling and a decrease in the infiltration capacity may occur (Martínez-Alvarez et al., 2016). On the long-term, irrigation with desalinated water will impact the deeper groundwater quality as well, as some of the water used will recharge the underlying aquifer.

Several environmental concerns are associated with desalination. The foremost is probably the enormous energy demand of desalination plants. The energy required to produce a cubic metre of freshwater from seawater by reverse osmosis is upward from 3 kWh, with an associated CO_2 emission of 1.4–1.8 kg per cubic metre of produced water (Elimelech and Phillip, 2011). The energy costs are significantly less when brackish water is used (Muñoz and Fernández-Alba, 2008). The disposal of large quantities of reject brine poses another environmental concern (Miller et al., 2015), also because the chemicals used during pre-treatment and cleaning become concentrated in the residual brine. The injection of brine into saline aquifers is considered one possible way to avoid disposal into the marine environment, but the associated risk to groundwater quality must be carefully considered (Stuyfzand and Raat, 2009).

The direct intake of seawater poses problems with entrainment of marine organisms and when the concentration of suspended particles is high, pre-treatment is necessary to prevent fouling of the membranes. Some plants, therefore, use saline groundwater instead, because of the natural filtration by the aquifer material (Figure 12.19). So-called beach wells are sometimes used for this purpose, but, among other limitations, the permeability of the subsurface may limit the amount of water that can be withdrawn, and there is the risk of abstracting too much freshwater. Intake pipes beneath the seafloor, drilled for example using directional horizontal drilling technologies, have been used with considerable success in some places (Peters and Pintó, 2008). The extraction of saline water through beach wells may be beneficial for the surrounding fresh groundwater resources by lowering saline water heads.

Brackish water is increasingly considered as feed water for desalination plants (Muñoz and Fernández-Alba, 2008; Greenlee et al., 2009; Stein et al., 2016). The main advantage is that it requires less energy to produce potable water from brackish water because of its lower salinity compared to seawater. Brackish groundwater is available in large quantities in many coastal regions, but when these are palaeo-waters, the supply may ultimately run

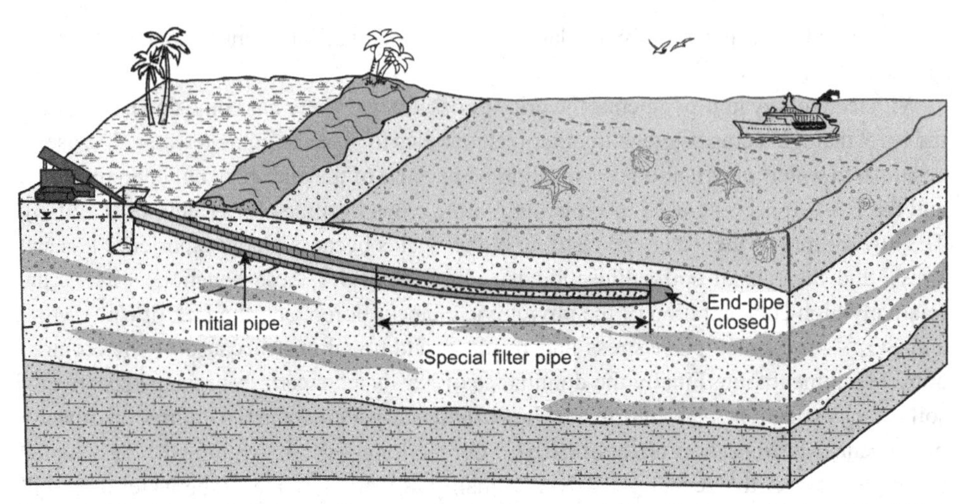

Figure 12.19 Schematic diagram of seawater intake from a coastal aquifer using a horizontally drilled well beneath the seafloor (modified from Peters and Pintó, 2008).

out (Stuyfzand and Raat, 2009). Moreover, the variations of the chemical composition within brackish groundwater bodies leads to temporal variability of the composition of the plant's feed water, which is undesirable from an operational point of view. While pumping brackish water can be a way to control seawater intrusion, there is also the risk that salinisation is exacerbated, or that uncontrolled pumping negatively impacts the overall condition of the aquifer (Hof et al., 2014).

Desalinated brackish groundwater has also been used for non-potable use. In London for example, brackish groundwater is pumped from an 87 m deep well to provide water for toilet flushing of a large entertainment complex (Smith et al., 2001). An additional benefit of the scheme was that the pumping lowers the water table, which had started to rise to undesirably high levels with the abandonment of more and more industrial wells. Some problems with the quality of the brackish groundwater were encountered, most notably the high hydrogen-sulphide concentrations, high iron content and aromatic hydrocarbons, which were leached from the former gasworks at the site. This shows that not only fresh groundwater reserves must be protected from contamination, also water bodies with a salinity traditionally perceived as being too high for beneficial use must be preserved as the advancement of treatment technologies may render their exploitation feasible in the future.

12.6 Socio-Economic Aspects

None of the technical solutions described in this chapter can be successfully applied if they are unacceptable to the community of stakeholders, or if they are economically unviable. Moreover, an effective legislative and institutional framework is required that allows seawater intrusion control to be fitted into a broader water management plan for a coastal jurisdiction. This also requires that measures can be enforced, which is often the major difficulty in successfully controlling water demand (Custodio, 2012b).

This is exemplified by the case of the urbanised coastal plain of Recife in Brazil described by Montenegro et al. (2006). In this area, deterioration of the limited groundwater resources has occurred due to an increase in exploitation and a reduction of recharge as a result of an expansion of paved surfaces. Overexploitation by a high number of private wells has caused a severe decline of the hydraulic head (up to 70 m in 20 years) and an increase of the salinity, and since the 1970s, multiple wells have been abandoned already. The centralised water supply is inadequate, and there is a large shortfall between the demand and the supply, which is compounded by losses of over 50% from the supply network. Groundwater resources are regulated by a state law that regulates ownership, well drilling and groundwater use. The implementation through the water management plan is based on a division of the region into zones and in the most critical ones, no new drilling is permitted, existing abstraction has to be severely reduced, and licences must be renewed every year. Initially, however, the management plan could not be introduced because at the time it was proposed, a drought occurred and to overcome shortages, a large number of wells were drilled. But also after the drought broke, and the plan came into effect, the enforcement of the regulations has been insufficient due to a lack of resources of the enforcing organisations.

Another case where the implementation of seawater intrusion control measures has proven to be challenging is the Batinah region on the coast of Oman, where irrigated agriculture relies fully on groundwater (Zekri, 2008). The reliance on groundwater was brought about by the widespread availability of diesel and later electrical pumps since the early 1970s. The large-scale use of groundwater led to a strong decline in water levels and seawater intrusion. The 1991 National Water Resources Master Plan was aimed at reducing usage by implementing a number of measures including a moratorium on new wells, improved irrigation efficiency, crop substitution, water tariffs for non-agricultural use and the re-use of treated wastewater. However, the cost of modern irrigation equipment was not offset by subsidies, which led to a limited uptake of more efficient irrigation practice. Moreover, while the date palm production remained at the same level, the area of vegetable crops increased and drip irrigation made it possible to grow summer vegetables, which led to an increased groundwater demand. To restrict groundwater abstraction in the Batinah region, with over 100 000 registered wells, Zekri (2009) proposed to implement annual electricity quota, as a proxy for water usage, using an online, prepaid metering system, and this system has been trialled in a pilot project. Surveys showed that the farmers were supportive of water quotas, provided that groundwater remains free of charge and that their crops' water requirements can be met (Zekri et al., 2014).

Controlling demand through water pricing can only be effective when the price of water is a major operational cost such as in irrigated agriculture. In the tourism industry, however, water forms only a small fraction of the total operational budget and pricing policies tend to be ineffective instruments to control demand. It was found, for example, that water-saving measures were introduced by hotel managers in the island of Mallorca because they were perceived positively by their guests and improved the quality of service, not because they led to direct cost savings (Deyà-Tortella et al., 2016). Understanding the drivers behind water use is thus another prerequisite for designing effective policies (Post et al., 2018b).

References

Abarca E (2006) Seawater intrusion in complex geological environments. Thesis, Technical University of Catalonia, UPC, Barcelona, Spain.

Abarca E, Carrera J, Sánchez-Vila X and Voss CI (2007) Quasi-horizontal circulation cells in 3D seawater intrusion. *J Hydrol* **339**: 118–129.

Abarca E, Karam H, Hemond HF and Harvey CF (2013) Transient groundwater dynamics in a coastal aquifer: The effects of tides, the lunar cycle, and the beach profile. *Water Resour Res* **49**: 2473–2488.

Abarca E, Vazquez-Sune E, Carrera J, Capino B, Gamez D and Batlle F (2006) Optimal design of measures to correct seawater intrusion. *Water Resour Res* **42**.

Abesser C, Shand P and Ingram J (2005) Millstone Grit of Northern England. Baseline Report Series 18. Environment Agency, Bristol, UK.

Addo KA, Larbi L, Amisigo B and Ofori-Danson PK (2011) Impacts of coastal inundation due to climate change in a CLUSTER of urban coastal communities in Ghana, West Africa. *Remote Sensing* **3**: 2029–2050.

Affholder M and Valiron F (2001) *Descriptive Physical Oceanography.* Boca Raton, FL: CRC Press.

Alcala FJ and Custodio E (2008) Using the Cl/Br ratio as a tracer to identify the origin of salinity in aquifers in Spain and Portugal. *J Hydrol* **359**: 189–207.

Alcolea A, Castro E, Barbieri M, Carrera J and Bea S (2007) Inverse modeling of coastal aquifers using tidal response and hydraulic tests. *Ground Water* **45**: 711–722.

Aliewi AS, Mackay R and Van Wonderen J (1993) Exploitation of Groundwater from Fresh Water Lenses in Saline Aquifers. Thesis, An-Najah National University

Allen DM (2004) Sources of ground water salinity on islands using 18O, 2H, and 34S. *Ground Water* **42**: 17–31.

Aller L, Bennett T, Lehr JH and Petty RJ (1987) *DRASTIC: A Standardized System for Evaluating Ground Water Pollution Potential Using Hydrogeologic Settings*. Ada, OK: Robert S. Kerr Environmental Research Laboratory.

Alley WM (2003) Desalination of ground water earth science perspectives. US Dept. of the Interior, USGS, Reston, VA.

Andersen M, Nyvang V, Jakobsen R and Postma D (2005) Geochemical processes and solute transport at the seawater/freshwater interface of a sandy aquifer. *Geochim Cosmochim Ac* **69**: 3979–3994.

Andersen MS (2001) Geochemical processes at a seawater–freshwater interface. Thesis, Technical University of Denmark, Lyngby.

Andersen MS, Jakobsen R, Nyvang V, Christensen FD, Engesgaard P and Postma D (2008) Density-driven seawater plumes in a shallow aquifer caused by a flooding event – field observations, consequences for geochemical reactions and potentials for remediation schemes. IAHS Publication 324: 483–490.

Andersen PF, Mercer JW and White HO (1988) Numerical modeling of salt-water-intrusion at Hallandale, Florida. *Ground Water* **26**: 619–630.

Anderson HR (1978) Hydrogeologic Reconnaissance of the Mekong Delta in South *Vietnam* and *Cambodia*. US Government Printing Office, Washington, DC.

Anderson MP, Woessner WW and Hunt RJ (2015) *Applied Groundwater Modeling: Simulation of Flow and Advective Transport*. Elsevier Science, New York.

Appelo CAJ (1994) Cation and proton exchange, pH variations, and carbonate reactions in a freshening aquifer. *Water Resour Res* **30**: 2793–2805.

Appelo CAJ and Postma D (2005) *Geochemistry, Groundwater and Pollution*. A. A. Balkema, London.

Appelo CAJ and Willemsen A (1987) Geochemical calculations and observations on salt water intrusions, I. A combined geochemical/minxing cell model. *J Hydrol* **94**: 313–330.

Appelo CAJ, Willemsen A, Beekman HE and Griffioen J (1990) Geochemical calculations and observations on salt water intrusions. II. Validation of a geochemical model with laboratory experiments. *J Hydrol* **120**: 225–250.

Aquilina L, Vergnaud-Ayraud V, Les Landes AA et al. (2015) Impact of climate changes during the last 5 million years on groundwater in basement aquifers. *Sci Rep* **5**: 14132.

Aravena R, Wassenaar LI and Plummer LN (1995) Estimating C-14 Groundwater Ages in a Methanogenic Aquifer. *Water Resour Res* **31**: 2307–2317.

Aristotle (350 BC) *Meteorology*. Vol. 2015.

Armandine Les Landes A, Aquilina L, Davy P, Vergnaud-Ayraud V and Le Carlier C (2015) Timescales of regional circulation of saline fluids in continental crystalline rock aquifers (Armorican Massif, western France). *Hydrol Earth Syst Sc* **19**: 1413–1426.

Armitage SJ, Jasim SA, Marks AE, Parker AG, Usik VI and Uerpmann H-P (2011) The southern route 'out of Africa': Evidence for an early expansion of modern humans into Arabia. *Science* **331**: 453–456.

Aslam M, Matsuno Y and Hatcho N (2016) Performance evaluation of fresh groundwater skimming wells in the Indus basin irrigation system of Pakistan: A selective review. *Mem Fac Agr Kindai Univ* **50**: 5–23.

Aswathanarayana U (2001) *Water Resources Management and the Environment*. A. A. Balkema, Lisse, Netherlands.

Ataie-Ashtiani B, Volker RE and Lockington DA (1999) Tidal effects on sea water intrusion in unconfined aquifers. *J Hydrol* **216**: 17–31.

Ataie-Ashtiani B, Werner AD, Simmons CT, Morgan LK and Lu CH (2013) How important is the impact of land-surface inundation on seawater intrusion caused by sea-level rise? *Hydrogeology Journal* **21**: 1673–1677.

Audra P, Mocochain L, Camus H, Gilli E, Clauzon G and Bigot JY (2004) The effect of the Messinian Deep Stage on karst development around the Mediterranean Sea. Examples from Southern France. *Geodin Acta* **17**: 389–400.

Bachu S (1995) Flow of variable-density formation water in deep sloping aquifers: review of methods of representation with case studies. *J Hydrol* **164**: 19–38.

Bachu S and Michael K (2002) Flow of variable-density formation water in deep sloping aquifers: minimizing the error in representation and analysis when using hydraulic-head distributions. *J Hydrol* **259**: 49–65.

Back W (1960) Origin of hydrochemical facies of ground water in the Atlantic Coastal Plain. Report, Part 1, pp. 87–95. Copenhagen.

(1966) *Hydrochemical Facies and Ground-Water Flow Patterns in Northern Part of Atlantic Coastal Plain*. US Government Printing Office, Washington, DC.

Back W, Baedecker MJ and Wood WW (1993) Scales in chemical hydrogeology – A historical perspective. Regional Ground-Water Quality (Alley WM, ed.). Van Nostrand Reinhold, New York.

Back W, Hanshaw BB, Herman JS and Van Driel JN (1986) Differential dissolution of a Pleistocene reef in the ground-water mixing zone of coastal Yucatan, Mexico. *Geology* **14**: 137–140.

Badaruddin S, Werner AD and Morgan LK (2015) Water table salinization due to seawater intrusion. *Water Resour Res* **51**: 8397–8408.

(2017) Characteristics of active seawater intrusion. *J Hydrol* **551**: 632–647.

Bakhtyar R, Brovelli A, Barry DA, Robinson C and Li L (2013) Transport of variable-density solute plumes in beach aquifers in response to oceanic forcing. *Adv Water Resour* **53**: 208–224.

Bakker M (2006) Analytic solutions for interface flow in combined confined and semi-confined, coastal aquifers. *Adv Water Resour* **29**: 417–425.

Bakker M, Kelson VA and Luther KH (2005) Multilayer analytic element modeling of radial collector wells. *Ground Water* **43**: 926–934.

Bakker M, Schaars F, Hughes JD, Langevin CD and Dausman AM (2013) Documentation of the seawater intrusion (SWI2) package for MODFLOW. Reston, VA.

Barker AP, Newton RJ, Bottrell SH and Tellam JH (1998) Processes affecting groundwater chemistry in a zone of saline intrusion into an urban sandstone aquifer. *Appl Geochem* **13**: 735–749.

Barlow PM (2003) *Ground Water in Freshwater-Saltwater Environments of the Atlantic Coast*. US Department of the Interior, USGS, Reston, VA.

Barnes RSK (1991) Dilemmas in the theory and practice of biological conservation as exemplified by British coastal lagoons. *Biol Conserv* **55**: 315–328.

Barros Grace V, Mas-Pla J, Oliveira Novais T, Sacchi E and Zuppi GM (2008) Hydrological mixing and geochemical processes characterization in an estuarine/mangrove system using environmental tracers in Babitonga Bay (Santa Catarina, Brazil). *Cont Shelf Res* **28**: 682–695.

Barry DA, Parlange JY, Haverkamp R and Ross PJ (1995) Infiltration under ponded conditions. 4. An explicit predictive infiltration formula. *Soil Science* **160**: 8–17.

Basu AR, Jacobsen SB, Poreda RJ, Dowling CB and Aggarwal PK (2001) Large groundwater strontium flux to the oceans from the bengal basin and the marine strontium isotope record. *Science* **293**: 1470–1473.

Bath AH and Edmunds WM (1981) Identification of connate water in interstitial solution of chalk sediment. *Geochim Cosmochim Ac* **45**: 1449–1461.

Bauer-Gottwein P, Gondwe BRN, Charvet G, Marín LE, Rebolledo-Vieyra M and Merediz-Alonso G (2011) Review: The Yucatán Peninsula karst aquifer, Mexico. *Hydrogeology Journal* **19**: 507–524.

Bear J (1972) *Dynamics of Fluids in Porous Media*. Elsevier, New York.

Hydraulics of Groundwater. McGraw-Hill, New York.

(1999) *Seawater Intrusion in Coastal Aquifers: Concepts, Methods and Practices*. Kluwer Academic, Dordrecht, Netherlands.

Bear J and Cheng AH-D (2010) *Modeling Groundwater Flow and Contaminant Transport*. Springer, Dordrecht, Netherlands.

Bear J and Verruijt A (1987) *Modeling Groundwater Flow and Pollution*. Springer, Dordrecht, Netherlands.

Beck AJ and Cochran MA (2013) Controls on solid-solution partitioning of radium in saturated marine sands. *Mar Chem* **156**: 38–48.

Beck AJ, Rapaglia JP, Cochran JK and Bokuniewicz HJ (2007) Radium mass-balance in Jamaica Bay, NY: Evidence for a substantial flux of submarine groundwater. *Mar Chem* **106**: 419–441.

Becker K and Fisher AT (2008) Borehole packer tests at multiple depths resolve distinct hydrologic intervals in 3.5-Ma upper oceanic crust on the eastern flank of Juan de Fuca Ridge. *J Geophys Res – Solid Earth* **113**.

Beebe CR, Ferguson G, Gleeson T, Morgan LK and Werner AD (2016) Application of an analytical solution as a screening tool for sea water intrusion. Groundwater **54**: 709–718.

Beekman HE (1991) Ion chromatography of fresh- and seawater intrusion: multicomponent dispersive and diffusive transport in groundwater. PhD thesis, Vrije Universiteit, Amsterdam.

Beekman HE, Anthony C and Appelo J (1991) Ion chromatography of fresh- and salt-water displacement: Laboratory experiments and multicomponent transport modelling. *J Contam Hydrol* **7**: 21–37.

Beekman HE, Eggenkamp HGM and Appelo CAJ (2011) An integrated modelling approach to reconstruct complex solute transport mechanisms – Cl and δ37Cl in pore water of sediments from a former brackish lagoon in The Netherlands. *Appl Geochem* **26**: 257–268.

Belitz K and Bredehoeft JD (1988) Hydrodynamics of Denver Basin – Explanation of subnormal fluid pressures. *Aapg Bull* **72**: 1334–1359.

Benkabbour B, Toto EA and Fakir Y (2004) Using DC resistivity method to characterize the geometry and the salinity of the Plioquaternary consolidated coastal aquifer of the Mamora plain, Morocco. *Environ Geol* **45**: 518–526.

Berg M, Tran HC, Nguyen TC, Pham HV, Schertenleib R and Giger W (2001) Arsenic contamination of groundwater and drinking water in Vietnam: A human health threat. *Environ Sci Technol* **35**: 2621–2626.

Berner RA (1984) Sedimentary pyrite formation – an update. *Geochim Cosmochim Ac* **48**: 605–615.

Bi XL, Liu FQ and Pan XB (2012) Coastal projects in China: From reclamation to restoration. *Environ Sci Technol* **46**: 4691–4692.

Bloomfield J, Williams R, Gooddy D, Cape J and Guha P (2006) Impacts of climate change on the fate and behaviour of pesticides in surface and groundwater – a UK perspective. *Sci Total Environ* **369**: 163–177.

Bostock J and Riley HT (1890) *The Natural History of Pliny*. HG Bohn.

Bower JW, Motz LH and Durden DW (1999) Analytical solution for determining the critical condition of saltwater upconing in a leaky artesian aquifer. *J Hydrol* **221**: 43–54.

Braithwaite F (1855) On the filtration of salt-water into the springs and wells under London and Liverpool. *Min Proc Inst Civ Eng* **14**: 507–523.

Brassington FC and Taylor R (2012) A comparison of field methods used to define saline–fresh groundwater interfaces at two sites in North West England. *Q J Eng Geol Hydroge* **45**: 173–181.

Brassington R (2007) *Field Hydrogeology*. 3rd edn. Wiley, Chichester, UK.

Bratton JF (2010) The three scales of submarine groundwater flow and discharge across passive continental margins. *J Geol* **118**: 565–575.

Bresciani E, Ordens CM, Werner AD, Batelaan O, Guan H and Post VEA (2014) Spatial variability of chloride deposition in a vegetated coastal area: Implications for groundwater recharge estimation. *J Hydrol* **519**: 1177–1191.

Brooks DA, Baca MW and Lo YT (1999) Tidal circulation and residence time in a macrotidal estuary: Cobscook Bay, Maine. *Estuar Coast Shelf S* **49**: 647–665.

Brooks H (1961) The submarine spring off Crescent Beach, Florida. *Q J Florida Acad Sci* **24**: 122–134.

Brown DL, Silvey WD and Norfolk Virginia. Dept. of Utilities (1977) Artificial *Recharge to a Freshwater-Sensitive Brackish-Water Sand Aquifer*. US Government Printing Office, Washington, DC.

Brown JS (1922) Relation of sea water to ground water along coasts. *Am J Sci, Ser 5* **4**: 274–294.

Bruington AE (1972) Saltwater intrusion into aquifers 1. *JAWRA* **8**: 150–160.

Bruun P (1962) Sea-level rise as a cause of shore erosion. *J. Waterw Harbors Div* **88**: 117–130.

Burnett WC, Aggarwal PK, Aureli A et al. (2006) Quantifying submarine groundwater discharge in the coastal zone via multiple methods. *Sci Total Environ* **367**: 498–543.

Burnett WC, Bokuniewicz H, Huettel M, Moore WS and Taniguchi M (2003) Groundwater and pore water inputs to the coastal zone. *Biogeochemistry* **66**: 3–33.

Burnett WC and Dulaiova H (2003) Estimating the dynamics of groundwater input into the coastal zone via continuous radon-222 measurements. *J Environ Radioact* **69**: 21–35.

Burnett WC, Peterson R, Moore WS and de Oliveira J (2008) Radon and radium isotopes as tracers of submarine groundwater discharge – Results from the Ubatuba Brazil SGD assessment intercomparison. *Estuar Coast Shelf S* **76**: 501–511.

Burnett WC, Taniguchi M and Oberdorfer J (2001) Measurement and significance of the direct discharge of groundwater into the coastal zone. *J Sea Res* **46**: 109–116.

Buschmann J and Berg M (2009) Impact of sulfate reduction on the scale of arsenic contamination in groundwater of the Mekong, Bengal and Red River deltas. *Appl Geochem* **24**: 1278–1286.

Bye JAT and Narayan KA (2009) Groundwater response to the tide in wetlands: Observations from the Gillman Marshes, South Australia. *Estuar Coast Shelf S* **84**: 219–226.

Cable JE, Bugna GC, Burnett WC and Chanton JP (1996) Application of Rn-222 and CH_4 for assessment of groundwater discharge to the coastal ocean. *Limnol Oceanogr* **41**: 1347–1353.

Calvache ML and Pulido-Bosch A (1991) Saltwater intrusion into a small coastal aquifer (Rio Verde, Almuñecar, southern Spain). *J Hydrol* **129**: 195–213.

(1994) Modeling the effects of salt-water intrusion dynamics for a coastal karstified block connected to a detrital aquifer. *Ground Water* **32**: 767–777.

Cardenas MB, Bennett PC, Zamora PB, Befus KM, Rodolfo RS, Cabria HB and Lapus MR (2015) Devastation of aquifers from tsunami-like storm surge by Supertyphoon Haiyan. *Geophys Res Lett* **42**: 2844–2851.

Carlston CW (1963) An early American statement of the Badon Ghyben-Herzberg principle of static fresh-water-salt-water balance. *Am Jour Sci* **261**: 89–91.

Carneiro JF, Boughriba M, Correia A, Zarhloule Y, Rimi A and El Houadi B (2010) Evaluation of climate change effects in a coastal aquifer in Morocco using a density-dependent numerical model. *Environ Earth Sci* **61**: 241–252.

Carr PA (1971) Use of harmonic analysis to study tidal fluctuations in aquifers near sea. *Water Resour Res* **7**: 632.

Carr PA and Van der Kamp G (1969) Determining aquifer characteristics by the tidal methods. *Water Resour Res* **5**.

Carrier GF (1958) The mixing of ground water and sea water in permeable subsoils. *J Fluid Mech* **4**: 479–488.

Carslaw HS and Jaeger JC (1959) *Conduction of Heat in Solids*. Clarendon Press, Oxford.

Chachadi AG and Lobo Ferreira JP (2001) Assessing aquifer vulnerability to sea-water intrusion using GALDIT method: Part 2 – GALDIT Indicators Description. p. 12. Guimarães, Portugal.

Chague-Goff C, Niedzielski P, Wong HKY, Szczucinski W, Sugawara D and Goff J (2012) Environmental impact assessment of the 2011 Tohoku-oki tsunami on the Sendai Plain. *Sediment Geol* **282**: 175–187.

Chan PSS (2001) A Survey Report of Historical Buildings and Structures Within the Project Area of the Central Reclamation Phase III. Antiquities and Monuments Office, Leisure and Cultural Service Department **7**.

Chang SW and Clement TP (2013) Laboratory and numerical investigation of transport processes occurring above and within a saltwater wedge. *J Contam Hydrol* **147**: 14–24.

Chapelle F (2001) *Ground-Water Microbiology and Geochemistry*. Wiley, New York.

Chapelle FH, Bradley PM and McManhon PB (1993) Subsurface microbiology. *Regional Ground-Water Quality* (Alley WM, ed.). Van Nostrand Reinhold, New York.

Chapelle FH and Lovley DR (1992) Competitive-exclusion of sulfate reduction by Fe(III)-reducing bacteria – a mechanism for producing discrete zones of high-iron groundwater. *Ground Water* **30**: 29–36.

Chapelle FH and McMahon PB (1991) Geochemistry of dissolved inorganic carbon in a coastal-plain aquifer.1. Sulfate from confining beds as an oxidant in microbial CO-2 production. *J Hydrol* **127**: 85–108.

Charette MA (2007) Hydrologic forcing of submarine groundwater discharge: Insight from a seasonal study of radium isotopes in a groundwater-dominated salt marsh estuary. *Limnol Oceanogr* **52**: 230–239.

Charette MA and Buesseler KO (2004) Submarine groundwater discharge of nutrients and copper to an urban subestuary of Chesapeake bay (Elizabeth River). *Limnol Oceanogr* **49**: 376–385.

Charette MA, Dulaiova H, Gonneea ME, Henderson PB, Moore WS, Scholten JC and Pham MK (2012) GEOTRACES radium isotopes interlaboratory comparison experiment. *Limnol Oceanogr Methods* **10**: 617–617.

Charette MA, Sholkovitz ER and Hansel CM (2005) Trace element cycling in a subterranean estuary: Part 1. Geochemistry of the permeable sediments. *Geochim Cosmochim Acta* **69**: 2095–2109.

Chen CX and Jiao JJ (1999) Numerical simulation of pumping tests in multilayer wells with non-Darcian flow in the wellbore. *Groundwater* **37**: 465–474.

Chen CX, Lin M and Cheng JM (2011) *Groundwater Dynamics*. Geological Publisher, Beijing.

Chen CX, Pei SP and Jiao JJ (2003a) Land subsidence caused by groundwater exploitation in Suzhou City, China. *Hydrogeol J* **11**: 275–287.

Chen CX, Wan JW and Zhan HB (2003b) Theoretical and experimental studies of coupled seepage-pipe flow to a horizontal well. *J Hydrol* **281**: 159–171.

Chen KP and Jiao JJ (2007) Seawater intrusion and aquifer freshening near reclaimed coastal area of Shenzhen. *Water Sci Technol* **7**: 137–145.

Chen KP, Jiao JJ, Huang JM and Huang RQ (2007) Multivariate statistical evaluation of trace elements in groundwater in a coastal area in Shenzhen, China. *Environ Pollut* **147**: 771–780.

Cheng AHD, Benhachmi MK, Halhal D, Ouazar D, Naji A and El Harrouni K (2003) Pumping optimization in saltwater-intruded aquifers. *Coastal Aquifer Management – Monitoring, Modeling, and Case Studies* 233–256. Lewis, Boca Ration, FL.

Cheng AHD, Halhal D, Naji A and Ouazar D (2000) Pumping optimization in saltwater-intruded coastal aquifers. *Water Resour Res* **36**: 2155–2165.

Cheng JM, Chen CX and Ji MR (2004) Determination of aquifer roof extending under the sea from variable-density flow modelling of groundwater response to tidal loading: case study of the Jahe River Basin, Shandong Province, China. *Hydrogeol J* **12**: 408–423.

Cherif S, El Ayni F, Jrad A and Trabelsi-Ayadi M (2013) Aquifer recharge by treated wastewaters: Korba case study (Tunisia). *Sustainable Sanitation Practice* **14**: 41–48.

Cherry J, Parker B, Bradbury K, Eaton T, Gotkowitz M, Hart D and Borchardt M (2004) *Role of Aquitards in the Protection of Aquifers from Contamination: A 'State of the Science' Report*. Awwa Research Foundation, Denver, CO.

Chesnaux R (2015) Closed-form analytical solutions for assessing the consequences of sea-level rise on groundwater resources in sloping coastal aquifers. *Hydrogeol J* **23**: 1399–1413.

Chua L, Lo EYM, Freyberg DL, Shuy EB, Lim TT, Tan SK and Ngonidzashe M (2007) Hydrostratigraphy and geochemistry at a coastal sandfill in Singapore. *Hydrogeol J* **15**: 1591–1604.

Chuang MH, Huang CS, Li GH and Yeh HD (2010) Groundwater fluctuations in heterogeneous coastal leaky aquifer systems. *Hydrol Earth Syst Sci* **14**: 1819–1826.

Church JA, Clark PU, Cazenave A et al. (2013) *Sea Level Change*, in *Climate Change 2013: The Physical Science Basis. Contribution of Working Group I to the Fifth Assessment Report of the Intergovernmental Panel on Climate Change* (Stocker TF, Qin D, Plattner G-K, Tignor M, Allen SK, Boschung J, Nauels A, Xia Y, V. B and Midgley PM, eds), 1137–1216. Cambridge, United Kingdom and New York, NY, USA.

Church TM (1996) An underground route for the water cycle. *Nature* **380**: 579–580.

Clark ID and Fritz P (1997) *Environmental Isotopes in Hydrogeology*. Lewis, Boca Raton, FL.

Clark PU, Dyke AS, Shakun JD, Carlson AE, Clark J, Wohlfarth B, Mitrovica JX, Hostetler SW and McCabe AM (2009) The Last Glacial Maximum. *Science* **325**: 710–714.

Clemmer ND (2003) Characterization of Chlorinated Solvent Degradation in a Constructed Wetland. Thesis, Air Force Inst. of Tech, Captain, USAF.

Clendenon C (2009) Ancient Greek hydromyths about the submarine transport of terrestrial fresh water through seabeds offshore of karstic regions. *Acta Carsologica* **38**: 293–302.

Coetsiers M and Walraevens K (2008) The Neogene Aquifer, Flanders, Belgium. *Natural Groundwater Quality* (Edmunds WM and Shand P, eds), 263. Blackwell, Oxford.

Cohen D, Person M, Wang P et al. (2010) Origin and extent of fresh paleowaters on the Atlantic continental shelf, USA. *Ground Water* **48**: 143–158.

Colombani N and Mastrocicco M (2016) Geochemical evolution and salinization of a coastal aquifer via seepage through peaty lenses. *Environ Earth Sci* **75**: 798.

Comte J-C, Cassidy R, Obando J et al. (2016) Challenges in groundwater resource management in coastal aquifers of East Africa: Investigations and lessons learnt in the Comoros Islands, Kenya and Tanzania. *J Hydrol Regional Stud* **5**: 179–199.

Comte J-C, Join J-L, Banton O and Nicolini E (2014) Modelling the response of fresh groundwater to climate and vegetation changes in coral islands. *Hydrogeol J* **22**: 1905–1920.

Comte JC, Wilson C, Ofterdinger U and González-Quirós A (2017) Effect of volcanic dykes on coastal groundwater flow and saltwater intrusion: A field-scale multiphysics approach and parameter evaluation. *Water Resour Res* **53**: 2171–2198.

Condesso de Melo MT, Carreira Paquete PMM and Marques Da Silva MA (2001) Evolution of the Aveiro Cretaceous aquifer (NW Portugal) during the Late Pleistocene and present day: evidence from chemical and isotopic data. Palaeowaters in Coastal Europe: Evolution of Groundwater since the Late Pleistocene. *Geological Society Special Publication*, Vol. 189 (Edmunds WM and Milne CJ, eds), pp. 49–70. Geological Society, London.

Cooper HH (1959) A hypothesis concerning the dynamic balance of fresh water and salt water in a coastal aquifer. *J Geophys Res* **64**: 461–467.

Cooper HH, Kohout FA, Henry HR and Glover RE (1964) *Sea Water in Coastal Aquifers*. US Government Printing Office, Washington, DC.

Cooper JAG and Pilkey OH (2004) Sea-level rise and shoreline retreat: time to abandon the Bruun Rule. *Global Planet Change* **43**: 157–171.

Coplen TB (1993) Uses of environmental isotopes. *Regional Ground-Water Quality* (Alley WM, ed.). Van Nostrand Reinhold, New York.

Corlis NJ, Herbert Veeh H, Dighton JC and Herczeg AL (2003) Mixing and evaporation processes in an inverse estuary inferred from δ2H and δ18O. *Cont Shelf Res* **23**: 835–846.

Cox D and Gordon Jr L (1970) *Estuarine Pollution in the State of Hawaii*. Vol. 1. Water Resources Research Center, Honolulu, HI.

Craig H (1961) Isotopic variations in meteoric waters. *Science* 1702–1703.

Craig RF (2004) *Craig's Soil Mechanics*. Spon Press, London.

Crook I, Miner T, Norman L, Dahlhaus P and Clarkson T (2008) *Acid Sulfate Soils, Landslides, Soil Erosion and Salinity for On-Ground Staff*. Vol. 2015. Corangamite Catchment Management Authority, Colac, VC.

Crossland CJ, Kremer HH, Lindeboom H, Crossland JIM and Le Tissier MD (2005) *Coastal Fluxes in the Anthropocene: The Land-Ocean Interactions in the Coastal Zone Project of the International Geosphere-Biosphere Programme*. Springer Science and Business Media, Berlin.

Cuffey KM and Marshall SJ (2000) Substantial contribution to sea-level rise during the last interglacial from the Greenland ice sheet. *Nature* **404**: 591–594.

Custodio E (2008) Acuíferos detríticos costeros del litoral Mediterráneo Peninsular: Valle bajo y delta del Llobregat (Coastal detrital aquifers in the peninsular Mediterranean littoral: Llobregat's lower valley and delta). *Enseñanza de las Ciencias de la Tierra*, pp. 295–304.

(2012a) Low Llobregat Aquifers: Intensive Development, Salinization, Contamination, and Management. The Llobregat – The Story of a Polluted Mediterranean River, Vol. 21 (Sabater S, Ginebreda A and Barcelo D, eds), 27–50. Springer, Berlin.

(2012b) *Coastal Aquifer Management in Europe*. Cassis, France.

Custodio E and Bruggeman GA (1987) *Groundwater problems in coastal areas: a contribution to the International Hydrological Programme*. UNESCO, Paris.

d'Andrimont R (1903) Notes sur l'hydrologie du littoral belge (Notes on the hydrology of the Belgium coast). *Ann Soc Géol Belgique* **XXIX**: M129–M144.

Dagan G and Bear J (1968) Solving the problem of local interface upconing in a coastal aquifer by the method of small perturbations. *J Hydraul Res* **6**: 15–44.

Dagan G and Zeitoun DG (1998) Seawater-freshwater interface in a stratified aquifer of random permeability distribution. *J Contam Hydrol* **29**: 185–203.

Dakin RA, Farvolden RN, Cherry JA and Fritz P (1983) Origin of dissolved solids in groundwaters of Mayne Island, British Columbia, Canada. *J Hydrol* **63**: 233–270.

Daniele L, Vallejos A, Sola F, Corbella M and Pulido-Bosch A (2011) Hydrogeochemical processes in the vicinity of a desalination plant (Cabo de Gata, SE Spain). *Desalination* **277**: 338–347.

Darling WG, Edmunds WM and Smedley PL (1997) Isotopic evidence for palaeowaters in the British Isles. *Appl Geochem* **12**: 813–829.

Darling WG and Gooddy DC (2006) The hydrogeochemistry of methane: Evidence from English groundwaters. *Chem Geol* **229**: 293–312.

Darwin C (1852) *Journal of Researches into the Natural History and Geology of the Countries Visited during the Voyage of H.M.S. Beagle Round the World, under the Command of Capt. Fitz Roy, R.N.* J. Murray, London.

Datta B, Vennalakanti H and Dhar A (2009) Modeling and control of saltwater intrusion in a coastal aquifer of Andhra Pradesh, India. *J Hydro-Environ Res* **3**: 148–158.

Davis R Jr (1955) Attempt to detect the antineutrinos from a nuclear reactor by the Cl 37 ($\bar{v}$, e–) A 37 reaction. *Phys Rev* **97**: 766.

Davis SN (1978) Floatation of fresh-water on sea-water, a historical note. *Ground Water* **16**: 444–445.

Davis SN, Whittemore DO and Fabryka-Martin J (1998) Uses of chloride/bromide ratios in studies of potable water. *Ground Water* **36**: 338–350.

De Franco R, Biella G, Tosi L et al. (2009) Monitoring the saltwater intrusion by time lapse electrical resistivity tomography: The Chioggia test site (Venice Lagoon, Italy). *J Appl Geophys* **69**: 117–130.

De Louw PGB, Eeman S, Oude Essink GHP, Vermue E and Post VEA (2013) Rainwater lens dynamics and mixing between infiltrating rainwater and upward saline groundwater seepage beneath a tile-drained agricultural field. *J Hydrol* **501**: 133–145.

De Louw PGB, Essink GHPO, Stuyfzand PJ and Van der Zee SEATM (2010) Upward groundwater flow in boils as the dominant mechanism of salinization in deep polders, The Netherlands. *J Hydrol* **394**: 494–506.

De Louw PGB, Van Berchum A, Buma J, Doornenbal P, Lorwa M, Oude Essink GHP, Pauw PS, Provoost Y and Visser M (2016) *A Self-Flowing Seepage System to Protect a Freshwater Lens from Local Sea Level Rise*. Cairns, Australia.

De Louw PGB, Vandenbohede A, Werner AD and Oude Essink GHP (2013) Natural saltwater upconing by preferential groundwater discharge through boils. *J Hydrol* **490**: 74–87.

De Montety V, Radakovitch O, Vallet-Coulomb C, Blavoux B, Hermitte D and Valles V (2008) Origin of groundwater salinity and hydrogeochemical processes in a confined coastal aquifer: Case of the Rhone delta (Southern France). *Appl Geochem* **23**: 2337–2349.

De Ruig JHM (1998) Coastline management in The Netherlands: human use versus natural dynamics. *J Coastal Conserv* **4**: 127–134.

De Vries J (1994) Willem Badon Ghijben and Johan MK Pennink: Pioneers of coastal-dune hydrology. *Hydrogeol J* **2**: 55–57.

De Wiest RJM (1965) *Geohydrology*. Wiley, New York.

Debuisson J and Moussu H (1967) *Une étude expérimentale de l'intrusion des eaux marines dans une nappe côtière du Sénégal sous l'effet de l'exploitation*. IAHS Publication 72.

DeConto RM and Pollard D (2016) Contribution of Antarctica to past and future sea-level rise. *Nature* **531**: 591–597.

Delsman JR (2015) Saline groundwater – surface water interaction in coastal lowlands. Thesis, IOS Press BV, Amsterdam.

Delsman JR, Hu-a-ng KRM, Vos PC, De Louw PGB, Oude Essink GHP, Stuyfzand PJ and Bierkens MFP (2014a) Paleo-modeling of coastal saltwater intrusion during the Holocene: an application to the Netherlands. *Hydrol Earth Syst Sci* **18**: 3891–3905.

Delsman JR, Waterloo MJ, Groen MMA, Groen J and Stuyfzand PJ (2014b) Investigating summer flow paths in a Dutch agricultural field using high frequency direct measurements. *J Hydrol* **519**: 3069–3085.

Deng J, Harff J, Li Y, Zhao Y and Zhang H (2016) Morphodynamics at the Coastal Zone in the Laizhou Bay, Bohai Sea. *J Coastal Res* **74**: 59–69.

Dent D (1986) *Acid Sulphate Soils: A Baseline for Research and Development*. International Institute for Land Reclamation and Improvement, Wageningen, Nethelands.

Deusdará K, Forti M, Borma L, Menezes R, Lima J and Ometto J (2017) Rainwater chemistry and bulk atmospheric deposition in a tropical semiarid ecosystem: the Brazilian Caatinga. *J Atmos Chem* **74**: 71–85.

Deutsch WJ (1997) *Groundwater Geochemistry: Fundamentals and Applications to Contamination*. Lewis, Boca Raton, FL.

Deyà-Tortella B, Garcia C, Nilsson W and Tirado D (2016) The effect of the water tariff structures on the water consumption in Mallorcan hotels. *Water Resour Res* **52**: 6386–6403.

Dickinson JE, Hanson RT, Ferre TPA and Leake SA (2004) Inferring time-varying recharge from inverse analysis of long-term water levels. *Water Resour Res* **40**.

Diersch HJG and Kolditz O (2002) Variable-density flow and transport in porous media: approaches and challenges. *Adv Water Resour* **25**: 899–944.

Dillon P, Pavelic P, Page D, Beringen H and Ward J (2009) *Managed Aquifer Recharge: An Introduction*. National Water Commission, Canberra, Australia.

Domenico PA and Schwartz FW (1990) *Physical and Chemical Hydrogeology*. Wiley, New York.

Dost H (1973) *Acid Sulphate Soils: Proceedings of International Symposium on Acid Sulphate Soils* International Institute for Land Reclamation and Improvement/ILRI, Wageningen, Netherlands.

Drabbe J and Badon Ghijben W (1889) Nota in verband met de voorgenomen putboring nabij Amsterdam (Note concerning the intended well drilling near Amsterdam). *Tijdschrift van het Koninklijk Instituut van Ingenieurs Verhandelingen* **1888/1889**: 8–22.

Dror I, Berkowitz B and Gorelick SM (2004) Effects of air injection on flow through porous media: Observations and analyses of laboratory-scale processes. *Water Resour Res* **40**.

Du Y, Ma T, Chen L, Shan H, Xiao C, Lu Y, Liu C and Cai H (2015) Genesis of salinized groundwater in Quaternary aquifer system of coastal plain, Laizhou Bay, China: Geochemical evidences, especially from bromine stable isotope. *Appl Geochem* **59**: 155–165.

Du Y, Ma T, Chen L, Xiao C and Liu C (2016) Chlorine isotopic constraint on contrastive genesis of representative coastal and inland shallow brine in China. *J Geochem Explor* **170**: 21–29.

Du Commun J (1828) On the cause of fresh water springs, fountains, etc. *Am J Sci Arts* **14**: 174–176.

Dulaiova H, Ardelan M, Henderson PB and Charette MA (2009) Shelf-derived iron inputs drive biological productivity in the southern Drake Passage. *Global Biogeochem Cy* **23**: 1–14.

Dupuit J (1863). Etudes Théoriques et Pratiques sur le mouvement des Eaux dans les canaux découverts et à travers les terrains perméables (Second ed.). Dunod, Paris.

Eckhardt K and Ulbrich U (2003) Potential impacts of climate change on groundwater recharge and streamflow in a central European low mountain range. *J Hydrol* **284**: 244–252.

Edmunds WM, Milne CJ and Geological Society of London. (2001) *Palaeowaters in Coastal Europe: Evolution of Groundwater since the Late Pleistocene*. Geological Society, London.

Eggenkamp HGM and Coleman ML (2009) The effect of aqueous diffusion on the fractionation of chlorine and bromine stable isotopes. *Geochim Cosmochim Acta* **73**: 3539–3548.

Eggenkamp HGM, Middelburg JJ and Kreulen R (1994) Preferential Diffusion of Cl-35 Relative to Cl-37 in Sediments of Kau Bay, Halmahera, Indonesia. *Chem Geol* **116**: 317–325.

Elberling B, Larsen F, Christensen S and Postma D (1998) Gas transport in a confined unsaturated zone during atmospheric pressure cycles. *Water Resour Res* **34**: 2855–2862.

Elimelech M and Phillip WA (2011) The future of seawater desalination: Energy, technology, and the environment. *Science* **333**: 712–717.

Ellerd MG, Massmann JW, Schwaegler DP and Rohay VJ (1999) Enhancements for passive vapor extraction: The Hanford study. *Ground Water* **37**: 427–437.

Erban LE, Gorelick SM, Zebker HA and Fendorf S (2013) Release of arsenic to deep groundwater in the Mekong Delta, Vietnam, linked to pumping-induced land subsidence. *Proc Natl Acad Sci USA* **110**: 13751–13756.

Erbs Poulsen S, Rømer Rasmussen K, Bøie Christensen N and Christensen S (2010) Evaluating the salinity distribution of a shallow coastal aquifer by vertical multielectrode profiling (Denmark). *Hydrogeol J* **18**: 161–171.

Erkens G, Van der Meulen MJ and Middelkoop H (2016) Double trouble: Subsidence and CO_2 respiration due to 1,000 years of Dutch coastal peatlands cultivation. *Hydrogeol J* **24**: 551–568.

Erskine AD (1991) The effect of tidal fluctuation on a coastal aquifer in the UK. *Ground Water* **29**: 556–562.

Essaid HI (1990) A multilayered sharp interface model of coupled freshwater and saltwater flow in coastal systems: Model development and application. *Water Resour Res* **26**: 1431–1454.

Eyrolle F, Ducros L, Le Dizes S, Beaugelin-Seiller K, Charmasson S, Boyer P and Cossonnet C (2018) An updated review on tritium in the environment. *J Environ Radioact* **181**: 128–137.

Fakir Y and Razack M (2003) Hydrodynamic characterization of a Sahelian coastal aquifer using the ocean tide effect (Dridrate Aquifer, Morocco). *Hydrolog Sci J* **48**: 441–454.

Fass T, Cook PG, Stieglitz T and Herczeg AL (2007) Development of saline ground water through transpiration of sea water. *Ground Water* **45**: 703–710.

Fasullo JT, Boening C, Landerer FW and Nerem RS (2013) Australia's unique influence on global sea level in 2010–2011. *Geophys Res Lett* **40**: 4368–4373.

Faye S, Maloszewski P, Stichler W, Trimborn P, Faye SC and Gaye CB (2005) Groundwater salinization in the Saloum (Senegal) delta aquifer: minor elements and isotopic indicators. *Sci Total Environ* **343**: 243–259.

Feldmann J and Levermann A (2015) Collapse of the West Antarctic Ice Sheet after local destabilization of the Amundsen Basin. *Proc Natl Acad Sci USA* **112**: 14191–14196.

Ferguson G and Gleeson T (2012) Vulnerability of coastal aquifers to groundwater use and climate change. *Nature Clim Change* **2**: 342–345.

Ferguson G, McIntosh JC, Grasby SE, Hendry MJ, Jasechko S, Lindsay MBJ and Luijendijk E (2018) The persistence of brines in sedimentary basins. *Geophys Res Lett* **45**: 4851–4858.

Ferrara, V and Pappalardo, G (2003) Salinisation of coastal aquifers in the southwestern Hyblean plateau (SE sicily, Italy). Coastal Aquifers Intrusion Technology: Mediterranean Countries, Vol. 1 (López-Geta JA, Orden JA, Gómez JD, Ramos G, Mejías M and Rodríguez L, eds). IGME, Madrid Spain, 103–112.

Ferris JG (1951) Cyclic fluctuations of water level as a basis for determining aquifer transmissibility. *IAHS Publ* **33**: 148–155.

Fetter CW (1972) Position of saline water interface beneath Oceanic Islands. *Water Resour Res* **8**: 1307.

(2001) *Applied Hydrogeology*. Prentice Hall, Upper Saddle River, NJ.

Field MS (2002) *A Lexicon of Cave and Karst Terminology with Special Reference to Environmental Karst Hydrology*. Vol. EPA/600/R-02/003. National Center for Environmental Assessment, Office of Research and Development, US Environmental Protection Agency, Washington, DC.

Fisher AT (2005) Marine hydrogeology: recent accomplishments and future opportunities. *Hydrogeol J* **13**: 69–97.

Fitzpatrick RW, Shand P and Merry RH (2009) Acid sulfate soils. Natural History of the Riverland and Murrayland (Jennings J, ed.), 65–111. Royal Society of South Australia.

Fletcher CJN (2004) *Geology of Site Investigation Boreholes from Hong Kong*. C. Fletcher, Hong Kong.

Fleury P, Bakalowicz M and De Marsily G (2007) Submarine springs and coastal karst aquifers: A review. *J Hydrol* **339**: 79–92.

Forchheimer P (1919) Zur Theorie der Grundwasserströmung. *Sitzungsberichten der Akademie der Wissenschaften in Wien, Mathem-natunw Klasse, Abteilung IIa* **128**: 7–14.

Foreman TL (2003) Management of seawater intrusion in the Los Angeles coastal basin, California: an evolution of practice. *Coastal Aquifers Intrusion Technology: Mediterranean Countries* (Lopez-Geta JA, De la Orden, JA, Dios Gomez J De Ramos G and Rodriguez L, eds). Instituto Geologico y Minero de Espana, Madrid.

Foster S, Garduno H, Evans R, Olson D, Tian Y, Zhang W and Han Z (2004) Quaternary Aquifer of the North China Plain: Assessing and achieving groundwater resource sustainability. *Hydrogeol J* **12**: 81–93.

Fratesi B (2013) Hydrology and geochemistry of the freshwater lens in coastal karst. Coastal Karst Landforms (Lace MJ and Mylroie JE, eds), 59–75. Springer, Dordrecht, Netherlands.

Freeze RA and Cherry JA (1979) *Groundwater.* Prentice Hall, Englewood Cliffs, NJ.

Frihy OE and Khafagy AA (1991) Climate and induced changes in relation to shoreline migration trends at the Nile delta promontories, Egypt. *Catena* **18**: 197–211.

Frind EO (1982) Seawater intrusion in continuous coastal aquifer-aquitard systems. *Adv Water Resour* **5**: 89–97.

Fukuda H (1955) Air and vapor movement in soil due to wind gustiness. *Soil Sci* **79**: 249–256.

Fukuo Y and Kaihotsu I (1988) A theoretical-analysis of seepage flow of the confined groundwater into the lake bottom with a gentle slope. *Water Resour Res* **24**: 1949–1953.

Galloway DL and Burbey TJ (2011) Review: Regional land subsidence accompanying groundwater extraction. *Hydrogeol J* **19**: 1459–1486.

Galvis-Rodriguez S, Post VEA, Werner AD, Sinclair P and Bosserelle A (2017) *Sustainable management of the Bonriki Water Reserve, Tarawa, Kiribati.* SPC Technical Report SPC00054. Pacific Community (SPC).

Garcia-Castellanos D, Estrada F, Jiménez-Munt I, Gorini C, Fernàndez M, Vergés J and De Vicente R (2009) Catastrophic flood of the Mediterranean after the Messinian salinity crisis. *Nature* **462**: 778–781.

García-Menéndez O, Morell I, Ballesteros BJ, Renau-Pruñonosa A, Renau-Llorens A and Esteller MV (2016) Spatial characterization of the seawater upconing process in a coastal Mediterranean aquifer (Plana de Castellón, Spain): Evolution and controls. *Environ Earth Sci* **75**: 728.

Garcia-Solsona E, Garcia-Orellana J, Masque P and Dulaiova H (2008) Uncertainties associated with Ra-223 and Ra-224 measurements in water via a Delayed Coincidence Counter (RaDeCC). *Mar Chem* **109**: 198–219.

Gat JR (1983) Precipitation, groundwater and surface waters: control of climate parameters on their isotopic composition and their utilization as palaeoclimatological tools. Palaeoclimates and Palaeowaters: A Collection of Environmental Isotope Studies 3–12. International Atomic Energy Agency, Vienna, Australia.

Gaur AS and Vora KH (1999) Ancient shorelines of Gujarat, India, during the Indus civilization (Late Mid-Holocene): A study based on archaeological evidences. *Curr Sci India* **77**: 180–185.

Geelen LHWT, Kamps PTWJ and Olsthoorn TN (2016) From overexploitation to sustainable use, an overview of 160 years of water extraction in the Amsterdam dunes, the Netherlands. *J Coastal Conserv* **21**: 1–12.

Gelhar LW, Welty C and Rehfeldt KR (1992) A critical review of data on field-scale dispersion in aquifers. *Water Resour Res* **28**: 1955–1974.

Geng X, Boufadel MC, Xia Y, Li H, Zhao L, Jackson NL and Miller RS (2014) Numerical study of wave effects on groundwater flow and solute transport in a laboratory beach. *J Contam Hydrol* **165**: 37–52.

Geotechnical Engineering Office (1993) *Guide to Retaining Wall Design.* Office, Hong Kong.

Gerstandt K, Peinemann KV, Skilhagen SE, Thorsen T and Holt T (2008) Membrane processes in energy supply for an osmotic power plant. *Desalination* **224**: 64–70.

Geyer W, Morris J, Prahl F and Jay D (2000) Interaction between physical processes and ecosystem structure: A comparative approach. Estuarine Science: A Synthetic Approach to Research and Practice (Hobbie JE, ed.). Island Press, New York.

Gimenez E and Morell I (1997) Hydrogeochemical analysis of salinization processes in the coastal aquifer of Oropesa (Castellon, Spain). *Environ Geol* **29**: 118–131.

Gingerich SB, Voss CI and Johnson AG (2017) Seawater-flooding events and impact on freshwater lenses of low-lying islands: Controlling factors, basic management and mitigation. *J Hydrol* **551**: 676–688.

Glaser R, Haberzettl P and Walsh R (1991) Land reclamation in Singapore, Hong Kong and Macau. *GeoJournal* **24**: 365–373.

Glover PWJ (2016) Archie's law – a reappraisal. *Solid Earth* **7**: 1157–1169.

Glover RE (1959) The pattern of fresh-water flow in a coastal aquifer. *J Geophys Res* **64**: 457–459.

Goebel M, Pidlisecky A and Knight R (2017) Resistivity imaging reveals complex pattern of saltwater intrusion along Monterey coast. *J Hydrol* **551**: 746–755.

Gogu RC and Dassargues A (2000) Current trends and future challenges in groundwater vulnerability assessment using overlay and index methods. *Environ Geol* **39**: 549–559.

Goldenberg LC (1985) Decrease of hydraulic conductivity in sand at the interface between seawater and dilute clay suspensions. *J Hydrol* **78**: 183–199.

Gonneea ME, Mulligan AE and Charette MA (2013) Seasonal cycles in radium and barium within a subterranean estuary: Implications for groundwater derived chemical fluxes to surface waters. *Geochim Cosmochim Ac* **119**: 164–177.

Grant CJ (1973) *Acid Sulphate Soils in Hong Kong*. Vol. 2. International Institute for Land Reclamation and Improvement/ILRI, Wageningen, Netherlands.

Green TR, Taniguchi M, Kooi H, Gurdak JJ, Allen DM, Hiscock KM, Treidel H and Aureli A (2011) Beneath the surface of global change: Impacts of climate change on groundwater. *J Hydrol* **405**: 532–560.

Greenlee LF, Lawler DF, Freeman BD, Marrot B and Moulin P (2009) Reverse osmosis desalination: Water sources, technology, and today's challenges. *Water Res* **43**: 2317–2348.

Greskowiak J (2014) Tide-induced salt-fingering flow during submarine groundwater discharge. *Geophys Res Lett* **41**: 6413–6419.

Griffioen J (1994) Uptake of phosphate by iron hydroxides during seepage in relation to development of groundwater composition in coastal areas. *Environ Sci Technol* **28**: 675–681.

Grinat M, Epping D and Meyer R (2018) Long-Time Resistivity Monitoring of a Freshwater/Saltwater Transition Zone Using the Vertical Electrode System. SAMOS, Gdansk, Poland.

Grinsted A, Moore JC and Jevrejeva S (2013) Projected Atlantic hurricane surge threat from rising temperatures. *Proc Natl Acad Sci USA* **110**: 5369–5373.

Groen J (2002) The effects of transgressions and regressions on coastal and offshore groundwater. Thesis, Vrije Universiteit, Amsterdam.

Groen J, Velstra J and Meesters AGCA (2000) Salinization processes in paleowaters in coastal sediments of Suriname: evidence from delta Cl-37 analysis and diffusion modelling. *J Hydrol* **234**: 1–20.

Guevara Morel CR, Van Reeuwijk M and Graf T (2015) Systematic investigation of non-Boussinesq effects in variable-density groundwater flow simulations. *J Contam Hydrol* **183**: 82–98.

Guha H and Panday S (2012) Impact of sea level rise on groundwater salinity in a coastal community of south Florida. *J Am Water Resour Assoc* **48**: 510–529.

Guo HP (2008) Groundwater movement and subsurface air flow induced by land reclamation and tidal fluctuation in coastal aquifers. Thesis, Hong Kong University.

Guo HP and Jiao JJ (2007) Impact of coastal land reclamation on ground water level and the sea water interface. *Ground Water* **45**: 362–367.

(2008a) Changes of coastal groundwater systems in response to large-scale land reclamation. New Topics in Water Resources Research and Management (Andreassen HM, ed.). Nova Science.

(2008b) Numerical study of airflow in the unsaturated zone induced by sea tides. *Water Resour Res* **44**.

(2010) Theoretical study of the impact of tide-induced airflow on hydraulic head in air-confined coastal aquifers. *Hydrolog Sci J* **55**: 435–445.

Guo HP, Jiao JJ and Li HL (2010) Groundwater response to tidal fluctuation in a two-zone aquifer. *J Hydrol* **381**: 364–371.

Guo HP, Jiao JJ and Weeks EP (2008) Rain-induced subsurface airflow and Lisse effect. *Water Resour Res* **44**.

Guo HP, Zhang ZC, Cheng GM, Li WP, Li TF and Jiao JJ (2015) Groundwater-derived land subsidence in the North China Plain. *Environ Earth Sci* **74**: 1415–1427.

Guo Q, Li HL, Boufadel MC, Xia Y and Li G (2007) Tide-induced groundwater head fluctuation in coastal multi-layered aquifer systems with a submarine outlet-capping. *Adv Water Resour* **30**: 1746–1755.

Hager WH (2004) Jules Dupuit – Eminent hydraulic engineer. *J Hydraul Eng* **130**: 843–848.

Han DM, Kohfahl C, Song X, Xiao G and Yang J (2011) Geochemical and isotopic evidence for palaeo-seawater intrusion into the south coast aquifer of Laizhou Bay, China. *Appl Geochem* **26**: 863–883.

Han DM, Post VEA and Song X (2015) Groundwater salinization processes and reversibility of seawater intrusion in coastal carbonate aquifers. *J Hydrol* **531**: 1067–1080.

Han DM, Song XF, Currell MJ and Tsujimura M (2012) Using chlorofluorocarbons (CFCs) and tritium to improve conceptual model of groundwater flow in the South Coast Aquifers of Laizhou Bay, China. *Hydrol Process* **26**: 3614–3629.

Hancock GJ, Webster IT and Stieglitz TC (2006) Horizontal mixing of Great Barrier Reef waters: Offshore diffusivity determined from radium isotope distribution. *J. Geophys. Res. Oceans* **111**. **C12**.

Hancock PL, Skinner BJ and Dineley DL (2000) *The Oxford Companion to the Earth*. Oxford University Press, Oxford.

Hanshaw BB and Back W (1980) Chemical mass-wasting of the northern Yucatan Peninsula by groundwater dissolution. *Geology* **8**: 222–224.

Hanshaw BB, Back W and Deike RG (1971) A geochemical hypothesis for dolomitization by ground water. *Econ Geol* **66**: 710–724.

Hanshaw BB, Back W and Rubin M (1967) Carbonate equilibria and radiocarbon distribution related to groundwater flow in the Floridian limestone aquifer. IAHS Publication 74: 601–614.

Hantush MS and Jacob CE (1955) Non-steady radial flow in an infinite leaky aquifer. *Trans AGU* **36**: 95–100.

Hardie LA (1996) Secular variation in seawater chemistry: An explanation for the coupled secular variation in the mineralogies of marine limestones and potash evaporites over the past 600 m.y. *Geology* **24**: 279–283.

Harris GD (1904) *Underground Waters of Southern Louisiana*. USGS Water-Supply Paper 101. US Government Printing Office, Washington, DC.

Hassanizadeh MS and Leijnse A (1995) A non-linear theory of high-concentration-gradient dispersion in porous media. *Adv Water Resour* **18**: 203–215.

Hatch CE, Fisher AT, Ruehl CR and Stemler G (2010) Spatial and temporal variations in streambed hydraulic conductivity quantified with time-series thermal methods. *J Hydrol* **389**: 276–288.

Hathaway JC, Poag CW, Valentine PC, Miller RE, Schultz DM, Manheim FT, Kohout FA, Bothner MH and Sangrey DA (1979) United-States-Geological-Survey core drilling on the Atlantic Shelf. *Science* **206**: 515–527.

Hathaway JC, Schlee JJ, Poag CW et al. (1976) *Preliminary Summary of the 1976 Atlantic Margin Coring Project of the US Geological Survey*. US Government Printing Office, Washington, DC.

He B, Takase K and Wang Y (2008) Numerical simulation of groundwater flow for a coastal plain in Japan: data collection and model calibration. *Environ Geol* **55**: 1745–1753.

Heiss JW and Michael HA (2014) Saltwater-freshwater mixing dynamics in a sandy beach aquifer over tidal, spring-neap, and seasonal cycles. *Water Resour Res* **50**: 6747–6766.

Heiss JW, Post VEA, Laattoe T, Russoniello CJ and Michael HA (2017) Physical controls on biogeochemical processes in intertidal zones of beach aquifers. *Water Resour Res* **53**: 9225–9244.

Heiss JW, Ullman WJ and Michael HA (2014) Swash zone moisture dynamics and unsaturated infiltration in two sandy beach aquifers. *Estuarine Coastal Shelf Sci* **143**: 20–31.

Hem JD (1985) *Study and Interpretation of the Chemical Characteristics of Natural Water*. USGS, Alexandria, VA.

Henderson RD, Day-Lewis FD, Abarca E, Harvey CF, Karam HN, Liu L and Lane JW (2010) Marine electrical resistivity imaging of submarine groundwater discharge: sensitivity analysis and application in Waquoit Bay, Massachusetts, USA. *Hydrogeol J* **18**: 173–185.

Hendizadeh R, Kompanizare M, Hashemi MR and Rakhshandehroo GR (2016) Steady critical discharge rates from vertical and horizontal wells in fresh–saline aquifers with sharp interfaces. *Hydrogeol J* **24**: 865–876.

Henry HR (1964) *Interfaces between Salt Water and Fresh Water in Coastal Aquifers*. USGS Water-Supply Paper C35-70. US Government Printing Office, Washington, DC.

Herczeg A, Dogramaci S and Leaney F (2001) Origin of dissolved salts in a large, semi-arid groundwater system: Murray Basin, Australia. *Mar Freshwater Res* **52**: 41–52.

Herzberg A (1888) *Erläuterungs-Bericht zum Entwässerungs- und Wasserversorgungs-Projekt des Inseldorfes Nordeney* (Explanatory report on the dewatering and water supply project of the island village Norderney). Vol. NLA AU. Rep. 16/3, no. 1212. Niedersächsisches Landesarchiv Aurich, Aurich, Germany.

(1901) Die Wasserversorgung einiger Nordseebäder (Water supply of some selected North Sea Spas). *J Gaslight Water Supply* **44**: 5.

Hicks WS, Bowman GM and Fitzpatrick RW (2009) Effect of season and landscape position on the aluminium geochemistry of tropical acid sulfate soil leachate. *Aust J Soil Res* **47**: 137–153.

Hill M (1988) A comparison of coupled freshwater-saltwater sharp-interface and convective-dispersive models of saltwater intrusion in a layered aquifer system. *Develop Water Sci* **35**: 211–216.

Hiscock KM (2005) *Hydrogeology: Principles and Practice*. Blackwell, Malden, MA.

Hoeksema RJ (2007) Three stages in the history of land reclamation in the Netherlands. *Irrig Drain* **56**: S113–S126.

Hof A, Blázquez-Salom M, Colom MC and Périz AB (2014) Challenges and solutions for urban-tourist water supply on Mediterranean tourist islands: The case of Majorca, Spain. The Global Water System in the Anthropocene: Challenges for Science and Governance (Bhaduri A, Bogardi J, Leentvaar J and Marx S, eds), 125–142. Springer International, Cham, Switzerland.

Holzbecher E (1998) *Modeling Density-Driven Flow in Porous Media – Principles, Numerics, Software*. Springer, Berlin.

(2005) Ghijben–Herzberg equilibrium. *Water Encyclopedia*. John Wiley, Hoboken, NJ.

Hori K and Saito Y (2007) An early Holocene sea-level jump and delta initiation. *Geophys Res Lett* **34**.

Houben G (2018) Annotated translation of 'Die Wasserversorgung einiger Nordseebäder [The water supply of some North Sea spas]' by Alexander Herzberg (1901). *Hydrogeol J* **26**: 1789–1799.

Houben GJ, Koeniger P and Sültenfuß J (2014) Freshwater lenses as archive of climate, groundwater recharge, and hydrochemical evolution: Insights from depth-specific water isotope analysis and age determination on the island of Langeoog, Germany. *Water Resour Res* **50**: 8227–8239.

Houben G and Post VEA (2017) The first field-based descriptions of pumping-induced saltwater intrusion and upconing. *Hydrogeol J* **25**: 243–247.

Houben GJ, Stoeckl L, Mariner KE and Choudhury AS (2018) The influence of heterogeneity on coastal groundwater flow – physical and numerical modeling of fringing reefs, dykes and structured conductivity fields. *Adv Water Resour* **113**: 155–166.

Hougham AL and Moran SB (2007) Water mass ages of coastal ponds estimated using Ra-123 and Ra-224 as tracers. *Mar Chem* **105**: 194–207.

Howard KWF (1987) Beneficial aspects of sea-water intrusion. *Ground Water* **25**: 398–406.

Howard KWF and Mullings E (1996) Hydrochemical analysis of ground-water flow and saline incursion in the Clarendon Basin, Jamaica. *Groundwater* **34**: 801–810.

Hsieh PA, Bredehoeft JD and Farr JM (1987) Determination of aquifer transmissivity from earth tide analysis. *Water Resour Res* **23**: 1824–1832.

Hu LT and Jiao JJ (2010) Modeling the influences of land reclamation on groundwater systems: A case study in Shekou peninsula, Shenzhen, China. *Eng Geol* **114**: 144–153.

Hu LT, Jiao JJ and Guo HP (2008) Analytical studies on transient groundwater flow induced by land reclamation. *Water Resour Res* **44**.

Hubbert MK (1940) The theory of ground-water motion. *J Geol* **48**: 785–944.

Hughes JD, Langevin CD and Brakefield-Goswami L (2009) Effect of hypersaline cooling canals on aquifer salinization. *Hydrogeol J* **18**: 25–38.

Hughes JD and Sanford WE (2004) *SUTRA-MS: A Version of SUTRA Modified to Simulate Heat and Multiple-Solute Transport*. USGS, Reston, VA.

Huizer S, Karaoulis MC, Oude Essink GHP and Bierkens MFP (2017) Monitoring and simulation of salinity changes in response to tide and storm surges in a sandy coastal aquifer system. *Water Resour Res* **53**: 6487–6509.

Humboldt AV and Thrasher JS (1856) The Island of Cuba. Derby and Jackson, New York.

Hwang DW, Kim G, Lee WC and Oh HT (2010) The role of submarine groundwater discharge (SGD) in nutrient budgets of Gamak Bay, a shellfish farming bay, in Korea. *J Sea Res* **64**: 224–230.

HydroMetrics (2008) *Seawater Intrusion Response Plan Seaside Basin, Monterey County California*. HydroMetrics.

Illangasekare T, Tyler SW, Clement TP et al. (2006) Impacts of the 2004 tsunami on groundwater resources in Sri Lanka. *Water Resour Res* 42.

Imbeaux M (1906) Les nappes aquifères au bord de la mer: salure de leurs eaux. *Bull Séances Soc Sci Nancy, Ser III* **VI**: 131–143.

Inglis G (1817) On the cause of ebbing and flowing springs. Philos Mag J **L**: 81–83.

IOC SCOR and IAPSO (2010) *The International Thermodynamic Equation of Seawater – 2010: Calculation and Use of Thermodynamic Properties*. Manuals and Guides No. 56. Intergovernmental Oceanographic Commission.

IPCC (2007) Climate change 2007: The physical science basis. *Agenda* **6**: 333.

Iribar V and Custodio E (1993) *Advancement of Seawater Intrusion in the Llobregat Delta Aquifer*. CIMNE, Barcelona, Spain.

Izuka SK and Gingerich SB (1998) Estimation of the depth to the fresh-water/salt-water interface from vertical head gradients in wells in coastal and island aquifers. *Hydrogeol J* **6**: 365–373.

Jacob CE (1940) On the flow of water in an elastic artesian aquifer. *Trans AGU* **21**: 574–586.

(1950) Flow of groundwater. Engineering Hydraulics (Rouse H, ed.), 321–386. John Wiley, New York.

Jakobsen R and Postma D (1994) In-situ rates of sulfate reduction in an aquifer (Romo, Denmark) and implications for the reactivity of organic-matter. *Geology* **22**: 1103–1106.

(1999) Redox zoning, rates of sulfate reduction and interactions with Fe-reduction and methanogenesis in a shallow sandy aquifer, Romo, Denmark. *Geochim Cosmochim Acta* **63**: 137–151.

Jakovovic D, Werner AD and Simmons CT (2011) Numerical modelling of saltwater up-coning: Comparison with experimental laboratory observations. *J Hydrol* **402**: 261–273.

Japan Green Resources Agency (2004) *Technical Reference for Effective Groundwater Development*. Kanagawa, Japan.

Jeen S-W, Kim J-M, Ko K-S, Yum B and Chang H-W (2001) Hydrogeochemical characteristics of groundwater in a mid-western coastal aquifer system, Korea. *Geosci J* **5**: 339–348.

Jeevanandam M, Kannan R, Srinivasalu S and Rammohan V (2007) Hydrogeochemistry and groundwater quality assessment of lower part of the ponnaiyar river Basin, Cuddalore district, South India. *Environ Monitor Assess* **132**: 263–274.

Jeng DS, Li L and Barry DA (2002) Analytical solution for tidal propagation in a coupled semi-confined/phreatic coastal aquifer. *Adv Water Resour* **25**: 577–584.

Jha MK, Namgial D, Kamii Y and Peiffer S (2008) Hydraulic parameters of coastal aquifer systems by direct methods and an extended tide – aquifer interaction technique. *Water Resour Manag* **22**: 1899–1923.

Jia GD, Duan GJ and Zhong ZS (2002) Groundwater exploitation, an important CO_2 source (in Chinese with English abstract). *Earth Sci* **27**: 4.

Jiang YN (2016) *A Great Jump in Land Reclamation in Shenzhen*. Vol. 2018. Initium Media, Hong Kong (https://theinitium.com/article/20160418-mainland-shenzhen-reclamation/).

Jiao JJ, Wang X and Nandy S (2005) Confined groundwater zone and slope instability in weathered igneous rocks in Hong Kong. *Eng Geol* **80**: 71–92.

Jiao JJ (2000) Modification of regional groundwater regimes by land reclamation. *Hong Kong Geol* **6**: 29–36.

(2002) Preliminary conceptual study on impact of land reclamation on groundwater flow and contaminant migration in Penny's Bay. *Hong Kong Geol* **8**: 14–20.

(2007) A 5,600-year-old wooden well in Zhejiang Province, China. *Hydrogeol J* **15**: 1021–1029.

Jiao JJ and Guo HP (2009) Airflow induced by pumping tests in unconfined aquifer with a low-permeability cap. *Water Resour Res* **45** (10).

Jiao JJ, Leung CM, Chen KP, Huang JM and Huang RQ (2005) Physical and chemical processes in the subsurface system in the land reclaimed from the sea. *Collections of Coastal Geo-Environment and Urban Development* P399–407. China Dadi, Beijing, China.

Jiao JJ and Li HL (2004) Breathing of coastal vadose zone induced by sea level fluctuations. *Geophys Res Lett* **31** (11).

Jiao JJ, Nandy S and Li HL (2001) Analytical studies on the impact of land reclamation on ground water flow. *Ground Water* **39**: 912–920.

Jiao JJ, Shi L, Kuang X, Lee CM, Yim WW-S and Yang S (2015a) Reconstructed chloride concentration profiles below the seabed in Hong Kong (China) and their implications for offshore groundwater resources. *Hydrogeol J* **23**: 277–286.

Jiao JJ and Tang ZH (1999) An analytical solution of groundwater response to tidal fluctuation in a leaky confined aquifer. *Water Resour Res* **35**: 747–751.

Jiao JJ, Wang Y, Cherry JA, Wang XS, Zhi BF, Du HY and Wen DG (2010) Abnormally high ammonium of natural origin in a coastal aquifer-aquitard system in the Pearl River Delta, China. *Environ Sci Technol* **44**: 7470–7475.

Jiao JJ, Zhang XT, Liu YL and Kuang XX (2015b) Increased water storage in the Qaidam Basin, the North Tibet Plateau from GRACE gravity data. *Plos One* **10** (10).

Johannes RE (1980) The ecological significance of the submarine discharge of groundwater. *Mar Ecol Prog Ser* **3**: 365–373.

Johnson AG, Glenn CR, Burnett WC, Peterson RN and Lucey PG (2008) Aerial infrared imaging reveals large nutrient-rich groundwater inputs to the ocean. *Geophys Res Lett* **35**.

Johnson TA and Whitaker R (2004) Saltwater intrusion in the coastal aquifers of Los Angeles County, California. Coastal Aquifer Management – Monitoring, Modeling, and Case Studies (Cheng AHD and Ouazar D, eds), Lewis, Boca Raton, FL.

Jones BF, Vengosh A, Rosenthal E and Yechieli Y (1999) Geochemical investigation. Seawater Intrusion in Coastal Aquifers: Concepts, Methods and Practices (Bear J, Cheng AH-D, Sorek S, Ouazar D and Herrebra I, eds). xv. Kluwer Academic, Dordrecht, Netherlands.

Jones RE, Beeman RE and Suflita JM (1989) Anaerobic metabolic processes in the deep terrestrial subsurface. *Geomicrobiol J* **7**: 117–130.

Jørgensen NO (2002) Origin of shallow saline groundwater on the Island of Læsø, Denmark. *Chem Geol* **184**: 359–370.

Jørgensen NO, Andersen MS and Engesgaard P (2008) Investigation of a dynamic seawater intrusion event using strontium isotopes (87Sr/86Sr). *J Hydrol* **348**: 257–269.

Judd AG and Hovland M (2007) *Seabed Fluid Flow: The Impact of Geology, Biology and the Marine Environment*. Cambridge University Press, Cambridge.

Kacimov A and Abdalla O (2010) Water table response to a tidal agitation in a coastal aquifer: The Meyer-Polubarinova-Kochina theory revisited. *J Hydrol* **392**: 96–104.

Kämpf J (2014) South Australia's large inverse estuaries: On the road to ruin. Estuaries of Australia in 2050 and Beyond (Wolanski E, ed.), 153–166. Springer, Dordrecht, Netherlands.

Kamps P, Nienhuis P, Van den Heuvel D and De Joode H (2016) *Monitoring Well Optimization for Surveying the Fresh/Saline Groundwater Interface in the Amsterdam Water Supply Dunes*. Cairns, Australia.

Kana TW, Michel J, Hayes MO and Jenson JR (1984) The physical impact of sea level rise in the area of Charleston, South Carolina. Greenhouse Effect and Sea Level Rise: A Challenge for This Generation (Barth M and Titus J, eds), xiii, 325. Van Nostrand Reinhold, New York.

Karlsen RH, Smits FJC, Stuyfzand PJ, Olsthoorn TN and Van Breukelen BM (2012) A post audit and inverse modeling in reactive transport: 50 years of artificial recharge in the Amsterdam Water Supply Dunes. *J Hydrol* **454**: 7–25.

Kashef AAI (1983) Harmonizing Ghyben-Herzberg interface with rigorous solutions. *Ground Water* **21**: 153–159.

Kass A, Gavrieli I, Yechieli Y, Vengosh A and Starinsky A (2005) The impact of freshwater and wastewater irrigation on the chemistry of shallow groundwater: A case study from the Israeli Coastal Aquifer. *J Hydrol* **300**: 314–331.

Kazemi GA (2008) Editor's message: Submarine groundwater discharge studies and the absence of hydrogeologists. *Hydrogeol J* **16**: 201–204.

Kench PS, Owen SD and Ford MR (2014) Evidence for coral island formation during rising sea level in the central Pacific Ocean. *Geophys Res Lett* **41**: 820–827.

Kerrou J and Renard P (2009) A numerical analysis of dimensionality and heterogeneity effects on advective dispersive seawater intrusion processes. *Hydrogeol J* **18**: 55–72.

Ketabchi H, Mahmoodzadeh D, Ataie-Ashtiani B and Simmons CT (2016) Sea-level rise impacts on seawater intrusion in coastal aquifers: Review and integration. *J Hydrol* **535**: 235–255.

Ketabchi H, Mahmoodzadeh D, Ataie-Ashtiani B, Werner AD and Simmons CT (2014) Sea-level rise impact on fresh groundwater lenses in two-layer small islands. *Hydrol Process* **28**: 5938–5953.

Khan A, Mojumder SK, Kovats S and Vineis P (2008) Saline contamination of drinking water in Bangladesh. *Lancet* **371**: 385–385.

Khoo TC (2009) Singapore water: yesterday, today and tomorrow. Water Management in 2020 and Beyond 237–250. Springer, Dordrecht, Netherlands.

Kidder R (1957) Unsteady flow of gas through a semi-infinite porous medium. *J Appl Mech* **27**: 329–332.

Kim G, Burnett WC, Dulaiova H, Swarzenski PW and Moore WS (2001) Measurement of Ra-224 and Ra-226 activities in natural waters using a radon-in-air monitor. *Environ Sci Technol* **35**: 4680–4683.

Kim G, Kim JS and Hwang DW (2011) Submarine groundwater discharge from oceanic islands standing in oligotrophic oceans: Implications for global biological production and organic carbon fluxes. *Limnol Oceanogr* **56**: 673–682.

Kim G, Ryu JW, Yang HS and Yun ST (2005) Submarine groundwater discharge (SGD) into the Yellow Sea revealed by Ra-228 and Ra-226 isotopes: Implications for global silicate fluxes. *Earth Planet Sc Lett* **237**: 156–166.

Kim KH, Heiss JW, Michael HA, Cai W-J, Laattoe T, Post VEA and Ullman WJ (2017) Spatial patterns of groundwater biogeochemical reactivity in an intertidal beach aquifer. *J Geophys Res Biogeosci* **122**: 2548–2562.

Kim KY, Chon CM and Park KH (2007) A simple method for locating the fresh water-salt water interface using pressure data. *Ground Water* **45**: 723–728.

Kim KY, Seong H, Kim T, Park KH, Woo NC, Park YS, Koh GW and Park WB (2006a) Tidal effects on variations of fresh-saltwater interface and groundwater flow in a multilayered coastal aquifer on a volcanic island (Jeju Island, Korea). *J Hydrol* **330**: 525–542.

Kim RH, Kim JH, Ryu JS and Chang HW (2006b) Salinization properties of a shallow groundwater in a coastal reclaimed area, Yeonggwang, Korea. *Environ Geol* **49**: 1180–1194.

Kim Y, Lee KS, Koh DC, Lee DH, Lee SG, Park WB, Koh GW and Woo NC (2003) Hydrogeochemical and isotopic evidence of groundwater salinization in a coastal aquifer: a case study in Jeju volcanic island, Korea. *J Hydrol* **270**: 282–294.

Kirkham D (1947) Field method for determination of air permeability of soil in its undisturbed state. *Soil Sci Soc Am J* **11**: 93–99.

(1967) Explanation of paradoxes in Dupuit-Forchheimer Seepage Theory. *Water Resour Res* **3**: 609–622.

Kiro Y, Weinstein Y, Starinsky A and Yechieli Y (2013) Groundwater ages and reaction rates during seawater circulation in the Dead Sea aquifer. *Geochim Cosmochim Acta* **122**: 17–35.

Kiro Y, Yechieli Y, Voss CI, Starinsky A and Weinstein Y (2012) Modeling radium distribution in coastal aquifers during sea level changes: The Dead Sea case. *Geochim Cosmochim Acta* **88**: 237–254.

Kirsch R (2006) *Groundwater Geophysics: A Tool for Hydrogeology*. Springer, Berlin.

Klassen J and Allen DM (2017) Assessing the risk of saltwater intrusion in coastal aquifers. *J Hydrol* **551**: 730–745.

Kohout FA (1960) Cyclic flow of salt water in the Biscayne aquifer of southeastern Florida. *J Geophys Res* **65**: 2133–2141.

(1961) Fluctuations of ground-water levels caused by dispersion of salts. *J Geophys Res* **66**: 2429–2434.

(1965) A hypothesis concerning cyclic flow of salt water related to geothermal heating in the Floridan aquifer. Trans NY Acad Sci **28**: 249–271.

(1966) *Submarine Springs: A Neglected Phenomena of Coastal Hydrology*. Central Treaty Organization, Ankara, Turkey.

Kohout FA and Hoy ND (1963) Some aspects of sampling salty ground water in coastal aquifers. *Ground Water* **1**: 28–43.

Kolditz O, Ratke R, Diersch H-JG & Zielke W (1998) Coupled groundwater flow and transport: 1. Verification of variable density flow and transport models. Adv Water Resour **21**: 27–46.

Konikow LF (2011a) The secret to successful solute – transport modeling. *Groundwater* **49**: 144–159.

(2011b) Contribution of global groundwater depletion since 1900 to sea-level rise. *Geophys Res Lett* 38.

Konikow LF, August LL and Voss CI (2001) Effects of clay dispersion on aquifer storage and recovery in coastal aquifers. *Transp Porous Media* **43**: 45–64.

Konikow LF and Bredehoeft JD (1978) Computer model of two-dimensional solute transport and dispersion in ground water. Techniques of Water Resources Investigation, Book 7. USGS, Reston, VA.

Kooi H and Groen J (2001) Offshore continuation of coastal groundwater systems; predictions using sharp-interface approximations and variable-density flow modelling. *J Hydrol* **246**: 19–35.

(2003) Geological processes and the management of groundwater resources in coastal areas. *Netherlands J Geosci* **82**: 31–40.

Kooi H, Groen J and Leijnse A (2000) Modes of seawater intrusion during transgressions. *Water Resour Res* **36**: 3581–3589.

Kopp RE, Kemp AC, Bittermann K, Horton BP, Donnelly JP, Gehrels WR, Hay CC, Mitrovica JX, Morrow ED and Rahmstorf S (2016) Temperature-driven global sea-level variability in the Common Era. *Proc Natl Acad Sci USA* **113**: E1434–E1441.

Kraemer TF and Reid DF (1984) The occurrence and behavior of radium in saline formation water of the United-States Gulf-Coast region. *Isot Geosci* **2**: 153–174.

Krijgsman W, Hilgen FJ, Raffi I, Sierro FJ and Wilson DS (1999) Chronology, causes and progression of the Messinian salinity crisis. *Nature* **400**: 652–655.

Krumbein WC and Aberdeen E (1937) The sediments of Barataria Bay. *J Sediment Res* **7**.

Kruseman GP and De Ridder NA (1994) *Analysis and Evaluation of Pumping Test Data*. 2nd edn. ILRI, Wageningen, Netherlands.

Kuan WK, Jin GQ, Xin P, Robinson CE, Gibbes B and Li L (2012) Tidal influence on seawater intrusion in unconfined coastal aquifers. *Water Resour Res* **48**.

Kuang XX, Jiao JJ and Liu K (2015) Numerical studies of vertical Cl-, H-2 and O-18 profiles in the aquifer-aquitard system in the Pearl River Delta, China. *Hydrol Process* **29**: 4199–4209.

Kuang XX, Jiao JJ and Wang Y (2016) Chloride as tracer of solute transport in the aquifer-aquitard system in the Pearl River Delta, China. *Hydrogeol J* **24**: 1121–1132.

Kuang XX, Jiao JJ and Li HL (2013) Review on airflow in unsaturated zones induced by natural forcings. *Water Resour Res* **49**: 6137–6165.

Kummer T and Spinelli GA (2009) Thermal effects of fluid circulation in the basement aquifer of subducting ocean crust. *J Geophys Res Solid Earth* **114**.

Kwong HT and Jiao JJ (2016) Hydrochemical reactions and origin of offshore relatively fresh pore water from core samples in Hong Kong. *J Hydrol* **537**: 283–296.

Kwong JSM (1997) *A Review of Some Drained Reclamation Works in Hong Kong*. Geotechnical Engineering Office, Civil Engineering Department, Hong Kong.

Lambeck K and Chappell J (2001) Sea level change through the last glacial cycle. *Science* **292**: 679–686.

Lambeck K, Rouby H, Purcell A, Sun Y and Sambridge M (2014) Sea level and global ice volumes from the Last Glacial Maximum to the Holocene. *Proc Natl Acad Sci USA* **111**: 15296–15303.

Land LS (1973) Holocene meteoric dolomitization of Pleistocene limestones, North Jamaica. *Sedimentology* **20**: 411–424.

Langaas K, Nilsen KI and Skjaeveland SM (2006) Tidal pressure response and surveillance of water encroachment. *Spe Reserv Eval Eng* **9**: 335–344.

Lange M, Burkhard B, Garthe S, Gee K, Kannen A, Lenhart H and Windhorst W (2010) Analyzing coastal and marine changes: Offshore wind farming as a case study (www.researchgate.net/profile/Marcus_Lange/publication/233932674_Analyzing_Coastal_and_Marine_Changes_Offshore_Wind_Farming_as_a_Case_Study/links/0fcfd50d1829574f8e000000/Analyzing-Coastal-and-Marine-Changes-Offshore-Wind-Farming-as-a-Case-Study.pdf?origin=publication_detail).

Langevin CD, Thorne DT Jr, Dausman AM, Sukop MC and Guo W (2008) SEAWAT Version 4: A Computer Program for Simulation of Multi-Species Solute and Heat Transport. USGS Techniques and Methods Book 6, Chapter A22, 39 pp. US Government Printing Office, Washington, DC.

Lebbe L (1981) The Subterranean Flow of Fresh and Salt Water underneath the Western Belgian Beach. *Sveriges Geologiska Undersökning Rapporter och meddelanden* **27**: 193–219.

(1983) Mathematical model of the evolution of the fresh water lens under the dunes and the beach with semi-diurnal tides. *Geol Appl Idrogeologia* **18**: 211–226.

(1999) Parameter identification in fresh-saltwater flow based on borehole resistivities and freshwater head data. *Adv Water Resour* **22**: 791–806.

Lee CH and Cheng RTS (1974) On seawater encroachment in coastal aquifers. *Water Resour Res* **10**: 1039–1043.

Lee CM, Jiao JJ, Luo X and Moore WS (2012) Estimation of submarine groundwater discharge and associated nutrient fluxes in Tolo Harbour, Hong Kong. *Sci Total Environ* **433**: 427–433.

Lee DR (1977) A device for measuring seepage flux in lakes and estuaries. *Limnol Oceanogr* **22**: 140–147.

Legresy B (2016) *Sea Level Data.* Vol. 2018 (www.cmar.csiro.au/sealevel/sl_data_cmar .html#).

Lei J, Men Q, Wang Y and Wang B (2002) Review of'green house+ deep well seawater'industrialized culture pattern of turbot (*Scophthalmus maximus*) (in Chinese). *Mar Fish Res* **23**: 7.

Lentz EE, Thieler ER, Plant NG, Stippa SR, Horton RM and Gesch DB (2016) Evaluation of dynamic coastal response to sea-level rise modifies inundation likelihood. *Nat Clim Change* **6**: 696–700.

Leung CM, Jiao JJ, Malpas J, Chan WT and Wang YX (2005) Factors affecting the groundwater chemistry in a highly urbanized coastal area in Hong Kong: An example from the Mid-Levels area. *Environ Geol* **48**: 480–495.

Leung WKL, Li DCH and Pickles AR (2007) *Heaving of Airfield Pavement at Hong Kong International Airport*. Atlantic City, NJ.

Levanon E, Yechieli Y, Gvirtzman H and Shalev E (2017) Tide-induced fluctuations of salinity and groundwater level in unconfined aquifers – Field measurements and numerical model. *J Hydrol* **551**: 665–675.

Li FL (2005) Monitoring and numerical simulation of saltwater intrusion in the Eastern Coast of Laizhou Bay, China. PhD thesis, China Ocean University.

Li GM and Chen CX (1991) Determining the length of confined aquifer roof extending under the sea by the tidal method. *J Hydrol* **123**: 97–104.

Li HL and Jiao JJ (2001a) Tide-induced groundwater fluctuation in a coastal leaky confined aquifer system extending under the sea. *Water Resour Res* **37**: 1165–1171.

(2001b) Analytical studies of groundwater-head fluctuation in a coastal confined aquifer overlain by a semi-permeable layer with storage. *Adv Water Resour* **24**: 565–573.

(2002) Analytical solutions of tidal groundwater flow in coastal two-aquifer system. *Adv Water Resour* **25**: 417–426.

(2003a) Tide-induced seawater-groundwater circulation in a multi-layered coastal leaky aquifer system. *J Hydrol* **274**: 211–224.

(2003b) Influence of the tide on the mean watertable in an unconfined, anisotropic, inhomogeneous coastal aquifer. *Adv Water Resour* **26**: 9–16.

(2005) One-dimensional airflow in unsaturated zone induced by periodic water table fluctuation. *Water Resour Res* **41**.

Li HL, Jiao JJ and Luk M (2004) A falling-pressure method for measuring air permeability of asphalt in laboratory. *J Hydrol* **286**: 69–77.

Li HL, Jiao JJ, Luk M and Cheung KY (2002) Tide-induced groundwater level fluctuation in coastal aquifers bounded by L-shaped coastlines. *Water Resour Res* **38**.

Li HL, Li GY, Cheng JM and Boufadel MC (2007) Tide-induced head fluctuations in a confined aquifer with sediment covering its outlet at the sea floor. *Water Resour Res* **43**.

Li HL, Li L and Lockington D (2005) Aeration for plant root respiration in a tidal marsh. *Water Resour Res* **41**.

Li HL, Xia YQ and Wang L (2011a) Tide-induced air pressure fluctuations in a coastal unsaturated zone: Effects of thin low-permeability pavements. *Ground Water Monitor* **31**: 40–47.

Li J, Zhan HB, Huang GH and You KH (2011b) Tide-induced airflow in a two-layered coastal land with atmospheric pressure fluctuations. *Adv Water Resour* **34**: 649–658.

Li L, Barry DA and Jeng DS (2001) Tidal fluctuations in a leaky confined aquifer: Dynamic effects of an overlying phreatic aquifer. *Water Resour Res* **37**: 1095–1098.

Li L, Barry DA, Cunningham C, Stagnitti F and Parlange JY (2000a) A two-dimensional analytical solution of groundwater responses to tidal loading in an estuary and ocean. *Adv Water Resour* **23**: 825–833.

Li L, Barry DA, Stagnitti F and Parlange JY (1999) Submarine groundwater discharge and associated chemical input to a coastal sea. *Water Resour Res* **35**: 3253–3259.

Li L, Barry DA, Stagnitti F, Parlange JY and Jeng DS (2000b) Beach water table fluctuations due to spring-neap tides: moving boundary effects. *Adv Water Resour* **23**: 817–824.

Li P and Qiao P (1982) The model of evolution of the Pearl River Delta during last 6,000 years (in Chinese with English abstract). *J Sediment Res* **3**: 33–42.

Li QF, Zhang Y, Juck D, Fortin N and Greer CW (2011c) Impact of intensive land-based fish culture in Qingdao, China, on the bacterial communities in surrounding marine waters and sediments. *Evid-Based Compl Alt*.

Li W (2006) Study on design theory of groundwater reservoir in porous medium. Thesis, Hehai University, Hehai University.

Lichtner PC, Steefel CI and Oelkers EH (1996) *Reactive Transport in Porous Media*. Mineralogical Society of America, Washington, DC.

Lin C and Melville MD (1994) Acid sulfate soil landscape relationships in the Pearl River Delta, southern China. *Catena* **22**: 105–120.

Lin C, Melville MD and Hafer S (1995) Acid sulfate soil-landscape relationships in an undrained, tide-dominated estuarine floodplain, eastern Australia. *Catena* **24**: 177–194.

Lin H-CJ, Richards DR, Yeh G-T, Cheng J-R and Cheng H-P (1997) *FEMWATER: A Three-Dimensional Finite Element Computer Model for Simulating Density-Dependent Flow and Transport in Variably Saturated Media*. DTIC Document.

Lin IT, Wang CH, You CF, Lin S, Huang KF and Chen YG (2010) Deep submarine groundwater discharge indicated by tracers of oxygen, strontium isotopes and barium content in the Pingtung coastal zone, southern Taiwan. *Mar Chem* **122**: 51–58.

Liu Y, Jiao JJ, Liang WZ and Luo X (2018) Using tidal fluctuation-induced dynamics of radium isotopes (Ra-224, Ra-223, and Ra-228) to trace the hydrodynamics and geochemical reactions in a coastal groundwater mixing zone. *Water Resour Res* **54**: 2909–2930.

Llopis-Albert C and Pulido-Velazquez D (2014) Discussion about the validity of sharp-interface models to deal with seawater intrusion in coastal aquifers. *Hydrol Process* **28**: 3642–3654.

Lloyd JW, Howard KWF, Pacey NR and Tellam JH (1982) The value of iodide as a parameter in the chemical characterization of groundwaters. *J Hydrol* **57**: 247–265.

Loke MH, Chambers JE, Rucker DF, Kuras O and Wilkinson PB (2013) Recent developments in the direct-current geoelectrical imaging method. *J Appl Geophys* **95**: 135–156.

Long RA (1965) Feasibility of a *Scavenger-Well Syste*m as a *Solution* to the *Problem* of *Vertical Salt-Water Encr*oachment. Department of Conservation, Louisiana Geological Survey, Baton Rouge.

Lovley DR and Goodwin S (1988) Hydrogen concentrations as an indicator of the predominant terminal electron-accepting reactions in aquatic sediments. *Geochim Cosmochim Acta* **52**: 2993–3003.

Lowe DR (1975) Water escape structures in coarse-grained sediments. *Sedimentology* **22**: 157–204.

Lu C, Kitanidis PK and Luo J (2009) Effects of kinetic mass transfer and transient flow conditions on widening mixing zones in coastal aquifers. *Water Resour Res* **45**.

Lu C, Shi L, Chen Y, Xie Y and Simmons CT (2016) Impact of kinetic mass transfer on free convection in a porous medium. *Water Resour Res* **52**: 3637–3653.

Lu N and Likos WJ (2004) *Unsaturated Soil Mechanics*. John Wiley, Hoboken, NJ.

Lu XX, Zhang SR, Xie SP and Ma PK (2007) Rapid channel incision of the lower Pearl River (China) since the 1990s as a consequence of sediment depletion. *Hydrol Earth Syst Sc* **11**: 1897–1906.

Lumb P (1976) Land Reclamation in Hong Kong, *Materials and methods for low cost road, rail, and reclamation works: proceedings of a residential workshop*, Leura, Australia, 6–10 September 1976. pp. 299–314.

(1980) *Thirty Years of Soil Engineering in Hong Kong*. Rupert H. Myers Lecture, Leura.

Luo X, Jiao JJ, Moore WS and Lee CM (2014) Submarine groundwater discharge estimation in an urbanized embayment in Hong Kong via short-lived radium isotopes and its implication of nutrient loadings and primary production. *Mar Pollut Bull* **82**: 144–154.

Luo X, Jiao JJ, Wang X-S and Liu K (2016) Temporal Rn-222 distributions to reveal groundwater discharge into desert lakes: Implication of water balance in the Badain Jaran Desert, China. *J Hydrol* **534**: 87–103.

Luo Y and Zhou X (2006) *Soil Respiration and the Environment*. Elsevier/Academic Press, Amsterdam.

Lusczynski NJ (1961) Head and flow of ground water of variable density. *J Geophys Res* **66**: 4247.

Lusczynski NJ and Swarzenski WV (1966) Salt-water encroachment in southern Nassau and southeastern Queens Counties, Long Island, New York USGS Water Supply Paper 1613-F, 76 pp.

Ma CX (2005) Study on mode of artificial recharge of groundwater in Shangdong Province (in Chinese). Thesis, Wuhan University.

Maas K (1998) Over grondwatergetijden (On groundwater tides). *Stromingen* **4**: 33–36.

Machel HG (2004) Concepts and models of dolomitization: A critical reappraisal. *Geol Soc London Spec Publ* **235**: 7–63.

MacIntyre S, Wanninkhof R and Chanton J (1995) Trace gas exchange across the air-water interface in freshwater and coastal marine environments. *Biogenic Trace Gases: Measuring Emissions from Soil and Water*, Vol. **52–97** (Matson PA and Harriss RC, eds). John Wiley, Hoboken, NJ.

Macpherson GL (2009) CO_2 distribution in groundwater and the impact of groundwater extraction on the global C cycle. *Chem Geol* **264**: 328–336.

Magaritz M, Goldenberg L, Kafri U and Arad A (1980) Dolomite formation in the seawater-freshwater interface. *Nature* **287**: 622–624.

Maimone M (2002) Developing an effective coastal aquifer management program. *Proceedings SWIM17 Delft 2002* (Boekelman RH, Hornschuh JCS, Olsthoorn TN, Oude Essink GHP, Peute L and Stark JM, eds), 327–336. Delft, Netherlands.

(2004) Defining and managing sustainable yield. *Ground Water* **42**: 809–814.

Malone MJ, Claypool G, Martin JB and Dickens GR (2002) Variable methane fluxes in shallow marine systems over geologic time: The composition and origin of pore waters and authigenic carbonates on the New Jersey shelf. *Mar Geol* **189**: 175–196.

Mantoglou A (2003) Pumping management of coastal aquifers using analytical models of saltwater intrusion. *Water Resour Res* **39**.

Manzano M, Custodio E and Carrera J (1993) *Fresh and Salt Water in the Llobregat Delta Aquitard: Application of the Ion Chromatography Theory to the Field Data*. CIMNE, Barcelona, Spain.

Margat J, Foster S and Droubi A (2006) Concept and importance of non-renewable resources. Non-renewable Groundwater Resources – A Guidebook on Socially-Sustainable Management for Water-Policy Makers (Foster S and Loucks DP, eds), 13–24. UNESCO, Paris.

Marimuthu S, Reynolds DA and La Salle CLG (2005) A field study of hydraulic, geochemical and stable isotope relationships in a coastal wetlands system. *J Hydrol* **315**: 93–116.

Marine IW (1974) Geohydrology of buried triassic basin at Savannah River Plant, South Carolina. *AAPG Bull* **58**: 1825–1837.

Marine IW and Fritz SJ (1981) Osmotic model to explain anomalous hydraulic heads. *Water Resour Res* **17**: 73–82.

Marlow DR, Moglia M, Cook S and Beale DJ (2013) Towards sustainable urban water management: A critical reassessment. *Water Res* **47**: 7150–7161.

Martens CS, Kipphut GW and Valklump J (1980) Sediment-water chemical-exchange in the coastal zone traced by insitu Rn-222 flux measurements. *Science* **208**: 285–288.

Martens K, Walraevens K and Al Farran N (2011) Hydrochemistry of the upper Miocene-Pliocene-Quaternary aquifer complex of Jifarah plain, NW-Libya. *Geol Belg* **14**: 159–174.

Martínez J, Benavente J, García-Aróstegui JL, Hidalgo MC and Rey J (2009) Contribution of electrical resistivity tomography to the study of detrital aquifers affected by seawater intrusion–extrusion effects: The river Vélez delta (Vélez-Málaga, southern Spain). *Eng Geol* **108**: 161–168.

Martínez ML, Intralawan A, Vázquez G, Pérez-Maqueo O, Sutton P and Landgrave R (2007) The coasts of our world: Ecological, economic and social importance. *Ecol Econ* **63**: 254–272.

Martínez-Alvarez V, Martin-Gorriz B and Soto-García M (2016) Seawater desalination for crop irrigation – A review of current experiences and revealed key issues. *Desalination* **381**: 58–70.

Marui A (2003) Groundwater conditions along the seawater/freshwater interface on a volcanic island and a depositional area in Japan. *Geol Q* **47**: 381–388.

Massmann JW (1989) Applying groundwater-flow models in vapor extraction system-design. *J Environ Eng ASCE* **115**: 129–149.

Massmann JW and Madden M (1994) Estimating air conductivity and porosity from vadose-zone pumping tests. *J Environ Eng ASCE* **120**: 313–328.

Mastrocicco M, Giambastiani BMS and Colombani N (2013) Ammonium occurrence in a salinized lowland coastal aquifer (Ferrara, Italy). *Hydrol Process* **27**: 3495–3501.

Matschoss C (1921) *Preussens Gewerbeförderung und ihre grossen Männer – Dargestellt im Rahmen der Geschichte des Vereins zur Geschichte des Gewerbfleises 1821–1921* (The Advancement of the Prussian Commerce and Industry and its great Men – illustrated within the framework of the Association of Commerce development). Verlag d. Vereins Dt. Ingenieure, Berlin, Germany.

Mazor IE (2004) *Chemical and Isotopic Groundwater Hydrology.* Marcel Dekker, New York.

McAnally WH and Pritchard DW (1997) Salinity control in Mississippi River under drought flows. *J Waterw Port C ASCE* **123**: 34–40.

McArthur JM, Ravenscroft P, Safiulla S and Thirlwall MF (2001) Arsenic in groundwater: Testing pollution mechanisms for sedimentary aquifers in Bangladesh. *Water Resour Res* **37**: 109–117.

McCain WD Jr. (1991) Reservoir-fluid property correlations – state of the art (includes associated papers 23583 and 23594).

McCarty PL (1997) Microbiology – breathing with chlorinated solvents. *Science* **276**: 1521–1522.

McClatchie S, Middleton JF and Ward TM (2006) Water mass analysis and alongshore variation in upwelling intensity in the eastern Great Australian Bight. *J. Geophys. Res* **111**.

McCobb TD and Weiskel PK (2003) *Long-Term Hydrologic Monitoring Protocol for Coastal Ecosystems*. USGS, Northborough, MA.

McElwee CD (1985) A model study of salt-water intrusion to a river using the sharp interface approximation. *Groundwater* **23**: 465–475.

McFarland ER, Bruce TS and Virginia Department of Environmental Quality (2006) *The Virginia Coastal Plain Hydrogeologic Framework.* US Department of the Interior, USGS, Reston, VA.

McGinnis RL (2002) Osmotic desalinization process. Google Patents.

McGranahan G, Balk D and Anderson B (2007) The rising tide: Assessing the risks of climate change and human settlements in low elevation coastal zones. *Environment and Urbanization* **19**: 17–37.

McIntosh JC, Grasby SE, Hamilton SM and Osborn SG (2014) Origin distribution and hydrogeochemical controls on methane occurrences in shallow aquifers, southwestern Ontario, Canada. *Appl Geochem* **50**: 37–52.

McMahon PB, Williams DF and Morris JT (1990) Production and carbon isotopic composition of bacterial CO_2 in deep coastal-plain sediments of South-Carolina. *Ground Water* **28**: 693–702.

McNeal BL (1968) Prediction of the effect of mixed-salt solutions on soil hydraulic conductivity. *Soil Sci Soc Am J* **32**: 190–193.

McWhorter DB (1971) *Infiltration Affected by Flow of Air.* Hydrology Papers no. 49. Colorado State University.

Meier HEM, Kjellström E and Graham LP (2006) Estimating uncertainties of projected Baltic Sea salinity in the late 21st century. *Geophys Res Lett* **33**.

Meinzer OE (1936) Movements of ground water. *AAPG Bull* **20**: 704–725.

Meisler H, Leahy PP and Knobel LL (1984) Effect of eustatic sealevel changes on saltwater–freshwater relations in the northern Atlantic coastal plain. USGS Water Supply Paper 2225, 28 pp. US Government Printing Office, Washington, DC.

Melchior PJ (1978) *The Tides of the Planet Earth*. Pergamon Press, Oxford.

Melloul A and Collin M (2006) Hydrogeological changes in coastal aquifers due to sea level rise. *Ocean Coast Manage* **49**: 281–297.

Merkel B, Planer-Friedrich B and Nordstrom DK (2008) *Groundwater Geochemistry: A Practical Guide to Modeling of Natural and Contaminated Aquatic Systems*. Springer, Berlin.

Merkens J-L, Reimann L, Hinkel J and Vafeidis AT (2016) Gridded population projections for the coastal zone under the Shared Socioeconomic Pathways. *Global Planet Change* **145**: 57–66.

Merritt ML (2004) *Estimating Hydraulic Properties of the Floridan Aquifer System by Analysis of earth-Tide, Ocean-Tide, and Barometric Effects, Collier and Hendry Counties, Florida*. US Department of the Interior, Branch of Information Services, Tallahassee, FL.

Micallef A, Mountjoy JJ, Schwalenberg K et al. (2018) How offshore groundwater shapes the seafloor. *Eos Trans AGU* **99**.

Michael HA, Charette MA and Harvey CF (2011) Patterns and variability of groundwater flow and radium activity at the coast: A case study from Waquoit Bay, Massachusetts. *Mar Chem* **127**: 100–114.

Michael HA, Lubetsky JS and Harvey CF (2003) Characterizing submarine groundwater discharge: A seepage meter study in Waquoit Bay, Massachusetts. *Geophys Res Lett* **30**.

Michael HA, Mulligan AE and Harvey CF (2005) Seasonal oscillations in water exchange between aquifers and the coastal ocean. *Nature* **436**: 1145–1148.

Michael HA, Post VEA, Wilson AM and Werner AD (2017) Science, society, and the coastal groundwater squeeze. *Water Resour Res* **53**: 2610–2617.

Michael HA, Scott KC, Koneshloo M, Yu X, Khan MR and Li K (2016) Geologic influence on groundwater salinity drives large seawater circulation through the continental shelf. *Geophys Res Lett* **43**: 10782–10791.

Milanovic P (2004) *Water Resources Engineering in Karst*. CRC Press, Boca Raton, FL.

Miller JA (2000) *Ground Water Atlas of the United States*. USGS, Reston, VA.

Miller S, Shemer H and Semiat R (2015) Energy and environmental issues in desalination. *Desalination* **366**: 2–8.

Millero FJ, Feistel R, Wright DG and McDougall TJ (2008) The composition of Standard Seawater and the definition of the Reference-Composition Salinity Scale. Deep Sea Res Part I **55**: 50–72.

Mitrovica JX and Peltier WR (1991) On postglacial geoid subsidence over the equatorial oceans. *J Geophys Res Solid Earth* **96**: 20053–20071.

Mok KM, Wong H and Fan XJ (2011) Water resources management in Macao SAR to tackle its sea water intrusion problem. Asian and Pacific Coasts 2011 (Lee JH-W and Ng C-O, eds), 1402–1409. World Scientific, Hong Kong.

Mondal NC, Singh VS, Saxena VK and Prasad RK (2008) Improvement of groundwater quality due to fresh water ingress in Potharlanka Island, Krishna delta, India. *Environ Geol* **55**: 595–603.

Monsen NE, Cloern JE, Lucas LV and Monismith SG (2002) A comment on the use of flushing time, residence time, and age as transport time scales. *Limnol Oceanogr* **47**: 1545–1553.

Montenegro S, Montenegro A, Cabral J and Cavalcanti G (2006) *Intensive Exploitation and Groundwater Salinity in Recife Coastal Plain (Brazil): Monitoring and Management Perspectives*. Cagliari-Chia, Laguna, Italy.

Moore WS (1976) Sampling radium-228 in the deep ocean. *Deep Sea Res* **23**: 647–651.

(1984) Radium isotope measurements using germanium detectors. *Nucl Instrum Methods Phys Res* **223**: 407–411.

(1996) Large groundwater inputs to coastal waters revealed by Ra-226 enrichments. *Nature* **380**: 612–614.

(1999) The subterranean estuary: A reaction zone of ground water and sea water. *Mar Chem* **65**: 111–125.

(2000) Determining coastal mixing rates using radium isotopes. *Cont Shelf Res* **20**: 1993–2007.

(2003) Sources and fluxes of submarine groundwater discharge delineated by radium isotopes. *Biogeochemistry* **66**: 75–93.

(2008) Fifteen years experience in measuring Ra-224 and Ra-223 by delayed-coincidence counting. *Mar Chem* **109**: 188–197.

(2010a) The effect of submarine groundwater discharge on the ocean. *Annu Rev Mar Sci* **2**: 59–88.

(2010b) A reevaluation of submarine groundwater discharge along the southeastern coast of North America. *Global Biogeochem Cy* **24**.

Moore WS and Arnold R (1996) Measurement of Ra-223 and Ra-224 in coastal waters using a delayed coincidence counter. *J Geophys Res Oceans* **101**: 1321–1329.

Moore WS, Blanton JO and Joye SB (2006) Estimates of flushing times, submarine groundwater discharge, and nutrient fluxes to Okatee Estuary, South Carolina. *J Geophys Res Oceans* **111**.

Moore WS and Cai P (2013) Calibration of RaDeCC systems for ^{223}Ra measurements. *Mar Chem* **156** 130–137 XX–XX.

Moore WS and Krest J (2004) Distribution of Ra-223 and Ra-224 in the plumes of the Mississippi and Atchafalaya Rivers and the Gulf of Mexico. *Mar Chem* **86**: 105–119.

Moore WS, Sarmiento JL and Key RM (2008) Submarine groundwater discharge revealed by Ra-228 distribution in the upper Atlantic Ocean. *Nat Geosci* **1**: 309–311.

Moosdorf N and Oehler T (2017) Societal use of fresh submarine groundwater discharge: An overlooked water resource. *Earth Sci Rev* **171**: 338–348.

Morgan LK, Stoeckl L, Werner AD and Post VEA (2013) An assessment of seawater intrusion overshoot using physical and numerical modeling. *Water Resour Res* **49**: 6522–6526.

Morgan LK and Werner AD (2016) Comment on 'Closed-form analytical solutions for assessing the consequences of sea-level rise on groundwater resources in sloping coastal aquifers', Hydrogeol J 23: 1399–1413.

Morgan LK, Werner AD, Ivkovic KM, Carey H and Sundaram B (2013) *A National-Scale Vulnerability Assessment of Seawater Intrusion: First-Order Assessment of Seawater Intrusion for Australian Case Study Sites*. Record 2013/19. Geoscience Australia, Canberra, and National Centre for Groundwater Research and Training, Adelaide.

Mörner NA (1996) Rapid changes in coastal sea level. *J Coastal Res* **12**: 797–800.

Mualem Y and Bear J (1974) The shape of the interface in steady flow in a stratified aquifer. *Water Resour Res* **10**: 1207–1215.
Muhammad A-K and Husam A-N (2011) Hydro-geochemical characteristics of groundwater beneath the Gaza Strip. *J Water Resour Protect* **2011**: 341–348.
Mulder A, Vandegraaf AA, Robertson LA and Kuenen JG (1995) Anaerobic ammonium oxidation discovered in a denitrifying fluidized-bed reactor. *Fems Microbiol Ecol* **16**: 177–183.
Mulligan AE and Charette MA (2006) Intercomparison of submarine groundwater discharge estimates from a sandy unconfined aquifer. *J Hydrol* **327**: 411–425.
Mulligan AE, Evans RL and Lizarralde D (2007) The role of paleochannels in groundwater/seawater exchange. *J Hydrol* **335**: 313–329.
Muñoz I and Fernández-Alba AR (2008) Reducing the environmental impacts of reverse osmosis desalination by using brackish groundwater resources. *Water Res* **42**: 801–811.
Murdoch LC and Kelly SE (2003) Factors affecting the performance of conventional seepage meters. *Water Resour Res* **39**.
Murtaza G, Saqib M, Ghafoor A, Javed W, Murtaza B, Ali MK and Abbas G (2015) Climate change and water security in dry areas dry areas climate change water security. *Handb Clim Change Adapt*: 1701–1730.
Muskat M (1934) The flow of compressible fluids through porous media and some problems in heat conduction. *J Appl Phys* **5**: 71–94.
(1946) *The Flow of Homogeneous Fluids through Porous Media*. J. W. Edwards, Ann Arbor, MI.
Mylroie JE (2013) Coastal karst development in carbonate rocks. Coastal Karst Landforms (Lace JM and Mylroie EJ, eds), 77–109. Springer, Dordrecht, Netherlands.
Nakada S, Umezawa Y, Taniguchi M and Yamano H (2012) Groundwater Dynamics of Fongafale Islet, Funafuti Atoll, Tuvalu. *Groundwater* **50**: 639–644.
Needham J and Wang L (1959) *Science and civilisation in China. Vol. 3, Mathematics and the sciences of the heavens and the earth*. Cambridge University Press, Cambridge.
Neeper DA (2002) Investigation of the vadose zone using barometric pressure cycles. *J Contam Hydrol* **54**: 59–80.
Nelms DL, Virginia Office of Drinking Water and Geological Survey (US) (2003) *Aquifer Susceptibility in Virginia, 1998–2000*. US Department of the Interior, USGS, Richmond, VA.
Neumann B, Vafeidis AT, Zimmermann J and Nicholls RJ (2015) Future coastal population growth and exposure to sea-level rise and coastal flooding–a global assessment. *PLoS One* **10**: e0118571.
Neuzil CE (1986) Groundwater-flow in low-permeability environments. *Water Resour Res* **22**: 1163–1195.
Ni JR, Borthwick AGL and Qin HP (2002) Integrated approach to determining postreclamation coastlines. *J Environ Eng ASCE* **128**: 543–551.
Nicholls RJ and Cazenave A (2010) Sea-level rise and its impact on coastal zones. *Science* **328**: 1517–1520.
Nielsen DM (1991) *Practical Handbook of Ground-water Monitoring*. Lewis, Chelsea, MI.
Nielsen P (1990) Tidal dynamics of the water-table in beaches. *Water Resour Res* **26**: 2127–2134.
Nilson RH, Peterson EW, Lie KH, Burkhard NR and Hearst JR (1991) Atmospheric pumping – a mechanism causing vertical transport of contaminated gases through fractured permeable media. *J Geophys Res Solid Earth* **96**: 21933–21948.

Nishikawa T, Siade AJ, Reichard EG, Ponti DJ, Canales AG and Johnson TA (2009) Stratigraphic controls on seawater intrusion and implications for groundwater management, Dominguez Gap area of Los Angeles, California, USA. *Hydrogeol J* **17**: 1699–1725.

Nocchi M and Salleolini M (2013) A 3D density-dependent model for assessment and optimization of water management policy in a coastal carbonate aquifer exploited for water supply and fish farming. *J Hydrol* **492**: 200–218.

Nomitsu T, Toyohara Y and Kamimoto R (1927) On the contact surface of fresh and salt-water under the ground near a sandy sea-shore. *Mem Coll Sci Kyoto Imperial Univ, Ser A* **X**: 281–302.

Noske RA (1995) the ecology of mangrove forest birds in Peninsular Malaysia. *Ibis* **137**: 250–263.

Oberdorfer JA, Hogan PJ and Buddemeier RW (1990) Atoll Island hydrogeology – flow and fresh-water occurrence in a tidally dominated system. *J Hydrol* **120**: 327–340.

Ogilvy RD, Meldrum PI, Kuras O et al. (2009) Automated monitoring of coastal aquifers with electrical resistivity tomography. *Near Surface Geophys* **7**: 367–375.

Oki DS, Souza WR, Bolke EL and Bauer GR (1998) Numerical analysis of the hydrogeologic controls in a layered coastal aquifer system, Oahu, Hawaii, USA. *Hydrogeol J* **6**: 243–263.

Olshausen J (1904) Flut und Ebbe in artesischen Tiefbrunnen in Hamburg. *J Gasbeleuch Wasserversor* **47**: 381–385, 412–415.

Ortoleva PJ (1994) *Basin Compartments and Seals*. American Association of Petroleum Geologists, Tulsa, OK.

Ortuño F, Molinero J, Garrido T and Custodio E (2012) Seawater injection barrier recharge with advanced reclaimed water at Llobregat Delta aquifer (Spain). *Water Sci Technol* **66**: 2083.

Otton JK (2006) *Environmental Aspects of Produced-Water Salt Releases in Onshore and Coastal Petroleum-Producing Areas of the Conterminous US – a Bibliography.* USGS Open-File Report 2006-1154, 223pp. US Government Printing Office, Washington, DC.

Oude Essink GHP (1996) Impact of sea level rise on groundwater flow regimes: A sensitivity analysis for the Netherlands. Thesis, TU Delft, Delft University of Technology.

(1999) Impact of sea level rise in the Netherlands. Seawater Intrusion in Coastal Aquifers: Concepts, Methods and Practices (Bear J, Cheng AH-D, Sorek S, Ouazar D and Herrebra I, eds), xv, 625. Kluwer Academic, Dordrecht, Netherlands.

(2000) *Groundwater Modelling Lecture Notes*. Utrecht, Netherlands.

(2001a) Salt water intrusion in a three-dimensional groundwater system in the Netherlands: A numerical study. *Transp Porous Media* **43**: 137–158.

(2001b) *Density Dependent Groundwater Flow: Salt water intrusion in coastal aquifers*. Utrecht, Netherlands.

(2001c) Improving fresh groundwater supply – problems and solutions. *Ocean Coast Manage* **44**: 429–449.

Oude Essink GHP, Van Baaren ES and De Louw PGB (2010) Effects of climate change on coastal groundwater systems: A modeling study in the Netherlands. *Water Resour Res* **46**.

PAGES PIWGo (2016) Interglacials of the last 800,000 years. *Rev Geophys* **54**: 162–219.

Palmer HS (1927) *Geology of the Honolulu Artesian System. Supplement of the Report of the Honolulu Sewer and Water Commission to the Legislature of the Territory of Hawaii*. Mercantile Press, Honolulu, HI.

(1957) Origin and diffusion of the Herzberg principle with especial reference to Hawaii. *Pac Sci* **11**: 181–189.

Park H-Y, Jang K, Ju JW and Yeo IW (2012a) Hydrogeological characterization of seawater intrusion in tidally-forced coastal fractured bedrock aquifer. *J Hydrol* **446**–447: 77–89.

Park J, Oh C and Kim J (2007) Three-dimensional numerical simulation of density-dependent groundwater flow and salt transport due to groundwater pumping in a heterogeneous and true anisotropic coastal aquifer. *A New Focus on Groundwater–Seawater Interactions (Proceedings of Symposium HS1001 at IUGG2007, Perugia, July 2007)*, IAHS Publ. 312 (Sanford W, Langevin C, Polemio M and Povinec P, eds), 294–395. IAHS.

Park J, Sanford RA and Bethke CM (2006) Geochemical and microbiological zonation of the Middendorf aquifer, South Carolina. *Chem Geol* **230**: 88–104.

Park Y, Lee JY, Kim JH and Song SH (2012b) National scale evaluation of groundwater chemistry in Korea coastal aquifers: Evidences of seawater intrusion. *Environ Earth Sci* **66**: 707–718.

Parker GG, Ferguson GE and Love SK (1955) Water resources of southeastern Florida, with special reference to geology and ground water of the Miami area. USGS Water Supply Paper 1255, 965 pp.

Parker JC (2003) Physical processes affecting natural depletion of volatile chemicals in soil and groundwater. *Vadose Zone J* **2**: 222–230.

Parkhurst DL and Appelo CAJ (2013) *Description of Input and Examples for PHREEQC Version 3: A Computer Program for Speciation, Batch-Reaction, One-Dimensional Transport, and Inverse Geochemical Calculations*. USGS Techniques and Methods, book 6, chap. A43, 497 pp.

Parlange JY, Stagnitti F, Starr JL and Braddock RD (1984) Free-surface flow in porous-media and periodic-solution of the shallow-flow approximation. *J Hydrol* **70**: 251–263.

Patel HM, Eldho TI and Rastogi AK (2010) Simulation of radial collector well in shallow alluvial riverbed aquifer using analytic element method. *J Irrig Drain E ASCE* **136**: 107–119.

Paul DB, Davidson RR and Cavalli NJ (1992) *Slurry Walls: Design, Construction, and Quality Control*. ASTM, Philadelphia, PA.

Paulsen RJ, Smith CF, O'Rourke D and Wong TF (2001) Development and evaluation of an ultrasonic ground water seepage meter. *Ground Water* **39**: 904–911.

Pauw PS, Groen J, Groen MMA, Van der Made KJ, Stuyfzand PJ and Post VEA (2017) Groundwater salinity patterns along the coast of the Western Netherlands and the application of cone penetration tests. *J Hydrol* **551**: 756–767.

Pedersen TA and Curtis JT (1991) *Soil Vapor Extraction Technology – Reference Handbook*. US Environmental Protection Agency, Washington, DC.

Pierson W, Bishop K, Van Senden D, Horton P and Adamantidis C (2002) *Environmental Water Requirements to Maintain Estuarine Processes Environmental Flows Initiative Technical Report 3*. Commonwealth of Australia, Canberra.

Peltier WR and Fairbanks RG (2006) Global glacial ice volume and Last Glacial Maximum duration from an extended Barbados sea level record. *Quat Sci Rev* **25**: 3322–3337.

Pennink JMK (1905) Investigations for ground-water supplies. *Trans Am Soc Civil Eng LIV D* 169–181.

Perlmutter NM and Geraghty JJ (1963) *Geology and Ground-Water Conditions in Southern Nassau and Southeastern Queens Counties, Long Island, N.Y.* US Government Printing Office, Washington, DC.

Person M, Dugan B, Swenson JB, Urbano L, Stott C, Taylor J and Willett M (2003) Pleistocene hydrogeology of the Atlantic continental shelf, New England. *Geol Soc Am Bull* **115**: 1324.

Person M, Marksamer A, Dugan B, Sauer PE, Brown K, Bish D, Licht KJ and Willett M (2012) Use of a vertical δ18O profile to constrain hydraulic properties and recharge rates across a glacio-lacustrine unit, Nantucket Island, Massachusetts, USA. *Hydrogeol J* **20**: 325–336.

Person M, Wilson JL, Morrow N and Post VEA (2017) Continental-shelf freshwater water resources and improved oil recovery by low-salinity waterflooding. *AAPG Bull* **101**: 1–18.

Peters T and Pintó D (2008) Seawater intake and pre-treatment/brine discharge – environmental issues. *Desalination* **221**: 576–584.

Peterson RN, Moore WS, Chappel SL, Viso RF, Libes SM and Peterson LE (2016) A new perspective on coastal hypoxia: The role of saline groundwater. *Mar Chem* **179**: 1–11.

Phien-wej N, Giao PH and Nutalaya P (2006) Land subsidence in Bangkok, Thailand. *Eng Geol* **82**: 187–201.

Philip JR (1973) Periodic nonlinear diffusion – integral relation and its physical consequences. *Aust J Phys* **26**: 513–519.

Pickles A and Tosen R (1998) Settlement of reclaimed land for the new Hong Kong International Airport. *Proc ICE Geotech Eng* **131**: 191–209.

Pierdinock MJ and Fedder RP (1997) *Bioslurping in a Tidally-Controlled Formation: A Case Study.* Battelle Press, New York.

Pinder GF and Cooper HH (1970) A numerical technique for calculating the transient position of the saltwater front. *Water Resour Res* **6**: 875–882.

Pinder GF and Gray WG (1977) *Finite Element Simulation in Surface and Subsurface Hydrology.* Academic Press, New York.

Piper AM (1944) A graphic procedure in the geochemical interpretation of water-analyses. Trans AGU **25**: 914–923.

Pit IR, Dekker SC, Kanters TJ, Wassen MJ and Griffioen J (2017) Mobilisation of toxic trace elements under various beach nourishments. *Environ Pollut* **231**: 1063–1074.

Plant GW, Covil CS, Hughes RA and Airport Authority Hong Kong (1998) *Site Preparation for the New Hong Kong International Airport.* Thomas Telford, London.

Plummer LN, Vacher HL, Mackenzie FT, Bricker OP and Land LS (1976) Hydrogeochemistry of Bermuda: A case history of ground-water diagenesis of biocalcarenites. *GSA Bull* **87**: 1301–1316.

Poch RM, Thomas BP, Fitzpatrick RW and Merry RH (2009) Micromorphological evidence for mineral weathering pathways in a coastal acid sulfate soil sequence with Mediterranean-type climate, South Australia. *Aust J Soil Res* **47**: 403–422.

Pokhrel YN, Hanasaki N, Yeh PJF, Yamada TJ, Kanae S and Oki T (2012) Model estimates of sea-level change due to anthropogenic impacts on terrestrial water storage. *Nat Geosci* **5**: 389–392.

Pool M and Carrera J (2010) Dynamics of negative hydraulic barriers to prevent seawater intrusion. *Hydrogeol J* **18**: 95–105.

(2011) A correction factor to account for mixing in Ghyben-Herzberg and critical pumping rate approximations of seawater intrusion in coastal aquifers. *Water Resour Res* **47**.

Pool M, Post VEA and Simmons CT (2014) Effects of tidal fluctuations on mixing and spreading in coastal aquifers: Homogeneous case. *Water Resour Res* **50**: 6910–6926.

(2015) Effects of tidal fluctuations and spatial heterogeneity on mixing and spreading in spatially heterogeneous coastal aquifers. *Water Resour Res* **51**: 1570–1585.

Poore J and Nemecek T (2018) Reducing food's environmental impacts through producers and consumers. *Science* **360**: 987–992.

Porcelli D (2008) Investigating groundwater processes using U- and Th-series nuclides. *Radioactivity Environ* **13**: 105–153.

Post VEA (2004) Groundwater salinization processes in the coastal area of the Netherlands due to transgressions during the Holocene. Thesis, Vrije Universiteit, Amsterdam.

Post VEA, Kooi H and Simmons C (2007) Using hydraulic head measurements in variable-density ground water flow analyses. *Ground Water* **45**: 664–671.

Post VEA (2012) Electrical conductivity as a proxy for groundwater density in coastal aquifers. *Ground Water* **50**: 785–792.

(2018) Annotated translation of 'Nota in verband met de voorgenomen putboring nabij Amsterdam (Note concerning the intended well drilling near Amsterdam)' by J. Drabbe and W. Badon Ghijben (1889). *Hydrogeol J* **26**: 1–18.

Post VEA, Banks E and Brunke M (2018a) Groundwater flow in the transition zone between freshwater and saltwater: A field-based study and analysis of measurement errors. *Hydrogeol J* **26**: 1771–1778.

Post VEA, Eichholz M and Brentführer R (2018b) *Groundwater Management in Coastal Zones*. Federal Institute for Geosciences and Natural Resources, Hannover, Germany.

Post VEA, Essink GO, Szymkiewicz A, Bakker M, Houben G, Custodio E and Voss C (2018c) Celebrating 50 years of SWIMs (salt water intrusion meetings). *Hydrogeol J* **26**: 1767–1770.

Post VEA, Groen J, Kooi H, Person M, Ge S and Edmunds WM (2013) Offshore fresh groundwater reserves as a global phenomenon. *Nature* **504**: 71–78.

Post VEA and Houben GJ (2017) Density-driven vertical transport of saltwater through the freshwater lens on the island of Baltrum (Germany) following the 1962 storm flood. *J Hydrol* **551**: 689–702.

Post VEA, Houben GJ and Van Engelen J (2018d) What is the Ghijben–Herzberg principle and who formulated it? *Hydrogeol J* **26**: 1801–1807.

Post VEA and Kooi H (2003) Rates of salinization by free convection in high-permeability sediments: Insights from numerical modeling and application to the Dutch coastal area. *Hydrogeol J* **11**: 549–559.

Post VEA, Plicht H and Meijer HAJ (2003) The origin of brackish and saline groundwater in the coastal area of the Netherlands. *Geologie en mijnbouw* **82**: 133–147.

Post VEA and Prommer H (2007) Multicomponent reactive transport simulation of the Elder problem: Effects of chemical reactions on salt plume development. *Water Resour Res* **43**.

Post VEA and Simmons CT (2010) Free convective controls on sequestration of salts into low-permeability strata: Insights from sand tank laboratory experiments and numerical modelling. *Hydrogeol J* **18**: 39–54.

Post VEA, Vandenbohede A, Werner AD and Teubner MD (2013) Groundwater ages in coastal aquifers. *Adv Water Resour* **57**: 1–11.

Povinec PP, Aggarwal PK, Aureli A et al. (2006) Characterisation of submarine groundwater discharge offshore south-eastern Sicily. *J Environ Radioactivity* **89**: 81–101.

Prieto C, Kotronarou A and Destouni G (2006) The influence of temporal hydrological randomness on seawater intrusion in coastal aquifers. *J Hydrol* **330**: 285–300.

Pulido-Leboeuf P (2004) Seawater intrusion and associated processes in a small coastal complex aquifer (Castell de Ferro, Spain). *Appl Geochem* **19**: 1517–1527.

Purdum J and Engel GR (2003) Reducing Seawater Intrusion via Low-tide Pumping. *Water Well Journal* **57**: 28–29.

Raj PP (1995) *Geotechnical Engineering*. Tata McGraw-Hill, New Delhi.

Ramasamy SM, Kumanan CJ, Saravanavel J, Rajawat AS, Tamilarasan V and Ajay (2010) Geomatics based analysis of predicted sea level rise and its impacts in parts of Tamil Nadu Coast, India. *J Ind Soc Remot* **38**: 640–653.

Ramsay D and Bell R (2008) *Coastal Hazards and Climate Change: A Guidance Manual for Local Government in New Zealand*. Ministry for the Environment, New Zealand.

Rangel-Medina M, Monreal Saavedra R, Morales Montaño M and Castillo Gurrola J (2003) *Caracterizacion geoquimica e isotopica del agua subterranea y determinacion de la migración de la intrusion marina en el acuifero de la Costa de Hermosillo, son., México*. IGME, Madrid.

Rao SN and Mathew PK (1995) Effects of exchangeable cations on hydraulic conductivity of a marine clay. *Clay Clay Miner* **43**: 433–437.

Rawlinson G (1889) *History of Phoenicia*. Library of Alexandria.

Re V and Zuppi GM (2011) Influence of precipitation and deep saline groundwater on the hydrological systems of Mediterranean coastal plains: A general overview. *Hydrol Sci J* **56**: 966–980.

Reager JT, Gardner AS, Famiglietti JS, Wiese DN, Eicker A and Lo M-H (2016) A decade of sea level rise slowed by climate-driven hydrology. *Science* **351**: 699–703.

Reeves HW, Thibodeau PM, Underwood RG and Gardner LR (2000) Incorporation of total stress changes into the ground water model SUTRA. *Ground Water* **38**: 89–98.

Reichard EG, Geological Survey (US) and Water Replenishment District of Southern California (2003) *Geohydrology, Geochemistry, and Ground-Water Simulation-Optimization of the Central and West Coast Basins, Los Angeles County, California*. US Department of the Interior, USGS, Information Services, Sacramento, CA.

Reilly TE and Goodman AS (1985) Quantitative-analysis of saltwater fresh-water relationships in groundwater systems – a historical-perspective. *J Hydrol* **80**: 125–160.

(1987) Analysis of saltwater upconing beneath a pumping well. *J Hydrol* **89**: 169–204.

Renken RA (2005) *Impact of Anthropogenic Development on Coastal Ground-Water Hydrology in Southeastern Florida, 1900–2000*. US Department of the Interior, USGS, Reston, VA.

Reuss FF (1809) Sur un nouvel effet de l'électricité galvanique. *Mem Soc Imp Natur Moscou* **2**: 327–337.

Revelle R (1941) Criteria for recognition of seawater in groundwaters. *Trans AGU* **22**: 593–597.

Rey J, Martínez J, Barberá GG, García-Aróstegui JL, García-Pintado J and Martínez-Vicente D (2013) Geophysical characterization of the complex dynamics of groundwater and seawater exchange in a highly stressed aquifer system linked to a coastal lagoon (SE Spain). *Environ Earth Sci* **70**: 2271–2282.

Rezaei M, Sanz E, Raeisi E, Ayora C, Vázquez-Suñé E and Carrera J (2005) Reactive transport modeling of calcite dissolution in the fresh-salt water mixing zone. *J Hydrol* **311**: 282–298.

Rhoads GH and Robinson ES (1979) Determination of aquifer parameters from well tides. *J Geophys Res* **84**: 6071–6082.

Richter BC and Kreitler CW (1991) *Identification of Sources of Ground-Water Salinization Using Geochemical Techniques*. US Environmental Protection Agency, Ada, OK.

Robins NS and Dance LT (2003) A new conceptual groundwater-flow system for the Central South Downs aquifer. *Water Environ J* **17**: 111–116.

Robins NS, Jones HK and Ellis J (1999) An aquifer management case study – the chalk of the English South Downs. *Water Resour Manag* **13**: 205–218.

Robinson CE, Li L and Barry DA (2007) Effect of tidal forcing on a subterranean estuary. *Adv Water Resour* **30**: 851–865.

Robinson CE, Xin P, Li L and Barry DA (2014) Groundwater flow and salt transport in a subterranean estuary driven by intensified wave conditions. *Water Resour Res* **50**: 165–181.

Robinson CE, Xin P, Santos IR, Charette MA, Li L and Barry DA (2018) Groundwater dynamics in subterranean estuaries of coastal unconfined aquifers: Controls on submarine groundwater discharge and chemical inputs to the ocean. *Adv Water Resour* **115**: 315–331.

Rosenberry DO and Morin RH (2004) Use of an electromagnetic seepage meter to investigate temporal variability in lake seepage. *Ground Water* **42**: 68–77.

Rosenthal E (1987) Chemical-composition of rainfall and groundwater in recharge areas of the Bet-Shean Harod multiple aquifer system, Israel. *J Hydrol* **89**: 329–352.

Rotzoll K (2010) Effects of groundwater withdrawal on borehole flow and salinity measured in deep monitor wells in Hawai'i-implications for groundwater management. USGS Scientific Investigations Report 2010-5058, 42 pp. US Government Printing Office, Washington, DC.

Rotzoll K, El-Kadi AI and Gingerich SB (2008) Analysis of an unconfined aquifer subject to asynchronous dual-tide propagation. *Ground Water* **46**: 239–250.

Rousseau-Gueutin P, Love AJ, Vasseur G, Robinson NI, Simmons CT and De Marsily G (2013) Time to reach near-steady state in large aquifers. *Water Resour Res* **49**: 6893–6908.

Rubin Y and Hubbard SS (2006) *Hydrogeophysics*. Springer, Netherlands.

Rushton KR (1980) Differing positions of saline interfaces in aquifers and observation boreholes. *J Hydrol* **48**: 185–189.

Russak A and Sivan O (2010) Hydrogeochemical tool to identify salinization or freshening of coastal aquifers determined from combined field work, experiments, and modeling. *Environ Sci Technol* **44**: 4096–4102.

Russak A, Sivan O and Yechieli Y (2016) Trace elements (Li, B, Mn and Ba) as sensitive indicators for salinization and freshening events in coastal aquifers. *Chem Geol* **441**: 35–46.

Russoniello CJ, Fernandez C, Bratton JF, Banaszak JF, Krantz DE, Andres AS, Konikow LF and Michael HA (2013) Geologic effects on groundwater salinity and discharge into an estuary. *J Hydrol* **498**: 1–12.

Saha D, Dwivedi SN and Singh RK (2014) Aquifer system response to intensive pumping in urban areas of the Gangetic plains, India: The case study of Patna. *Environ Earth Sci* **71**: 1721–1735.

Sanchez-Martos F, Pulido-Bosch A, Molina-Sanchez L and Vallejos-Izquierdo A (2002) Identification of the origin of salinization in groundwater using minor ions (Lower Andarax, Southeast Spain). *Sci Total Environ* **297**: 43–58.

Sanford WE (1997) Correcting for diffusion in carbon-14 dating of ground water. *Ground Water* **35**: 357–361.

Sanford WE, Doughten MW, Coplen TB, Hunt AG and Bullen TD (2013) Evidence for high salinity of Early Cretaceous sea water from the Chesapeake Bay crater. *Nature* **503**: 252–256.

Sanford WE and Konikow LF (1989) Simulation of calcite dissolution and porosity changes in saltwater mixing zones in coastal aquifers. *Water Resour Res* **25**: 655–667.

Sanford WE and Pope JP (2009) Current challenges using models to forecast seawater intrusion: lessons from the Eastern Shore of Virginia, USA. *Hydrogeol J* **18**: 73–93.

Sanial V, Buesseler KO, Charette MA and Nagao S (2017) Unexpected source of Fukushima-derived radiocesium to the coastal ocean of Japan. *Proc Natl Acad Sci USA* **114**: 11092–11096.

Santos IR, Erler D, Tait D and Eyre BD (2010) Breathing of a coral cay: Tracing tidally driven seawater recirculation in permeable coral reef sediments. *J Geophys Res* **115**.

Santos IR, Eyre BD and Huettel M (2012) The driving forces of porewater and groundwater flow in permeable coastal sediments: A review. *Estuar Coast Shelf Sci* **98**: 1–15.

Santos IR, Niencheski F, Burnett W, Peterson R, Chanton J, Andrade CFF, Milani IB, Schmidt A and Knoeller K (2008) Tracing anthropogenically driven groundwater discharge into a coastal lagoon from southern Brazil. *J Hydrol* **353**: 275–293.

Saye SE and Pye K (2007) Implications of sea level rise for coastal dune habitat conservation in Wales, UK. *J Coastal Conserv* **11**: 31–52.

Scheible O, Mulbarger M, Sutton P, Simpkin T and Daigger G (1993) *Manual: Nitrogen Control*. Environmental Protection Agency, Cincinnati, OH.

Schmorak S and Mercado A (1969) Upconing of fresh water – sea water interface below pumping wells, field study. *Water Resour Res* **5**: 1290–1311.

Scientific Committee on Oceanic Research (2004) *Submarine Groundwater Discharge: Management Implications, Measurements and Effects*. IHP-VI, Series on Groundwater No. 5, IOC Manuals and Guides No. 44 (http://unesdoc.unesco.org/images/0013/001344/134436e.pdf).

Seasholes NS (2003) *Gaining Ground: A History of Landmaking in Boston*. MIT Press, Cambridge, MA.

Sebben ML, Werner AD and Graf T (2015) Seawater intrusion in fractured coastal aquifers: A preliminary numerical investigation using a fractured Henry problem. *Adv Water Resour* **85**: 93–108.

SEDAC (2007) *Percentage of Total Population Living in Coastal Areas*. Vol. 2016 Socioeconomic Data and Applications Center (SEDAC), NASA, Houston, TX.

Segeren WA (1983) Introduction to polders of the world. *Water Int* **8**: 51–54.

Segol G, Pinder GF and Gray WG (1975) A Galerkin-finite element technique for calculating the transient position of the saltwater front. *Water Resour Res* **11**: 343–347.

Selim HME and Amacher MC (1997) *Reactivity and Transport of Heavy Metals in Soils*. CRC/Lewis, Boca Raton, FL.

Selivanov A (1993) Modelling response of accumulative coasts to the possible future sea-level rise: General approach and examples from the former USSR coasts. pp. 63–74.

Senio K (1951) *On the Ground Water Near the Seashore*. Vol. 33. Association Internationale d'Hydrologie Scientifique, Brussels, Belgium.

Servan-Camas B and Tsai FTC (2009) Saltwater intrusion modeling in heterogeneous confined aquifers using two-relaxation-time lattice Boltzmann method. *Adv Water Resour* **32**: 620–631.

(2010) Two-relaxation-time lattice Boltzmann method for the anisotropic dispersive Henry problem. *Water Resour Res* **46**.

Shackleton NJ (1987) Oxygen isotopes, ice volume and sea-level. *Quat Sci Rev* **6**: 183–190.

Shalem Y, Weinstein Y, Levi E, Herut B, Goldman M and Yechieli Y (2014) The extent of aquifer salinization next to an estuarine river: An example from the eastern Mediterranean. *Hydrogeol J* **23**: 69–79.

Shalev E, Lazar A, Wollman S, Kington S, Yechieli Y and Gvirtzman H (2009) Biased monitoring of fresh water-salt water mixing zone in coastal aquifers. *Ground Water* **47**: 49–56.

Shan C (1995) Analytical solutions for determining vertical air permeability in unsaturated soils. *Water Resour Res* **31**: 2193–2200.

Sharqawy MH, Lienhard JH and Zubair SM (2012) Thermophysical properties of seawater: a review of existing correlations and data. *Desalination Water Treatment* **16**: 354–380.

Shaw R, Clasen T, Smith L, Albert J, Bastable A and Fesselet J-F (2006) The drinking water response to the Indian Ocean tsunami, including the role of household water treatment. *Disaster Prevent Manage* **15**: 190–201.

Shennan I and Horton B (2002) Holocene land- and sea-level changes in Great Britain. *J Quat Sci* **17**: 511–526.

Sherrod BL, Bucknam RC and Leopold EB (2000) Holocene relative sea level changes along the Seattle Fault at Restoration Point, Washington. *Quat Res* **54**: 384–393.

Shi L and Jiao JJ (2014) Seawater intrusion and coastal aquifer management in China: A review. *Environ Earth Sci* **72**: 2811–2819.

Shiklomanov IA (1998) World water resources: A new appraisal and assessment for the 21st century.

Shugar DH, Walker IJ, Lian OB, Eamer JBR, Neudorf C, McLaren D and Fedje D (2014) Post-glacial sea-level change along the Pacific coast of North America. *Quat Sci Rev* **97**: 170–192.

Siemon E, Christiansen AV and Auken E (2009) A review of helicopter-borne electromagnetic methods for groundwater exploration. *Near Surface Geophys* **7**: 629–646.

Silva B, Rivas T, Garcia-Rodeja E and Prieto B (2007) Distribution of ions of marine origin in Galicia (NW Spain) as a function of distance from the sea. *Atmos Environ* **41**: 4396–4407.

Silvestru E (2001) The riddle of paleokarst solved. *J Creation* **15**: 105–114.

Simmons CT, Pierini ML and Hutson JL (2002) Laboratory investigation of variable-density flow and solute transport in unsaturated–saturated porous media. *Transp Porous Media* **47**: 215–244.

Simpson MJ and Clement TP (2004) Improving the worthiness of the Henry problem as a benchmark for density-dependent groundwater flow models. *Water Resour Res* **40**.

Singh A (2014) Optimization modelling for seawater intrusion management. *J Hydrol* **508**: 43–52.

Sivan O, Yechieli Y, Herut B and Lazar B (2005) Geochemical evolution and timescale of seawater intrusion into the coastal aquifer of Israel. *Geochim Cosmochim Acta* **69**: 579–592.

Slomp CP and Van Cappellen P (2004) Nutrient inputs to the coastal ocean through submarine groundwater discharge: controls and potential impact. *J Hydrol* **295**: 64–86.

Slooten LJ, Carrera J, Castro E and Fernandez-Garcia D (2010) A sensitivity analysis of tide-induced head fluctuations in coastal aquifers. *J Hydrol* **393**: 370–380.

Small C and Nicholls RJ (2003) A global analysis of human settlement in coastal zones. *J Coastal Res* **19**: 584–599.

Smart PL, Dawans JM and Whitaker F (1988) Carbonate dissolution in a modern mixing zone. *Nature* **335**: 811–813.

Smiles DE and Stokes AN (1976) Periodic-solutions of a nonlinear diffusion equation used in groundwater flow theory – examination using a Hele-Shaw model. *J Hydrol* **31**: 27–35.

Smith AJ (2004) Mixed convection and density-dependent seawater circulation in coastal aquifers. *Water Resour Res* **40**.

Smith AJ, Khow J, Lodge B and Bavister G (2001) Desalination of poor quality brackish groundwater for non-potable use. *Desalination* **139**: 207–215.

Smith AJ and Turner JV (2001) Density-dependent surface water-groundwater interaction and nutrient discharge in the Swan-Canning Estuary. *Hydrol Process* **15**: 2595–2616.

Smith CG, Cable JE, Martin JB and Roy M (2008) Evaluating the source and seasonality of submarine groundwater discharge using a radon-222 pore water transport model. *Earth Planet Sci Lett* **273**: 312–322.

Sola F, Vallejos A, Daniele L and Pulido-Bosch A (2014) Identification of a Holocene aquifer–lagoon system using hydrogeochemical data. *Quat Res* **82**: 121–131.

Solomon DK and Cook PG (2000) 3H and 3He. *Environmental Tracers in Subsurface Hydrology* (Cook PG and Herczeg AL, eds), 397–424. Springer, Boston, MA.

Sonenshein RS, US Geological Survey, Metropolitan Dade County Department of Environmental Resources Management and Miami-Dade Water and Sewer Department (1997) *Delineation and Extent of Saltwater Intrusion in the Biscayne Aquifer, Eastern Dade County, Florida, 1995*. The Survey, Tallahassee, FL.

Song SH, Lee GS, Kim JS, Seong B, Kim Y, Woo MH and Park N (2006) Electrical resistivity survey for delineating seawater intrusion in a coastal aquifer. pp. 289–293. Proceedings 1st SWIM-SWICA Joint Saltwater Intrusion Conference, Cagliari-Chia Laguna, Italy, September 24–29.

Song ZY, Li L, Kong J and Zhang HG (2007) A new analytical solution of tidal water table fluctuations in a coastal unconfined aquifer. *J Hydrol* **340**: 256–260.

Sous D, Petitjean L, Bouchette F, Rey V, Meulé S, Sabatier F and Martins K (2016) Field evidence of swash groundwater circulation in the microtidal rousty beach, France. *Adv Water Resour* **97**: 144–155.

Sparks DL (1999) Kinetics and mechanisms of chemical reactions at the soil mineral/ water interface. *Soil Physical Chemistry* (Sparks DL, ed.), 409. CRC Press, Boca Raton, FL.

Spechler RM (1994) *Saltwater Intrusion and Quality of Water in the Floridan Aquifer System, Northeastern Florida*. USGS, Washington, DC.

Stamkart FJ and Matthes CJ (1851) Verslag van de Heeren F.J. Stamkart en C.J. Matthes, over den stand van het water in den put op de Noordermarkt, ingediend in de Vergadering van den 21sten Junij j.l. (Report by the gentlemen F.J. Stamkart and C.J. Matthes about the water level in the well on the Noordermarkt, submitted to the assembly on the 21st of June). Tijdschrift voor Wis- en Natuurkundige Wetenschappen, uitgegeven door de Eerste Klasse van het Koninklijk Nederlandsche Instituut van Wetenschappen, Letterkunde en Schoone Kunsten, Vierde deel 301–324.

Stanley DJ and Warne AG (1994) Worldwide initiation of holocene marine deltas by deceleration of sea-level rise. *Science* **265**: 228–231.

Steggewentz JH (1933) De invloed van de getijbeweging van zeeën en getijrivieren op de stijghoogte van grondwater. Thesis, Technische Hoogeschool, Delft.

Stieglitz TC, Cook PG and Burnett WC (2010) Inferring coastal processes from regional-scale mapping of 222Radon and salinity: Examples from the Great Barrier Reef, Australia. J Environ Radioactivity **101**: 544–552.

Stein S, Russak A, Sivan O, Yechieli Y, Rahav E, Oren Y and Kasher R (2016) Saline groundwater from coastal aquifers as a source for desalination. *Environ Sci Technol* **50**: 1955–1963.

Steuer A, Siemon B and Auken E (2009) A comparison of helicopter-borne electromagnetics in frequency- and time-domain at the Cuxhaven valley in Northern Germany. *J Appl Geophys* **67**: 194–205.

Stewart MT (1999) Geophysical investigations. *Seawater Intrusion in Coastal Aquifers – Concepts, Methods and Practices* (Bear J, Cheng AHD, Sorek S, Ouazar D and Herrera I, eds), 9–50. Springer, Dordrecht, Netherlands.

Stiff HA Jr. (1951) The interpretation of chemical water analysis by means of patterns. *J Petrol Technol* **3**: 15–17.

Stigter TY, Van Ooijen SPJ, Post VEA, Appelo CAJ and Carvalho Dill AMM (1998) A hydrogeological and hydrochemical explanation of the groundwater composition under irrigated land in a Mediterranean environment, Algarve, Portugal. *J Hydrol* **208**: 262–279.

Stocker T (2014) *Climate Change 2013: The Physical Science Basis: Working Group I Contribution to the Fifth Assessment Report of the Intergovernmental Panel on Climate Change*. Cambridge University Press, Cambridge.

Stoeckl L, Walther M and Graf T (2016) A new numerical benchmark of a freshwater lens. *Water Resour Res* **52**: 2474–2489.

Storlazzi CD, Gingerich SB, Van Dongeren A et al. (2018) Most atolls will be uninhabitable by the mid-21st century because of sea-level rise exacerbating wave-driven flooding. *Sci Adv* 4.

Strack ODL (1976) A single-potential solution for regional interface problems in coastal aquifers. *Water Resour Res* **12**: 1165–1174.

Stringfield VT and LeGrand HE (1971) Effects of karst features on circulation of water in carbonate rocks in coastal areas. *J Hydrol* **14**: 139–157.

Stumm W and Morgan JJ (1995) *Aquatic Chemistry: Chemical Equilibria and Rates in Natural Waters*. 3rd ed. Wiley-Interscience, Hoboken, NJ.

Stuyfzand PJ (1988) *Hydrochemie en hydrologie van duinen en aangrenzende polders tussen Noordwijk en Zandvoort aan Zee*. KIWA, Nieuwegein, Netherlands.

(1989) *A New Hydrochemical Classification of Water Types*. IAHS, Baltimore, MD.

(1993) Hydrochemistry and hydrology of the coastal dune area of the Western Netherlands. Thesis, Vrije Universiteit, Amsterdam.

(1999) Patterns in groundwater chemistry resulting from groundwater flow. *Hydrogeol J* **7**: 15–27.

(2017) Observations and analytical modeling of freshwater and rainwater lenses in coastal dune systems. *J Coastal Conserv* **21**: 577–593.

Stuyfzand PJ and Raat KJ (2009) Benefits and hurdles of using brackish groundwater as a drinking water source in the Netherlands. *Hydrogeol J* **18**: 117–130.

Stuyfzand PJ, Schaars F and Van der Made KJ (2012) *Multitracing the Origin of Brackish and Saline Groundwaters near a Dune Catchment Area with Beach Nourishment (Monster, Netherlands)*. Buzios, Brazil.

Sufi AB, Latif M and Skogerboe GV (1998) Simulating skimming well techniques for sustainable exploitation of groundwater. *Irrigation Drainage Syst* **12**: 203–226.

Suh JY (2004) Hydrogeochemical studies of groundwater from reclaimed land adjacent to Rozelle Bay, Sydney, Australia. *Geosci J* **8**: 301–312.

Sukop MC and Thorne DT Jr (2006) *Lattice Boltzmann Modeling – An Introduction for Geoscientists and Engineers*. Springer, Berlin.

Sumner ME and Naidu R (1998) *Sodic Soils: Distribution, Properties, Management and Environmental Consequences*. Oxford University Press, Oxford.

Sun HB (1997) A two-dimensional analytical solution of groundwater response to tidal loading in an estuary. *Water Resour Res* **33**: 1429–1435.

Sun Y and Torgersen T (1998) The effects of water content and Mn-fiber surface conditions on 224Ra measurement by 220Rn emanation. *Mar Chem* **62**: 299–306.

Swarzenski P, Dulaiova H, Dailer M, Glenn C, Smith C and Storlazzi C (2013) A geochemical and geophysical assessment of coastal groundwater discharge at select sites in Maui and O'ahu, Hawai'i. Groundwater in the Coastal Zones of Asia-Pacific, 27–46. Springer, New York.

Swarzenski PW, Reich CD, Spechler RM, Kindinger JL and Moore WS (2001) Using multiple geochemical tracers to characterize the hydrogeology of the submarine spring off Crescent Beach, Florida. *Chem Geol* **179**: 187–202.

Sweet W, Park J, Marra J, Zervas C and Gill S (2014) *Sea Level Rise and Nuisance Flood Frequency Changes around the United States*. NOAA Technical Report NOS CO-OPS 073. Silver Spring, MD.

Talley LD (2011) *Descriptive Physical Oceanography: An Introduction*. Academic Press, New York.

Tambach TJ, Veld H and Griffioen J (2009) Influence of HCl/HF treatment on organic matter in aquifer sediments: A Rock-Eval pyrolysis study. *Appl Geochem* **24**: 2144–2151.

Tang ZH and Jiao JJ (2001) A two-dimensional analytical solution for groundwater flow in a leaky confined aquifer system near open tidal water. *Hydrol Process* **15**: 573–585.

Taniguchi M, Burnett WC, Cable JE and Turner JV (2002) Investigation of submarine groundwater discharge. *Hydrol Process* **16**: 2115–2129.

Taniguchi M, Burnett WC, Smith CF, Paulsen RJ, O'Rourke D, Krupa SL and Christoff JL (2003) Spatial and temporal distributions of submarine groundwater discharge rates obtained from various types of seepage meters at a site in the Northeastern Gulf of Mexico. *Biogeochemistry* **66**: 35–53.

Taniguchi M and Fukuo Y (1993) Continuous measurements of groundwater seepage using an automatic seepage meter. *Ground Water* **31**: 675–679.

Telford WM, Telford WM, Geldart LP and Sheriff RE (1990) *Applied Geophysics*. Cambridge University Press, Cambridge.

Tellam JH, Lloyd JW and Walters M (1986) The morphology of a saline groundwater body: Its investigation, description and possible explanation. *J Hydrol* **83**: 1–21.

Terawaki T, Yoshikawa K, Yoshida G, Uchimura M and Iseki K (2003) Ecology and restoration techniques for Sargassum beds in the Seto Inland Sea, Japan. *Mar Pollut Bull* **47**: 198–201.

Terry JP and Falkland AC (2010) Responses of atoll freshwater lenses to storm-surge overwash in the Northern Cook Islands. *Hydrogeol J* **18**: 749–759.

Terzaghi K (1925) *Erdbaumechanik auf Bodenphysikalischer Grundlage*. Franz Deuticke, Leipzig, Germany.

Timms WA and Acworth RI (2005) Propagation of pressure change through thick clay sequences: an example from Liverpool Plains, NSW, Australia. *Hydrogeol J* **13**: 858–870.

Todd DK (1980) *Groundwater Hydrology*. 2nd ed. John Wiley, New York.

Todd DK and Mays LW (2005) Groundwater Hydrology. John Wiley, Hoboken, NJ.

Tokunaga T, Shimada J, Kimura Y, Inoue D, Mogi K and Asai K (2010) A multiple-isotope (δ37Cl, 14C, 3H) approach to reveal the coastal hydrogeological system and its temporal changes in western Kyushu, Japan. *Hydrogeol J* **19**: 249–258.

Tolman C (1937) *Ground Water*. McGraw-Hill, New York.

Tornqvist TE and Hijma MP (2012) Links between early Holocene ice-sheet decay, sea-level rise and abrupt climate change. *Nature Geosci* **5**: 601–606.

Tóth J (2009) *Gravitational Systems of Groundwater Flow: Theory, Evaluation, Utilization*. Cambridge University Press, Cambridge.

Touma J, Vachaud G and Parlange JY (1984) Air and water-flow in a sealed, ponded vertical soil column – experiment and model. *Soil Sci* **137**: 181–187.

Townley LR (1995) The response of aquifers to periodic forcing. *Adv Water Resour* **18**: 125–146.

Toyohara Y (1935) A study on the coastal ground water at Yumigahama, Tottori. *Mem Coll Sci Kyoto Imperial Univ, Ser A* **XVIII**: 295–309.

Trefry MG (1999) Periodic forcing in composite aquifers. *Adv Water Resour* **22**: 645–656.

Trefry MG and Bekele E (2004) Structural characterization of an island aquifer via tidal methods. *Water Resour Res* **40**.

Trefry MG and Johnston CD (1998) Pumping test analysis for a tidally forced aquifer. *Ground Water* **36**: 427–433.

Tse KC and Jiao JJ (2008) Estimation of submarine groundwater discharge in Plover Cove, Tolo Harbour, Hong Kong by Rn-222. *Mar Chem* **111**: 160–170.

Tung S, Leung JKC and Jiao JJ (2010) Geogenic study of soil radon concentration in Hong Kong under a ten-point system. *Environ Earth Sci* **68**: 679–689.

Turner SF and Foster MD (1934) A study of salt-water encroachment in the Galveston Area, Texas. *Eos Trans AGU* **15**: 432–435.

Ullman WJ, Chang B, Miller DC and Madsen JA (2003) Groundwater mixing, nutrient diagenesis, and discharges across a sandy beachface, Cape Henlopen, Delaware (USA). *Estuarine Coastal Shelf Sci* **57**: 539–552.

Underwood MR, Peterson FL and Voss CI (1992) Groundwater lens dynamics of Atoll Islands. *Water Resour Res* **28**: 2889–2902.

USEPA (1999) *The Class V Underground Injection Control Study: Vol. 20. Salt Water Intrusion Barrier Wells*. US Environmental Protection Agency, Washington, DC.

Vachaud G, Vauclin M, Khanji D and Wakil M (1973) Effects of air pressure on water flow in an unsaturated stratified vertical column of sand. *Water Resour Res* **9**: 160–173.

Vacher HL (1988) Dupuit–Ghyben–Herzberg analysis of strip-island lenses. *Geol Soc Am Bull* **100**: 580–591.

(2004) Introduction: Varieties of carbonate islands and a historical perspective. Developments in Sedimentology, Vol. 54 (Vacher HL and Quinn TM, eds), 1–33. Elsevier, New York.

Valle-Levinson A, Marino-Tapia I, Enriquez C and Waterhouse AF (2011) Tidal variability of salinity and velocity fields related to intense point-source submarine groundwater discharges into the coastal ocean. *Limnol Oceanogr* **56**: 1213–1224.

Van Beek P, Souhaut M and Reyss JL (2010) Measuring the radium quartet (Ra-228, Ra-226, Ra-224, Ra-223) in seawater samples using gamma spectrometry. *J Environ Radioactivity* **101**: 521–529.

Van Dam JC (1976) Possibilities and limitations of the resistivity method of geoelectrical prospecting in the solution of geo-hydrological problems. *Geoexploration* **14**: 179–193.

(1999) Exploitation, restoration and management. Seawater Intrusion in Coastal Aquifers – Concepts, Methods and Practices (Bear J, Cheng AHD, Sorek S, Ouazar D and Herrera I, eds), 73–125. Springer, Dordrecht, Netherlands.

Van der Kamp G (1972) Tidal fluctuations in a confined aquifer extending under the sea. Int Geol Cong 24: 101–106.

Van der Kamp G and Gale JE (1983) Theory of earth tide and barometric effects in porous formations with compressible grains. *Water Resour Res* **19**: 538–544.

Van der Veer P (1977a) Analytical solution for steady interface flow in a coastal aquifer involving a phreatic surface with precipitation. *J Hydrol* **34**: 1–11.

(1977b) Analytical solution for a 2-fluid flow in a coastal aquifer involving a phreatic surface with precipitation. *J Hydrol* **35**: 271–278.

Van de Ven GP, Drainage ICoIa and ICID (2004) *Man Made Lowlands – History of Water Management and Land Reclamation in the Netherlands*. Uitgeverij Matrijs, Utrecht, Netherlands.

Van Geldern R, Hayashi T, Böttcher ME, Mottl MJ, Barth JAC and Stadler S (2013) Stable isotope geochemistry of pore waters and marine sediments from the New Jersey shelf: Methane formation and fluid origin. *Geosphere* **9**: 96–112.

Van Sambeek MHG, Eggenkamp HGM and Vissers MJM (2000) The groundwater quality of Aruba, Bonaire and Curaçao: A hydrogeochemical study. *Netherlands J Geosci* **79**: 459–466.

Van Weert F, Van der Gun J and Reckman J (2009) Global overview of saline groundwater occurrence and genesis. IGRAC Report GP 2009-1.

Vandenbohede A and Lebbe L (2006) Occurrence of salt water above fresh water in dynamic equilibrium in a coastal groundwater flow system near De Panne, Belgium. *Hydrogeol J* **14**: 462–472.

(2011) Heat transport in a coastal groundwater flow system near De Panne, Belgium. *Hydrogeol J* **19**: 1225–1238.

(2012) Groundwater chemistry patterns in the phreatic aquifer of the central Belgian coastal plain. *Appl Geochem* **27**: 22–36.

Vandenbohede A, Lebbe L, Gysens S, Delecluyse K and DeWolf P (2008) Salt water infiltration in two artificial sea inlets in the Belgian dune area. *J Hydrol* **360**: 77–86.

Veatch AC (1906) *Fluctuations of the Water Level in Wells, with Special Reference to Long Island, New York*. US Government Printing Office, Washington, DC.

Velstra J, Groen J and De Jong K (2011) Observations of salinity patterns in shallow groundwater and drainage water from agricultural land in the northern part of the Netherlands. *Irrig Drain* **60**: 51–58.

Vengosh A and Rosenthal E (1994) Saline groundwater in Israel – its bearing on the water crisis in the country. *J Hydrol* **156**: 389–430.

(1998) Chloride/bromide and chloride/fluoride ratios of domestic sewage effluents and associated contaminated ground water. *Ground Water* **36**: 815–824.

Verruijt A (1971) Steady dispersion across an interface in a porous medium. *J Hydrol* **14**: 337–347.

(2013) *Theory and Problems of Poroelasticity*. Delft University of Technology, Delft, Netherlands.

Villholth KG and Neupane B (2011) Tsunamis as long-term hazards to coastal groundwater resources and associated water supplies. *Tsunami – A Growing Disaster* (Mokhtari M, ed.). InTechOpen, 87–104.

Vink A, Steffen H, Reinhardt L and Kaufmann G (2007) Holocene relative sea-level change, isostatic subsidence and the radial viscosity structure of the mantle of north-west Europe (Belgium, the Netherlands, Germany, southern North Sea). *Quat Sci Rev* **26**: 3249–3275.

Vithanage M, Villholth KG, Mahatantila K, Engesgaard P and Jensen KH (2009) Effect of well cleaning and pumping on groundwater quality of a tsunami-affected coastal aquifer in eastern Sri Lanka. *Water Resour Res* **45**.

Volker A and Van der Molen WH (1991) The influence of groundwater currents on diffusion-processes in a lake bottom – an old report reviewed. *J Hydrol* **126**: 159–169.

Volker RE and Rushton KR (1982) An assessment of the importance of some parameters for sea-water intrusion in aquifers and a comparison of dispersive and sharp-interface modeling approaches. *J Hydrol* **56**: 239–250.

Vos P (2015) *Origin of the Dutch Coastal Landscape: Long-Term Landscape Evolution of the Netherlands during the Holocene, Described and Visualized in National, Regional and Local Palaeogeographical Map Series*. Barkhuis, Groningen, Netherlands.

Voss CI (1999) USGS SUTRA code – history, practical use, and application in Hawaii. Seawater Intrusion in Coastal Aquifers – Concepts, Methods and Practices (Bear J, Cheng AHD, Sorek S, Ouazar D and Herrera I, eds), 249–313. Springer, Dordrecht, Netherlands.

Voss CI and Wood WW (1994) Synthesis of geochemical, isotopic and groundwater modelling analysis to explain regional flow in a coastal aquifer of southern Oahu, Hawaii. Mathematical Models and Their Applications to Isotope Studies in Groundwater Hydrology, IEAE-TECDOC-777, 147–178. International Atomic Energy Agency, Vienna, Austria.

Wada Y, Van Beek LPH, Weiland FCS, Chao BF, Wu YH and Bierkens MFP (2012) Past and future contribution of global groundwater depletion to sea-level rise. *Geophys Res Lett* **39**.

Waelbroeck C, Labeyrie L, Michel E, Duplessy JC, McManus JF, Lambeck K, Balbon E and Labracherie M (2002) Sea-level and deep water temperature changes derived from benthic foraminifera isotopic records. *Quat Sci Rev* **21**: 295–305.

Wallis I, Prommer H, Pichler T, Post V, B. Norton S, Annable MD and Simmons CT (2011) Process-based reactive transport model to quantify arsenic mobility during aquifer storage and recovery of potable water. *Environ Sci Technol* **45**: 6924–6931.

Walraevens K, Cardenal-Escarcena J and Van Camp M (2007) Reaction transport modelling of a freshening aquifer (Tertiary Ledo-Paniselian Aquifer, Flanders-Belgium). *Appl Geochem* **22**: 289–305.

Walraevens K, Van Camp M, Lermytte J, Van der Kemp WJM and Loosli HH (2001) Pleistocene and Holocene groundwaters in the freshening Ledo-Paniselian aquifer in Flanders, Belgium. Palaeowaters in Coastal Europe: Evolution of Groundwater since the Late Pleistocene, Geological Society Special Publication 189 (Edmunds WM and Milne CJ, eds), 49–70. Geological Society, London.

Walther M, Bilke L, Delfs J-O, Graf T, Grundmann J, Kolditz O and Liedl R (2014) Assessing the saltwater remediation potential of a three-dimensional, heterogeneous, coastal aquifer system. *Environ Earth Sci* **72**: 3827–3837.

Walther M, Graf T, Kolditz O, Liedl R and Post V (2017) How significant is the slope of the sea-side boundary for modelling seawater intrusion in coastal aquifers? *J Hydrol* **551**: 648–659.

Wang F and Jiao JJ (2005) *Preliminary Numerical Study on Groundwater System at the Hong Kong International Airport*. Hong Kong.

Wang W, Liu HI, Li YQ and Su JL (2014) Development and management of land reclamation in China. *Ocean Coast Manage* **102**: 415–425.

Wang Y (2011) Isotopic and hydrogeochemical studies of the coast aquifer-aquitard system in the Pearl River Delta, China. Thesis, University of Hong Kong.

Wang Y and Jiao JJ (2012) Origin of groundwater salinity and hydrogeochemical processes in the confined Quaternary aquifer of the Pearl River Delta, China. *J Hydrol* **438**: 112–124.

Wang Y, Jiao JJ and Cherry JA (2012) Occurrence and geochemical behavior of arsenic in a coastal aquifer-aquitard system of the Pearl River Delta, China. *Sci Total Environ* **427**: 286–297.

Wang Y, Jiao JJ, Cherry JA and Lee CM (2013) Contribution of the aquitard to the regional groundwater hydrochemistry of the underlying confined aquifer in the Pearl River Delta, China. *Sci Total Environ* **461**: 663–671.

Ward JD, Simmons CT, Dillon PJ and Pavelic P (2009) Integrated assessment of lateral flow, density effects and dispersion in aquifer storage and recovery. *J Hydrol* **370**: 83–99.

Warren J (2000) Dolomite: Occurrence, evolution and economically important associations. *Earth Sci Rev* **52**: 1–81.

Watson RT (2001) *Climate Change 2001: Synthesis Report: Contribution of Working Groups I, II, and III to the Third Assessment Report of the Intergovernmental Panel on Climate Change*. Cambridge University Press, Cambridge.

Watson TA, Werner AD and Simmons CT (2010) Transience of seawater intrusion in response to sea level rise. *Water Resour Res* **46**.

Webb AP and Kench PS (2010) The dynamic response of reef islands to sea-level rise: Evidence from multi-decadal analysis of island change in the Central Pacific. *Global Planet Change* **72**: 234–246.

Weeks EP (1978) *Field Determination of Vertical Permeability to Air in the Unsaturated Zone*. Department of the Interior, USGS, Washington, DC.
(2002) The Lisse effect revisited. *Ground Water* **40**: 652–656.

Wentworth CK (1948) Growth of the Ghyben–Herzberg transition zone under a rinsing hypothesis. *Eos Trans AGU* **29**: 97–98.

Werner AD (2017) On the classification of seawater intrusion. *J Hydrol* **551**: 619–631.

Werner AD, Alcoe DW, Ordens CM, Hutson JL, Ward JD and Simmons CT (2011) Current practice and future challenges in coastal aquifer management: Flux-based and trigger-level approaches with application to an Australian case study. *Water Resour Manag* **25**: 1831–1853.

Werner AD, Bakker M, Post VEA, Vandenbohede A, Lu CH, Ataie-Ashtiani B, Simmons CT and Barry DA (2013) Seawater intrusion processes, investigation and management: Recent advances and future challenges. *Adv Water Resour* **51**: 3–26.

Werner AD, Jakovovic D and Simmons CT (2009) Experimental observations of saltwater up-coning. *J Hydrol* **373**: 230–241.

Werner AD, Sharp HK, Galvis SC, Post VEA and Sinclair P (2017) Hydrogeology and management of freshwater lenses on atoll islands: Review of current knowledge and research needs. *J Hydrol* **551**: 819–844.

Werner AD and Simmons CT (2009) Impact of sea-level rise on sea water intrusion in coastal aquifers. *Groundwater* **47**: 197–204.

Werner AD, Ward JD, Morgan LK, Simmons CT, Robinson NI and Teubner MD (2012) Vulnerability indicators of sea water intrusion. *Ground Water* **50**: 48–58.

White I and Falkland T (2009) Management of freshwater lenses on small Pacific islands. *Hydrogeol J* **18**: 227–246.

White JK and Roberts TOL (1994) *The Significance of Groundwater Tidal Fluctuations* (Wilkinson WB, ed.). Thomas Telford, London.

Wigley TML and Plummer LN (1976) Mixing of carbonate waters. *Geochim Cosmochim Acta* **40**: 989–995.

Wilson AM (2005) Fresh and saline groundwater discharge to the ocean: A regional perspective. *Water Resour Res* **41**.

Wilson AM and Gardner LR (2006) Tidally driven groundwater flow and solute exchange in a marsh: Numerical simulations. *Water Resour Res* **42**.

Winslow AG, Doyel WW and Wood LA (1957) *Salt Water and Its Relation to Fresh Ground Water in Harris County, Texas*. USGS Water Supply Paper 1360.

Winter TC (1998) *Ground Water and Surface Water: A Single Resource*. USGS, Denver, CO.

Wong PP, Losada IJ, Gattuso JP, Hinkel J, Khattabi A, McInnes KL, Saito Y and Sallenger A (2014) Coastal systems and low-lying areas. Climate Change 2014: Impacts, Adaptation, and Vulnerability Part A: Global and Sectoral Aspects Contribution of Working Group II to the Fifth Assessment Report of the Intergovernmental Panel on Climate Change (Field CB, Barros VR, Dokken DJ et al., eds), 361–409. Cambridge University Press, Cambridge.

Wood WW and Hyndman DW (2017) Groundwater depletion: A significant unreported source of atmospheric carbon dioxide. *Earth's Future* **5**: 1133–1135.

Wood WW, Sanford WE and Frape SK (2005) Chemical openness and potential for misinterpretation of the solute environment of coastal sabkhat. *Chem Geol* **215**: 361–372.

Wooding RA (1969) Growth of fingers at an unstable diffusing interface in a porous medium or Hele–Shaw cell. *J Fluid Mech* **39**: 477–495.

Woodroffe SA and Horton BP (2005) Holocene sea-level changes in the Indo-Pacific. *J Asian Earth Sci* **25**: 29–43.

World Health Organization (2006) *Guidelines for Drinking-Water Quality: First Addendum to Volume 1, Recommendations*. World Health Organization, Geneva, Switzerland.

Wozencraft J and Millar D (2005) Airborne LIDAR and integrated technologies for coastal mapping and nautical charting. *Mar Technol Soc J* **39**: 27–35.

Wriedt G and Bouraoui F (2009) *Large Scale Screening of Seawater Intrusion Risk in Europe – Methodological Development and Pilot Application along the Spanish Mediterranean Coast*. European Communities, Luxembourg.

Wyatt DE, Richers DM and Pirkle RJ (1995) Barometric pumping effects on soil-gas studies for geological and environmental characterization. *Environ Geol* **25**: 243–250.

Xia YQ, Li HL, Boufadel MC, Guo Q, Li GH and Magos A (2007) Tidal wave propagation in a coastal aquifer: Effects of leakages through its submarine outlet-capping and offshore roof. *J Hydrol* **337**: 249–257.

Xin P, Robinson CE, Li L, Barry DA and Bakhtyar R (2010) Effects of wave forcing on a subterranean estuary. *Water Resour Res* **46**.

Xiong X (2011) Numerical simulation of underground reservoir aquifer recharge based on prevention and control of seawater intrusion. MSc thesis, Shandong Jianzhu University.

Xu YS, Ma L, Du YJ and Shen SL (2012) Analysis of urbanisation-induced land subsidence in Shanghai. *Nat Hazards* **63**: 1255–1267.

Xue YQ, Wu J, Liu PM, Wang JJ, Jiang QB and Shi HW (1993) Sea-water intrusion in the coastal area of Laizhou Bay, China. 1. Distribution of sea-water intrusion and its hydrochemical characteristics. *Ground Water* **31**: 532–537.

Yang J, Graf T and Ptak T (2015) Sea level rise and storm surge effects in a coastal heterogeneous aquifer: A 2D modelling study in northern Germany. *Grundwasser* **20**: 39–51.

Yang SY, Yim WWS and Huang GQ (2008) Geochemical composition of inner shelf Quaternary sediments in the northern South China Sea with implications for provenance discrimination and paleoenvironmental reconstruction. *Global Planet Change* **60**: 207–221.

Yechieli Y (2000) Fresh-saline ground water interface in the western Dead Sea area. *Ground Water* **38**: 615–623.

Yechieli Y, Gavrieli I, Berkowitz B and Ronen D (1998) Will the Dead Sea die? *Geology* **26**: 755–758.

Yechieli Y, Kafri U and Sivan O (2009a) The inter-relationship between coastal sub-aquifers and the Mediterranean Sea, deduced from radioactive isotopes analysis. *Hydrogeol J* **17**: 265–274.

Yechieli Y, Kafri U, Wollman S, Shalev E and Lyakhovsky V (2009b) The effect of base level changes and geological structures on the location of the groundwater divide, as exhibited in the hydrological system between the Dead Sea and the Mediterranean Sea. *J Hydrol* **378**: 218–229.

Yechieli Y, Shalev E, Wollman S, Kiro Y and Kafri U (2010) Response of the Mediterranean and Dead Sea coastal aquifers to sea level variations. *Water Resour Res* **46**.

Yechieli Y and Sivan O (2011) The distribution of saline groundwater and its relation to the hydraulic conditions of aquifers and aquitards: Examples from Israel. *Hydrogeol J* **19**: 71–81.

Yeung AT (2016) Geotechnical works of the Hong Kong–Zhuhai–Macao Bridge Project. *Jpn Geotech Soc Spec Publ* **2**: 109–121.

Younger PL (1996) Submarine groundwater discharge. *Nature* **382**: 121–122.

Zagwijn WH (1989) The Netherlands during the Tertiary and the Quaternary: A case history of Coastal Lowland evolution. Coastal Lowlands: Geology and Geotechnology (Van der Linden WJM, Cloetingh SAPL, Kaasschieter JPK, Van de Graaff WJE, Vandenberghe J and Van der Gun JAM, eds), 107–120. Springer, Dordrecht, Netherlands.

Zekri S (2008) Using economic incentives and regulations to reduce seawater intrusion in the Batinah coastal area of Oman. *Agr Water Manage* **95**: 243–252.

(2009) Controlling groundwater pumping online. *J Environ Manage* **90**: 3581–3588.

Zekri S, Madani K, Kalbus E, Jayasuriya H and Zaier R (2014) *Groundwater Management in Oman by Water Metering: Adoptive Research*. Muscat, Sultanate of Oman.

Zektser IS, Dzhamalov RG and Everett LG (2007) *Submarine Groundwater*. CRC/Taylor and Francis, Boca Raton, FL.

Zhang SQ and Dai FC (2001) Isotope and hydrochemical study of seawater intrusion in Laizhou Bay, Shandong Province. *Sci China Ser E* **44**: 86–91.

Zhang ZK (2004) Research on the seawater intrusion control and treatment for the city of Laizhou the Wanghe River down stream region (in Chinese). Thesis, China Ocean University.

Zhang ZH, Shi DH, Ren FH, Yin ZZ, Sun JC and Zhang CY (1997) Evolution of Quaternary groundwater system in North China Plain. *Sci China Ser D* **40**: 276–283.

Zheng C and Bennett GD (2002) *Applied Contaminant Transport Modeling*. Wiley-Interscience, New York.

Zhou X and Wang Y (2009) Brief review on methods of estimation of the location of a fresh water-salt water interface with hydraulic heads or pressures in coastal zones. *Ground Water Monit R* **29**: 77–84.

Zhou X, Zhou HY and Zhang L (2008) Characteristics of Piezometric heads and determination of fresh water-salt water interface in the coastal zone near Beihai, China. *Environ Geol* **54**: 67–73.

Zimmerman JTF (1988) Estuarine residence times. Hydrodynamics of Estuaries 1, Estuarine Physics (Kjerfve B, ed.), 76–84. CRC Press, Boca Raton, FL.

Zong YQ, Huang G, Switzer AD, Yu F and Yim WWS (2009) An evolutionary model for the Holocene formation of the Pearl River Delta, China. *Holocene* **19**: 129–142.

Zong YQ, Lloyd JM, Leng MJ, Yim WWS and Huang G (2006) Reconstruction of Holocene monsoon history from the Pearl River Estuary, southern China, using diatoms and carbon isotope ratios. *Holocene* **16**: 251–263.

Zong YQ, Yu FL, Huang GQ, Lloyd JM and Yim WWS (2010) The history of water salinity in the Pearl River Estuary, China, during the Late Quaternary. *Earth Surf Proc Land* **35**: 1221–1233.

Zuurbier KG, Hartog N and Stuyfzand PJ (2016) Reactive transport impacts on recovered freshwater quality during multiple partially penetrating wells (MPPW-)ASR in a brackish heterogeneous aquifer. *Appl Geochem* **71**: 35–47.

Zuurbier KG, Raat KJ, Paalman M, Oosterhof AT and Stuyfzand PJ (2017) How subsurface water technologies (SWT) can provide robust, effective, and cost-efficient solutions for freshwater management in coastal zones. *Water Resour Manage* **31**: 671–687.

Zwemer SM and Zwemer AE (1911) *Zigzag Journeys in the Camel Country: Arabia in Picture and Story*. New York.

Index

Print
By Bo he United States
sters